DEVELOPMENTAL RESEARCH METHODS

THIRD EDITION

SCOTT A. MILLER

University of Florida

For information:

Sage Publications, Inc.
2455 Teller Road
Thousand Oaks, California 91320
E-mail: order@sagepub.com

Sage Publications India Pvt. Ltd.
B-42, Panchsheel Enclave
Post Box 4109
New Delhi 110 017 India

Sage Publications Ltd.
1 Oliver's Yard
55 City Road
London EC1Y 1SP
United Kingdom

Sage Publications Asia-Pacific Pvt. Ltd
33 Pekin Street #02-01
Far East Square
Singapore 048763

Printed in the United States of America

Library of Congress Cataloging-in-Publication Data

Miller, Scott A., 1944-
Developmental research methods / Scott A. Miller. — 3rd ed.
p. cm.
Includes bibliographical references and indexes.
ISBN-13: 978-1-4129-5029-9 (pbk.)
1. Developmental psychology—Research—Methodology. I. Title.

BF713.M56 2007
155.072—dc22

2006029740

Printed on acid-free paper

07 08 09 10 11 10 9 8 7 6 5 4 3 2 1

Acquiring Editor: Cheri Dellelo
Editorial Assistant: Anna Mesick
Production Editor: Sarah K. Quesenberry
Copy Editor: Teresa Wilson
Proofreader: Kevin Gleason
Typesetter: C&M Digitals (P) Ltd.

Contents

Preface

This book is intended for anyone who wants to learn more about how to do research in developmental psychology. It does not teach everything about how to do research—no book could. But it does, I hope, provide a helpful basis, a set of guidelines and principles that can aid in both the execution of one's own research and the evaluation of the research of others.

I have tried to strike a balance between the general and the specific. This balance is reflected in the book's organization: an initial 10 chapters that discuss general matters of research methodology, followed by 4 (somewhat lengthier) chapters devoted to specific research topics in developmental psychology. The balance is also reflected in the approach that is taken to discussing research. This book is neither an abstract "design and analysis" treatise on the one hand, nor a cookbook of hands-on exercises on the other. It is written instead to reflect the way that I believe most of us actually go about doing research—to convey an appreciation of the issues that must be addressed, the decisions that must be made, and the obstacles that must be overcome at every phase in a research project. I hope that the book captures something of both the excitement and the challenge of doing good research on topics that really do matter.

The primary audience for this book will doubtless come from laboratory or research methods courses in developmental or child psychology. I have assumed that any student in such courses will have had at least one prior course in developmental or child psychology. Other kinds of course work (e.g., statistics, general research methods) would be helpful but are not necessary. With suitable adjustments by the instructor, the book should be appropriate for both advanced undergraduates and beginning graduate students.

This edition differs in a number of ways from the previous edition of the book. One major change is the addition of qualitative research, an interesting and increasingly important approach to study that provides a valuable complement to the more traditional quantitative methods in which most of us were trained. A number of other topics have been added as well; among the topics new to this edition are archival research, meta-analysis, microgenetic methods, conceptual development, and executive functioning. Even when the topics remain the same, the discussion has been thoroughly updated; almost half of the References are new to the third edition.

This edition retains the features (glossary, exercises, summaries) that have proved useful in previous editions of the book. It also adds a new feature: a series of Focus On boxes directed either to particular methodologies in developmental psychology (in the early chapters of the book) or to particular topics of current interest (in the later chapters). The purpose of the boxes is the usual purpose of such a feature: They permit a more extended treatment of some interesting and informative contemporary material than would coverage in the regular text.

For the most part, the chapter divisions and the sequence of chapters are the same as in the second edition of the text. I realize that different instructors organize this material in different ways, and indeed I have used different approaches myself over my years of teaching methods courses (and across the first two editions of this book). Although the different chapters are of course linked, to a good extent each is also a stand-alone unit. Thus the chapters—and often even particular sections within them—can be assigned in any order that an instructor prefers.

I am grateful to many people for various kinds of help during the writing of this edition of the book. General support was provided by the Department of Psychology, University of Florida. Feedback from student users of the book was a valuable source of ideas—as indeed has been true for each edition of the book. I am grateful for the help and support provided by the editorial team at Sage: Cheri Dellelo, Jim Brace-Thompson, Sarah Quesenberry, Karen Ehrmann, Deya Saoud, Teresa Wilson, and Anna Mesick. Among their services was the recruitment of an outstanding group of reviewers, to whom I express my gratitude. Finally, I thank Ira Fischler for advice on IRB issues and James Algina for his help with my numerous statistical questions.

—Scott A. Miller

SAGE Publications gratefully acknowledges the contributions of the following reviewers:

Libby B. Blume
University of Detroit Mercy

Shannon Casey
University of California, San Diego

Christy Jensen
Center for Public Policy Research, Aging & Health
College of William & Mary

1

Introduction

One of the classic topics in developmental psychology is young children's egocentrism. Egocentrism refers to an inability to break away from one's own perspective to take into account the perspectives of others. That others do not experience the world in the same way that we do—see what we see, know what we know, wish what we wish—seems obvious to older children or adults. This fact is not always obvious to young children, however. Instead, young children often seem to operate as though they assume that everyone shares their own particular point of view upon the world—hence the label "egocentric."

Consider how a young boy responds to an experimental task in which he is asked to imagine that he is buying a birthday present for his mother (Flavell, Botkin, Fry, Wright, & Jarvis, 1968). An array of gifts, selected to vary in both age-appropriateness and sex-appropriateness, is laid out before the child. Does the child head immediately for the silk stockings or grown-up books? While such a response is possible, it is not very common among 3- or 4-year-olds. A more likely response is selection of one of the shiny new toy trucks. The young child knows what he wants; how could mother not want the same?

Textbook writers are in some respects similar to the young child standing before the array of gifts. To them, the interest and importance, even the beauty, of their subject are self-evident. If asked to justify why anyone else should care about the topic, the response may be one of bewilderment or frustration: How could anyone *not* see that this is a fascinating and vitally important subject? One might as well question the value of a shiny new truck!

Nevertheless (and here comes the egocentrism), it is difficult to see that any justification is needed for an interest in research in developmental psychology. What could be more obvious than the need to study how people develop? If some further justification *is* requested, it is easy to provide. Certainly no branch of psychology is broader in scope than developmental. And certainly no branch of psychology addresses more fundamental scientific issues than does developmental, because developmental psychology simultaneously encompasses all the other areas of the field (perception, thinking, personality, etc.) and adds to them a single basic question: How do people get to be the way they are? How is it, for example, that virtually all people come to understand and use an incredibly complex language system? Where do individual differences in intelligence or personality come from? What are the effects of early child-rearing practices on later development? Questions such as these cut to the heart of what psychology as a science can potentially tell us.

Such questions are not only of scientific interest. More obviously than any other branch of the field, developmental psychology speaks to issues that make a difference in the lives of everyone. Consider again some of the questions posed in the preceding paragraph. The issue of early experience and later development may be a fascinating scientific problem for the researcher, but it is a matter of urgent practical importance for any parent concerned with the optimal development of his or her children. That people differ in intelligence may raise a number of intriguing theoretical questions, but this fact also has enormous interpersonal and societal consequences. One of the exciting things about being a developmental psychologist is this feeling that one is dealing with questions that really matter.

Answers to such questions do not come easily, however. Indeed, it often seems that the most basic and important questions are the hardest to resolve. The difficulty of doing good research is a continuing theme throughout the book and hence need not be documented here. But let us briefly consider one example to introduce some general points. It is a problem that has already been touched on twice: drawing a relation between parental child-rearing practices and child development. How might we study this problem scientifically?

To anyone with even a rudimentary background in scientific methods, the general answer to this question is obvious: through controlled experimental study (if this answer is *not* obvious, it should become so in chapter 2). What the researcher might do, for example, is randomly assign infants at birth to families of different backgrounds and different child-rearing philosophies. Effects of child-rearing practice could then be determined apart from the contributions of the parents' genes to the children's development. Or the researcher might decide to assign different child-rearing practices on a random basis to different families. This procedure would avoid the confounding factor of parental choice in child rearing and allow a clear focus on the child-rearing methods themselves. The researcher might even decide, for purposes of comparison, to include a group of parents who rear their own children in whatever way they wish. In any case, the researcher would study the children as they grew and take extensive measures of their development. If such research could be carried out for even a few years, we would know much more about the consequences of different methods of child rearing than we do now.

Needless to say, the research program just outlined is the stuff of science fiction (or of methods textbooks), not fact. We do not have experiments of this sort, and it is to be hoped that we never will.[1] In this case the ethical problems are clearly sufficient to prohibit the research. If they were not, the practical difficulties in actually carrying out such studies would be staggering. These two factors—ethical limitations and practical constraints—act to rule out many well-designed, "textbook-like" experiments that any developmental psychologist could easily dream up. The result is that we have to fall back upon less scientifically satisfactory methods of gathering the desired information. That such methods do exist, and that they lead to genuine gains in knowledge, will be another continuing theme. But the appropriate methods and the resulting knowledge often do not come easily.

The main points of the discussion thus far are easy to summarize. Developmental psychology addresses questions that are of both great scientific and great practical importance. Studying such questions is often very difficult, and these difficulties place serious constraints on what can be known. Nevertheless, methods of study do exist, and gains in knowledge are being made literally every day. What we have, then, is a field of study in which the potential benefits of research are great, the challenges to successful study formidable, and the progress in knowledge slow but meaningful—in short, an ideal place for an ambitious researcher.

Goals of the Book

This book has three general goals. The first and most obvious is to help promote the skills necessary to do good research in developmental psychology. To this end, principles and precepts of various sorts are presented. Some of these principles are specific to issues of development; others are more general to the field of psychology. Some, indeed, are not even specific to psychology but reflect applications of the general scientific method. Whenever possible, however, I embed the discussion within the context of developmental issues. And, as already suggested, developmental psychology presents enough methodological problems of its own to challenge any researcher.

A second goal is to provide exposure to important research areas within the field. No one, after all, does "research in development"; studies are always directed to some particular content area, and every content area presents its own set of methodological challenges. It is impossible in one book to cover every interesting topic in the field or to convey everything about any given topic. But we can make a start on some of the most interesting and well-studied topics.

The third goal is to foster skills necessary for critically evaluating research and the conclusions that can be drawn from research. Such skills, of course, are not separate from those needed to carry out studies, but most of us are likely to use them far more often. Not everyone is going to do research in developmental psychology, but everyone is a consumer of the results of such research. Consider again some of the practical issues for which research in developmental psychology is relevant. Is physical punishment ever justified when disciplining children, or should such techniques be avoided altogether? Does violence on TV promote aggression in children? Should early enrichment programs be provided for children at risk for school failure? Are mandatory retirement ages ever justified, and if so, for what kinds of occupations? And what sort of research programs, if any, should the federal government support? Questions such as these are of interest to every parent, taxpayer, or voter. Intelligent answers to the questions are most likely if one knows the conclusions that have been drawn from relevant research. Intelligent answers are even more likely if one knows the methodology behind the research and can sensibly weigh the various strengths, weaknesses, and uncertainties when evaluating the conclusions.

Steps in a Research Program

What are the things that must go right in the course of a study if the final product is to be an increment in knowledge? The answer is quite a number of things, most of which are discussed at length in later sections. The purpose of the present section is simply to provide an introductory orientation to the skills necessary to do good research—a brief overview and a preview of topics to come.

The starting point for any successful program of research is *good ideas.* This is at once the most obvious and the least teachable of the various requirements. Because it is both obvious and difficult to teach, the criterion of good ideas tends to be neglected in discussions of how to do research, the focus instead being on the skills necessary to implement whatever ideas one may have (but see Leong & Muccio, 2006, and McGuire, 1997, for exceptions). This neglect will hold true here as well. It is important to remember, however, that all the technical skill in the world will not save a study if the ideas behind it are not any good. It is important to realize too that the really important differences among researchers—the factors that separate the average researcher from the one whose research shakes the field—lie less in the technical skill with which they execute studies than in their abilities to think in truly original and penetrating ways about an issue.

A second criterion is *knowledge of past work.* Anyone embarking on a program of research must have a thorough knowledge of what has already been done on the topic in question. Indeed, this step might logically be listed as the first, because really good ideas probably cannot be generated without knowledge of what has gone before. In any case, knowledge of the literature is essential when researchers come to evaluate just how testworthy their ideas are. There is little point in executing a brilliant idea for a study if someone else has already done the same thing. More common, perhaps, is the case in which certain important points of procedure would have been decided differently if the researcher had only known about similar work by others. Few things can be more depressing to an investigator than going to all the effort of carrying out a study and only then learning that the findings of some earlier study render the effort pointless.

Keeping abreast of the literature is no easy task at a time when professional journals publish thousands of articles in developmental psychology annually. Luckily, helpful sources do exist. There have long been journals whose sole purpose is to provide cross-referenced abstracts of articles and books published in psychology—in particular, *Psychological Abstracts* for research in psychology in general and *Child Development Abstracts and Bibliography* for research in child psychology. In recent years such hard-copy abstracting systems have been supplanted by various computer-based versions. Probably the most helpful of the electronic search engines for the psychology researcher is PsycINFO. Both Reed and Baxter (2006) and Rosnow and Rosnow (2006) provide helpful guides with respect to how to use PsycINFO. An online source of help is an American Psychological Association Web site devoted to PsycINFO (www.apa.org/psycinfo/training).

Another valuable aid to literature searches are the various books and journals devoted to review articles on major topics. Table 1.1 lists and briefly describes some of the most helpful of these sources. Table 1.2 provides a list of some of the major empirical journals that publish research in developmental psychology. It is good practice to scan the most recent volumes of these journals before making a final decision about procedures. Note that almost all journals are now available in electronic form. Finally, the best guide to past work may often come not from written sources but from consultations with an experienced researcher in the field. And bibliographic assistance aside, discussing one's ideas with others is generally a helpful part of the problem-solving process.

Once the ideas for the study have been generated, the next step is to translate them into an *adequate experimental design.* It was suggested earlier that a technically perfect design is of little value if the ideas being tested do not merit study. I now add the converse point: that a brilliant idea may come to nothing if it cannot be embodied in a scientifically testable form. Matters of experimental design are a central topic in the coming chapters. For now, two points can be made. The first is a reiteration of a point made earlier. Very often in developmental psychology, ethical or practical constraints rule out research designs that, from a purely scientific point of view, would be ideal for studying an issue. The challenge then becomes to devise alternative procedures that can lead to valid conclusions. The second point is that designs in developmental psychology are often complicated by the fact that age is included as a variable of primary interest. As we see later, age is in some ways an especially difficult variable with which to work. But, of course, changes with age are of great interest for most developmental psychologists.

Our hypothetical researchers have now reached the point at which they have an idea for

Table 1.1 Sources for Review Articles in Developmental Psychology

Advances in Child Development and Behavior (Volume 1 published in 1963, new volumes at an almost annual rate since)

Annals of Child Development (published annually from 1984 to 1998)

Annual Review of Psychology (published annually since 1950)

Blackwell Handbooks of Developmental Psychology (volumes devoted to Infancy, Early Childhood, Adolescence, Cognitive Development, and Social Development, published from 2001 to 2005)

Handbook of Child Psychology, 6th edition, 2006

Handbook of the Psychology of Aging, 5th edition, 2001

Minnesota Symposia on Child Psychology (published annually since 1967)

New Directions for Child and Adolescent Development (multiple volumes each year since 1978)

Developmental Review (journal of reviews of research in developmental psychology)

Psychological Bulletin (journal of reviews of research in all areas of psychology)

Table 1.2 Sources for Empirical Reports in Developmental Psychology

Adolescence	*Human Development*	*Journal of Experimental Child Psychology*
Applied Developmental Science	*Infancy*	*Journal of Genetic Psychology*
British Journal of Developmental Psychology	*Infant and Child Development*	*Journal of Research on Adolescence*
Child Development	*Infant Behavior and Development*	*Journals of Gerontology*
Cognitive Development	*International Journal of Behavioural Development*	*Merrill-Palmer Quarterly*
Developmental Psychology	*Journal of Applied Developmental Psychology*	*Monographs of the Society for Research in Child Development*
Developmental Science	*Journal of Applied Gerontology*	*Parenting*
Experimental Aging Research	*Journal of Cognition and Development*	*Psychology and Aging*
Genetic, Social, and General Psychology Monographs	*Journal of Early Adolescence*	*Research in Human Development*
The Gerontologist		*Social Development*

a study, have surveyed the relevant literature, and have decided (at least tentatively) on an experimental design. The next step is to seek *human subjects approval* for the research—that is, to submit a proposal to the university committee responsible for monitoring the ethical conduct of research. Ethics is the subject of chapter 9, and there will be much to say then about both procedures to follow and criteria to consider when evaluating ethics. For now, I settle for one basic point—the need for independent determination of the ethics of research. Researchers must, of course, do everything possible to ensure that their own research projects are ethical. They may not make this decision alone, however; rather, research can proceed only after an independent committee has been satisfied that the research is ethically sound.

Although the next step is not a necessary one in all research projects, in particular cases it may be essential. It is to carry out a *pilot study*—that is, to do some preliminary testing and practicing before beginning the experiment proper. There are two general reasons for pilot testing, both of which may be especially important in work with children. One reason is to give the tester practice in working with the particular procedures and subject groups, the goal being to minimize experimenter error once the real study starts. The second is to test out any uncertain aspects of the procedure to make certain that they work as intended. Are the instructions clear? Is the test session a reasonable length? Is a particular experimental manipulation convincing? The actual questions will vary from project to project, but the general issue is the same: Is the study ready to go?

Assuming a positive answer to the ready-to-go question, the next step is again an obvious one: *obtaining research participants.* Obvious though this step may seem, it is not discussed in many textbooks on methodology, in which experimental designs somehow magically eventuate in data without the messy intermediate step of finding people on whom to try them. In fact, many researchers spend a good portion of their professional careers, not in the interesting business of thinking up research, but in the much more tedious business of finding participants with whom to do the research. This situation is especially true for developmental psychologists, who do not have readily available populations such as college sophomores or laboratory rats with which to work. Researchers of infancy cannot post sign-up sheets on which babies can volunteer for experiments; they must somehow locate parents with infants and induce them to bring their babies in for testing. The researcher who wishes to study large samples of 5-, 7-, and 9-year-olds will almost certainly need to work through a school system to find sufficient numbers to test. The investigator of possible changes in functioning with old age will need to locate and recruit elderly participants, possibly through contacts with various organizations that serve the elderly. All of these groups can present special problems of access.

It is difficult to offer specific guidelines with respect to obtaining participants, because procedures may vary from one locality to another. A few pieces of general advice can be offered, however. One is to allow plenty of time. Research almost always takes longer than the beginning researcher expects it to, and one common contributor to the delays is difficulty in obtaining participants. A second piece of advice is to be as persuasive as possible when presenting one's proposed research to those (principals, teachers, parents, children themselves) who must decide about participation. As is stressed in chapter 9 on the topic of ethics, the primary consideration when presenting research to prospective participants is to be honest and informative, so that decisions about participation can be fairly made. It is also important, however, to be clear about the value of the research, or else no one may decide to participate. Finally, perhaps the most helpful course, once again, is to find an experienced investigator of the subject group in question and solicit advice about how to proceed.

Note that the primary problem posed by difficulties in obtaining participants is not the inconvenience or loss of time suffered by the experimenter. The problem, rather, is that one aspect of proper experimental design is selection of appropriate subject groups. As we see later, an otherwise well-conducted experiment may be of little value if the experimenter has failed to obtain the right kind of participants.

Once participants have been recruited, the testing can begin. At this point the experimenter's *testing skills* become important. The phrase *testing skills* refers here to all of the abilities needed in actually working with research participants, whether in face-to-face interactions or in observing and measuring behavior. At issue, then, are questions of the following sort: Have the instructions clearly conveyed what is required? Has the tester biased performance through facial cues or inadvertent reinforcement? Have the responses been accurately recorded? The issue, in short, is whether the on-paper study that has been worked out in advance can be adequately realized in the actual experimental setting. Again, it is clear that a successful passage through the earlier steps of the research program will be of no avail if the present step is not also negotiated successfully. A researcher, for example, may have devised a beautiful plan for studying problem solving in 5-year-olds, but the results are not likely to mean much if the researcher has no conception of how to talk to 5-year-olds and consequently leaves the children either frightened or bewildered.

Discussions of testing skills occur at various places throughout the text, with the most concentrated coverage coming in chapter 5. As noted, some of the points made are general ones that apply to psychology as a whole, and others are specific to developmental psychology. Although any kind of research can be difficult, the researcher in developmental psychology often faces special problems that stem from the special nature of the subject groups tested. Skills that are sufficient when testing a college student may not be sufficient when working with a crying infant, a shy preschooler, or a suspicious octogenarian. The challenge is even greater if several distinct age groups must be accommodated within the same study.

No aspect of research methodology can be conveyed in a totally adequate fashion through a textbook alone. In no case, however, is a textbook treatment less adequate than for the question of how to work with participants. Although various guidelines can be given verbally, the only real way to become skilled in working with infants, preschoolers, or elderly people is to spend considerable amounts of time actually working with infants, preschoolers, or the elderly.

The conclusion of the testing does not mean that the researcher's job is done. The next step is the *statistical analysis* of the data. The question that must be answered now is whether the various factors under study have or have not produced a consistent and meaningful pattern of results. For the majority of studies, the accepted way to answer this question is through application of certain well-developed statistical procedures to the data. This statistical analysis will not, in itself, answer deeper questions about the theoretical or practical significance of the results. It does, though, set constraints within which such interpretations must operate.

Statistical analysis is a large topic, the subject of separate courses and books. This book does not cover it at any length. chapter 8, however, does provide a summary of certain general principles of statistics.

The final phase of a research program is the *communication* of what has been done and found. Science is a matter of shared information, and a research finding is simply not a finding until it has been communicated to others. The usual way to communicate findings in developmental psychology is through publication in a professional journal. Such publication requires that the researcher prepare a clear, accurate, and concise written report of the study. chapter 10 gives advice about how to prepare such reports.

Plan of the Book

The next six chapters deal with general principles of design and procedure. Chapter 2 is, in fact, entitled "General Principles"; it considers such basic concepts as independent and dependent variables, experimental control, and various forms of validity. Chapter 3 is entitled "Design"; its concern is the different ways in which studies can be constructed and the comparisons and conclusions that are possible given the different approaches. Because of the book's developmental focus, special attention is paid to designs for comparing different age groups.

In chapter 4, "Measurement," the focus shifts from the independent variable to the dependent variable, the concern being the ways in which we measure the outcomes in research. Decisions about both design and measurement must actually be implemented, and this is the topic of chapter 5, "Procedure": challenges that can arise in translating an on-paper study into an actual study, as well as ways to overcome the challenges. Chapter 6, "Contexts for Research," concerns the kinds of settings (e.g., structured lab environment, natural "field" setting) in which developmental research occurs, with a consideration of the advantages and disadvantages of each of the various possibilities. Finally, chapter 7, "Qualitative and Applied Research," provides, as its title suggests, a focus on two kinds of research whose importance and methodological challenges justify a separate consideration: qualitative research and applied forms of study.

The next three chapters consider three of the essential steps in executing research. Chapter 8 presents some general principles of statistical tests and statistical reasoning. Chapter 9 discusses ethical issues in research in developmental psychology, and chapter 10 presents guidelines for writing papers in psychology.

The final section of the book is devoted to specific research areas in developmental psychology. Chapter 11 is concerned with methods of studying development in infancy. The next two chapters are topically defined. In chapter 12 the focus is on ways to study cognitive development, especially during early and middle childhood; in chapter 13 the concern is the study of social development. Finally, in chapter 14 the focus is again chronological, with a discussion of methods of studying development in old age.

Summary

The chapter begins with a discussion of both the importance of research in developmental psychology and the challenges in doing such research well. This discussion leads to an overview of the three general goals of the book: to foster the skills necessary to carry out research in developmental psychology, to provide an introduction to interesting and important research topics in the field, and to promote the critical-evaluative skills that will allow readers to become intelligent consumers of research.

The middle section of the chapter provides an orientation to the steps that must be successfully negotiated if a research project is to be informative. The starting point is both the most important and the least teachable of the steps: generating *good ideas* that are worthy of empirical study. A closely related and perhaps even prior step is *knowledge of past work,* for research always grows out of what has gone before. Good ideas must be translated into an *adequate experimental design,* from which clear and valid conclusions can be drawn. Before research can begin, it is necessary to receive *human subjects approval,* and it is often useful to carry out a *pilot study,* both to refine uncertain aspects of the procedure and to sharpen testing skills. Another important and often difficult preliminary to research is *obtaining participants:* identifying the appropriate subject group and then securing its cooperation. Once the study begins, the experimenter's *testing skills* become important—that is, all the skills needed to interact with participants and observe behavior in nonbiasing ways. The conclusion of the data collection is followed by *statistical analysis* to determine what reliable and potentially informative patterns are identifiable in the results. The final step is the *communication* of one's research to others, usually in the form of publication in a professional journal.

The run-through of steps in a research program serves to introduce topics considered at various points throughout the book. Further introduction is provided by the concluding section of the chapter, which briefly previews each of the remaining chapters.

Exercises

1. The text suggests that research in developmental psychology speaks to many everyday, real-world issues of personal, social, or political importance. Spend a week or so thinking about this claim whenever you read the newspaper or listen to the news. For how many of the topics that you encounter in the news might an understanding of principles of developmental psychology be in some way valuable?

2. One way to gain a quick impression of the topics of current interest within a field of study is by scanning recent issues of some of the field's major research journals. This is also a good way to get ideas for your own research. Select at least three of the journals listed in Table 1.2 and locate their most recent volumes either at your library or, if they are available electronically, online (note that a volume, which typically spans an entire year, includes several separate issues). Read the titles of each of the articles, and for any that you find intriguing, read the abstract (which provides a brief summary of the article) as well.

3. Select two topics in developmental psychology that especially interest you, and conduct a PsycINFO search for each.

Note

1. There *are* accounts, perhaps apocryphal, of ancient rulers who carried out quite systematic experiments in child rearing. One story concerns a 13th-century king named Frederick II whose experiment, like many a contemporary study, produced some interesting data but not of the sort that Frederick intended:

> He bade foster mothers and nurses to suckle the children, to bathe and wash them, but in no way to prattle with them, or to speak to them, for he wanted to learn whether they would speak the Hebrew language, which was the oldest, or Greek, or Latin, or Arabic, or perhaps the language of their parents, of whom they had been born. But he laboured in vain because the children all died. For they could not live without the petting and joyful faces and loving words of their foster mothers. (Ross & McLaughlin, 1949, p. 366)

2

General Principles

Having some specific studies to refer to will help clarify the discussion that follows. Described next, therefore, are two examples of research in developmental psychology. Both studies have been simplified somewhat to make the points drawn from them easier to follow.

Dufresne and Kobasigawa (1989) were interested in the determinants of improvement in memory across childhood. Why do older children generally remember better than younger children? The particular determinant on which the research focused is labeled *study time.* The issue with respect to study time is what children do when they have a chance to study some set of to-be-remembered-items prior to being tested for memory. How much time, for example, do the children spend in studying the material before deciding that they are ready, and how sensibly do they distribute their effort across the various items? Perhaps one reason older children remember better than younger ones is that they make better use of their study time.

The memory task that Dufresne and Kobasigawa selected to test this hypothesis is called a *paired-associates* task. A paired-associates task consists of two phases: an initial phase during which pairs of words are presented together, followed by a second phase during which only one member of the pair is presented and the child must attempt to remember the matching item. In Dufresne and Kobasigawa's study, the pairs were of two sorts: an "easy" set in which the paired items were highly related (e.g., cat-dog, shoe-sock) and a "hard" set in which the items were not related (e.g., book-frog, skate-baby). All of the participants (children from the first, third, fifth, and seventh grades) received both sets of items, and for both they were allowed to study the material for as long as they wished before being tested.

Table 2.1 shows the average study time for each pair of items. Several conclusions are suggested by the values in the table. As would be expected, older children generally studied the items longer than did younger children. As would also be expected, hard items elicited more study time than did easy items. Finally, this easy-hard differentiation was not apparent among the youngest participants; rather, it was only the two older age groups who directed more attention to the hard items. I return to this last finding in particular later in the chapter.

Table 2.1 Mean Study Time (in Seconds) for the Child Participants in the Dufresne and Kobasigawa Study

	Type of item		
Group	*Easy*	*Hard*	*Combined*
Grade 1	5.40	5.20	5.30
Grade 3	5.53	6.96	6.25
Grade 5	4.23	8.42	6.33
Grade 7	4.45	12.48	8.47
Combined	4.90	8.27	

SOURCE: Adapted from "Children's Spontaneous Allocation of Study Time: Differential and Sufficient Aspects," by A. Dufresne and A. Kobasigawa, 1989, *Journal of Experimental Child Psychology, 47*, p. 282. Copyright © 1989, Academic Press.

The second study was also concerned with memory but of a different sort and at a different phase of the life cycle. Cherry and Park (1993) examined memory for spatial locations in samples of young (mean age = 21 years) and older (mean age = 69 years) adults. Participants first viewed a spatial array of 24 common objects. The objects were then scrambled, and the participants' task was to reproduce the original spatial arrangement.

Two presentation conditions were compared. For half of the participants, the objects were presented within the context of a colored, three-dimensional model with a number of distinctive landmarks. For the other half, the background consisted of a two-dimensional, black- and-white map of the three-dimensional model. Two questions were of interest. Would spatial memory be better when the locations to be remembered were embedded within a natural and distinctive context? And (as some studies had suggested) would any facilitative effect of a helpful context be greater for older than for younger participants?

Table 2.2 shows the results. It can be seen that the context did indeed make a difference: Performance was better with the model background than with the map background. Age also made a difference, with the younger participants outperforming the older ones. Finally, although the two age groups differed in overall performance, they did not differ in response to the context manipulation. Both the young adults and the older adults did better with the model than with the map.

Variables

I begin the discussion of general principles with some terminology. Research in psychology involves variables and the relations that hold among variables. The variables are of two sorts: dependent and independent. **Dependent variables** are outcome variables—those measures whose values constitute the results of a study. In the first example the dependent variable was the number of seconds that the child studied each pair of items; in the second example the dependent variable was the number of objects that the adult was able to place correctly. Such variables are dependent in the sense that variation in them follows from or depends on other factors. A central job for the researcher is to

Table 2.2 Mean Number of Items Correctly Placed by the Adult Participants in the Cherry and Park Study

	Context		
Group	*Map*	*Model*	*Combined*
Young	14.5	15.9	15.2
Old	11.1	14.5	12.8
Combined	12.8	15.2	

SOURCE: Adapted from "Individual Differences and Contextual Variables Influence Spatial Memory in Younger and Older Adults," by K. E. Cherry and D. C. Park, 1993, *Psychology and Aging*, 8, p. 520. Copyright © 1993, American Psychological Association.

determine what these other factors are. They are variable necessarily: If there were no possibility of variation in the dependent measure, there would be no point in doing the study.

The dependent variable is something that the researcher measures but does not directly control. **Independent variables**, in contrast, are variables that are under the control of the researcher. The object of the study is to determine whether the particular independent variables chosen do in fact relate to variations in the dependent variable. The independent variables in the Dufresne and Kobasigawa study were the age of the child and the hard-easy contrast, whereas those in the Cherry and Park study were age and type of context. Such variables are independent in the sense that their values are decided on in advance rather than following as results of the study. The "variable" part is again necessary: If there were no variation in the independent variable, there would be no possibility of determining whether that factor has an effect. Variation and comparison are intrinsic parts of all research.

The description of research as divisible into independent and dependent variables is valid for many but not for all studies. Suppose, for example that you wish to know whether there is a relation between a child's IQ and how well that child does in school. You might test a sample of grade-school children and collect two measures: performance on an IQ test and grades in school. Your interest would be in whether variations in one measure relate to variations in the other; for example, do children with high IQs tend to do well in school? A study like this does not have an independent variable whose values are under the experimenter's control; rather, IQ, grades, and the relation between them are all outcome variables in the study. "Correlational" research of this sort is discussed at length later. The point for now is simply that not all studies fit the independent variable–dependent variable mold.

The example studies can serve to illustrate a further point about independent variables. The contrasts that define an independent variable can be created in two ways. One way is through an experimental manipulation that literally creates the variable. This is what Dufresne and Kobasigawa did when they constructed their easy and hard sets of items and what Cherry and Park did when they designed their map and model backgrounds. This was not the approach, however, for the other independent variable in both studies: chronological age. Clearly, investigators cannot create an age contrast in the same way that they can create an easy-hard contrast. In the case of a variable like age, the control occurs not through manipulation but through *selection:* choosing people for study who are at the desired levels of the variable (e.g., 20 years old or 70 years old). Because selection is the only control possible, age and other "subject variables" can present special problems of interpretation—an issue to which I return later in the chapter.

A bit more terminology is necessary before proceeding. Independent variables are also referred to as **factors**, and the particular values that the variables take are referred to as **levels**. The Dufresne and Kobasigawa study, therefore, can be described as a 4 × 2 factorial design—that is, an experiment with two factors, one of which (age) has four levels and one of which (condition) has two levels. Similarly, the Cherry and Park study can be described as a 2 (age) × 2 (condition) factorial design. Note that symbolizing the design in this way serves to tell us the number of distinct cells or groups in the experiment. For example, in the Cherry and Park study there are four (2 × 2) distinct groups: young adults in the model condition, young adults in the map condition, old adults in the model condition, and old adults in the map condition.

Validity

All research involves variables and the relations that hold among variables. When we wish to describe research, therefore, the construct of variables is central: What kinds of contrasts are being examined, and what forms do the examinations take? When we wish to move beyond description to *evaluation* of research, the central construct becomes that of **validity**. The question of validity is the question of accuracy: Has the study in fact demonstrated what it claims to demonstrate? All of the specific methodological points discussed throughout the book come down to this one basic question of the accuracy of the conclusions that we draw from research.

Various forms of validity can be distinguished (Shadish, Cook, & Campbell, 2002). In this chapter I discuss three forms: *internal, external,* and *construct.* Chapter 8 will add a fourth form: *statistical conclusion validity.*

Internal validity applies within the context of the study itself. The issue in question is whether the independent variables really relate to the dependent variables in the manner claimed. Have we drawn the correct conclusions about the causal impact (or lack of causal impact) of one set of variables on the other set? Let us take the Dufresne and Kobasigawa study as an example. Their conclusions are internally valid if the hard items really did produce longer study times than the easy items, if the average study time really did increase as a function of age, and if the ability to differentiate between easy and hard also really increased with age. If there is a plausible alternative explanation for any of these findings, then the internal validity of the study is thrown in doubt. Suppose, for example, that seventh graders in the study had been selected primarily from "gifted" classes, whereas the younger children represented more average ability levels. If so, we would have an alternative explanation for the seeming improvements with age: The differences reflect not natural changes with age but rather differences in ability level. (I discuss this problem, labeled *selection bias,* more fully later.)

The question of **external validity** is the question of generalizability. It applies, therefore, once we move outside the immediate context of the study. The question now is whether we can generalize the findings of the study to other samples, situations, and behaviors—not just any samples, situations, and behaviors, of course, but those for which we wish the study to be predictive. In this case let us take the Cherry and Park study as the example. Their findings would have external validity if young adults really do in general have better spatial memory than older adults, if distinctive contextual cues really do in general facilitate spatial memory, and if both young and old really do in general benefit equally from such cues. In each case the "in general" refers to what is found across a variety of samples of young and old, a variety of measures of spatial memory, and a variety of contextual cues. If any one of the findings fails to generalize across these dimensions, then that finding lacks external validity. Perhaps, for example, contextual cues make a difference only for "small-scale" environments such as Cherry and Park's model and map, and there is no comparable effect in full-size, real-life settings. If this limitation actually held (other research makes clear that it does not), then the Cherry and Park study would have limited external validity.

Exactly what forms of generalizability are important varies to some extent across studies. Table 2.3 lists and briefly describes the most common dimensions that are relevant to external validity.

A satisfactory study must have both internal validity and external validity. As Campbell and Stanley (1966) observe, "internal validity is the basic minimum without which any experiment is uninterpretable" (p. 5). Logically the internal validity question is the primary one, because findings can hardly be generalized if there are no valid findings in the first place. External validity is also critical, however. Internally valid conclusions do not mean much if they cannot be generalized beyond the study in which they occur.

Internal validity is also a prerequisite for the third form of validity: **construct validity**. Construct validity has to do with theoretical accuracy: Have we arrived at the correct explanation for any cause-and-effect relations that the study has demonstrated? We assume, in other words, that we have internally valid conclusions; the question now is whether we know *why* the results have occurred.

Suppose, for example, that we are confident that the context manipulation in the Cherry and Park study really did cause variations in memory performance. Why did the context make a difference? Probably the most obvious explanation—and the one that has guided most such research—is that it is the distinctiveness of the visual information that is important: Locations are easiest to remember when they are embedded within a well-differentiated spatial surround. But perhaps there is a different basis. Maybe the model was more interesting and engaging than the map, resulting in

Table 2.3 Dimensions of External Validity

Dimension	*Issue*
Sample	Do the results generalize beyond the sample tested to some broader population of interest?
Setting	Do the results generalize beyond the setting used in the research (e.g., a structured laboratory environment) to the real-life settings of interest (e.g., behavior at home or at school)?
Researcher	Are the results specific to the research team that collects the data, or would the same results be obtained by any team of investigators?
Materials	Are the results specific to the particular materials used to represent the constructs of interest, or would the same results be obtained with any appropriate set of materials?
Time	Are the results specific to the particular time period during which the data were collected, in either a short-term sense (e.g., a measure administered in late afternoon) or a long-term sense (e.g., a measure affected by historical events)?

closer attention and hence better memory. By this view, any manipulation that heightens attention should improve performance, quite apart from the spatial distinctiveness of the cues. Or maybe the participants were more confident when confronted with the relatively familiar model than when confronted with the abstract map, and it was this heightened confidence that led to better memory. By this view, any manipulation that increases confidence should improve performance. If plausible competing explanations for the results cannot be ruled out, then the study lacks construct validity.

The preceding discussion has been just a first pass at constructs that will recur in various contexts throughout the book. For now, let us settle for one more point with respect to validity. It concerns the difficulty of simultaneously achieving the various forms of validity in the same study. This difficulty exists because often research decisions that maximize one form of validity work against another form. The trade-off is most obvious with regard to internal and external validity. In general, the more tightly controlled an experiment is, the greater its internal validity—that is, the more certain the experimenter can be that the variables really do relate in the manner hypothesized. At the same time, the artificiality of a tightly controlled experiment may make generalization to the nonlaboratory world hazardous. Conversely, research conducted in natural settings with naturally occurring behaviors may pose little problem of generalizability, because the situations to which the researcher wishes to generalize are precisely those under study. The lack of experimental control, however, may make the establishment of valid relationships very difficult.

Sampling

Decisions about variables have to do with the *what* of research: What independent variables am I going to manipulate, and what potential outcomes of these variables am I going to measure? Also important are decisions about *who:* With what sorts of participants am I going to explore these independent variable–dependent variable links?

The selection of participants for research is referred to as **sampling**. Sampling is important because of the constraints on the scope of research. With very rare exceptions, psychologists are not able to study all of the people in whom they are interested. The researcher of infancy, for example, is not going to test all of the world's babies, or even all those in the United States, or (probably) even all those in one specific geographical community. Instead, what researchers do is to test **samples**, from which they hope to generalize to the larger **population** of interest. The generalization is legitimate if the sample is *representative* of the larger population. This, clearly, is an issue of external validity.

How can researchers ensure that a sample is representative of the population to which they wish to generalize? A logical first step is to define what the population of interest is. It need not be as broad as all of the world's infants; more likely, perhaps is something like "all full-term, healthy 3-month-olds growing up in the United States." Once the desired population has been defined, the next step is **random sampling** from that population. As the term implies, random sampling means that every member of the population has an equal chance of being selected for the research. If all members of the population really are equally likely to be selected, then the most probable outcome of the sampling process is that the characteristics of the sample will mirror those of the population. Note, however, that the likelihood that this desired outcome will in fact be achieved varies directly with the size of the sample. A random sample of 100 is a good deal more likely to be representative than a random sample of 10. This principle is just one of a number of arguments (we will encounter some others in chapter 8) for using large rather than small sample sizes.

In some instances researchers may use modified forms of random sampling, especially when the intended sample size is limited and pure random selection might therefore not produce the desired outcome. In **stratified sampling** researchers first identify the subgroups within the population that they want to be sure are represented in their correct proportions in the final sample. A researcher might want to be sure that males and females are represented equally, for example, or that different ethnic groups appear in proportions that match their numbers in the general population, or that freshmen are just as common as seniors in a college student sample. Samples are then drawn in the desired proportions from the identified subgroups—thus, equal numbers of males and females, 25% of the participants from each year in college, and so forth.

The goal of stratified sampling is to ensure that different members of the population are represented in their actual proportions in the sample selected. In contrast, with **oversampling** the researcher deliberately samples one or more subgroups at rates *greater* than their proportion in the target population, the goal being to achieve a sufficiently large sample of the subgroup to permit conclusions. Suppose, for example, that we plan to conduct a survey of high school students in which comparisons among ethnic groups are one of the issues of interest, and suppose also that Asian Americans constitute 3% of the high school population in the city in which we are working. Even with a total sample of 1,000 students, a random sampling approach will give us only about 30 Asian American participants, which may not be enough to draw conclusions. If we deliberately oversample Asian Americans, however (say at a 6% rather than a 3% rate, thus giving 60 students total), we can end up with a sufficient subsample for analysis, while still achieving adequate numbers in the other groups of interest.

How often do psychologists in fact draw their samples in the textbook-perfect fashion just described? The answer is: not very often. Random sampling and its variants are occasionally found in psychological research—perhaps most commonly in large survey projects in which it is important that the sample match some target population. More generally, most researchers undoubtedly start with at least an implicit notion of the population to which they wish to generalize, and most would certainly avoid selecting a sample that is clearly nonrepresentative of this population. Nevertheless, true random sampling from some target population is rare. The most obvious and frequent deviation from randomness is geographical. Researchers tend to draw samples from the communities in which they themselves live and work. Often, moreover, they may sample from only one or a few of the available hospitals, day care centers, or schools within the community. Such selection of samples primarily on the basis of availability or cooperation is referred to as **convenience sampling**. Samples obtained in this way may not be representative of the broader population with respect to variables such as social class and race, and they *cannot* be completely representative with respect to variables like region of the country or size of the community.

How important are these deviations from random sampling? There is no simple answer to this question; among the dimensions that are relevant are the topic under study; what the researcher wishes to conclude about the topic; and, of course, just how nonrandom and potentially nonrepresentative the sample is. We will revisit issues of sampling throughout the book in the context of particular kinds of research. For now, I settle for two pieces of advice, one directed to the reader of research reports and the other to the author of such reports.

The advice for the reader is to make a careful reading of the Participants section an important part of the critical evaluation of any research project. However satisfactory the other elements of a study may be, the results do not mean much if the sample is not representative of some larger population of interest. One question concerns the standing of the sample on the demographic characteristics that may affect response. At the least, these characteristics will include age, sex, and race; for particular studies additional dimensions (e.g., income level, geographical region, health status) may also be important. Another question concerns the method of recruitment. What was the initial pool from which participants were drawn, how many of these potential participants actually made it into the study, and (if there was any dropout) how many stayed in the study until the end? Finding a representative pool of potential participants is a good starting point for research, but it is not sufficient; the real question is how well the final sample reflects the starting point.

The advice for the author follows from the points just made. Readers cannot critically evaluate the samples for research if Participants sections do not tell them enough about the samples. It is the author's responsibility to make sure that all of the necessary information is conveyed to the reader. Helpful further sources with respect to what sorts of information to convey include the *APA Publication Manual* (APA, 2001), Hartmann (2005), and Rosnow and Rosnow (2006).

Control

The notion of control was touched on in each of the preceding sections. Recall that the independent variable is defined as a variable that is under the control of the researcher. Control is central to the establishment of validity, especially internal validity. And selection of the right participants is one sort of control that a researcher must exercise. The purpose of the present section is to discuss the further sorts of control that become important once participants are in hand.

As Table 2.4 indicates, three forms of control are important in the execution of studies. The table summarizes the forms and gives examples of how each type applies or might apply to the illustrative studies. Both the forms of control and the examples are elaborated and should become clearer as we go. The table is intended simply as a guide to help keep track of the distinctions to be made.

One type of control concerns the exact form of the independent variable. If the interest, for example, is in the effects of a certain kind of reinforcement, then the researcher must be able to deliver exactly this kind of reinforcement to the participants. If any unintended deviations occur—in form, timing, consistency, or whatever—the researcher can no longer be certain what the independent variable is. Or consider again the Dufresne and Kobasigawa examination of study time. Because the researchers' interest was in possible effects of item difficulty, it was critical that they present the same easy-hard contrast to all of the children.

The point being made about this first form of control is hardly an esoteric one. It is simply that if one wants to study the possible effects of something, one must first be able to produce that something. Note, however, that doing so is not always as easy as in the two example studies, in which the levels of the independent variables were defined simply by the different stimulus materials that were presented. When the experimental manipulation is more complicated, delivering the variable in the same form to all participants can become a challenge. The challenges, moreover, are often multiplied when children are the participants, a point to which I return later.

A second form of control has to do with factors in the experimental setting other than the independent variable. Independent variables do not occur in a vacuum; there must always be a context for them, and it is the job of the researcher to determine exactly what this context will be. In giving a memory test, for example (as in the two example studies), the researcher must decide not only what test to use but also what the immediate environment for the testing will be like. One easy decision in this particular case is to make the environment as quiet as

Table 2.4 Forms of Control in Experimental Research

Type of control	*Methods of achieving*	*Examples from illustrative studies*
Over the independent variable	• Make the critical elements of the experimental manipulation the same for all participants	• In Dufresne and Kobasigawa, present the same sets of easy and hard items in the same way to all the children
Over other potentially important factors in the experimental setting	• Hold the factors constant for all participants • Disperse variations in the other factors randomly across participants	• In Cherry and Park, use the same quiet testing room for all participants • In Dufresne and Kobasigawa, vary the time of testing randomly across children
Over preexisting differences among the participants	• Randomly assign participants to experimental conditions • Match participants on potentially important attributes prior to experimental conditions • Test each participant under every experimental condition	• In Cherry and Park, randomly assign half of the participants at each age to the model condition and half to the map condition • In Cherry and Park, measure the participants' IQs and assign equal-IQ participants to the different conditions (not actually done) • In Dufresne and Kobasigawa, test every child with both the hard and the easy items

possible, in order to minimize distractions. Once the experimenter has made this decision, it is then his or her job to ensure that each participant receives the same quiet environment.

Let us introduce some further terminology at this point. Differences in scores on the dependent variable are referred to as the *variance* of the study. Those differences that can be attributed to the independent variables are called **primary variance**; those that result from other factors are called *secondary variance* or *error variance.* By controlling the level of other potential variables, experimenters attempt to maximize the proportion of primary variance in the study. Perhaps even more important, they attempt to make sure that other sources of variance are not systematically associated with any of the independent variables. Suppose, for example, that Cherry and Park had tested all of their young adult participants in a quiet laboratory on campus but all of their older participants in a noisy room at a senior citizens' center. Clearly, in this case there would have been two independent variables—age and testing environment—when only one had been intended. Any such unintended conjunction of two potentially important variables is referred to as **confounding**. A major goal of good research design is to rule out confounding.

As Table 2.4 indicates, control of unwanted variables can take a couple of forms. Often it is possible to control the variable by making it the same for all participants. This is the case in the memory example, in which the noise level of the testing environment is held constant for all participants. Sometimes, however, such literal equating is not practical. We can return to the Dufresne and Kobasigawa study for an example. In research with school-aged children, a plausible contributor to how the children respond is the time of day at which the testing occurs. Cooperation and attentiveness are not necessarily the same late in the school day as they are first thing in the morning, or immediately before recess as compared with immediately after, or on a Friday as compared with a Monday. Clearly, Dufresne and Kobasigawa would have introduced a potentially important confounding if they had tested all of their first-graders early in the day and all of their seventh-graders in the afternoon. One way to avoid this problem would be to test all of the children at the same point in the day, say at one o'clock on Wednesday. With this approach, however, most studies would take months to complete, and even then only time of day and not time of year (which also can be important) would be held constant across participants. A sensible alternative would be to let the time of testing vary across children but to make sure that the variations are the same for the different groups being compared—in this case, first-, third-, fifth-, and seventh-graders. In this case the control of the time-of-testing variable would lie not in its equation but in its randomization—that is, by dispersing differences in it equally across the groups of interest.

Shorn of certain specifics, the discussion thus far should have a familiar sound to it. What has been presented here is simply the classic scientific method: to determine the effects of some factor, systematically vary that factor (the first form of control) while holding other potentially important factors constant (the second form of control).

There is still a third form of control that is essential. Thus far, the "other potentially important factors" that have been discussed have been factors within the experimental setting—for example, the noise level of the testing room. Another important source of variance in any experiment stems from individual differences among the participants. Participants are not identical at the start of an experiment, and differences among them contribute error variance to the final results. Because there is no way to rule out such differences, the method of control must again be through dispersion rather than equation. What the experimenter must ensure is that the differences are spread equally across the different treatment groups—or, to make the same point in different words, that the groups are equivalent prior to the application of the treatment. Doing so requires that the experimenter have control not only over the form of the treatment but also over who gets what treatment.

How can the experimenter assign people to groups in a way that will ensure that the groups all are initially equivalent? The answer is that although there is no way literally to ensure equivalence, there are ways to come as close as can reasonably be expected. The most common method is through **random assignment** of participants to the different groups. Random assignment means that each participant has an equal chance of being assigned to each group. If each participant has an equal chance of being assigned to each group, then the characteristics associated with each participant (IQ, sex, relevant past experience—whatever might affect the results) have an equal chance of falling in each group. It follows that the most probable outcome of the assignment process is that these characteristics will end up equally distributed in the different groups, a result that is, of course, the researcher's goal. The logic of random assignment is clearly the same as the logic of random sampling, and the success of the process shows a similar dependence on sample size. One could not randomly divide 8 participants into two groups and conclude with any confidence that the randomization had produced equivalent groups. With a sample of 80 participants, the odds are much better.

Random assignment is a much more frequent component of research than is true random sampling. Indeed, random assignment has been referred to as "the key defining attribute of the experimental method" (McCall & Green, 2004, p. 4).

Powerful though random assignment is, it does have a limitation. At best, random assignment makes it *probable* that the groups being compared are equivalent; it cannot guarantee this outcome. An obvious question follows: Why settle for probability? Why not identify the dimensions on which we wish the groups to be equal (e.g., intelligence, SES, health status—the list will vary across studies) and then assign participants based on these dimensions—thus, the same proportion of high-intelligence participants in each group, the same proportion of middle-class participants in each group, and so forth? Why, in short, not do the assignment in a way that *ensures* equivalence?

The general answer to this question is that such matching is more difficult than might at first appear and that the attempt to achieve it can sometimes create more problems than it solves. A more specific answer is given in chapter 3, when we return to the issue of selecting and assigning participants. Also discussed in chapter 3 is the third general technique for achieving equivalence: testing every participant under each experimental condition.

Subject Variables

Manipulable Versus Nonmanipulable Variables

Thus far the discussion of experimental control has focused on the ideal situation for research: the case in which the researcher can systematically manipulate the independent variables of interest while holding all other variables constant, and can assign participants to the different treatment groups either randomly or randomly within certain desired constraints. With many variables such control is not only desirable but quite feasible. We saw examples of this kind of control in both of the cited studies: the easy-hard contrast in the Dufresne and Kobasigawa study, and the model-map contrast in the Cherry and Park study.

The developmental psychologist's life is complicated, however, by the fact that not all variables of interest lend themselves to the kind of manipulation that good research design demands. Again, both of the cited studies provide examples, and in this case it is the same example: chronological age. Clearly, age is not something that the researcher randomly assigns to people; rather it is a characteristic that people bring to the experimental setting. Age is just one example of what are called **subject** (or *classification* or *attribute)* **variables**: intrinsic properties of individuals that cannot be experimentally manipulated but must be taken as they naturally are. Other common examples are race and sex. The researcher who wishes to work with such characteristics as independent variables forgoes the possibility of control through manipulation. The only control possible in such cases is control through selection of people who already possess the characteristic.

A number of other variables of interest, although not literally nonmanipulable, are never in fact the subject of controlled experiments with humans. From a theoretical perspective, for example, it would be very interesting to know whether infants deprived of mothers develop in the same way as infants who have mothers. Despite the early work of Frederick II (noted in chapter 1), we do not have manipulative studies of this issue. Yet there has long been a literature on "maternal deprivation" and its effects on the child. What researchers have done is to identify situations in which infants have already been left motherless (usually in orphanages) and then take advantage of these "natural experiments" by studying how the infants develop. And there are numerous similar examples of psychologists' ability to capitalize upon naturally occurring events—studies of malnutrition in infancy, of father absence during childhood, of social isolation in old age, and so forth. In each case the independent variable is created through selection rather than experimental manipulation.

Research with nonmanipulated variables does not attain the status of the "true experiment," because the controlled manipulation that constitutes the heart of an experiment is not possible. For this reason such research is labeled as *preexperimental* in Campbell and Stanley's (1966) influential discussion of experimental design. Because of the lack of control, such studies can never establish cause-and-effect conclusions with the certainty that is possible in a manipulative experiment.

What exactly are the limitations of such research? The problems are of two main sorts. First, it is impossible to assign participants randomly to groups. Because random assignment is impossible, there is no way to be sure that the groups under study are equivalent except for the variable of interest (e.g., presence or absence of mother), and therefore no way to be sure that any differences between groups are caused by that variable. This, in fact, was one criticism of the early maternal deprivation studies. Perhaps babies who grow up in orphanages are a nonrandom subset of the general population of babies, a subset that includes an unusually high proportion of genetic or organic problems. If so, then differences between orphanage babies and other babies could not be attributed with any confidence to the effects of the orphanage rearing. In a well-designed experiment, such confounding would be ruled out by random assignment. This, it should be clear, is a problem with internal validity: We cannot be certain that our independent variable is really the causal factor.

The other problem concerns the broad-scale and longstanding nature of most subject variables. Orphanage rearing, father absence, social isolation, growing up Black (or White), and growing up male (or female) all encompass a host of factors that can affect an individual's development. Thus, even if we find a significant effect associated with a particular subject variable, we still do not know what the specific causal factors are. This, too, has been a problem in research on maternal deprivation. Although the damaging effects of certain kinds of orphanage rearing are not in dispute, there has long been debate about whether the effects result from lack of normal mothering or from more general cognitive-perceptual deprivation. Even if we could conclude that mothering per se is important, we still would not know which of the many things that mothers normally do with infants are critical to the effect. Again, there is a confounding of factors that a well-designed experiment would keep separate. A researcher with control over variables is unlikely to set up an independent variable that is so global that its effects cannot be interpreted. This, it should be clear, is a problem with construct validity: We do not know whether we have arrived at the correct theoretical interpretation of the results.

This discussion is not meant to suggest that there is no value in demonstrating that a variable like maternal deprivation or sex or age is associated with important outcomes in the child. But it should be realized that such a demonstration is merely the first step in a research program.

Age as a Variable

Because of its importance in developmental research, the variable of chronological age deserves a somewhat fuller consideration. Much research in developmental psychology has as one of its points a demonstration that participants of different ages either are or are not similar on the dependent variables being studied. Even studies with a single age group may have age comparisons at their core, for often the comparison is implicit rather than explicit. A researcher of neonates, for example, may not include a comparison group of older children in the study, but findings about how neonates function can nevertheless be interpreted in light of a large body of information about the functioning of older children. To take a very simple example, one would hardly do research to determine whether young infants have color vision (e.g., Adams, 1995) unless one already knew that color vision is eventually part of the human competence.

Developmental psychologists are sometimes apologetic about the "merely age differences" nature of much research in developmental psychology. But the identification of genuine changes with age is clearly a valid part of a science of development. Not only is description a legitimate part of any science, but accurate description provides the phenomena to which explanatory models must speak. It is only when we know, for example, that young children do not understand conservation (Piaget & Szeminska, 1952) that we can begin to build a model of why this is so and of where eventual understanding comes from.

Although we may agree that the study of age changes is legitimate, it is important to be clear about exactly what is meant by a "genuine change with age." What is *not* meant, certainly, is that chronological age in any direct sense causes the change. What *is* meant is that variables that are regularly and naturally associated with age produce the change. It is then the job of the researcher to determine which of the potentially important variables are in fact important.

The earlier discussion stressed that a primary goal of experimental control is the creation of groups that are equivalent in every way except for the independent variable being examined. This goal takes on special meaning in the case of a broad subject variable like age. Imagine that you are interested in comparing 7-year-olds and 12-year-olds. If you wish to make the groups equivalent in every way except age, then you will have to find 7- and 12-year-olds whose levels of biological maturation are the same, who have been going to school for the same number of years, whose general experiences in the world are equivalent, and so forth. Clearly, such a goal is not only impossible but quite misguided. Variables like biological maturation, years of schooling, and general experience are among the variables that are "regularly and naturally associated with age." As such, they are factors to be studied, not ruled out through experimental control.

On the other hand, there are other potentially important factors that must not be allowed to confound the age comparison. An obvious kind of confounding would occur if all of the 7-year-olds were boys and all of the 12-year-olds girls. Maleness is not an intrinsic part of being 7, nor is femaleness an intrinsic part of being 12; hence, this factor must not be allowed to covary with age. A somewhat less obvious confounding might occur if all of the 7-year-olds were drawn from one school and all of the 12-year-olds from another school. The mere fact of attending different schools is probably not important, and in any case this difference may be unavoidable for the particular age range studied. Nevertheless, it will be important for the researcher to select schools that are as comparable as possible on dimensions such as educational philosophy, geographical location, and socioeconomic status of the population served. If this criterion is not met, then an apparent age change may not in fact be genuine.

As these examples suggest, decisions about what to match and what not to match when comparing different ages are generally straightforward. As we will see, however, such decisions are not always straightforward, nor is it always easy to achieve whatever matching one has decided on. We will return to the issue of age comparisons in chapter 3.

Outcomes

Researchers manipulate independent variables in order to examine effects on dependent variables. But what are the possible effects? In a factorial study—that is, a study with two or more independent variables—the possible effects are of two sorts: main effects and interactions.

Main Effects

A main effect is a direct effect of an independent variable on a dependent variable. It is what researchers examine when they compare the levels of a single independent variable independent of (or summed across) the other independent variables in the study. Both of the illustrative studies provide examples of main effects. In the Cherry and Park study there was a main effect of age: Young participants performed better than older ones. The means for this effect are shown in the rightmost column of Table 2.2; they are the values for all the young participants and all the old participants in the study, summed across the levels of the other independent variable (the model-map contrast). Similarly, there was a main effect of experimental condition, and the values for this effect are shown at the bottom of the table: the means for all participants in the model and all those in the map condition, summed across the two levels of age.

The Dufresne and Kobasigawa study also produced main effects of age and experimental condition. The means for these effects appear in the "combined" portions of Table 2.1. In both studies, therefore, we can say that both independent variables had an effect: Scores on the dependent variable varied as a function of age and experimental condition. Note, however, that the effect of age in the Dufresne and Kobasigawa study is more complicated than the other main effects, because in this case there are four levels of the independent variable rather than simply two. Main effects for variables with more than two levels pose some special statistical and interpretive complexities—an issue to which I return in chapter 8.

Interactions

A main effect is an effect of a single independent variable considered in isolation. An **interaction**, in contrast, becomes possible when we consider two or more independent variables simultaneously. An interaction occurs whenever the effect of one independent variable varies with the level of another independent variable.

The Dufresne and Kobasigawa study produced an interaction in addition to its two main effects. In this study, the effects of the easy-hard comparison varied with the level of age—little effect at the two younger grade levels, a marked effect at the two older grade levels. As with any interaction, the results can also be stated with the opposite emphasis: The effects of grade varied with the level of item difficulty—no differences with the easy items, strong differences with the hard items. This two-way ("two-way" because two independent variables are involved) interaction is graphed in Figure 2.1. The data are the same as those presented in Table 2.1; the graphical presentation, however, makes the nature of the interaction more visible. Note, in particular, the nonparallel nature of the lines. Graphically, an interaction is always signaled by some deviation from parallelism—some spreading apart or crossing over of lines that reflects the differential effects of one variable across the levels of the other. Note also, however, that graphs are not sufficient to determine that an interaction has occurred; rather, there must be a statistical test of the data to identify both main effects and interactions. I discuss the most common such test, the analysis of variance or ANOVA, in chapter 8.

What would the graph look like if there were no interaction? Figure 2.2, which plots the means from the Cherry and Park study, provides an answer. Recall that their study found equivalent benefits from the model condition for both the younger and the older participants—thus, two main effects (age and condition) but no interaction. This situation is reflected in the essentially parallel lines of Figure 2.2. (That the lines are not perfectly parallel reflects the fact that there was a slight trend toward an interaction—a slight

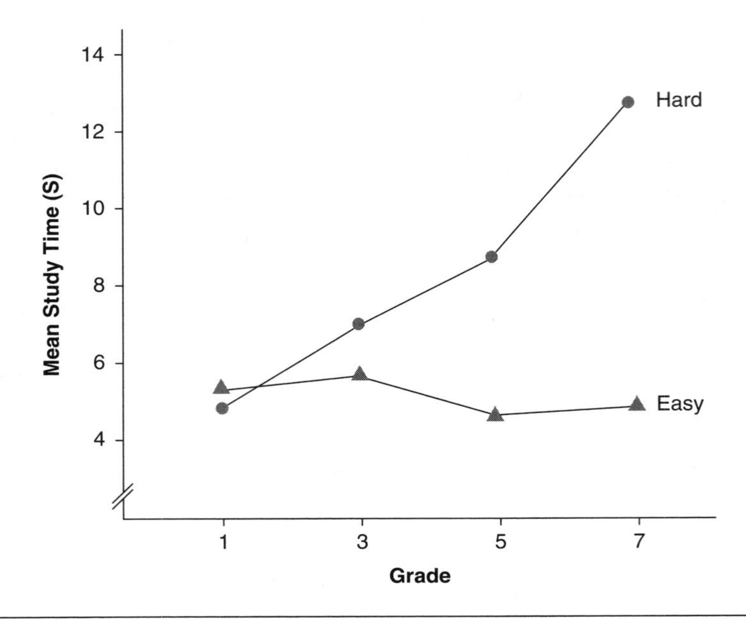

Figure 2.1 Interaction of age and experimental condition in the Dufresne and Kobasigawa study

SOURCE: Adapted from "Children's Spontaneous Allocation of Study Time: Differential and Sufficient Aspects," by A. Dufresne and A. Kobasigawa, 1989, *Journal of Experimental Child Psychology, 47*, 274–296. Copyright © 1989, Academic Press.

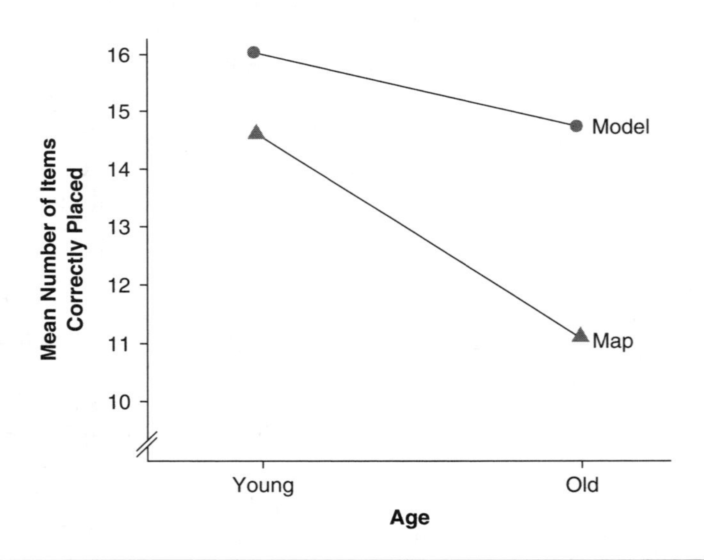

Figure 2.2 Main effects in the Cherry and Park study

SOURCE: Adapted from "Individual Differences and Contextual Variables Influence Spatial Memory in Younger and Older Adults," by K. E. Cherry and D. C. Park, 1993, *Psychology and Aging, 8*, 517–526. Copyright © 1993, American Psychological Association.

tendency for older participants to benefit more from the model.)

The interaction in the Dufresne and Kobasigawa study was between a subject variable and an experimentally manipulated variable. Interactions are not limited to such designs, however; rather, they can occur between independent variables of any sort. Interactions are possible, therefore, in any multiple-factor experiment. Figure 2.3 shows an interaction between two experimentally manipulated variables, and Figure 2.4 shows an interaction between two subject variables. The main finding of the Patterson and Carter (1979) study, pictured in Figure 2.3, was that the presence of a desired reward lessened children's self-control when they were simply waiting for the reward but enhanced self-control when they were working to complete a task rather than simply waiting. One finding of the Underwood, Coie, and Herbsman (1992) study, pictured in Figure 2.4, was that children's tendency to use "display rules" to mask feelings of sadness varied as a function of both age and sex. At the two younger grade levels, girls were slightly more likely than boys to report that they would attempt to disguise the fact that they were sad; by seventh grade, however, a clear difference had emerged in favor of boys.

As a comparison of Figures 2.1, 2.3, and 2.4 suggests, interactions can take a variety of forms. They can also become exceedingly complicated when more than two independent variables are involved. Although some researchers try, it is seldom possible to make sense of a four- or five-way interaction.

Interpreting any sort of interaction can be a complex matter, both statistically and theoretically (Levin, 1985; Rosnow & Rosenthal, 1995). I settle here for one basic point about interactions. The most general implication of a significant interaction between two variables is that interpretations of main effects involving those variables must be made with caution. In the Dufresne and Kobasigawa study, for example, there were main effects of both age and item difficulty; as Figure 2.1 reveals, however, the age effect was limited to the hard items, and the item effect was limited to the older children. In the Patterson and Carter study, in contrast, the main effect of the reward

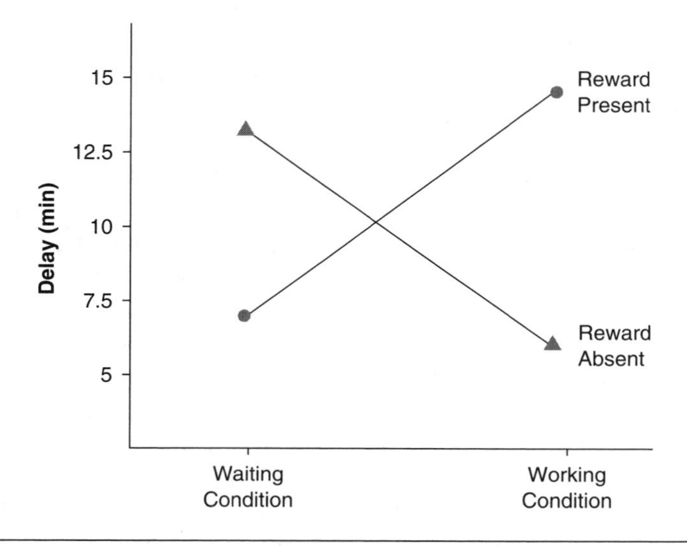

Figure 2.3 Interaction of experimental conditions in the Patterson and Carter study

SOURCE: Adapted from "Attentional Determinants of Children's Self-Control in Waiting and Working Situations," by C. J. Patterson and D. B. Carter, 1979, *Child Development, 50*, 272–275. Copyright © 1979, Blackwell Publishing, Inc.

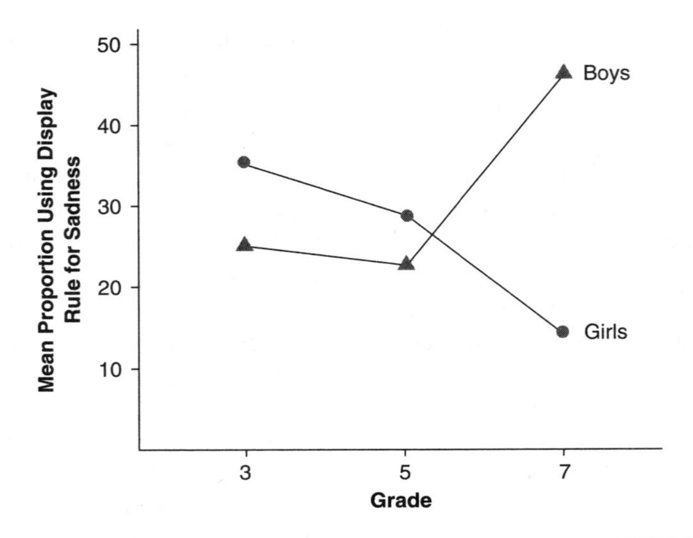

Figure 2.4 Interaction of age and sex in the Underwood, Coie, and Herbsman study

SOURCE: Adapted from "Display Rules for Anger and Aggression in School-Age Children," by M. K. Underwood, J. D. Coie, and C. R Herbsman, 1992, *Child Development, 63*, 366–380. Copyright © 1992, Blackwell Publishing, Inc.

present–reward absent manipulation was *not* significant, a finding that would suggest that this variable had no effect. Such a conclusion, however, is clearly contradicted by a separate analysis of the working and waiting conditions. An interaction, then, is a signal that the world is more complicated than we might have expected. Studying an independent variable in isolation cannot give us a full picture of the way in which that variable operates.

Note that the point just made can also be put in the context of external validity. An interaction implies a limitation in the generality of conclusions about the independent variables that enter into the interaction. In the Dufresne and Kobasigawa study, for example, the effects of item difficulty did not generalize across age, and the effects of age did not generalize across item difficulty. Conversely, the absence of an interaction is evidence in support of the external validity of conclusions regarding the variables in question—at least across the particular dimensions and levels that are sampled.

Threats to Validity

As we have seen, the ultimate goal in designing research is always to arrive at valid conclusions about the phenomena being studied. The converse to successful research design comes when there are threats to validity—uncertainties or limitations in what can be concluded that the design has failed to rule out. Several threats to validity were touched on in this chapter, and many more are discussed in the coming chapters. It will be helpful for the coming discussion to have a brief overview of the factors to be considered—an overall list and a set of definitions that can be referred to as necessary. This is the purpose of Table 2.5.

Table 2.5 is derived from an influential monograph by Campbell and Stanley (1966) that was subsequently elaborated by Cook and Campbell (1979) and Shadish et al. (2002). It does not provide an exhaustive list of things that can go wrong in research (Shadish et al. discuss 37 threats to validity!); it does, however, include many of the problems that are discussed later in

Table 2.5 Threats to Validity

Source	*Description*
Selection bias	Assignment of initially nonequivalent participants to the groups being compared
Selective drop-out	Nonrandom, systematically biased loss of participants in the course of the study
History	Potentially important events occurring between early and later measurements in addition to the independent variables being studied
Maturation	Naturally occurring changes in the participants as a function of the passage of time during the study
Testing	Effects of taking a test upon performance on a later test
Reactivity	Unintended effects of the experimental arrangements upon participants' responses
Instrumentation	Unintended changes in experimenters, observers, or measuring instruments in the course of the study
Statistical regression	Tendency of initially extreme scores to move toward the group mean upon retesting
Low reliability	Errors of measurement in the assessment of the dependent variable
Low statistical power	Low probability of detecting genuine effects because of characteristics of the design and statistical tests
Mono-operation bias	Use of a single operationalization of either the independent or dependent variable
Mono-method bias	Use of a single experimental method for examining possible relations between the independent and dependent variables

the text. Again, there is no expectation that the table is completely self-explanatory; its purpose, rather, is as a preliminary guide to concepts that will receive further attention as we go along.

Summary

This chapter begins with some basic terms and concepts. All research involves variables. *Dependent variables* are the outcome variables in research—for example, the number of aggressive acts in a study of aggression. *Independent variables* are potential causal factors that are controlled by the researcher—for example, reinforcement for aggression. The goal of most research is to determine whether variations in the independent variable relate to variations in the dependent variable—for example, does aggression increase following reinforcement?

The basic issue with respect to all research is validity. *Validity* refers to the accuracy with which conclusions can be drawn from research. Three forms are discussed in this chapter: *internal validity,* which concerns the accuracy of cause-and-effect conclusions within the context of the study; *external validity,* which concerns the generalizability of the conclusions beyond the study; and *construct validity,* which concerns the accuracy of the theoretical interpretation of the conclusions.

An important decision that the researcher must make concerns the participants for research. The goal in sampling participants is to obtain a *sample* that is representative of the larger *population* to which the researcher wishes to generalize. The common prescription for achieving representativeness is to do *random sampling* from the target population. In fact, most research in developmental psychology employs sampling procedures that are less than totally random. In some instances the deviations are intentional and systematic, the goal being to ensure that the sample possesses certain characteristics; *stratified sampling* and *oversampling* are examples. More commonly, the deviations reflect the use of samples that are readily available, an approach known as *convenience sampling.* How important such departures from randomness are varies across different topics. Nevertheless, representativeness and external validity remain important questions to examine for any study.

The discussion turns next to the construct of control. Three kinds of control are important if clear cause-and-effect conclusions are to be drawn. A first is over the exact form of the independent variable. A second is over other potentially important factors in the situation. Two methods of achieving this second form of control are discussed: holding the other factors constant and randomly dispersing variations in them across participants. The third kind of control is over preexisting differences among participants. One method of achieving this form of control, random assignment, is discussed in the present chapter; two others (matching and within-subject testing) are deferred for later consideration.

In some kinds of research, the degree of control is limited by the nature of the variables. The term *subject variable* refers to preexisting differences among people that are not experimentally manipulable; examples include age, sex, and race. The only control possible with such variables is through selection, a point that applies also to situations (e.g., maternal deprivation) whose experimental induction would be unethical. Although such variables are often of great interest to the developmental psychologist, cause-and-effect conclusions are difficult to establish in the absence of experimental manipulation. Specifying the exact basis for an effect can be a problem with a broad and multifaceted variable; ruling out other possible causal factors can also be difficult.

Subject variables are often of special interest when they enter into interactions. An *interaction* occurs whenever the effects of one independent variable depend on the level of another variable. In contrast, a *main effect* refers to an effect of an independent variable that is independent of the other factors in the study. Interactions can occur with independent variables of any sort, and they can take a variety of forms. Their most general message is that relations are complicated and that conclusions about any one variable must be made with caution.

The chapter concludes with a brief return to the concept of validity and an overview of some of the major threats to validity that are considered throughout the book.

Exercises

1. Find at least three recent summaries of developmental psychology research in the popular press (newspapers, magazines). For each, generate a list of possible threats to the validity of the research. If the description of the research is not complete enough for you to evaluate some forms of validity, specify what further information you would need.

2. Consider the task of recruiting research participants of the following ages: 6 months, 4 years, 12 years, 70 years. For each age group, generate a list of ways in which you might recruit prospective participants. For each method of sampling, discuss the likely representativeness of your final sample.

3. A particular construct can serve as either an independent or a dependent variable, depending on the way it is used in research. Consider the following constructs: anxiety, activity level, academic readiness. For each, generate a study in which the construct serves as (a) a dependent variable, (b) an experimentally manipulated independent variable, (c) a subject variable, and (d) a correlational variable.

3

Design

We saw in chapter 2 that all research involves comparison. In most cases, the comparison is between different levels of an independent variable. If the independent variable is a nonmanipulable subject characteristic such as age, then the researcher must select participants who already possess different levels of the characteristic. If the independent variable is an experimentally manipulable factor, then the researcher must assign participants to conditions that embody the desired levels of the factor. In either case, the researcher must do the selecting and assigning in a way that will allow a clear, nonconfounded comparison of the different levels being studied (the internal validity question), that will permit generalization to other samples and situations of interest (the external validity question), and that will allow identification of the causal bases for any relations that are found (the construct validity question).

The steps and the goals just sketched are issues of *experimental design.* Design, in the words of Kerlinger and Lee (2000, p. 449), is "the plan and structure of investigation"—the way in which studies are put together. Although the overall goal—valid conclusions—is always the same, studies can in fact be put together in a variety of ways. This chapter considers some of the most important dimensions along which research designs vary.

Once again, the sample studies described in chapter 2 can serve to illustrate some general points and standard terminology. Both the Dufresne and Kobasigawa and the Cherry and Park studies included two levels of an experimentally manipulated variable: easy versus hard items in Dufresne and Kobasigawa, and the model versus map context in Cherry and Park. Cherry and Park assigned separate participants to their two experimental conditions; hence their approach can be labeled a **between-subject design**. Dufresne and Kobasigawa tested all of their participants in both the easy and hard conditions; hence their approach can be labeled a **within-subject design**. One basic decision that a researcher must make is whether to use the same or different participants when comparing the effects of two or more experimental treatments. Strengths and weaknesses of both approaches are discussed later in this chapter.

Both of the sample studies also included the nonmanipulable variable of chronological age. In this case the methodological decision was the same: Both sets of researchers tested separate participants at the different ages. The strategy of testing different groups of people at different ages is referred to as a **cross-sectional design**. It is not the only possible approach to studying differences with age. Dufresne and Kobasigawa, for example, might have tested a sample of first-graders, waited 2 years and tested the children again as third-graders, waited another 2 years and tested the children in fifth grade, and finally tested the now seventh-graders after one last 2-year wait. The strategy of repeatedly testing the same sample of participants across the ages of interest is referred to as a **longitudinal design**.

It should be clear that there is a basic similarity between the between-versus-within contrast and the cross-sectional versus longitudinal contrast. In both cases the central issue is whether to examine effects within the same people or across different people. The relative merits of cross-sectional and longitudinal approaches are also discussed shortly.

Although the Dufresne and Kobasigawa and the Cherry and Park studies differed on the between-versus-within dimension, they were similar in another, perhaps more basic, respect. The similarity is that both studies did include an experimentally manipulated independent variable: easy-hard in Dufresne and Kobasigawa, and model-map in Cherry and Park. As we saw in chapter 2, not all studies include true independent variables of this sort. In so-called correlational or nonexperimental designs, the variables are simply measured, not controlled, and the researcher then searches for relations among the measures. Correlational designs are the third major topic considered in this chapter.

Because age comparisons are central to research in developmental psychology, the chapter begins with a consideration of designs for studying age. The discussion then moves to methods for comparing experimental conditions, and the chapter concludes with a consideration of the strengths and limitations of correlational research.

Age Comparisons

As noted earlier, age is just one of a number of subject variables that can be examined in research. Because the focus here is on age, it is worth noting an important difference between age and most other subject variables—a difference with implications for choice of research design. The investigator of a variable like sex or race does not have the option of waiting for the participants to change from one level of the variable to another; rather, studies of these variables must necessarily involve separate groups of people. In the case of age, however, today's 6-year-old is tomorrow's 8- or 10- or 20-year-old. It is because of this natural change along the age dimension that the researcher of age differences has the option of adopting either a within-subjects or a between-subjects approach.

There is a further point here as well. If we do a study to compare boys and girls, then our interest clearly is in differences (or, of course, lack of differences) between boys and girls. If we do a study to compare 6- and 10-year-olds, our interest may be partly in differences between 6- and 10-year olds, but it is likely to go deeper as well. What we may really be interested in is the possibility that the 6-year-old *will become like* the 10-year-old, or, equivalently, that the 10-year-old *was once like* the 6-year-old. Our interest, in short, may be not just in age *differences* but in age *changes*. As we will see, one of the thorny problems for developmental research is to determine when differences between age groups really reflect natural changes with age as people develop.

Longitudinal Designs

A *longitudinal study* tests the same sample at least twice across some period of time.

Although there are no clear-cut rules for deciding when a study with repeated testing becomes "longitudinal," at least two rough criteria seem to govern use of the label. First, the reference is usually to the study of naturally occurring rather than experimentally induced changes. Thus, the use of delayed follow-up tests in intervention or training research is not usually classified as longitudinal, even though the same children may be tested several times. Second, the reference is typically to repeated tests that span an appreciable period of time. Thus, simply testing the same people several times at 1-week intervals is not likely to earn a study the designation "longitudinal." Note, however, that what constitutes "an appreciable period of time" will vary with the developmental level of the sample. A series of 1-week retests probably *would* be considered longitudinal if the participants were only a few days old at the time of the initial testing.

Longitudinal studies are a good deal less common than are cross-sectional studies. It is not difficult to see why. Longitudinal studies are more time-consuming, more expensive, and more difficult to bring to successful completion than are cross-sectional studies. Consider as examples the two illustrative studies from chapter 2. The Dufresne and Kobasigawa experiment probably took a few weeks to complete. Had they opted for longitudinal rather than cross-sectional testing, the minimum time period for the study would have been 6 years. The contrast is, of course, even clearer for the Cherry and Park study. If these authors had decided on a longitudinal approach, they would have had to wait 40 or 50 years for their young adults to turn into elderly adults.

In itself, the extended time frame of the longitudinal approach is simply a practical problem—bothersome certainly, but not a threat to the validity of the conclusions. There are other problems associated with the extended time, however, that do threaten validity. One is the possible obsolescence of the tests and instruments being used. Because the essence of the longitudinal design is the earlier time–later time comparison, the researcher is committed to continued use of whatever measures were selected at the beginning of the project. Often, however, a test may become outmoded or lose its theoretical interest in the course of a long study; conversely, new tests and new issues will almost certainly arise. Thus, what one wants to know in 2010 may not be what one wished to know in 1980. This problem of test obsolescence is especially great in very long-term studies, such as some of the life-span studies begun in the 1920s (Kagan, 1964). It need not be a problem in more short-term longitudinal efforts.

Other problems relate to the nature of the sample in longitudinal research. Any at all long-term longitudinal study requires a substantial commitment of time and effort on the part of its participants (and, in the case of child samples, the parents of the participants as well). Samples may be selected, therefore, at least partly on the basis of factors such as belief in the value of research or probable geographical stability. If so, they may not be representative of the population to which the researcher wishes to generalize. Furthermore, any single longitudinal sample, all born at about the same time, constitutes but a single generation or **cohort**, and any findings may be at least somewhat specific to this one generation. We may be interested, for example, in how people change across the first 30 years of life. If all of our sample were born in 1940, however, then all we know with any certainty is how people born in 1940 changed as they encountered the changing world of the 1940s, 1950s, and 1960s. Had our sample been born either earlier or later, we might have obtained somewhat different results.

Although longitudinal samples may be nonrepresentative in various ways, they do at least avoid Campbell and Stanley's (1966) problem of **selection bias**—that is, the selection of initially nonequivalent groups for comparison. There can be no problem of selection bias when each participant is being compared with himself or herself. There can, however, be **selective dropout** (also labeled *attrition* or *mortality*), and such dropout does in fact occur. People can be lost from longitudinal samples for a variety of reasons—change of residence, unwillingness to continue to participate, or (especially in elderly samples) mortality in its literal sense. If such dropouts were random, then the only problems would be the reduction in sample size and the waste of effort in collecting early measures for which there turns out to be no later counterpart. Often, however, the dropout is not random but selective—that is, participants who are lost from the study are systematically different from those who remain. In longitudinal studies of IQ, for example, participants who drop out tend to have lower scores on the initial tests than do participants who continue (e.g., Siegler & Botwinick, 1979). Because the lower-competence dropouts contribute scores at the younger but not the older ages, the result is a "positive bias" in favor of the older groups. It is possible, of course, to limit the younger-older comparison to people who remain in the study and thus contribute scores at all ages. In this case, however, the initially nonrepresentative sample becomes even more nonrepresentative.

There is still one further way in which the participants in a longitudinal study differ from the broader population to which the researcher wishes to generalize. The difference is an obvious one: The participants in a longitudinal study undergo repeated psychological testing of a kind that most of the population escapes. Two of Campbell and Stanley's (1966) threats to validity are therefore potentially relevant. One is **testing**: the effects on later test performance of having taken the same or a similar test earlier. It seems likely, for example, that taking the same IQ test repeatedly at fairly close intervals could eventually begin to affect responses, and indeed research demonstrates that practice effects do occur (e.g., Rabbitt, Diggle, Holland, & McInnes, 2004). The second problem is the more general one of **reactivity**. Knowledge that one is the subject of research can affect anyone's responses, and such knowledge is probably especially salient for the participants in long-term, frequent-measurement longitudinal studies. Responses obtained from such participants, therefore, may not be representative of the typical course of development.

The final problem to be noted is an elaboration of the earlier point about the one-generational nature of many longitudinal samples. In longitudinal research there is an inevitable confounding between the age of the participants and the historical time of testing. This confounding follows from the fact that the age comparisons are all within subject; if we want different ages, therefore, we must test at different times. Suppose, for example, that we wish to examine possible changes between age 15 and age 20. We select a sample born in 1985 whom we test at age 15 and again at age 20. Should the second measure differ from the first, we would have two possible explanations for the difference: the fact that the participants are 5 years older, or the fact that one test was given in 2000 and the other in 2005. Age can never be disentangled from time in a longitudinal design.

How likely is it that this potential problem will actually be a problem? One determinant is undoubtedly the nature of the phenomena being studied. Let us move to the elderly years for an example of this point. Imagine that your interest is in changes in visual acuity as people age. You test a sample of 65-year-olds in 2000 and the same people at age 70 in 2005. Although historical time is a logically possible explanation for any changes you find, it is not a very plausible explanation in the case of a dependent variable like acuity. What is more likely, should you find differences, is that the visual system really undergoes natural changes between age 65 and age 70. Imagine, however, that instead of visual acuity you had tested attitudes toward airport security. You find that people are more concerned about security and more accepting of strong security measures at age 70 than at age 65. A clear case of increased caution with increasing age? Hardly, given that the events of September 11 intervened between the two measurements (I draw this example from Schaie & Caskie, 2005). In this case the historical-cultural explanation seems the more plausible one. In both cases, however, the standard longitudinal design permits conclusions that are at best plausible, not certain. The confounding of age and time can never be removed.

This catalog of the woes that can beset the longitudinal researcher raises the question of why anyone other than a confirmed masochist would ever attempt a longitudinal study. The answer, as might be expected, is that the longitudinal approach has a number of compensating virtues (Bullock, 1995; Hartmann, 2005; Jordan, 1994). It is to the more positive side of longitudinal studies that I turn next.

I noted earlier the distinction between age changes and age differences. As long as different samples are studied at different ages, the only direct measure a study can provide is of age differences; it is a further inference that any differences found reflect changes from the earlier to the later age. In longitudinal studies, however, the measure of age changes is direct rather than inferred. As we have seen, there can be questions about why the changes occur or how generalizable they are. But at least the focus is squarely on the central question of developmental psychology: that of intraindividual development over time.

The focus on intraindividual development makes the longitudinal approach uniquely suited to questions of individual consistency or individual change. Suppose that you wish to know whether a child's IQ tends to remain the same or to go up or down as the child develops. Clearly, you cannot answer this question by testing different children at different ages; rather, you must follow the same child as he or she develops. Whenever the interest is in individual consistency or change, then the longitudinal approach is not merely a nicety; it is a must.

The value of the longitudinal approach is not limited to tracing the course of a single trait or a single behavioral system over time. The value, rather, is much broader, for potentially *any* interesting cross-age patterning can be examined if we only obtain the measures of interest. In some cases the focus may be on the relation between one aspect of the child's development early in life and some other aspect later in life. We might seek to determine, for example, whether speed of skeletal maturation in the first 2 years relates to age of onset of puberty at adolescence. In other cases the interest may be in the relation between some aspect of the environment early in life and some aspect of development later in life. Thus, we might try to determine whether the parents' child-rearing practices during the child's first 2 years relate to measures of the child's personality at middle childhood or adolescence. Whenever the interest is in the relation between something early and something later, then the longitudinal approach is again a must.

Longitudinal research is also especially suited for tracing the continuous and progressive transformations that certain very general behavioral systems undergo as the child develops. This rather murky statement needs to be clarified by examples, and two examples will in fact readily occur to anyone familiar with research in developmental psychology. One is Piaget's research on the development of intelligence in infancy (Piaget, 1952). Piaget studied each of his three children longitudinally from birth to about age 2, painstakingly charting the sequences within and relations among various domains of intelligent behavior. The result was a conception of infant intelligence that in scope and insight surpassed anything that had come before and has served as a model for much that has been done since. It is possible that at least some of the same insights might have been derived from a judicious cross-sectional study of different babies at different ages; it is doubtful, however, that the full picture of infant intelligence and how it develops could ever have emerged without the intensive, almost day-to-day study of changes within a single child over time.

A similar argument can be made for research on early language development (e.g., Brown, 1973). In much the same way as Piaget, researchers of child language have used the longitudinal approach to trace gradual changes in language across the early years of language learning. What, for example, is the earliest form that negation takes in the child's speech, and how does this rudimentary form eventually turn into the complex rule system of the older child or adult? Again, the intensive longitudinal study, in which changes can be charted within a single child, has made possible a view of early language and how it evolves that probably could not have been gleaned from cross-sectional study alone.

Clearly, longitudinal research of the sort just described involves more than simply testing the same child at least twice; such research becomes, rather, an extended case history of individual development. When is such intensive longitudinal study likely to prove most fruitful? Certainly a prime rationale for such research lies in its application to new research terrain in which many of the basic phenomena still remain to be discovered. "New terrain" was certainly an accurate description of the field of infant intelligence when Piaget began his work. Once some idea of the general form and salient landmarks of development has emerged, more focused cross-sectional studies can be profitably applied. Longitudinal study is also especially suited to tracing the gradual construction of new abilities, the slow evolution of initially primitive forms through various intermediate steps to full maturity. How, for example (to add a Piagetian instance to the earlier example of negation), does the neonate's primitive grasping reflex eventually become the skilled, visually directed reaching of the older infant? Finally, the intensive study of the same children over time may be especially helpful when it comes to interpreting behavior—that is, attempting to move beyond the surface behavior itself to some conception of the underlying basis for it (a cognitive structure? linguistic rule? individually learned response? or what?). In most research the investigator sees the participants for the first and only time when they appear for testing, and the investigator's ability to make sense of their behavior is dependent on this very brief interaction. Piaget, however, had been studying the same children literally since birth, and his extensive knowledge of each child's background gave him an excellent basis for interpreting any particular behavior from the child.[1]

A last argument in support of the longitudinal approach is of a more negative sort. The main alternative to the longitudinal design is the cross-sectional design, yet the cross-sectional design is also subject to a number of criticisms. Possible problems with cross-sectional studies are the subject of the next section.

Cross-Sectional Designs

A cross-sectional study tests different people at different ages. For this reason, the cross-sectional approach cannot measure age changes directly, nor can it answer questions about individual stability over time. As we saw, these limitations of the cross-sectional approach provide a primary motivation for longitudinal study.

There are other possible problems. Because cross-sectional studies test different samples at different ages, the possibility of *selection bias* arises. Perhaps the groups being compared differ not just on the independent variable of interest (in this case age) but in other ways as well, and it is these other differences that produce differences on the dependent measures.

The issue of selection bias was discussed briefly in chapter 2 when I considered the special nature of age as an independent variable. As noted there, the goal is not to rule out all differences between groups other than chronological age, but just those differences that are not naturally associated with age. I noted too that in most cases the decision about what to match is fairly obvious—for example, sex, race, social class, IQ. What must be added now, however, is that actually achieving the desired matching may not always be easy. Developmental researchers typically draw samples of different ages from quite different sources—newborns from a hospital nursery, infants from parents who respond to solicitations to participate, preschoolers from preschools or day care centers, children between 5 and 11 from elementary schools, adolescents from junior high schools or high schools, adults from college classes. The populations served by these different settings may differ in a number of ways. Thus, even though the researcher may realize the importance of matching, selecting groups that are in fact comparable may prove difficult.

Bias can also occur in the form of *selective dropout* from the study. An initial equivalence between groups may quickly vanish if some participants drop out before testing is completed. The problem is not simply that there may be more dropouts at one age than another. The problem, rather, is the same one identified for longitudinal studies: People who drop out may be different from those who remain in. Thus, once again it is the "selective" part of selective dropout that threatens validity.

It is not hard to imagine situations in which selective dropout might bias comparisons between different ages. Suppose that we are doing a study of preschool children. We divide our sample into younger (2½ to 4) and older (4 to 5½) children, thus giving us two groups to compare. Our procedure is a fairly demanding one, requiring the child to process a variety of instructions and to continue to respond appropriately for a lengthy period. Not all preschool children are capable of such responding, and some are therefore lost from the study. The odds are strong that more children will be lost from the younger group than from the older group. The odds are strong also that those children who are lost will be, on the average, the less competent members of the sample. If so, we will end up with two noncomparable groups: a fairly representative sample of older children, and a nonrepresentative, biased-toward-superior sample of younger children. Clearly, any such differential dropout would decrease the chances of finding an improvement in performance with age.

Let us return to the issue of initial selection of participants. I have twice stated that decisions about what to match when comparing different age groups are generally straightforward. It is time now to consider the exceptions implied by the qualifier "generally."

Uncertainties about what should be matched are most likely when there is a wide separation between the ages being compared and thus many ways in which the groups potentially differ. They loom largest, therefore, in research comparing elderly adults with younger samples. Perhaps the most obvious example of this point is the variable of educational level. The average amount of schooling completed is greater now than it was 50 or 60 years ago. Suppose, then, that we wish to do a study comparing 25-year-olds and 75-year-olds. If we sample randomly at each age, our younger sample will be more highly educated than our older sample. We would have, then, a confounding of age and educational level. If we restrict our older sample to the more highly educated individuals, we will achieve comparability in educational level, but at the cost of selecting a nonrepresentative, positively biased older group. Neither solution is very satisfactory; perhaps the best course, if the researcher has the resources, is to incorporate both approaches. The main point, however, is that age and educational level, at this point in history, are unavoidably confounded in any attempt to compare adults of different ages.

The point just made about matching is actually part of a larger point concerning cross-sectional designs. I noted earlier that the longitudinal approach to studying age differences involves an inevitable confounding of age and time of testing. I will add now that the cross-sectional approach involves an inevitable confounding of age and generation or cohort. The samples in a cross-sectional study, being different ages, must necessarily be born at different times and grow up under at least somewhat different sets of circumstances. The disparity in educational opportunities between today's 25-year-old and today's 75-year-old is just one example of such generational differences. Many other examples could easily be cited. Today's 75-year-olds lived through the Great Depression as children, encountered a world war in adolescence and another war in young adulthood, somehow survived into adulthood without TV or computers or many other commonplaces of modern life, and so forth. Suppose, then, that we find that 25-year-olds and 75-year-olds differ on our dependent measure. Should we attribute the difference to differences in age or differences in generation?

As with the other threats to validity discussed in this chapter, the extent to which the age-cohort confound is in fact a problem depends on the particular kind of study being done. Two factors are important in assessing the likelihood of cohort effects. One factor is the dependent variable under study. If our focus is on political attitudes or IQ test performance, then cohort effects may be quite important; indeed, such effects have been clearly demonstrated in the study of IQ (e.g., Schaie, 2005). If our focus is on heart-rate change or visual acuity, then cohort effects are much less likely to be important. In general, the more "basic" and "biological" a dependent variable appears, the less likely it is to vary across cohorts. Note, however, that there can almost always be dispute about how "basic" and cohort-general a particular variable is. Perhaps, for example, visual acuity actually does vary across generations as a function of changes in factors such as adequacy of artificial lighting or presence of TV during the formative years.

The other factor to consider is the age spread of the sample. Cohort effects are most obviously a problem in studies with widely separated age groups. Indeed, the issue of cohort differences first arose in research comparing young adult and old adult samples, and it is still most often discussed in that context. At the other extreme, the child psychologist who compares 3- and 4-year-olds probably does not need to worry about the fact that one group was born in 2000 and the other in 2001. Samples within the span of childhood can usually be assumed to belong to the same generation. Even here, however, doubts may arise. What about a comparison between pre–Sesame Street and post–Sesame Street generations? What about a comparison between a computers-since-kindergarten grade-schooler and a computers-since-sixth-grade adolescent? We live in a time of rapid cultural and educational change, and these changes may affect at least some between-age comparisons even among child samples.

The final problem to be noted is that of **measurement equivalence**. If we wish to compare the level of a particular behavior or particular ability in different age groups, then we need a procedure that can accurately tap the behavior or ability at each of the ages being studied. Often, however, a test that is appropriate for one age may not be appropriate for another age. A test of classification skills, for example, may be a fine indicator of such skills among 7-year-olds but may be too verbally demanding for many 4-year-olds. If so, the test may measure different things at the two ages: classification at age 7 and vocabulary at age 4. Note that the test would still reveal a genuine and perhaps important difference between the two ages: 7-year-olds really do perform better on this measure than do 4-year-olds. But the basis for the difference might not be the one that the investigator is seeking to study.

The problem of measurement equivalence is not limited to cross-sectional studies. The issue arises in any comparison of different ages; thus it applies with equal force to the longitudinal approach. The particular form of the equivalence problem, however, is likely to be different in longitudinal than in cross-sectional studies. Consider the longitudinal study of aggression (e.g., Cairns, Cairns, Neckerman, Ferguson, & Gariepy, 1989). The investigator who studies aggression in a group of children at age 4 and again at age 12 is unlikely to be interested simply in comparing levels of aggression at the two ages. If levels of aggression *were* the focus, then serious problems would arise from the fact that the forms that aggression takes and the circumstances under which it occurs are quite different at age 12 than at age 4. The fact that the same children are being studied over time, however, probably means that the real interest of the longitudinal investigator is the *stability of individual differences* in aggression as children develop. The question, in other words, is whether children who are relatively high or low in aggression at age 4 are also relatively high or low in aggression at age 12. A child may be high in aggression at both 4 and 12 even though the frequency and forms of the behavior have changed greatly. This focus on relative standing within a group, rather than absolute level of response, provides a partial solution to the measurement-equivalence problem. Note, however, that it is still necessary to have valid measures of aggression at both ages.

More Complicated Designs

A clear message from the preceding discussion is that both the longitudinal approach and the cross-sectional approach suffer from various limitations. Table 3.1 summarizes the problems that have been discussed. Some of these problems are at least in principle avoidable—for example, the possibility of selection bias in cross-sectional research. Some of the problems, however, are intrinsic to the longitudinal and cross-sectional designs and hence can never be ruled out. Specifically, it is impossible ever to avoid the confounding of age with generation in the cross-sectional approach or the confounding of age with time of measurement in the longitudinal approach.

These limitations of the traditional longitudinal and cross-sectional designs have been much discussed in recent years, and they have motivated the development of several new procedures for studying changes with age. Because these new procedures have thus far been applied most often in studies of old age, I defer the main discussion of them until the chapter on aging (chapter 14). A brief introduction is possible here, however.

Figure 3.1 provides a schematic summary of the two designs discussed thus far. The body of the figure shows the ages that would be obtained from the various combinations of date of birth and year of measurement. A longitudinal design would be represented by any of the rows in the figure. In this case, a sample of people born at the same time is studied

Table 3.1 Problems With Longitudinal and Cross-Sectional Designs

Longitudinal	*Cross-Sectional*
• Practical difficulties (expensive, time-consuming)	• No direct measure of age changes
• Possible obsolescence of measures	• Inapplicable to issues of individual stability
• Possible nonrepresentative samples	• Possible selection bias
• Limitation to a single cohort	• Possible selective drop-out
• Possible selective drop-out	• Difficulty in establishing equivalent measures
• Effects of repeated testing	• Confounding of age and time of birth (cohort)
• Difficulty in establishing equivalent measures	
• Confounding of age and time of measurement	

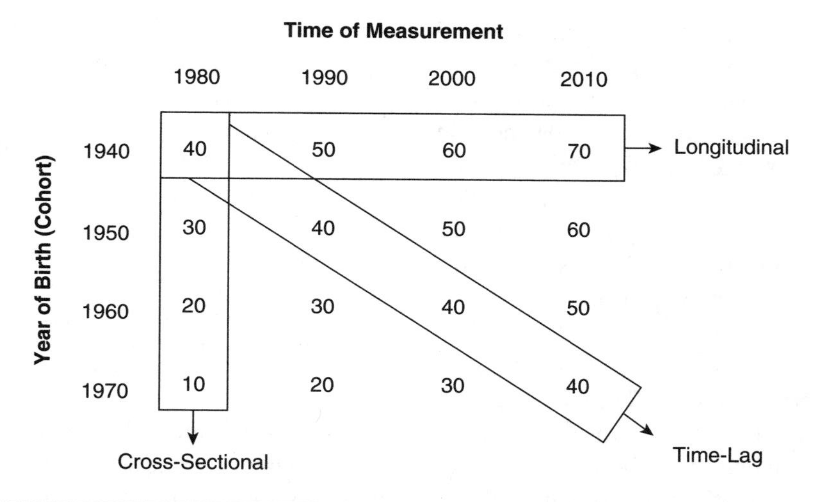

Figure 3.1 Examples of longitudinal, cross-sectional, and time-lag designs

NOTE: Numbers in the body of the figure indicate ages.

repeatedly across a span of ages. A cross-sectional design would be represented by any of the columns in the figure. In this case, separate samples born in different years are studied at the same time.

Figure 3.1 also includes a third design not yet discussed: the **time-lag design**. A time-lag design would be represented by any of the diagonals in the figure. Thus, we might study a sample of 40-year-olds in 1980, another sample of 40-year-olds in 1990, another sample of 40-year-olds in 2000, and another sample in 2010. Clearly, the time-lag design cannot give us direct information about age changes or age differences, because only one age group is studied. What it can do, however, is provide information about factors that may confound the age comparisons in longitudinal or cross-sectional designs. Specifically, if we find differences among our samples of 40-year-olds, then we know that these differences must reflect either generational factors (the main confound in the cross-sectional design) or time-of-measurement factors (the main confound in the longitudinal design) or, of course, some combination of the two factors. The fact that we cannot be certain which factor is important indicates that the time-lag design suffers from its own brand of confounding: a confound between generation and time of measurement.

Time-lag designs are not used very often. Occasionally, however, time-lag comparisons become available simply through the natural historical course of research. Piaget, for example, first studied young children's understanding of conservation during the 1930s and 1940s. When the Piagetian approach became popular decades later, the result was a second wave of conservation studies during the 1960s and 1970s. The conjunction of original research and newer research constitutes a time-lag comparison: two groups of children of the same age but born at different times and tested at different times. In this case neither cohort nor time of measurement appears important, since children of the 1970s responded to conservation tasks in essentially the same way as had children of the 1930s.

IQ tests provide a second, and contrasting, example. As I discuss more fully in chapter 12, when a child takes an IQ test, the child's performance is compared with that of children of the same age who formed the normative sample when the test was first developed. Thus, a

10-year-old (say) taking an IQ test in 2005 might be compared with 10-year-olds who took the same test in 1985. A common finding from such comparisons is a slight improvement in average performance over time, a phenomenon known as the *Flynn Effect* (Flynn, 1998). Because age is held constant, we know that the improvement must reflect effects of either time of measurement or (more probably) cohort.

Longitudinal, cross-sectional, and time-lag designs are sometimes referred to as the "simple" developmental designs. They are simple in comparison to the alternative, the decidedly *not* simple sequential design. A **sequential design** consists of a combination of longitudinal, cross-sectional, and time-lag components within a single research design, the goal being to tease apart effects of age, cohort, and time of measurement. The components can be combined and analyzed in different ways, and thus there are several different kinds of sequential design. In this chapter I discuss, briefly and generally, two of the kinds. In chapter 14 I return to the topic of sequential designs, but in this case in concrete rather than hypothetical form with a discussion of the major research program to date to employ such designs.

A word first about the logic of sequential designs. Ideally, what we would like to be able to do is to examine the contribution of all three potentially important factors—age, cohort, and time of measurement—within a single analysis. Unfortunately, doing so is precluded by the interdependencies among the three factors; as soon as any two of them are set, the levels of the third follow automatically. Once we have decided, for example, that we wish to study particular ages and particular cohorts, the times of measurement are necessarily fixed; they are whatever values we need to get the desired conjunction of ages and cohorts. The result of these interdependencies is that only two of the three factors can function as independent variables within the same analysis. The various sequential designs vary in which of the factors they concentrate on. In the first of the examples

Year of Birth (Cohort)	Time of Measurement 1980	1990	2000	2010
1940	40	50	60	
1950		40	50	60

Figure 3.2 Example of a cohort-sequential design

NOTE: Numbers in the body of the figure indicate ages.

discussed next, the independent variables are age and cohort; in the second, the independent variables are age and time of measurement.

Figure 3.2 presents a *cohort-sequential design*. A cohort-sequential design selects samples from different cohorts (i.e., years of birth) and tests them repeatedly across the same span of ages. It consists, therefore, of two (or more) overlapping longitudinal studies. In the example shown in the figure, groups born in 1940 and 1950 are tested three times across a 20-year span. Such a design offers several advantages in comparison to a standard longitudinal or cross-sectional approach: (a) Because different times of measurement are used, the age variable is not confounded with the cohort variable (the prime confound in cross-sectional studies). (b) Because samples are drawn from different years of birth, the longitudinal comparisons are not limited to a single generation or cohort. (c) Because different age groups are tested at each time of measurement, there is a cross-sectional as well as a longitudinal dimension. (d) Because the same age group is represented at different times of measurement, there is a time-lag dimension as well. There is, in short, more information than in a standard design, and thus more chance to disentangle the contributions of various factors.

Figure 3.3 shows a *time-sequential design*. A time-sequential design consists of two (or more) cross-sectional studies carried out at different times of measurement. In the example, samples of 40-, 50-, and 60-year-olds are compared in 1990,

Year of Birth (Cohort)	Time of Measurement: 1990	2000	2010
1930	60		
1940	50	60	
1950	40	50	60
1960		40	50
1970			40

Figure 3.3 Example of a time-sequential design

NOTE: Numbers in the body of the figure indicate ages.

in 2000, and in 2010. The samples at the different times may be either independent (i.e., different people at the three test occasions) or the same (if the original participants are followed longitudinally). This design has the same general virtue as the cohort-sequential design: It provides considerably more information than the simpler designs. A specific strength is that it unconfounds the variables of age and time of measurement (the prime confound in longitudinal studies). If independent samples are studied at the different times, this method also avoids some of the problems found with longitudinal designs (selective dropout, effects of repeated testing).

As noted, I consider these designs more fully in chapter 14. Two points can be made here, however. First, it is obvious that sequential designs, though more informative, are also considerably more costly—in time, effort, and money—than the simpler cross-sectional and longitudinal designs. Execution of the design pictured in Figure 3.3, for example, would require 20 years and (in the independent-samples version) nine groups of participants. In any research project there are a large number of things that would be desirable to do, only a subset of which it is actually possible to do. The best designs are always those that can actually be carried out.

The second point concerns the threats to validity that apply in the traditional designs—namely, the confound of age with cohort in the cross-sectional design and the confound of age with time in the longitudinal design. It is a point that applies to threats to validity in general. Although it is true that the goal of good research design is always to minimize threats to validity, the fact is that it is never possible to rule out every conceivable alternative explanation for one's findings. The question then becomes how plausible the alternative explanations are. And for much developmental research, especially within the span of childhood, the possibility of cohort or time-of-measurement effects is simply not plausible. In such cases, cross-sectional or longitudinal methods may produce—and indeed have produced—data about changes with age that are of considerable validity and use.

FOCUS ON

Box 3.1. The Microgenetic Method

As we have seen, one argument in support of longitudinal studies is that they provide a direct measure of change that is lacking in the cross-sectional approach. Most longitudinal studies, however, are limited to documenting the results or products of change. That is, they tell us what the individual is like at time 1 and time 2 and time 3. But they do not tell us *how* the changes from 1 to 2 to 3 come about, and they do not tell us about any intermediate states between 1 and 3 that are not represented in the times of measurement. In Robert Siegler's words, longitudinal studies provide snapshots of development (Siegler, 1996).

The microgenetic method, in contrast, is intended to provide something more akin to a movie of development. The **microgenetic method** refers to repeated, high-density observations of the

(Continued)

(Continued)

behaviors being studied across a period when change is occurring. It is therefore a form of longitudinal research, in that repeated observations are made of the same individuals across the time period of interest. In contrast to a standard longitudinal study, however, the observations are both more frequent and more closely spaced, and there is an emphasis on capturing not just levels of performance but processes of change.

Let us consider an example. Siegler and Jenkins (1989) were interested in how children develop strategies to solve simple arithmetic problems, such as 3 plus 5. Table 3.2 shows some of the strategies that children might use. To test these possibilities, Siegler and Jenkins performed a microgenetic study with 10 four- and five-year-olds who had not yet developed any of the more advanced strategies. The children participated in three experimental sessions per week across a period of 11 weeks. During each session they attempted to solve seven problems, and across sessions the problems gradually increased in complexity. Videotapes were made of the children's performance, and they were also directly questioned at various points about the strategies they were using. Through this approach, Siegler and Jenkins were able not only to document the gradual and often halting emergence of new strategies but also to identify precursors to and conditions for strategy change.

Table 3.2 Children's Strategies for Solving Simple Addition Problems

Strategy	*Typical use of strategy to solve 3 + 5*
Sum	Put up 3 fingers, put up 5 fingers, count fingers by saying "1, 2, 3, 4, 5, 6, 7, 8."
Finger recognition	Put up 3 fingers, put up 5 fingers, say "8" without counting.
Short-cut sum	Say "1, 2, 3, 4, 5, 6, 7, 8," perhaps simultaneously putting up one finger on each count.
Count-from-first-addend	Say "3, 4, 5, 6, 7, 8" or "4, 5, 6, 7, 8," perhaps simultaneously putting up one finger on each count.
Min (count-from-larger-addend)	Say "5, 6, 7, 8," or "6, 7, 8," perhaps simultaneously putting up one finger on each count beyond 5.
Retrieval	Say an answer and explain it by saying "I just knew it."
Guessing	Say an answer and explain it by saying "I guessed."
Decomposition	Say "3 + 5 is like 4 + 4, so it's 8."

SOURCE: From *How Children Discover New Strategies* (p. 59), by R. S. Siegler and E. Jenkins, 1989, Hillsdale, NJ: Erlbaum. Copyright © 1989 by Lawrence Erlbaum. Adapted with permission.

In discussing the results of this and other microgenetic studies, Siegler (1996) identifies five issues related to cognitive change for which microgenetic techniques can provide valuable data. Such techniques can inform us about the *path* of cognitive change: the sequences and levels through which children move in acquiring new knowledge. They can provide information about the *rate* of change: how quickly or slowly children master different forms of knowledge. They speak to the issue of *breadth* of change: when children acquire a new competency (such as a particular arithmetical strategy), how narrowly or broadly they apply it. They are relevant to the question of possible *variability* in the pattern of change: Do all children follow the same route in mastering a new concept? Finally, microgenetic methods can provide information about the *sources* of change: the experiences and processes through which children construct new knowledge.

The discussion to this point may have given the impression that the microgenetic approach is specific to the study of cognitive development. In fact, most applications of the approach to date have been in the cognitive realm. Among the other topics that have been studied are mnemonic strategies (e.g., Coyle & Bjorklund, 1997) scientific reasoning (e.g., Kuhn, 1995), problem solving (e.g., Chen & Siegler, 2000), and language (e.g., Ruhland & van Geert, 1998). The approach is not limited to cognitive outcomes, however; it has been applied, for example, to the study of early mother-infant interaction (Lavelli, Pantoja, Hsu, Messinger, & Fogel, 2005). Nor, as this last example illustrates, is it limited to older children.

As with all methods, the microgenetic approach is subject to possible criticisms and threats to validity (Miller & Coyle, 1999; Pressley, 1992). Perhaps the major concern is that the frequent, high-density observations may in themselves change the phenomenon being studied. Most 4-year-olds, after all, do not spend dozen of hours solving arithmetic problems and responding to explicit questions about what they are doing. Perhaps what occurs under such circumstances is in some ways different from the real-life processes that we are trying to capture.

As with any threat to validity, it is an empirical question whether this potential problem actually applies, and microgenetic researchers cite evidence that at least in some instances it does not (Kuhn, 1995; Siegler & Crowley, 1992). Perhaps the major argument in support of the approach is a more conceptual one, however. Understanding how change occurs is both one of the most fundamental and one of the most challenging questions in developmental psychology, and it is clearly desirable to have as many methods as possible with which to attack it. Microgenetic techniques are one such method.

Condition Comparisons

Within-Subject Versus Between-Subject Designs

I turn now to the question of how to make comparisons between two or more tests or experimental conditions. I noted earlier that two general approaches are possible: administering all tasks or conditions to the same participants or assigning different participants to different experimental groups. The former is labeled a *within-subject design;* the latter, a *between-subject design.* Because the discussion of these two approaches will involve much back-and-forth comparison, it is simpler to consider them together rather than separately.

How does an investigator decide whether to make comparisons within or between subjects? Just as in the longitudinal versus cross-sectional decision, matters of convenience may often play a role. Usually (with a qualifier to be noted shortly), a within-subject approach means that fewer participants are needed. Suppose, for example, that we have three tasks whose difficulty we wish to compare, and we know that we will need at least 20 respondents on each task to determine whether any differences in difficulty are present. If we opt for a between-subject approach, we will need at least 60 people to complete the study; with a within-participant approach, however, a mere 20 may suffice. Whenever the pool of possible participants is limited, the economy of a within-subject design may be attractive.

Considerations of convenience do not always fall on the side of the within-subject approach, however. The smaller sample size in a within-subject study is bought at an obvious price—namely, more time spent with each participant, either in longer experimental sessions or in a greater number of sessions. Especially in work with young children, lengthy or repeated sessions may tax the child's motivation or endurance. Even if the investigator is not concerned about such demands on the child, the parents or school authorities may be. In such situations, a between-subject design, which minimizes the demands on any one child, may be the most sensible approach.

Statistical considerations may also affect the within versus between decision. The statistical tests appropriate for within-subject comparisons are somewhat different from those appropriate for between-subject comparisons. Furthermore, within-subject tests are often more powerful than between-subject tests—that is, more likely to reveal a significant difference if a difference does in fact exist (I discuss the notions of significance and power more fully in chapter 8). This greater power stems from the reduction in unwanted variance afforded by the within-subject design. Recall the earlier discussion of primary variance compared with secondary or error variance. As I noted then, a goal of good experimental design is to maximize primary variance, or variance associated with the independent variable, and to minimize unwanted variance from other sources. I noted too that the inevitable differences that exist among different participants are one source of unwanted variance. Use of the same participants for all experimental conditions reduces such variance and hence enhances the power of any comparisons made. The result is a greater likelihood that a difference of a given magnitude will achieve statistical significance.

Both between-subject and within-subject designs are subject to their own particular forms of bias. The obvious threat in between-subject designs is selection bias. Because different people are assigned to different conditions, the possibility will always exist that any differences that are found between conditions reflect preexisting differences among the participants and not a true effect of the experimental manipulations. This possibility does not arise in a within-subject design, in which each participant responds under each condition. Note that this advantage of within-subject over between-subject designs parallels an advantage discussed earlier for longitudinal compared to cross-sectional approaches.

There are two ways to try to rule out possible selection biases in a between-subject design (recall Table 2.4). One is to match participants on variables of potential importance. I consider the pros and cons of matching shortly. The other is the approach discussed in chapter 2: random assignment of participants to different groups. If the sample size is sufficiently large and if the assignment to conditions is truly random, then preexisting differences among participants should be controlled and confounding of subject and condition avoided. As argued in chapter 2, the logic of the random-assignment approach is impeccable; the challenge is to ensure that the two "if" questions really do receive positive answers.

The most obvious threat to the validity of within-subject designs concerns the possible effects of extended testing. Consider a study in which the researchers wish to compare the relative difficulty of several cognitive tasks. They decide to use a within-subject design, in which every child receives every task. Because presenting several tasks takes time, the children may well become increasingly tired or bored as they move through the series of problems. If so, performance may be poorer on later tasks than on earlier ones. Alternatively, the children may be somewhat timid or confused at the start of the study but become increasingly relaxed and confident as the testing proceeds. In this case, performance may be better on later tasks than on earlier ones. In either case, the effects stemming from the repeated testing would cloud the intertask comparison that is the researcher's real interest.

"Warm-up" or "fatigue" effects of the sort just described fall under the general heading of order effects. The term **order effect** refers to any general tendency for response to change in a systematic fashion from early in a session to later in a session. Usually, the systematic change is either a general improvement or a general deterioration in performance.

Another potential problem in within-subject designs is the possibility of carryover effects. A **carryover effect** occurs whenever response to one task or condition varies as a function of whether another task or condition precedes or follows it. Let us try a simple example to clarify this rather forbidding definition. Imagine that we wish to compare the relative difficulty of two tasks: A and B. We will suppose that either task, presented in isolation, elicits 50% correct responses from our sample. It turns out, however, that when task A is presented first, experience with A suggests a helpful means of attacking task B; correct responses to B consequently rise to 70%. In contrast, when task B is presented first, experience with B suggests a means of solution that is maladaptive for task A; correct responses to A consequently fall to 30%. Note that in this case there is no general improvement or decline across the experimental session; rather, the finding is that response to one task depends on whether that task is presented before or after the other task. Although the specific mechanism may differ, the general import of order effects and carryover effects is the same: complications in the interpretation of task or condition comparisons.

Problems created by order effects are most likely when the experimenter adopts a constant order of presentation for the different tasks or conditions. An obvious prescription follows: Whenever comparisons among tasks or conditions are of interest, a single order of presentation should be avoided. There are two alternatives to constant order. One alternative is to randomize the order of tasks or conditions. In certain cases, perhaps especially when the number of tasks is large, randomization may be the most sensible approach. Generally, however, a better alternative than randomization is **counterbalancing** of the order of presentation. Counterbalancing is conveyed more easily through example than through definition; a simple example is given in the upper left portion of Table 3.3. As can be seen, counterbalancing is a method for distributing a particular task or condition equivalently across the various possible ordinal positions. Thus, in the example, task A occurs equally often in the first, second, and third positions; furthermore, it precedes and follows tasks B and C equally often in each position. The counterbalancing in this case is complete—that is, all possible permutations of the three tasks are used. Clearly, with more tasks the number of possible permutations increases; with four tasks there are 24 permutations (these are shown in the upper right part of Table 3.3), and with five tasks there are 120 permutations. In such cases, complete counterbalancing may not be feasible; it is still possible, however, to select a subset of orders that will provide a reasonable degree of balancing. Examples of such orders for four-task and five-task studies are shown in the bottom part of Table 3.3.

Counterbalancing has two advantages over randomization. First, it ensures that there is no confounding of task and order of presentation, an outcome that cannot be ensured by randomization alone. Second, because confounding has been ruled out, it permits the researcher to compare the different orders of presentation and tease out any order effects or carryover effects that may be present in the data. Note, however, that such effects are likely to be identifiable only if the sample size is reasonably large and each order is represented sufficiently often. This point provides the qualifier for the earlier statement that within-subject designs require fewer participants than between-subject designs: Whenever possible effects of order are of interest, then the *N* necessary for a within-subject study may increase substantially.

Table 3.3 Examples of Complete and Partial Counterbalancing

Complete balancing	
Three tasks	*Four tasks*
ABC	ABCD BACD CABD DABC
ACB	ABDC BADC CADB DACB
BAC	ACBD BCAD CBAD DBAC
BCA	ACDB BCDA CBDA DBCA
CAB	ADBC BDAC CDAB DCAB
CBA	ADCB BDCA CDBA DCBA
Partial balancing	
Four tasks	*Five tasks*
ABCD	ABCDE
BDAC	BEDAC
CADB	CAEBD
DCBA	DCBEA
	EDACB

Thus far I have discussed a number of factors that a researcher can weigh in deciding between a within-subject and a between-subject design. In some cases, however, there is no decision to make; the nature of the research question dictates the design to be used. Specifically, whenever the interest is in within-subject patterning of performance, then a within-subject design is necessary. Whenever the interest is in definite and persistent change as a result of the experimental manipulation, then a between-subject design is necessary. I now elaborate on both of these points.

The argument with respect to within-subject patterning parallels an argument that was made earlier in support of longitudinal designs. There, we saw that questions concerning individual consistency or individual change over time require a longitudinal approach that studies the same people as they develop. Similarly, questions concerning the relation between two or more measures at any given time require a within-subject approach that studies the same people across the different measures. Suppose, for example, that we wish to know whether children's social skills relate to their popularity with peers (Cillessen & Bellmore, 2002). Clearly, we cannot assess social skills in one group of children and popularity in another group; rather, we must have both measures for all children. Or suppose (to return to an earlier example) that we wish to know whether children's IQs relate to their grades in school. We cannot assess IQ in one sample and grades in another sample; again, we must have both measures for all children. Examples such as these illustrate a prime rationale for within-subject study: to identify interrelations and patterning in development.

The argument with respect to manipulations that produce change is in some respects similar to points made earlier concerning testing effects in longitudinal designs and carryover effects in within-subject designs. The essential point is that administering one task or experimental condition may change participants in a way that makes them unusable for other tasks or conditions. Suppose that we wish to compare the effectiveness of several different methods of training conservation concepts (e.g., Smith, 1968). We select a group of nonconservers and administer training condition A. We can hardly then take the same children and administer condition B, for if condition A is at all effective, many of the children will no longer be nonconservers! The same argument applies to any research whose goal is to bring about lasting change in its participants—intervention programs for so-called disadvantaged children, therapy programs for disturbed children, parent-education programs for new parents, and so forth. In each case, if we wish to compare the effectiveness of different programs we need a between-subject design that assigns different participants to the different approaches. Note too that the argument is not limited to attempts to produce sweeping changes à la intervention or therapy; the argument may apply to more focused, short-term changes as well. Suppose, for example, that we wish to know whether inducing children to use verbal rehearsal helps them on a short-term memory task (e.g., Ferguson & Bray, 1976). We cannot expect that children who have been taught such a strategy will necessarily abandon it once we remove the instruction to verbalize; rather, if we want a rehearsal–no rehearsal comparison we need to test separate groups of participants.

There is a possible objection to this last example and the conclusion drawn from it. In the verbal-rehearsal case our interest is not in the relative effectiveness of several different treatments; the interest, rather, is in whether a single treatment will lead to improvement over a no-treatment baseline. It is true that we cannot apply the treatment and later expect to get a measure of performance in its absence. But why not proceed in the opposite order—that is, first measure the children's natural level of memory performance, apply the treatment, and then measure memory performance again? Doing so would give us an example of what Campbell and Stanley (1966) labeled a One-Group Pretest-Posttest Design. The rationale would be that any improvement in performance from the pretest to the posttest would reflect the effects of the intervening treatment. If this rationale is valid, then there is no need to set up separate groups of participants.

In certain simple situations this kind of One-Group design may be sufficient for the researcher's purposes. Generally, however, it is not. The weakness of such a design should be evident from the earlier discussion of experimental control: It permits a confounding of the experimental treatment with a number of other factors that might produce a pretest-to-posttest change.

Let us take intervention programs as the example to make this point. Imagine that we find a group of at-risk 4-year-olds, give them a test of "academic readiness," subject them to a 1-year intervention program designed to enhance academic skills, readminister the academic-readiness test at the end of the program, and find that scores have improved significantly. Evidence for the effectiveness of our program? Not necessarily. It may be that the improvement results from natural biological-maturational changes as the children age from 4 to 5—Campbell and Stanley's **maturation** variable. It may be that the improvement results from other events in the children's lives during the course of the program—Campbell and Stanley's **history** variable. It may be that the improvement results from practice effects gained from taking the initial pretest—Campbell and Stanley's *testing* variable. Or it may be that the improvement results from the natural upward movement of initially low scores upon retesting—Campbell and Stanley's *regression* variable. None of these rival hypotheses can be ruled out with a One-Group design; all could be ruled out if we included a separate, no-treatment control group.

This comparison of within-subject and between-subject approaches could use some summarizing. Table 3.4 lists the various pros and cons that have been discussed as relevant to the within-subject versus between-subject decision.

Both between-subject and within-subject approaches come in a variety of forms. I turn next to two of the most important variants:

Table 3.4 Relative Merits of Within-Subject and Between-Subject Designs

Factor	*Comparison of designs*
Convenience	Fewer participants with within; less time per participant with between
Statistical tests	Generally more powerful with within
Order or carryover effects	A problem with within; not a problem with between
Possible selection bias	A problem with between; not a problem with within
Focus on within-subject patterning	Must have within; impossible with between
Focus on procedures that produce lasting change	Must have between; impossible with within

matched-groups designs (a form of between-subject research) and time-series designs (a form of within-subject research).

Matched-Groups Designs

A clear comparison of different experimental conditions requires that the participants assigned to the different conditions be equivalent at the start of the study. I have discussed two methods for creating such equivalence: random assignment of different participants to different conditions, and repeated testing of the same participants across all conditions. I now add a third possibility: use of a **matched-groups design**, in which participants are matched prior to their assignment to conditions.

The notion of matching was introduced briefly in chapter 2 in the discussion of random assignment. The question was posed then: Why settle for random assignment; why not ensure equivalence by matching the participants assigned to different groups on all the characteristics that might be important? A little thought will suggest an answer: We can never identify all the characteristics that might be important variables, and even if we could do so, we could never get the necessary data and achieve the necessary matching. Matching is always necessarily partial matching. Still, partial matching is presumably better than none; why not utilize it? It turns out that doing so has both advantages and disadvantages.

Since the most often matched-for variable in research with children is probably IQ, I will use IQ as our example. If we wish to match children on IQ, we must first administer IQ tests to all of our potential participants (or, perhaps, go to the school files and obtain already-collected IQ data). We then group together children with identical or close-to-identical IQ scores. The number of children in a group will depend on the number of experimental conditions—pairs of children if there are two conditions, trios if there are three conditions, and so forth. Working within these same-IQ groups, we then randomly assign different children to the different experimental conditions. Note, therefore, that random assignment remains important even in a matched-group design. Note also, however, that the initial matching on IQ ensures what randomization alone cannot ensure: that the experimental groups end up with equal IQs.

The great strength of the matching approach is that it does provide such exact and certain control for variables that might otherwise bias results. If IQ really does relate to performance on our dependent variable, then it is critical that there be no confounding of IQ and experimental condition. Matching also has certain statistical advantages. In much the same way as within-subject designs, a matched-group design reduces unwanted variance and hence increases the power of the statistical tests. Matching may be especially helpful, therefore, when the power to detect effects is known to be low—when the sample size is limited, for example, or when the expected differences between groups are small.

The main disadvantages of matching revolve around the question: Is it worth it? Matching typically requires a substantial investment of effort on the investigators' part, especially if they must pretest all of the potential participants (as opposed to relying on already-existing data). If the matched-for variable is not in fact related to performance on the dependent variable, then the matching will have added nothing. If the sample size is large and random assignment is used, the groups will probably end up equivalent anyway, and again matching will add nothing. The point here has to do with efficiency of effort. Any research project involves selection of a relatively few specific procedures from a much larger pool of potentially informative procedures. To devote a portion of one's limited time and effort to procedures that do not enhance the study is simply bad research practice.

In addition to the possible waste of effort, matching can sometimes create particular problems. In some cases, administering the matching pretest may bias the participants' responses to the later test of interest (Campbell and Stanley's reactivity variable). Perhaps, for example, being taken out of their classroom and given an IQ test is anxiety-arousing for some children and makes them suspicious of the friendly tester who later invites them to "come play a game." The tester's attempt to create a game-like atmosphere for the measures may therefore come to naught, and the validity of the study may be affected. Matching can also sometimes result in loss of participants. If participants are matched in the manner described earlier, then the unit becomes the matched group rather than the individual child—for example, the trios of matched-for-IQ children in a study with three experimental conditions. If any member of a trio is lost from the study for any reason, then the other two must be eliminated as well. Whenever dropout seems likely, matching may turn out to be a costly decision.

There is one situation in which matching is a tempting but usually unsound procedure. It is the case in which the investigator wishes to bring about equality in initially unequal groups of subjects. We saw an earlier example in the discussion of the differing educational levels of young adult and elderly adult samples. Consider another example, drawn from Neale and Liebert (1986). Imagine that you are interested in determining whether high school graduates achieve greater economic success in later life than do people who drop out of school. You are concerned, however, that the two groups differ in average IQ—perhaps a mean of 105 for the graduates and 90 for the dropouts. This IQ discrepancy provides an alternative explanation for any group differences you may find: Perhaps differences in economic success are simply reflections of differences in cognitive ability and have nothing to do with completing or not completing high school. You decide, therefore, to match the graduate and dropout groups in IQ. With IQ ruled out as a possible cause, you can more confidently attribute differences in economic success to the benefits of finishing school.

There are at least three problems with such a procedure, two of which I discuss here and one of which I defer for later consideration. First, the procedure guarantees some limit in external validity, since at least one of the two groups will not be completely representative of its parent population (i.e., either unusually high-IQ dropouts or unusually low-IQ graduates). Second, matching the groups on one dimension may systematically unmatch them on other dimensions that are related to finishing school.

Suppose, for example, that you decide to set the mean IQ for both groups at 90. In this case you will have a typical group of dropouts, but your group of graduates—precisely because they have succeeded despite mediocre IQs—is likely to be above average in other characteristics (e.g., motivation, family support) that contribute to school success. Conversely, setting the mean IQ at 105 will yield a typical group of graduates; now, however, the dropouts will be below average on the other determinants of performance in school. Matching the groups on one dimension may thus have the unintended effect of making them in general less similar rather than more similar.

The third problem with matching unequal groups is that it can lead to effects associated with statistical regression. In chapter 4 I will discuss how regression can be a problem in matched-groups designs in the context of a general consideration of statistical regression as a threat to validity.

Time-Series Designs

Time-series designs are easiest to introduce with an example. The goal of a project by Hall et al. (1971) was to reduce the disruptive "talking-out" behavior of a 10-year-old boy in special education classes. Their study, like all time-series research, proceeded in several phases.

The first phase was a **baseline** period: measurement of the initial frequency of the target behavior under normal classroom conditions. As Figure 3.4 shows, the behavior was indeed frequent—three to five outbursts across each of five 15-minute sessions. Following the baseline came the first application of the experimental treatment: The teacher ignored instances of talking-out behavior but paid increased attention when the child was behaving productively. The apparent result of this "contingent attention" regimen was a dramatic drop in talking out, as shown in the second part of Figure 3.4. The experimental treatment phase was followed by a reinstatement of the baseline conditions, during which the talking-out behavior shot back up in frequency. Finally, in a fourth and last phase the contingent attention treatment was restored, and talking out returned to a low level.

The Hall et al. study (1971) is an example of an *A-B-A-B* time-series design: an initial baseline phase (the first A), followed by an initial application of the experimental treatment (the first B), followed by a second baseline (the second A), followed by a second experimental treatment (the second B). Let us work through the rationale for each of these phases. The first baseline is clearly necessary—we need to know the initial, preintervention level of the target behavior to determine any effects of the experimental treatment. The first intervention phase is of course also necessary. But why not stop once we have shown that the experimental treatment reduces the behavior—that is, why go beyond an A-B design? The answer is that a simple A-B design would be subject to all the threats to validity (maturation, history, etc.) discussed earlier with respect to within-subject designs in general. These threats are especially difficult to rule out when we have only one participant for the research, as is the possibility that the change is simply random fluctuation that would have occurred even without the treatment. If we can show that the target behavior reemerges when we withdraw the treatment, we can be more certain that the treatment really was responsible for the decline. If we can show that a second administration of the treatment is associated with a second decline, then we can be even more certain that the treatment is the causal agent. And, of course, there are also pragmatic and ethical reasons for adding the final B phase in the A-B-A-B design. The goal, after all, is to reduce the undesirable behavior; thus, we do not want to end the study with the behavior still at its height.

It should be clear from this description that a **time-series design** is a special form of within-subject research. It is within-subject in the sense that each participant receives every level of the independent variable and comparisons are made

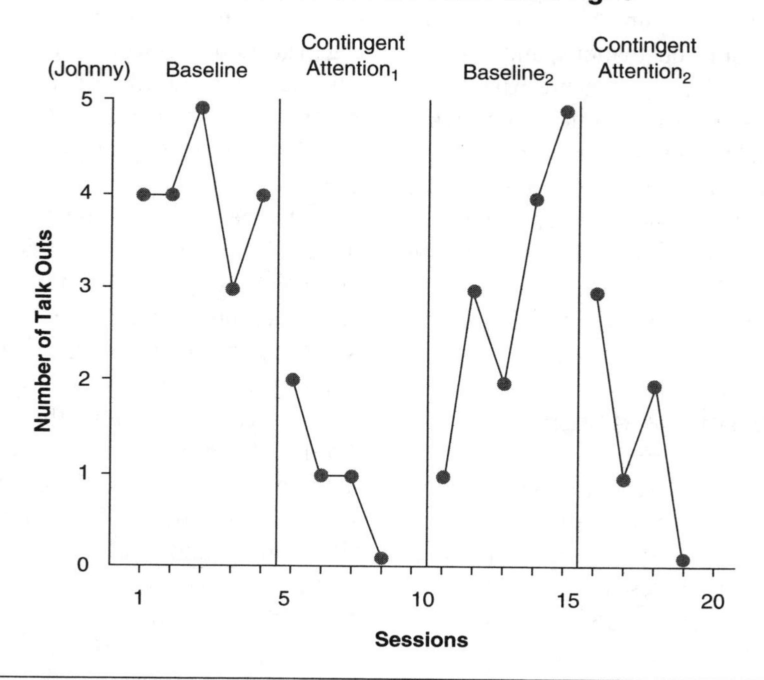

Figure 3.4 Example of a time-series design

NOTE: The level of the target behavior (talking out) varies as a function of the presence or absence of the experimental treatment.

SOURCE: From "The Teacher as Observer and Experimenter in the Modification of Disputing and Talking-Out Behaviors," by R. V. Hall, R. Fox, D. Willard, L. Goldsmith, M. Emerson, M. Owen, T. Davis, & E. Porcia, 1971, *Journal of Applied Behavior Analysis, 4*, p. 143. Copyright © 1971 by the Society for the Experimental Analysis of Behavior. Reprinted with permission.

within rather than between people. A time-series study also differs from the kinds of within-subject research discussed earlier in several respects, however. In most within-subject research, the levels of the independent variable represent different forms of some task or treatment (e.g., the easy-hard comparison in the Dufresne and Kobasigawa study); in a time-series study the levels are the presence or absence of the experimental treatment. In most within-subject studies, the comparisons are made within a single experimental session; in a time-series study, the comparisons are spread out over repeated sessions. Most within-subject studies sample and analyze groups of participants; many time-series studies (like the Hall et al., 1971, study) involve but a single participant. The time-series approach, in fact, constitutes the major source of designs for *single-subject research*—that is, research that attempts to identify effects of an experimental manipulation within a single participant. Finally—and related to these other points—time-series studies are often carried out for pragmatic purposes, the goal being to demonstrate the effectiveness of some intervention in ameliorating a problem behavior (as was true in the Hall et al. study). They are most often seen, therefore, in clinical or educational settings.

Time-series designs can involve complexities of both implementation and interpretation that I have not attempted to discuss. They can also encompass naturally occurring time series (for example, variations in purchasing behavior in response to economic fluctuations) and not just, as in Hall et al. (1971), those that are experimentally created. And whether naturally occurring or experimentally induced, they can come in many forms in addition to the A-B-A-B design described here. More extensive discussions of time-series research can be found in Barlow and Hersen (1984), Cook and Campbell (1979), Kazdin (1998), and Velicer and Fava (2003).

Correlational Research

In chapter 1 I noted several contemporary, socially important issues for which research in developmental psychology can be informative. Let us return to one of these issues for an example of correlational research. McLeod, Atkin, and Chaffee (1972) were interested in possible effects of TV violence on aggression in children. They collected various measures of aggression in a sample of 6th-graders through 10th-graders. They also measured how much violent TV each child in the sample typically watched. Their interest was in whether there was a relation between watching violent TV and being aggressive—that is, did the children who watched the most violence on TV also tend to be the most aggressive? In their study (as in many other similar studies) there *was* such a relationship, an outcome compatible with the hypothesis that watching violent television promotes aggression.

The study by McLeod et al. is an example of **correlational research**. It is correlational because there was no manipulation of an independent variable. McLeod et al. did not experimentally control the type of TV that their sample watched, nor did they control the level of aggression that the children showed. Instead, both TV viewing and aggression were *measured* as they naturally occurred, the intent being to see whether scores on one index covaried with scores on the other. Such a relation might be positive, with high scores on one measure tending to go with high scores on the other. This was the case in the study by McLeod et al. Or the relation might be negative, with high scores on one measure tending to go with low scores on the other.

Outcomes in correlational research are often assessed through use of a **correlational statistic**, which will be discussed more fully in chapter 8. For now, let us note that a correlation statistic is a measure of the degree of relation between two variables; it ranges from −1 (a perfect negative relation) through 0 (no relation) to +1 (a perfect positive relation). In the study by McLeod et al., the correlations varied to some extent, depending on the age and sex of the sample and the particular measure of aggression used; most of the values, however, fell in the range of .2 to .3. Such correlations indicate a modest positive relation between TV violence and aggression.

Although correlation statistics are typically associated with correlational research designs, it is important to note that the statistic and the design are separable. Statistics other than correlations can be used to examine the results of correlational research. McLeod et al., for example, might have divided their sample into high, medium, and low TV watchers and then used *t* tests or analysis of variance to compare levels of aggression across the three groups. In this case the statistic would be different, but the design would remain correlational. Because of this independence of design and statistic, some researchers prefer the term *nonexperimental* for the kind of research at issue here. Whatever the label, the defining aspect of such research is that variables are simply measured, not experimentally controlled.

Correlation and Causation

One of the truisms of research is that correlation does not imply causation. That is, simply from knowing that two variables are correlated, we cannot establish what causal relation, if any, holds between them. Thus, the results of the McLeod et al. study are compatible with the hypothesis that TV violence causes aggression, but the results cannot prove that this hypothesis is true.

Before discussing why correlation does not imply causation, it is worth noting that the reverse direction does hold true: Causation does imply correlation. That is, if two variables are causally related, we should expect (except in unusual circumstances) to find a correlation between them. Thus, correlation is a necessary basis for inferring causality, but it is not sufficient.

This basic limitation in correlational research stems from the absence of experimental control. As has been stressed repeatedly, it is control—control over the nature of the independent variable, control over the assignment of participants to conditions, control over other potentially important variables—that makes internally valid conclusions about cause and effect possible. Because correlational research lacks all these forms of control, the best that such studies can do is to demonstrate that two or more measures covary. They cannot tell us why.

Consider the McLeod et al. study. There are in this study, as in most correlational studies, three possible explanations for the correlation. One possibility is that watching violent TV causes children to be more aggressive. Had McLeod et al. experimentally manipulated TV viewing, they might have established this conclusion with some confidence. But because there was no experimental manipulation, there is a second possibility: Perhaps children who are already aggressive seek out violent TV. In this case it is the aggressive tendency that causes the TV viewing, not the reverse. Finally, there is still a third possibility: Perhaps TV viewing and aggression are both caused by some third factor but are not themselves causally related. It may be, for example, that certain parents' child-rearing practices promote both aggressive behavior and a liking for violent TV; the two measures thus covary, but neither one has any causal effect on the other.

This argument can be put in more general terms. Whenever there is a correlation between variable A and variable B, three possible explanations must be considered: A causes B, B causes A, or some third factor C causes both A and B.

The inability to establish causal relations is obviously a critical limitation to correlational designs. Why, then, are such designs used? The basic reason is that such designs are often the best that we can do. Many variables cannot be experimentally manipulated for ethical or practical reasons—parental child-rearing practices, for example, or exposure to drugs during the prenatal period. In such cases the only approach possible is a correlational one. In other cases experimental manipulation is possible but difficult, especially if the goal is to combine experimental control with a natural setting. The topic of TV violence and aggression provides an example of this point. It *is* possible to manipulate TV viewing experimentally and to measure subsequent aggression, and a large number of studies have done so; to varying degrees, however, all such studies are subject to criticisms of artificiality and lack of external validity. In a study like that of McLeod et al., however, the focus is squarely on the two variables of interest: naturally occurring TV viewing and naturally occurring aggression. A final virtue is that the correlational approach may allow us to sample a wider range of variation than is possible with an experimental design. In an experimental study of TV and aggression, we would probably have to limit ourselves to presenting two or three different types of TV experience. With a correlational approach, however, we can encompass the whole range of naturally occurring experiences, from 2 or 3 hours per week viewing on the one end to perhaps 40 or 50 hours on the other.

Ways to Strengthen Causal Inferences

Causality cannot be established with certainty from a correlational design. There are techniques, however, for heightening the plausibility of any causal inferences that might be drawn. In this section we will consider several such techniques.

A first strategy is quite commonsensical but still worth noting. In some cases one of the A-B causal directions is ruled out by the nature of the variables. Suppose that we find a positive correlation between body size and level of aggression. It is plausible that body size in some way affects aggression (although we would still need to specify exactly how). It is not plausible, however, that level of aggression has any causal effect on body size. In cases such as this, we need to entertain just two hypotheses: A causes B, or C causes A and B. The B to A link, however, is not a concern.

The logical-argument approach is relevant to the issue of the directionality of a causal relation between A and B. A second method is especially appropriate for eliminating third-factor C explanations. It makes use of a statistical procedure called the **partial correlation technique**. Partial correlation is a procedure for statistically removing, or "partialing out," the contribution of one variable from a correlation between two other variables. Essentially, what the partial-correlation technique does is to hold the potentially troublesome third variable constant while examining the relation between the two variables of interest. It is equivalent to asking how A and B relate in a sample in which everyone has the same score on variable C. The issue, of course, is whether the A-B correlation remains significant even when we control for C.

Suppose that we find a positive correlation between TV viewing and aggression but suspect that some third factor, such as methods of child rearing, actually produces the correlation. Assuming that we can obtain acceptable measures of child rearing, we could then use the partial-correlation technique to eliminate the contribution of child rearing from the TV-aggression correlation. If we find that the correlation remains as large or about as large as it was originally, we could conclude that child rearing was not an important confounding factor. Conversely, a substantial drop in the size of the correlation would indicate that child rearing does make an important contribution to the TV-aggression correlation.

Although the specific procedures differ, the goal behind the partial-correlation technique is the same as that for the matching technique discussed earlier in the chapter. In both cases the researcher seeks to eliminate confounding factors by equalizing them across the groups being compared. With matching, the equalization comes before the fact, in the assignment of participants to groups; with partial correlation, it comes after the fact, in the statistical removal of the confounding factors. Partial correlation also shares the same basic limitation that we saw in the case of matching: It is impossible through such techniques ever to remove *all* possible confounding factors. There are, in other words, lots of variable Cs, and no researcher is ever able to measure and control for them all.

A third approach to extracting causality from correlational data is concerned with the directionality issue—that is, does A cause B or does B cause A? The starting point is a basic fact about causality: Causes must come before their effects. Thus, if we can chart variations in the relation between A and B over time, we can come closer to determining whether it is A that leads to B or the reverse.

Charting across-time relations requires longitudinal study. Imagine that we conduct a study with 5-year-olds in which we measure both how much violent TV each child watches and how aggressive each child is. Imagine also that we study the same children 3 years later at age 8, again measuring both TV viewing and aggression. Clearly, we would then have two standard correlational studies, one with 5-year-olds and one with 8-year-olds. But we would also have an across-time correlational study, in that we could examine the correlations between the measures at age 5 and the measures at age 8. Suppose (to posit a simple case) we find that TV viewing at age 5 correlates with aggression at age 8 but that aggression at age 5 does not correlate with TV viewing at 8. In other words, early variations in TV experience are predictive of later variations in aggression; early variations in aggression, however, do not predict later TV viewing. Such an outcome would be compatible with the hypothesis that TV is the causal agent in the TV-aggression relation. It would not prove the hypothesis, even in the relatively simple case in which the results come out as clearly as in our example. But it would add support to the argument.

Discussion of a final method of strengthening causal inferences will serve also to introduce a basic point about research methods. Sometimes it is possible to complement the correlational approach with an experimental examination of the same issue. What we can do, in other words, is manipulate the variable that we believe to be causal and measure the effects on the other variable, thus creating a true independent variable–dependent variable relationship. As I noted earlier, the literature on TV violence includes a number of such experimental studies, in which TV viewing has been manipulated and subsequent aggression measured. Such studies provide exactly the forms of control that are lacking in correlational research. Because we have experimentally manipulated variable A, there can be no uncertainty about the causal direction between A and B—variations in B must follow from variations in A rather than the reverse. And because we can control factors other than our independent variable, there can be no third factor C to confound any A-B relation. We can be much more certain, therefore, about any cause-and-effect conclusions that we might draw.

The general point that this example illustrates concerns the value of converging operations when investigating complex, hard-to-study topics. The term **converging operations** (also labeled the *multimethod approach*—e.g., Brewer & Hunter, 1989) refers to the use, either within or across studies, of a variety of distinct methods of studying a particular topic (the converse—the exclusive use of one method—results in the threat to validity that Cook and Campbell, 1979, labeled *mono-method bias*). The basic idea is that the strengths of one method can, to at least some extent, compensate for the weaknesses of another method, and that conclusions based upon a convergence of evidence from different methods can be held with a greater certainty than can conclusions based on one approach alone.

This argument certainly applies to the issue of TV violence and aggression. Experimental approaches to this issue are uniquely suited for the identification of causality; at the same time, such studies may suffer from a variety of problems (artificiality, reactivity, etc.) that make their external validity doubtful. Correlational designs avoid many of the pitfalls of manipulative studies; as we have seen, however, the correlational approach is intrinsically limited in what it can tell us about cause and effect. It is precisely because of these limitations of any one method that we need a convergence of evidence from different methods. Thus, the correlational studies of TV viewing give us more confidence that experimental demonstrations of the impact of TV violence really do have some real-life generalizability. Correspondingly, the fact that experimental manipulations of TV violence affect aggression gives us a basis for arguing that TV viewing really is the causal factor in the TV-aggression correlation.[2]

Summary

This chapter addresses three issues that fall under the heading of experimental design:

comparison of different age groups, comparison of different experimental conditions, and contrasts between experimental and correlational designs.

Two designs have been most common in examinations of different age groups: the longitudinal and the cross-sectional. In a *longitudinal study* the same participants are studied across some span of time. Such an approach provides the only direct measure of age changes as opposed to age differences; it also provides the only way to study individual stability or individual change over time. On the negative side, longitudinal research is costly and time-consuming, factors that undoubtedly contribute to its relative infrequency. Longitudinal research is also subject to a number of biases. These biases include *selective drop-out* of participants, *testing* effects stemming from repeated exposure to the same measures, and the inevitable confounding between the age of the participant and the time of measurement.

In a *cross-sectional study* different participants are studied at different ages. The cross-sectional approach is generally more economical than the longitudinal approach, it avoids many of the problems of longitudinal study, and for many research questions it is perfectly adequate. Cross-sectional designs also have their limitations, however. Because each participant is studied just once, a cross-sectional study cannot provide direct evidence of changes with age. *Selection bias* in the formation of the different age groups may hamper the age comparisons. A further problem, which applies to both cross-sectional and longitudinal designs, is that of *measurement equivalence:* selecting measuring instruments that are equally appropriate for the age groups being compared. Finally, cross-sectional designs also contain an inevitable confounding: between the age of the participants and the generation or cohort to which they belong.

Limitations of the classic longitudinal and cross-sectional approaches have led in recent years to the development of alternative designs. In a *time-lag design,* age is held constant while generation and time of measurement are varied. Such designs provide an estimate of the importance of factors that are confounded with age in the traditional designs. More ambitious are the various *sequential designs,* which involve combinations of the simpler longitudinal, cross-sectional, and time-lag approaches. Sequential designs are unquestionably more informative than the simpler approaches; they are also more costly, however, and they still do not remove all possible sources of confounding.

The second section of the chapter is devoted to designs for comparing different tasks or experimental conditions. Two main approaches exist: *within-subject designs,* in which every participant responds to every task or condition, and *between-subject designs,* in which different participants are assigned to the different tasks or conditions. The within-subject approach is sometimes more economical, often affords greater statistical power, and is free of some of the problems (such as selection bias) that can affect between-subject designs. A within-subject approach is also essential when the interest is in within-subject patterning of performance. A between-subject approach, in turn, avoids many of the problems of within-subject testing—in particular, *order* or *carryover effects* stemming from the repeated testing. A between-subject approach is also essential when the experimental manipulation is intended to produce definite and lasting change.

The discussion turns next to specific variants of the between-subject and within-subject approaches. With a *matched-groups design,* participants are matched prior to assignment to experimental conditions. The advantage of matching is that it ensures that groups are equivalent on variables that might affect performance. Possible disadvantages include the increased time and effort, the potentially biasing effects of taking a matching pretest, the increased subject attrition if any matched-for participant is lost from the study, and the possibility that the groups will be systematically unmatched on variables other than the matched-for variable. In a *time-series design,* an experimental treatment is repeatedly administered and withdrawn, and changes in behavior are charted as a function of the treatment's presence or absence. Such designs are used most frequently in clinical or educational settings, often in the form of single-subject research.

The chapter concludes with a discussion of *correlational research.* In a correlational study there is no control of an independent variable; rather, two or more variables are measured, and the interest is in whether scores on the different measures covary. Correlational designs may be the only research option available for variables whose experimental manipulation is either impossible or very difficult. Furthermore, correlational research has the advantage of encompassing more levels of a variable than is possible in a controlled experimental study. On the negative side, the absence of experimental control means that correlational designs are intrinsically limited in what they can tell us about cause and effect. Methods useful for reducing the uncertainty and moving closer to the determination of causality include logical analysis of which causal directions are possible; *partial correlation,* in which the contribution of third-factor variables is statistically removed; longitudinal study to trace the pattern of correlations over time; and experimental manipulation of one of the variables.

Exercises

1. One theme of the chapter concerns the difficulty of distinguishing age effects from cohort or generational effects. Consider your own cohort. Are there experiences your generation has had that are at least somewhat different from those of other generations? What sorts of effects might these generational differences have in cross-sectional comparisons?

2. One way to think through the complexities of sequential designs is to imagine specific outcomes and what they would mean. Consider the cohort-sequential design schematized in the table below. The dependent variable is IQ, and we will assume that the means for the various groups range from 90 to 110. For each of the following outcomes, generate means that would be consistent with the specified result: (a) an effect of age alone, (b) an effect of cohort alone, (c) effects of both age and time of measurement.

			Time	
		1990	*2000*	*2010*
		Age M	Age M	Age M
	1930	60	70	80
Cohort	1940	50	60	70
	1950	40	50	60

3. This chapter stresses both the value and the difficulty of longitudinal research. One alternative approach to the study of across-time stability or change is the *retrospective method.* The retrospective method goes backward in time, typically beginning with some

adult outcome of interest and then attempting to recapture the important precursors or contributors from earlier in development. The obvious challenge in such research lies in obtaining accurate measures of events from the past. Pick some salient set of experiences from your own developmental history—performance in school, for example, or relations with friends, or trips with your family. Reconstruct, as best you are able, the developmental picture across some span of your childhood. Ask your parents to do the same, and compare your reconstruction with theirs. If possible, compare both accounts with objective records from the time periods at issue (e.g., report cards, family photo albums).

4. As the text discusses, a correlation between variables A and B is potentially susceptible to several causal interpretations: A causes B, B causes A, some third factor C causes both A and B, or some combination of all of these possibilities. Listed here are several examples of positive correlations that are in fact obtained in developmental research. For each, (a) generate as many plausible explanations for the correlation as you can, and (b) indicate the kinds of evidence you could collect to decide among the possibilities.

physical punishment by parents and aggression in children

reasoning by parents and prosocial behavior in children

physical attractiveness and popularity

IQ and school performance

academic self-concept and school performance

activity level and mental competence in old age

Notes

1. Ginsburg and Opper (1988) provide a good discussion of both this point and other virtues of Piaget's approach.

2. Modern statistical and analytical techniques to test causal inferences from correlational data go well beyond what I am able to cover here. Particularly influential is an approach labeled *structural equation modeling.* Brief (although still challenging) introductory treatments can be found in MacCallum and Austin (2000) and Ullman and Bentler (2003), as well as a special issue of the journal *Child Development* (Connell & Tanaka, 1987).

4

Measurement

Chapter 2 introduced the basic distinction between independent variables and dependent variables: Independent variables are factors that we control; dependent variables are outcomes that we measure. Most of chapter 3 on Design was concerned with the first of these topics: various ways to create and combine the independent variables in research. In this chapter the focus shifts to the dependent variable end: ways in which to measure the outcomes of the research process.

Measurement is a large topic, and discussions of measurement will consequently recur at various points throughout the book. The chapters devoted to specific kinds of developmental research (chapters 11 to 14), in particular, will contain much discussion of how constructs of interest to developmental psychologists are measured. The goal of the present chapter is simply to provide an overview of some basic points about measurement—points that can be returned to and elaborated on as necessary.

The organization for the chapter is as follows. An initial section serves to introduce a number of the basic concepts that are important for an understanding of measurement—the concept of operational definition, for example, and the distinctions among various levels of measurement, and the central notions of the reliability and the validity of measurement. The remainder of the chapter is then devoted to two important forms of measurement: standardized tests designed to assess some psychological attribute, and observational assessments of behavior. Some further general points will emerge in the context of these two types of measurement.

Some Basic Concepts

Once again, an example will provide a helpful context for many of the points to be made. I draw the example from a content area that has already been touched on in various places: the issue of TV violence and aggression. Liebert and Baron (1972) used an experimental approach to examine this issue, exposing half of their participants (5- to 9-year-old children) to a violent TV segment (3½ minutes of clips from *The Untouchables)* and half to a nonviolent segment and then measuring subsequent aggression in both groups. Several measures of aggression were taken, but one measure was the basis for most of their conclusions, and it is the one on which I concentrate here. Following the TV viewing, the children were seated in front of a response box that contained two buttons: a red button labeled "hurt" and a green button labeled "help." Wires ran from the box through an opening in the wall, and the children were told that the box was connected to a game that a child in the adjacent room was about to play. The game involved turning a handle, and the connection between box and game allowed the participants to affect the ease of play: Pushing the "help" button would loosen the handle and thus make the game easier, whereas pushing the "hurt" button would heat up the handle, hurt the other child's hands, and thus make the game harder.

A series of trials followed on which the children had a chance to push either of the buttons for as long as they wished. The question was whether the children who had watched the violent TV segment would show greater aggression by spending more time with the "hurt" button depressed. This was exactly what happened: Total duration of "hurt" responses was almost 50% greater for the violent TV group than for the nonviolent group. The results thus supported the proposition that exposure to TV violence increases aggression in children. (I will add a point that has probably already occurred to you, and that is that there was not really a game-playing child in the room next door.)

Operationalization

Consider two ways of summarizing the results of the Liebert and Baron study: (a) "Watching violent TV increases aggression in children," and (b) "Watching 3½ minutes of excerpts from *The Untouchables* in a laboratory setting increases the likelihood that 5 minutes later children will push a button that supposedly delivers a painful stimulus to an unseen child in another room." Clearly, the first statement has a more interesting and generalizable sound to it. But the second statement is more certainly accurate, for it describes exactly what was done and found, whereas the first statement involves a general conclusion that goes beyond the actual data.

The two ways of summarizing the Liebert and Baron data raise the important distinction between the conclusions that a researcher wishes to draw from a study and the actual manipulations and measurements of the study. "TV violence" and "aggression" are interesting constructs that are clearly worth knowing about. So too are any number of other things that developmental psychologists study—intelligence, creativity, self-concept, sex typing. The problem is that attributes like intelligence or creativity are not in fact "things" that are immediately and automatically observable; rather, if they are to be studied, they must somehow be *operationalized*—that is, translated into a specific and measurable form. All research requires measurement, and all measurement requires that some more general notion be made specific.

The verb *operationalize* has a noun form: **operational definition**. The notion of operational definition derives from work in physics in the1920s by P. W. Bridgman (1927). An operational definition defines a variable in terms of the operations used to produce or measure that variable. Thus, temperature might be defined as the displacement of mercury within a certain kind of container. Intelligence might be defined as performance on the Stanford-Binet IQ test. Or, to return to the Liebert and Baron study, aggression might be defined along the lines of the second of the two summary statements. In each case, there would be a clear tie to the measurement operations actually used.

A strict interpretation of the operational approach is that researchers are not allowed to make statements about their variables that go beyond the operations used to produce or measure those variables. In fact, few researchers today adhere to such a strict conception of what it means to be operational. Nevertheless, the operational movement has had a lasting and beneficial impact on research in psychology. What it has done is to set a generally accepted framework within which the task of measurement proceeds. The hallmark of this approach is the insistence that measurement operations be clearly specified, objective, and repeatable by any investigator in any appropriately equipped laboratory.

Let us consider how the business of translating theoretical construct into specific measurement might proceed for both the researcher and the reader of a research report. Imagine an observational study of aggression in preschool children. The researcher is interested in the possibility that social reinforcement promotes aggression in the preschool setting. The first task for the researcher is to decide on an operational definition for each of these rather global constructs. Because there are a large number of ways in which either construct could be operationalized, this decision will involve selecting particular indices from a larger pool of possibilities. Our researcher might decide, for example, to define social reinforcement as consisting of certain verbalizations (e.g., "good," "OK"), certain facial expressions (e.g., a smile directed toward the child), and certain nonverbal behaviors (e.g., pats or hugs). Aggression might be defined as consisting of various physical actions (e.g., hits, kicks, pinches) whose intent seems to be to injure another. Whatever the specific indices selected, it is then the researcher's job to carry out the measurements as accurately as possible and to convey to the reader exactly what was done.

The eventual reader of such a study has a job to do as well. The reader must begin by recognizing the point just made: that constructs like social reinforcement and aggression have many possible operational definitions, and that any one study will necessarily include only a subset of these possibilities. This point means that the particular operational definitions used may not correspond to the reader's own preconceptions about the meanings of social reinforcement and aggression, and the definitions may not correspond to those that the reader has encountered in other studies of these constructs. What the reader must do, therefore, is to set aside, at least temporarily, whatever preexisting notions he or she may have and focus instead on what was actually done in the study under consideration. Probably the most important skill that the reader of psychology reports must cultivate is the ability to move beyond the nice-sounding summaries found in Abstracts and Discussions (e.g., "social reinforcement promotes aggression") to evaluate research in terms of the specific operations actually used. If the specific operations are not satisfactory, then the general conclusions can hardly be compelling.

Quantification

The movement from global to specific is one of the hallmarks of measurement. Another hallmark is *quantification.* In the words of one of the pioneers of measurement theory, S. S. Stevens, "measurement is the assignment of numbers to aspects of objects or events according to one or another rule" (Stevens, 1968, p. 850). The nature of the numbers and rules varies, however, across different forms of measurement. The conclusions that can be drawn from the outcomes of the measurements vary accordingly.

Variations in the kinds of quantities involved define so-called *levels* or *scales* of measurement. Following Stevens (e.g., 1968), it is traditional to distinguish four levels of measurement. Each level fulfills the basic functions of any system of measurement; that is, a value is assigned to each observation, and the values serve to differentiate among the observations. They do so, however, in somewhat different ways.

The simplest form of measurement is referred to as a **nominal scale**. *Nominal* means "naming": assigning some qualitative label to each observation in the sample. Imagine that you are interested in studying toy preferences among preschool children. You present four different toys and allow each child to pick one to play with. Your measurement consists of recording which toy the child selects. The measurement in this case would be nominal, for all you are doing is assigning a label to each response. You could, of course, make the labels numbers—for example, write down a 1 each time a child selects the truck, a 2 each time a child picks the teddy bear, and so forth. Such numbers, however, would be merely substitute labels without any quantitative significance. And this is the distinguishing characteristic of nominal scales: They deal with qualitative classifications rather than quantitative differences.

We can return to aggression for an example of the second level of measurement. Suppose that we have the preschool teacher rate how aggressive the different children in the classroom are. We use a 5-point rating scale ranging from "very aggressive" through "moderately aggressive" to "very nonaggressive." The level of measurement in such a study would constitute an **ordinal scale**, for what we are doing is ordering the observations in terms of magnitude. In this case, in contrast to a nominal scale, there *is* a quantitative dimension to our observations, and the measurement serves to place each observation along this dimension. Thus, we can say that the "very aggressive" child is more aggressive than the "moderately aggressive" one, who is in turn more aggressive than the "very nonaggressive" one—or that category 5 is indeed greater than 3, which is in turn greater than 1. Note, however, that we still cannot talk about the size of the differences. We do not know, for example, whether the difference between a rating of 5 and a rating of 3 is the same as that between a rating of 3 and a rating of 1. Nor, of course, do we have any warrant for saying that a child who earns a rating of 5 is five times as aggressive as the child whose rating is 1. All that we can talk about is the ordering.

This restriction is lifted in the third scale of measurement, the interval scale. In an **interval scale** the points of the scale are not only ordered but also equidistant. A common (albeit nonpsychological) example is a thermometer. Measurements of temperature are clearly ordered: 40° is hotter than 30° is hotter than 20°. Furthermore, the points of the thermometer are equally spaced. Thus, we can say that the difference between 40° and 30° is precisely the same as the difference between 30° and 20° (in a physical, if not a psychological, sense). As we saw, this sort of quantitative precision is impossible with an ordinal scale.

There is still one limitation to an interval scale: There is no true zero point. Thermometers do, of course, include a zero, but the zero on the thermometer is simply an arbitrary point with values on both sides, not a true bottom to the scale. It does not signify a total absence of the characteristic being measured. Scales of measurement that meet all of the criteria for an interval scale and also contain a true zero point are called **ratio scales**. A common example of a ratio scale concerns the measurement of physical characteristics such as height or weight. A balance scale comprises not only equal intervals of weight but also a true zero—that is, the absence of any weight on the balance. Because of the zero point, a ratio scale permits proportional statements that are not possible with an interval scale. We can say, for example, that 40 pounds is twice as heavy as 20 pounds. We cannot say that 40° is twice as hot as 20°.

Table 4.1 summarizes the distinctions just discussed. We will return to the topic of measurement scales in chapter 8 on Statistics. As we will see then, the level of measurement is one factor that determines which statistical tests are appropriate to use.

Facets of Measurement

One emphasis of the discussion thus far has been on the need for choice when translating some general construct (such as aggression) into some specific and measurable form (such as hits and kicks). Not yet discussed, however, are the general dimensions along which choices about measurement are made. Measurement theorists identify a number of such dimensions or "facets" of measurement (Messick, 1983). I mention a few here and add some others later.

One basic decision concerns the specific aspect of the target behavior on which to concentrate. Imagine that our hypothetical researcher of preschool aggression has chosen hits as a measure of aggression. Still to be

Table 4.1 Levels of Measurement

Level	*Characteristics*	*Examples*
Nominal	Assignment to qualitatively distinct categories without quantitative significance	Gender, race, political party
Ordinal	Ordering in terms of magnitude along a quantitative dimension	Friendship rankings, class standing
Interval	Placement along a quantitative dimension with equal intervals	IQ scores, grade equivalents
Ratio	Placement along a dimension with equal intervals and a true zero point	Reaction time, number of errors

decided is exactly what it is about hits that will be measured. The researcher might decide, for example, to work with the *frequency* of the behavior—that is, how many hits a particular child delivers. Such a direct frequency count is probably the most obvious index of what we normally mean by "level of aggression." An alternative possibility, however, is to work not with the frequency of the behavior but with its *intensity*—that is, not how many hits a child delivers but how hard each hit is. Intensity also has an obvious tie to what we normally mean by "aggression." Still another possibility is to focus not on frequency or intensity but on the *timing* of the behavior. The researcher might decide, for example, to concentrate on the *latency* or quickness with which hits are elicited, or perhaps the total *duration* of the hitting episode. This frequency-intensity-timing trichotomy does not apply to all response measures that developmentalists examine; it does apply to many, however, and when it does not, there are generally other dimensions and thus other measurement decisions. Few outcomes of interest present but a single possible target for measurement.

The discussion of which aspect of a behavior to measure presupposes a prior measurement decision, and that is to concentrate on overt behavior. Not all measurements have overt behavior as their target. A researcher of aggression might be interested instead in aggressive thoughts or fantasies—thus, the underlying mental content, rather than the actual behavior. Such a researcher would presumably still elicit some measurable behavior from which to infer this mental content (e.g., self-reports of aggressive fantasies); in this case, however, the behavior would be simply a means to the desired end. Similarly, a researcher of aggressive emotions might elicit verbal reports of such emotions; again, however, the actual measurement target would be something other than the overt behavior. Alternatively, the researcher of emotions might bypass overt behavior altogether and instead measure physiological change (e.g., accelerated heart rate, increased blood pressure) in response to some aggression-eliciting situation. Emotions are just one of a number of constructs that lend themselves to such beneath-the-surface measurement. We will encounter various examples in the later chapters of this book.

Note that the possibilities sketched in the preceding paragraph embody several different dimensions of measurement. One dimension, clearly, is an overt to covert one. In some cases our interest is in actual behavior; in other cases it is in some less visible and more general construct (thoughts, motives, desires, etc.) that presumably underlies the observed behavior.

In some cases the specific target for our measurement operation is an overt behavior (such as hits); in other cases it is some covert, invisible-to-the-naked-eye process (such as heart rate). Finally, in some cases the actual and the specific target match, and in some cases they do not. With hits, for example, there is a match—what we are interested in is how often the child hits other children, and what we measure is how often the child hits other children. The researcher of emotions, however, is probably not interested in heart rate per se; rather, heart rate is simply a clue from which to infer the nature of the emotion.

The contrast just discussed is sometimes referred to as the *sign-sample* distinction. Sometimes our measures are samples of the target construct—that is, specific instances (hits, cries, smiles, etc.) of the behavior of interest. Sometimes, however, the measures are simply signs—not the target itself, but indices (a raised eyebrow, a heart rate change, etc.) from which the target can be inferred. And, of course, the same measure can function as either sign or sample depending on its use in a study. In a study of crying, cries are clearly samples. In a study of attachment, cries are signs—typically one of a number of signs—from which to determine the nature of the attachment relation.

A final distinction concerns the goals of the measurement operation—that is, what it is that we wish to do with the scores we obtain. In some instances the goal is to identify individual differences among the participants—to measure the extent to which members of the sample differ in aggression or attachment or whatever. This goal is especially likely in correlational research, in which the researcher seeks to determine whether variations in one set of scores (e.g., individual differences in children's aggression) relate to variations in some other set of scores (e.g., differences in parents' child rearing). In other research the interest is less in individual differences than in the immediate determinants of the behavior in question. Consider, for example, the hypothetical study of social reinforcement and aggression. The goal in such a study would be to relate fluctuations in aggression to the presence or absence of reinforcement, not to measure which children are in general most or least aggressive. Or consider experimental studies of TV violence and aggression, such as the Liebert and Baron (1972) experiment. Again, the goal is not to identify individual differences in aggression; the goal is to see whether aggression varies as a function of the kind of TV content to which the children are exposed.

The contrast just discussed is sometimes labeled the *trait-state* distinction. With the trait approach to measurement, our interest is in what people are like *in general*—the usual goal being to relate the measured characteristic to some other measure or measures on the same sample. With the state approach to measurement, our interest is in what people are like *at the moment*—the usual goal being to relate the variations in immediate response to some potential determinant of the target behavior. With either approach it is of course important to select a good operationalization of the target construct. Furthermore, the specific measurement operations may well be the same whether the focus is on states or traits. If, for example, some sort of composite index of physical acts (hits, kicks, pinches, etc.) makes sense as a measure of individual differences in aggression, then such a measure will probably also be appropriate in an experimental study of effects of TV violence. Even if the measures are the same, however, the ways in which they are collected and used are likely to differ across the two kinds of study. When the focus is on individual differences among people, issues of the sampling of behavior become critical: Our ability to order participants along the dimension of interest is dependent on our ability to obtain representative samples of each person's behavior. When the focus is on the effects of an experimental manipulation on the target behavior, issues of sampling and individual differences recede in importance: Now what we typically

Table 4.2 Facets of Measurement

Dimension	*Description*
Aspect of behavior	Which aspect of behavior (e.g., frequency, intensity, timing) is measured
Overt-covert	Whether the measurement target is overt (an actual behavior) or covert (e.g., a physiological change)
Sign-sample	Whether the measurement is an index from which the construct of interest can be inferred or an example of the target construct
Trait-state	Whether the focus is on persistent individual differences among participants or on the immediate effects of an experimental manipulation

aim for is a single, comparable sampling of behavior that will reveal the effects that we are seeking. Indeed, in this case preexisting individual differences become an impediment rather than a goal, because such differences contribute error variance that may obscure the effects of interest.

This section has considered a number of dimensions along which measurements vary. Table 4.2 summarizes the dimensions that have been discussed. As noted, fuller treatments of the measurement process (e.g., Messick, 1983) add a number of other facets as well. Despite this diversity of dimensions and choices, however, at some level the point to be taken away is a simple one: We choose measurements that will satisfy the specific goals of the research.

Sources of Measurement

Imagine that you are a researcher of aggression, and imagine also that you have successfully navigated all of the decision-making steps that we have considered up to this point. Thus, you know (let us say) that you plan to record a certain set of physical acts as instances of aggression, that you will be analyzing the frequency with which these acts occur, and that your goal is to identify individual differences in aggression in the sample being studied. You wish to be able to say, therefore (with appropriate numbers attached), that Johnny is highly aggressive, Billy moderately aggressive, Tommy nonaggressive, and so forth. Yet to be addressed, however, is a basic question about measurement: How are you going to get evidence about how Johnny, Billy, and Tommy actually behave? This is the issue that Hartmann and Pelzel (2005) label "sources of data": Through what contexts and procedures can we translate general measurement decisions into actual data about behavior?

The answer is that psychologists obtain their evidence in three general ways: through observation of behavior in the natural setting (the *naturalistic observation approach*), through creation of special environments or circumstances to elicit the behavior (the *laboratory approach*), and through reports about behavior from some knowledgeable source (the *verbal report approach*). Each of these general categories has a number of different possibilities embedded within it. Laboratory studies, for example, may elicit some sort of overt behavior, or they may measure some underlying physiological process, or they may collect their information through paper-and-pencil tests. Verbal reports may take the form of either questionnaires or face-to-face interviews, and they may be provided by children's parents, by children's teachers, or (in older children) by children themselves. Measurements take dozens of different specific forms; all, however, fall within one of the three general approaches.

Table 4.3 summarizes the distinctions just made and provides a brief overview of the strengths and weaknesses of each approach. There is no expectation that the points made will be fully comprehensible from this brief initial treatment. The goal, rather, is simply to introduce distinctions and issues that will recur throughout the book. As we encounter the various topics that developmental psychologists study, we can see the relative balance of the three approaches of study and some of the specific forms they take, as well as some of the strengths and limitations of each.

Table 4.3 Sources of Measurement

Source	*Strengths and limitations*
Naturalistic observation	The only direct source of information for how children naturally behave in the natural setting. The presence of observers, however, may alter the setting and thus the behavior. Observers may not be able to record some behaviors accurately. Finally, some behaviors (e.g., perceptual processes) are difficult if not impossible to observe in the natural setting.
Laboratory study	The controlled laboratory environment ensures that the behaviors of interest will occur, and it maximizes both accuracy of measurements and comparability across participants. In addition, for some dependent variables (e.g., physiological changes) laboratory study is the only option. To the extent that the lab environment differs from the real-life settings of interest, however, behaviors measured in the lab may not be generalizable.
Verbal report	The scope of information provided is greater with verbal reports than with any other method of study. As children develop, multiple sources become available, including peer reports and self-reports. Only some behaviors are accessible through verbal reports, however. In addition, verbal reports are not direct measures of behavior, and for various reasons they may sometimes be inaccurate.

FOCUS ON

Box 4.1. Looking Inside the Brain: Methods of Measuring Brain Activity

As the text notes, not all measurements in psychology have overt behavior as their target. Measures of underlying physiological processes can also be informative, especially in the infant and young child, for whom overt behaviors are often limited. We will see a number of examples in the later chapters of the book.

Although physiological measures date back to the birth of psychology, the kinds of processes that can be examined have expanded greatly as technology has advanced. In particular, recent years have seen some exciting advances in techniques to study the brain (Casey & de Haan, 2002; Nelson & Luciano, 2001). Such techniques have allowed scientists for the first time to examine not only the anatomy of the brain but also brain *activity* as people perform different tasks. A recent special issue of the journal *Developmental Science* (Casey & de Haan, 2002) discusses seven such **functional imaging techniques**. Here we sample three.

One such technique is labeled *positive emission tomography*, or *PET*. The PET procedure measures metabolic activity in the brain through the injection of a radioactive isotope into the bloodstream. Brain areas that are especially active "light up" as the radiation is emitted, and the PET scanner provides pictures of the activation. In studies of language processing, for example, areas within the left hemisphere tend to be especially active, providing evidence that particular brain areas are specialized for language processing.

Because it requires the injection of a radioactive substance, PET is regarded as a relatively "invasive" procedure, and as such it cannot be employed routinely in research with children (indeed, the Food and Drug Administration explicitly forbids such use). To date, PET studies with children have been limited to clinical cases in which diagnostic needs have justified use of the procedure.

A somewhat more widely applicable technique is *functional magnetic resonance imaging*, or *fMRI*. When a brain area is active, there is an increase in blood flow and blood oxygenation within the area, and the oxygenated blood is more highly magnetized than nonoxygenated blood. The fMRI procedure uses a powerful magnet to record the changes in oxygen level. Figure 4.1 provides an example of the sort of images that are produced. The question under study was whether different parts of the brain were implicated in face perception and location matching. As the figure shows, these tasks did in fact involve different parts of the brain, and this was true in both children and adults. The research also revealed a developmental difference, however, in that adults showed a more focused pattern of activation than did children.

Although fMRI does not carry the potential hazards of PET, its use does require that the participant remain very still for an extended period—something that is difficult for young children to do. For this reason, the procedure is seldom used below about age 5 or 6.

Figure 4.2 shows the apparatus that underlies a third technique: measurement of *event-related potentials*, or *ERPs*. The electrodes attached to the scalp record the neuronal activity that follows presentation of a particular stimulus. This technique is applicable to a wider range of ages than is either PET or fMRI; indeed, as the figure shows, it can be used even with infants. It is also applicable to a wide range of topics; among the topics covered in a recent review of ERP studies (Taylor & Baldeweg, 2002) are attention, memory, language, reading, and face perception. The fMRI procedure, we might add, has also been applied to a wide range of issues. Indeed, a number of studies have made use of both procedures (de Haan & Thomas, 2002).

Although their methods vary, it should be clear that each of these techniques has the same general goal: to provide pictures of underlying brain activity as different tasks are performed. A basic question is that of task comparison: To what extent are similar or different brain regions involved in the performance of different activities? For the developmental psychologist, there is a second basic question as well: To what extent are conclusions about brain activity similar or different as we move from infant to child to adolescent to adult? Often, as illustrated in the discussion of Figure 4.1, the answer turns out to be a mixture of similarities and differences: Basic aspects of brain organization and functioning are present quite early in life, but further developments occur with maturation and experience.

A final question—of interest at any point in the developmental span—concerns individual differences. On many tasks there are differences in quality of performance across individuals, and we can ask about the brain measures that distinguish successful from less successful performance. We can also ask about the brain patterns that characterize more extreme and pervasive individual differences—in particular, various clinical syndromes or developmental disorders. A variety of disorders have now been explored with the techniques described here, including dyslexia, autism, and attention-deficit hyperactivity disorder (ADHD). The new imaging techniques are thus clearly of applied as well as scientific value.

(Continued)

(Continued)

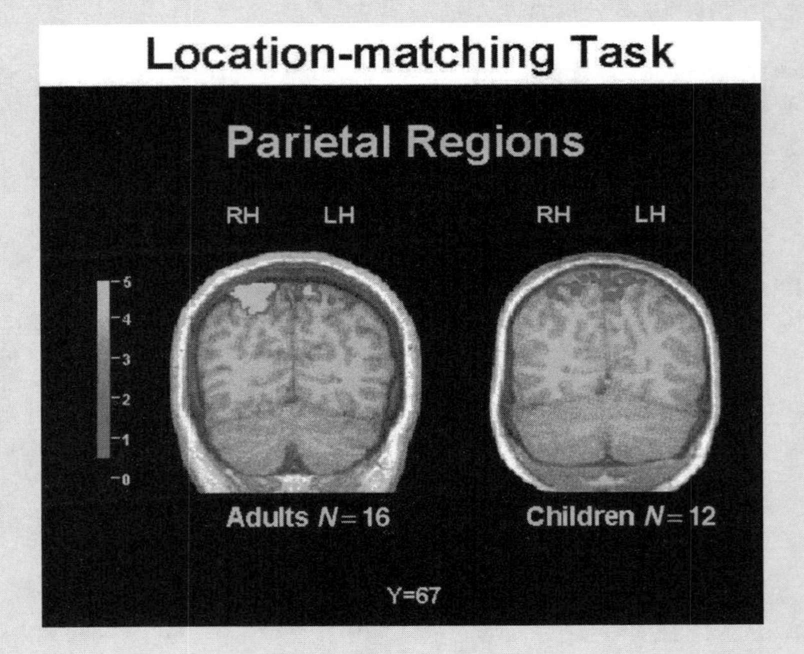

Face-matching Task

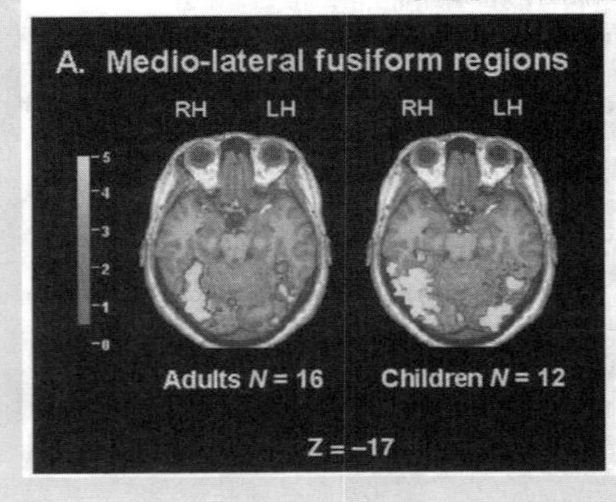

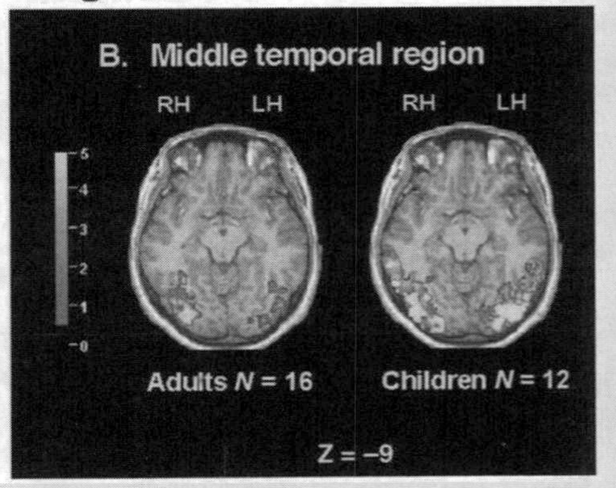

Figure 4.1 Examples of functional magnetic resonance imaging. Radio waves and a strong magnetic field are used to provide pictures of blood flow and chemical changes in the brain as different cognitive tasks are performed. In both children and adults, different regions of the brain are activated for different tasks. (Note: RH = right hemisphere and LH = left hemisphere.)

SOURCE: From "The Development of Face and Location Processing: An FMRI Study," by A. M. Passarotti, B. M. Paul, J. R. Bussiere, R. B. Buxton, E. C. Wong, and J. Stiles, 2003, *Developmental Science, 6*, 108, 109. Copyright 2003 © by Blackwell Publishers. Reprinted with permission.

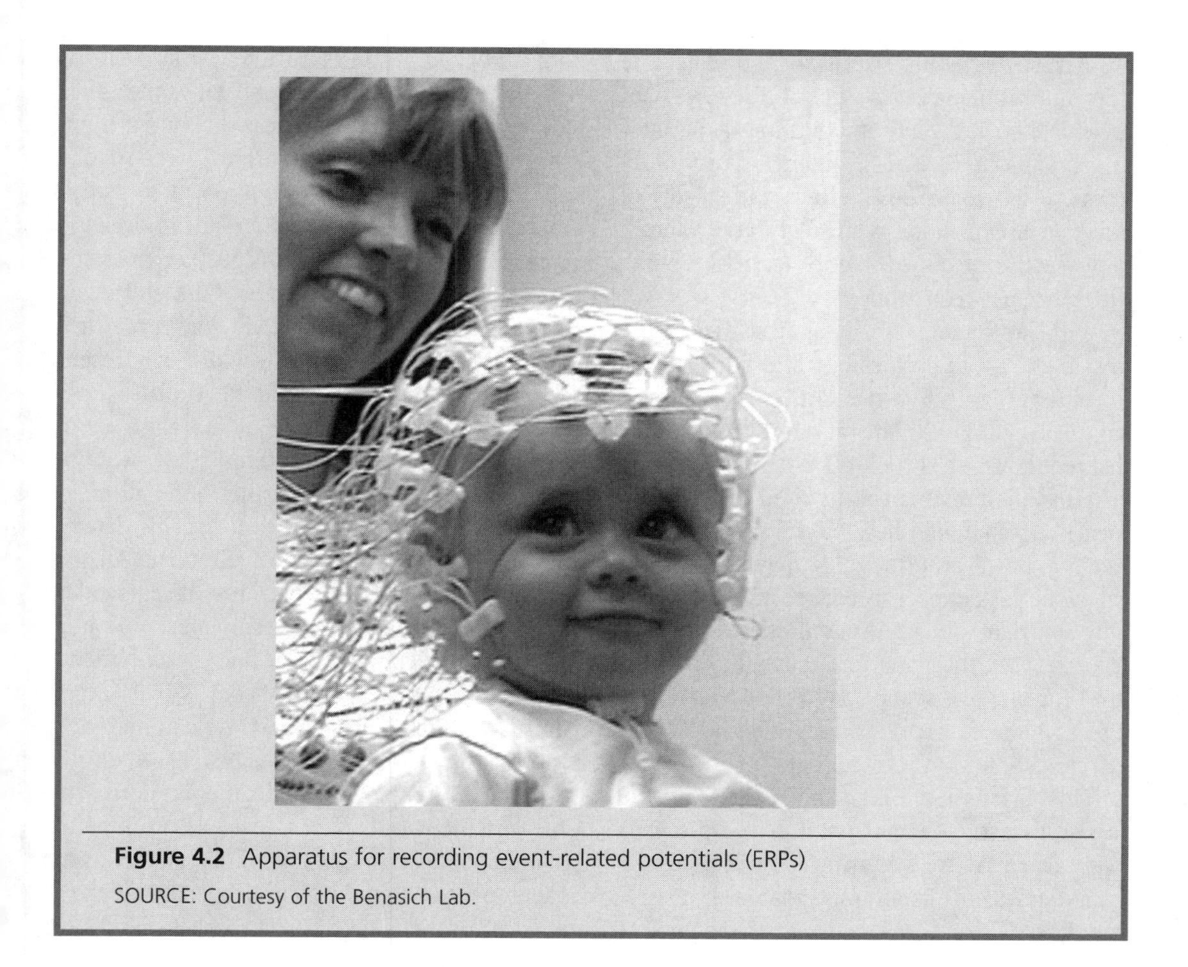

Figure 4.2 Apparatus for recording event-related potentials (ERPs)

SOURCE: Courtesy of the Benasich Lab.

Quality of Measurement

As we have seen, measurement is always a decision-making process: selection of a specific method of assessing a target construct from a much larger pool of possibilities. In this section we will consider some of the factors that determine how satisfactory a particular measurement choice is likely to be.

As indicated in chapter 2, the "variable" part of "dependent variable" implies that the scores we obtain must have some chance of varying as a function of the factors that the research is examining. One criterion of good measurement, therefore, is that it yields the desired level and range of values for the dependent variable. The converse occurs when scores are so bunched together that no effects could possibly emerge. This issue is most often discussed with respect to correlational research and the problem of restriction of range (see chapters 3 and 8); it applies as well, however, to experimental designs. Consider again the Cherry and Park (1993) study discussed in chapter 2. Imagine that instead of 24 objects, they had presented just 4 objects as stimuli for their memory test. In this case most of the participants would probably have performed perfectly, and there would have been no chance of finding the context effects (the map-model contrast) that the study was designed to explore.

This hypothetical reworking of the Cherry and Park study illustrates one possible kind of restriction of range: a so-called **ceiling effect**. A ceiling effect occurs when a task is too easy and most participants therefore earn scores that are at or close to the upper limit of the scale. The opposite problem is also possible: scores that are clustered at the low end of the scale because a task is too hard. This problem, as you might guess, is labeled a **floor effect**. Either sort of effect, ceiling or floor, is a threat to the validity of research, for either may obscure a genuine difference between groups.

With sufficient care, it should usually be possible to avoid both ceiling and floor effects. Previous research with similar tasks and samples can guide the attempt to find the desired level of difficulty. When doubt still remains, pilot testing can be used to refine the measures. No matter how careful the researcher may be, however, difficulties may still arise if the same tasks and procedures are to be used with participants of different developmental levels. A task that is at just the right difficulty level for one age group may produce a ceiling effect in somewhat older children and a floor effect in somewhat younger ones.

This discussion of cross-age matching of tasks raises the general issue of *measurement equivalence.* As we saw in chapter 3, measurement equivalence is an issue whenever a study compares distinct groups of participants. The question is whether the procedures and measures are equally fair and valid for the different groups. In developmental research the issue arises most often in the comparison of different ages, but it can apply to any comparison between naturally formed groups. In a study of sex differences in cognitive abilities, for example, it would be important to select materials that are equally familiar and interesting for both boys and girls. In research that compares different ethnic or cultural groups it is important—and often very difficult—to ensure that the measures are equally appropriate for the different groups. It can be difficult enough to devise satisfactory measures for one set of participants. The challenges of measurement are often multiplied when group comparisons are the focus.

One message from this consideration of measurement equivalence is that a single operationalization of a target construct is often not sufficient. A measure that works well for one group may not work well for some other group. This point, however, is actually a more general one: Even when group comparisons are not an issue, a single method of measurement can be a hazardous basis for drawing conclusions. Shadish et al. (2002) refer to the exclusive use of one measurement operation as **mono-operation bias**. As they note, any single operation will almost certainly underrepresent the target construct—that is, capture only some aspects of whatever it is (aggression, intelligence, sex typing, etc.) that we are attempting to measure. Furthermore, all measurements include some task-specific irrelevancies (e.g., the particular stimuli used, the exact wording of the instructions, the particular individual who acts as experimenter), and it is therefore difficult, with just one measurement, to know how much of the score is true target and how much is irrelevancy. We can be much more certain about our conclusions when we are able to employ several different forms of measurement rather than just one. This point should sound familiar, for it is the same argument made in chapter 3 with respect to the dangers of a mono-method approach to issues of cause and effect. The prescription in both cases is the use of *converging operations,* rather than just one method of study. (See also Eid and Diener, 2006.)

This discussion of the quality of measurement has yet to consider the two most important constructs in the evaluation of measurement: reliability and validity. **Reliability** refers to the consistency or the repeatability of a measurement. The issue is whether repeated applications of a measuring technique will yield identical or at least highly similar values. The higher the degree of agreement across the repeated applications, the higher the reliability. One goal of measurement is always to maximize reliability. *Validity* refers, as it always does, to accuracy—in this case the accuracy of a measurement. The question is whether the values that the measurement yields accurately reflect the target construct—that is, have we really measured what we think we have measured? Validity, clearly, is *the* overall issue with respect to the evaluation of any measurement. It is a general goal that subsumes all of the more specific considerations discussed throughout this section.

Questions of reliability and validity apply whenever we measure anything. In practice, however, the two constructs are most often discussed with respect to two kinds of measurement: standardized tests and observational assessments. Standardized tests and observations are the subject matter for the remaining sections of the chapter, and I will consequently address issues of reliability and validity more fully in the context of these two forms of assessment.

Tests

There are no generally agreed-upon criteria for deciding when some measurement operation merits the label *test.* As the term is used here, tests are standardized measuring instruments whose purpose is to assess some important psychological attribute. There are many psychological attributes of interest and even more tests that purport to measure them; compendia of such measures (e.g., Goldman & Mitchell, 2003; Plake, Impara, & Spies, 2003) may list up to several thousand entries. We will see a number of examples in the later chapters: for example, measures of temperament in infancy (chapter 11), self-concept in preschoolers (chapter 13), sex-role development in older children (chapter 13), and intelligence at various points in the lifespan (chapters 12 and 14). The examples, like tests in general, will vary in the specific content that is measured, the age group that is the target, the format that is used to elicit responses, and the uses to which the test scores are put. All, however, are similar in that they must satisfy the twin criteria of validity and reliability.

Validity

The issue of test validity is straightforward: Does the test in fact measure what it claims to measure? If the test is an IQ test, for example, does it really measure individual differences in intelligence, or do differences in people's scores have some other basis? Clearly, the mere fact that the test is labeled "IQ" cannot decide this issue; we need other criteria. In general, three main sorts of validity criteria are possible.

A first possibility is that the test has **content validity**. The term *content validity* refers to the adequacy with which the test items represent the conceptual domain of interest. Does the test include all of the important aspects of the target that we wish to measure, and are the various aspects properly weighted? Suppose that our test is designed to tap knowledge of fourth-grade arithmetic. A test that consisted solely of addition problems would have poor content validity. A test composed of a representative sampling of addition, subtraction, multiplication, and division problems would have much better content validity.

Content validity is usually desirable, but it is not always easy to achieve. Even with a target as circumscribed as elementary arithmetic, disagreements about the adequacy of the sampling may arise. How many two-digit and how many three-digit problems should there be, for example, and in what context or contexts should the problems be embedded? When the target is more complex than elementary arithmetic, content validity may be virtually impossible to demonstrate. Consider a very global construct like intelligence. No matter how wide ranging the sampling of items, it is doubtful that any test could ever be shown, through content analysis alone, to comprise a complete and representative sampling of every possible aspect of "intelligence." In such cases other indices of validity are necessary.

A second form of validity is **criterion validity**. The issue in criterion validity is whether a participant's performance on the test relates to some measure of the attribute in question—to some external criterion. In the case of a knowledge-of-arithmetic test, a reasonable criterion might be grades in arithmetic across the school year. A test that correlated highly with such grades would have good criterion validity. In the case of intelligence, a common criterion has been the ability of IQ tests to predict performance in school or on standardized achievement tests; indeed, historically it was the need to predict school performance that led to the development of the first successful IQ test (that of Binet and Simon in 1905). More generally, criterion validity is the main form of validity for any test whose primary function is prediction for pragmatic purposes. Thus, it is criterion validity that underlies the use of SATs or GREs to predict success in college, the use of times in the 40-yard dash to predict success as a professional football player, and so forth.

A distinction is sometimes made between two forms of criterion validity. A test that correlates with some contemporaneous external criterion is said to have *concurrent validity.* If, for example, we showed that IQ assessed during the second grade correlated with academic performance in the second grade, we would be demonstrating the concurrent validity of our IQ measure. A test that correlates with some *future* external criterion is said to have *predictive validity.* If we showed that IQ in the second grade relates to performance in high school, we would be demonstrating predictive validity.

The final form of validity to be considered is **construct validity**. Among psychometricians, construct validity is generally regarded as the most important form of test validity. Unfortunately, it is also the most difficult form for the researcher to achieve, as well as the most difficult form for the textbook author to convey. I will settle here for a brief introduction to this complex notion. Fuller discussions can be found in many sources, including Borsboom, Mellenbergh, and van Heerden (2004); Cronbach (1990); Kerlinger and Lee (2000); and Nunnally (1978).

The distinguishing aspect of construct validity is its theoretical grounding. As Kerlinger and Lee (2000) note, "It is not simply a question of validating a test. One must try to validate the theory behind the test" (p. 671). The starting point, therefore, is some theory of the construct (intelligence, creativity, self-concept, anxiety, etc.) that we wish to measure. From this theory various predictions can be drawn. These predictions may include hypotheses about the effects of certain kinds of experimental manipulations on the test scores. Suppose, for example, that we are attempting to validate a measure of anxiety. In this case we might predict that heightening the demand pressures of a testing situation would lead to higher scores on the measure; conversely, minimizing apparent pressure should lead to lower scores. Obtaining such a pattern of results would provide one piece of evidence for the construct validity of the test.

In addition to experimental tests, correlational data are important in the establishment of construct validity. Predicted correlations are typically of two sorts. Some predictions reflect hypotheses about which measures of the construct should correlate positively with which other measures. If we were devising a test of anxiety, for example, we might predict that self-reports of anxiety would correlate with physiological changes thought to indicate anxiety (e.g., increased heart rate). The demonstration of such expected correlations among theoretically linked measures is referred to as *convergent validity.* Other predictions might concern hypotheses about which measures should *not* correlate with each other. In validating a test of anxiety, for example, it might be important to demonstrate that some kinds of physiological changes do not relate to self-reported anxiety, thus ruling out generalized arousal as an explanation for the results. The demonstration of such differentiation among theoretically distinct measures is referred to as *divergent* (or *discriminant) validity.* Both sorts of demonstration are important. As Wilkinson (Wilkinson and the Task Force on Statistical Inference, 1999) puts it, "It is just as important to establish that a measure does *not* measure what it should not measure as it is to show that it *does* measure what it should" (p. 596).

As the preceding discussion indicates, construct validity, like criterion validity, is established largely through the demonstration of expected correlations among measures. There are, however, important differences between the two kinds of validity. With criterion validity there is usually a single external target, such as performance in school, to which we wish to predict; with construct validity there is a whole network of hypothesized interrelations. With criterion validity the goal is generally prediction for pragmatic purposes; with construct validity the goal is validation of an underlying theory. It is not accidental, therefore (although it can be confusing), that the label for this final form of test validity is the same as the label used in chapter 2 for one form of experimental validity. In both cases the issue is theoretical accuracy—in one case with respect to measurement and in one case with respect to the study as a whole.

Reliability

In addition to validity, standardized tests must possess satisfactory reliability. The issue of reliability as applied to tests is also straightforward: Does the test consistently measure whatever it is that it does measure? Imagine, for example, that we give an IQ test to the same child on consecutive days and compare the scores. Highly similar scores would indicate good reliability; markedly different scores would indicate poor reliability.

The IQ example illustrates one common form of reliability: **test-retest reliability**. There are two ways to assess test-retest reliability. One is to give literally the same test on two separate occasions. Clearly, however, if the tests are identical, the children may remember many of the responses, and this fact could artificially inflate the reliability (it could also *deflate* the reliability, if the child perceives the readministration of the test as a signal that the answers should be changed). To avoid this problem, retest reliability is sometimes assessed via an *alternative-forms procedure.* As the name implies, the alternative-forms approach requires two different but equivalent versions of the test, with one version given at time 1 and the other version at time 2. Again, high agreement in response would indicate high reliability.

The second main type of test reliability is labeled **internal consistency reliability**. The question at issue now is the consistency of response across the different items of a single test given at a single time. A common procedure, labeled the "split-half" approach, is to divide the test into the odd-numbered items and the even-numbered items and then to compare responses between these two categories. Once again, high reliability would be indicated by a high agreement in response. Note, of course, that the calculation of such reliability makes sense only if the different items are all meant to be measures of the same construct. If the items on a test are intended to measure different things, then we would not necessarily expect them to correlate.

It is important to distinguish reliability from certain other constructs that also deal with the consistency of independently obtained measurements. Suppose that our administrations of the IQ test are separated not by 1 day but by 2 years. If we find that the two scores differ substantially, should we conclude that the IQ test is unreliable, or should we conclude that the child's IQ has really changed in 2 years? Suppose that instead of intelligence we decided to measure weight. If our measurements show that the child weighs 15 pounds more at age 9 than at age 7, should we conclude that our scale is unreliable? A much more likely conclusion, of course, is that the child's weight has really changed in 2 years—that is, that the stability of weight is less than perfect as children grow. Many aspects of the child's functioning (including performance on IQ tests) are less than perfectly stable as the child develops. It is important, therefore, to distinguish between the reliability of a measurement and the stability of a behavior.

It is important, too, to distinguish between the reliability of a measurement and the extent to which a behavior generalizes. The issue of generalization is the issue of consistency in behavior across different situations. Suppose that our interest now is in aggression in preschool children. We go into a preschool and make recordings of the many acts of aggression there on display. From these we derive a level-of-aggressiveness measure for each child. We then go into the children's homes and measure aggressiveness there. We find that our measure of school aggression is only weakly related to our measure of home aggression—that, in short, there is a good deal of inconsistency between the two scores. Do we conclude that one or both of the measures are unreliable? Although this is a possible conclusion, it may be more reasonable to conclude that aggressive behavior is simply not very consistent from one environmental setting to another quite different environmental setting. If so, our finding has to do with generalization, not reliability.

It may be helpful to summarize the distinctions just made. Reliability is a property of a measurement; stability and generalization are properties of behavior. Reliability is something that the researcher always seeks to maximize. Stability and generalization, however, are phenomena to be studied, not maximized. Finally, these phenomena *can* be studied only if we have first achieved a satisfactory degree of reliability. It is only if we can be sure that our measures are reliable for a particular time and situation that we can hope to study consistency in behavior over time (the stability question) or situations (the generalization question).

This section of the chapter has discussed both test validity and test reliability. The relation between these two concepts should be easy to discern. Reliability is a necessary but not sufficient condition for validity. If the scores on a test have little or no consistency, then the test can hardly provide a valid measure of the attribute in question. Consistency alone, however, is no guarantee that a test is valid. We might design an "intelligence" test that consists of the ability to hop on one foot. Scores on this test might be perfectly consistent across items (the two feet) and across time; this perfect reliability, however, would not mean that we had succeeded in measuring intelligence.

Regression

Let us return for a moment to test-retest reliability. Less-than-perfect reliability means that scores on the second administration of a test will tend to differ from those on the first administration. Is it possible to go beyond this general statement to say something about the *direction* of the difference—that is, whether scores are likely to go up or down on the second test? In the case of any individual participant it is impossible to predict with certainty whether the score will be higher or lower the second time around than the first. At the level of group averages, however, a prediction *can* be made: On the average, participants with low scores on the first test will have higher scores on the second test, and participants with high scores on the first test will have lower scores on the second test. This tendency for initially extreme scores to move toward the group mean upon retesting is referred to as **regression toward the mean**.

Let us try a concrete example of this phenomenon before considering why it occurs. Suppose we give an IQ test to a sample of children and obtain the distribution of scores shown in Figure 4.3. Some children (the open circles) score distinctly below average, some (the closed circles) score distinctly above average, and some (the crossed circles) score within a range of average performance. Now suppose we administer the same test to the same sample a week later and obtain the distribution of scores shown in Figure 4.4. The figure shows that the initially low-scoring children have, on the average, moved up, and that the initially high-scoring children have, on the average, moved down. Thus, both groups have

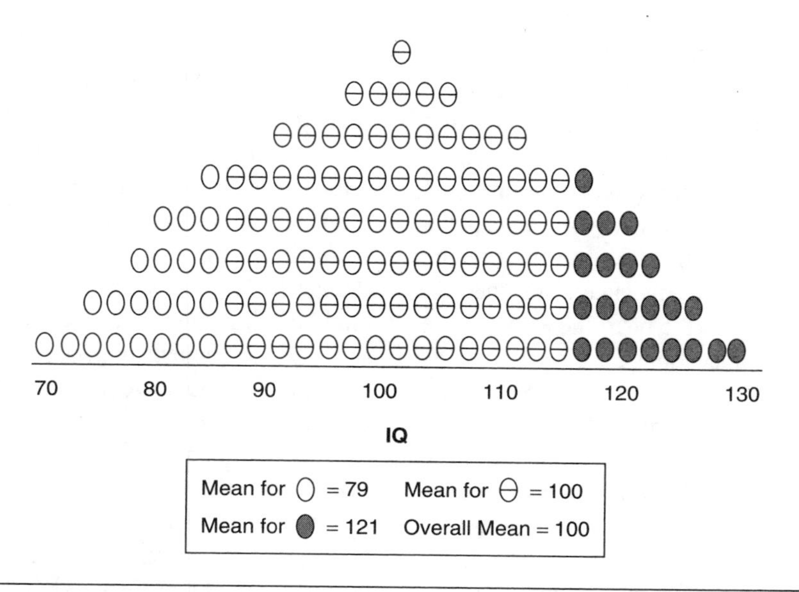

Figure 4.3 Hypothetical distribution of scores upon initial administration of an IQ test

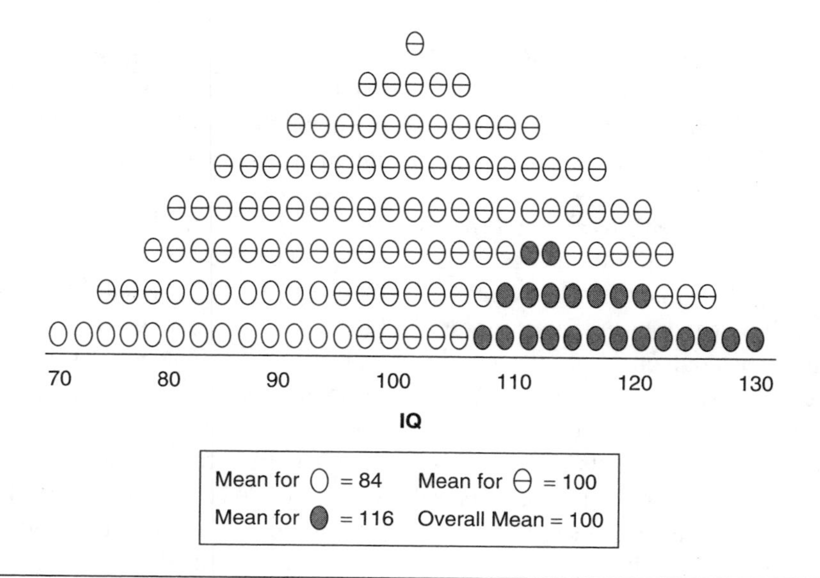

Figure 4.4 Hypothetical distribution of scores upon readministration of the IQ test

"regressed toward the mean." Because some of the initially average scorers have gone up or down, however, the overall range and mean remain about what they were originally.

Why does regression occur? Any participant's score on a test can be thought of as consisting of two components: the "true score," or the actual value on the dimension being measured, and error stemming from imperfect measurement. Clearly, "error" is just another way to talk about reliability: Perfect reliability means an absence of error; conversely, the greater the error, the lower the reliability. Two assumptions are generally valid concerning error. One assumption is that errors are normally distributed about the true score. This means that small errors are more likely than large errors; it also means that errors are equally likely to inflate or to deflate a particular individual's score. The second assumption is that errors occur randomly across participants and across testing occasions. This means that the errors for a particular participant are uncorrelated from one test to the next; the error on the first test has no implications for the error on the second test.

Consider now the scores in Figure 4.3. How have errors of measurement affected these scores? In particular, how have the relatively extreme errors, those that substantially inflate or deflate a score, affected the obtained distribution? It is reasonable to assume that the low-scoring (open circle) children, on the average, have suffered from a disproportionate number of negative errors; this is one reason that their scores are low. Similarly, it is reasonable to assume that the high-scoring children, on the average, have benefited from a disproportionate number of positive errors; this is one reason that their scores are high. But what happens when we retest the children? Recall that errors are uncorrelated from one test to the next. It is quite unlikely, therefore, that the extreme errors will affect the same children in the same way. The most probable outcome for any child, rather, is a relatively small error that is equally likely to inflate or deflate the true score. This "evening out" of errors over tests ensures that low scores will tend to move up and high scores tend to move down—ensures, in short, that we will find regression toward the mean.

The basic problem posed by regression should be obvious. Like poor reliability in general, regression is a threat to the validity of research. Furthermore, because regression is a systematic phenomenon, it can produce systematically incorrect conclusions. Suppose, for example, that in the IQ study described earlier we had not simply retested the children but also had introduced a new educational program between the first and second tests. Given the results in Figures 4.3 and 4.4, we might then have concluded that the effects of our program depend on the initial level of ability: The program enhances IQ for low-ability children but actually depresses IQ for high-ability children. Clearly, in such a case regression could produce a spurious impression of change where no change had occurred. Alternatively, regression might mask a true change; perhaps the program actually does benefit high-ability children, but the gains in true score are offset by the losses via regression.

Intervention research such as the example just sketched is perhaps the most common context for possible regression effects, given the focus of such research on children whose initial performance is low. Certain kinds of matched-group designs are also vulnerable to the problem. Let us consider a slightly reworked version of the dropout-graduate example from chapter 3. Imagine now that your interest is in the maintenance of IQ scores over time: Do graduates maintain their abilities better than dropouts? You match your two groups at an initial IQ level halfway between those for the parent populations (say 97) and then retest them 10 years later. Using regression alone, we can predict that the mean for the graduates will have gone up (given the selection of relatively low-scoring members from that population) and that the mean for the dropouts will have gone down (given the selection of relatively high-scoring members from that population). Again, regression may produce apparent effects that really have nothing to do with the independent variable being examined.

Observational Methods

Direct observation of behavior is both one of the most valuable and one of the most challenging approaches to measurement in psychology. I conclude the chapter, therefore, with a discussion of some of the complexities involved in observing behavior.

Some definition is in order at the start. In a sense, all research involves observation of behavior; how else would we get our dependent variables? In some cases, however, the recording of behavior is essentially, if not literally, automatic. Heart-rate responses may be recorded via an electrocardiogram. Judgments on a problem-solving task may be indicated by pushing a button. Questionnaires may be used to elicit a variety of responses from developmentally mature participants. Whatever the other problems in such research, the accuracy of the behavioral recording is not usually an issue.

In observational research the accuracy of the recording definitely *is* an issue. In observational studies the focus is generally on fairly large chunks of natural, ongoing behavior. Such behavior is not automatically recordable; rather, a human observer is required to make judgments about the occurrence and meaning of the behavior. The central question then is how accurately the observer can make these decisions.

Let us divide the discussion of observational methods into three general issues: deciding what to observe, deciding how to observe it, and determining the accuracy of the observations. Useful further sources include Bakeman, Deckner, and Quera (2005); Hartmann, Barrios, and Wood (2004); Odom and Ogawa (1992); and Pellegrini (2004).

Deciding What to Observe

At one level, answers to the "what" question are straightforward. Clearly, the researcher's general interests provide an initial delimiting of the behaviors that might be observed. The nature of the behavior then determines whether observational assessment is a reasonable option or not. Some behaviors lend themselves more readily to observation than do others. Aggression, for example, is a natural candidate for observational assessment: a frequent, overt, readily "see-able" behavior. Although other techniques of measurement (e.g., rating scales, contrived experimental tests) do exist, they may often be less satisfactory for the researcher's purpose. Conversely, heart-rate changes, or physiological responses in general, are poor candidates for observational assessment. Such responses are difficult and in some cases impossible to see, and alternative methods of assessment are both available and far more sensible.

Things become more complicated once the researcher moves beyond the initial decision to use observational techniques and attempts to determine exactly which aspects of the behavior to record. Suppose that we are studying mothers' styles of interacting with their infants. We know to start with that we cannot record *everything;* observation always involves some abstraction from the moment-to-moment specifics of the behavior. But at what level of specificity should we abstract? Should we record the fact that the mother raised her eyebrows, opened her eyes wider, turned up the corners of her mouth, and emitted a vocalization? Or should we work at a more global and interpretive level and record the fact that the mother smiled at and talked to her baby? Should we become more global and interpretive still and indicate that the mother seemed to encourage the behavior just produced by the baby? Or should we move to an even more global level and record that the mother acted in a warm, positive manner toward her child?

The distinctions just raised are often discussed under the heading of molecular as opposed to molar approaches to observation. A **molecular observational system** focuses on relatively fine-grained details of behavior, staying close to the actual behavior and to an essentially neutral depiction of exactly what occurs. There is, of course, still some loss of detail and some interpretation involved; even so, the attempt is to be relatively complete, specific, and nonevaluative. A **molar observational system**, in contrast, involves some stepping back from the actual behavior, some summing together of molecular units to arrive at a more evaluative overall category. "Smiles" or "hugs" would be examples of molar categories at a relatively specific level; "encourages" or "distracts" would be examples at a more global and interpretive level.

As the preceding may suggest, the molecular versus molar contrast is really more a continuum than a dichotomy. Observational systems may embody various degrees of specificity and interpretation. Tables 4.4 and 4.5 give examples of systems for the study of mother–infant interaction that fall fairly clearly into the molecular and molar categories. In the Als, Tronick, and Brazelton (1979) research, described in Table 4.4, the interest was in very basic forms of organization and adaptation shown by young infants in interaction with their mothers. In the Lamb (1976) research, described in Table 4.5, the interest was in infants' methods of maintaining contact with their parents in a strange situation. The main contrast drawn concerned the two global and clearly molar categories, distal/affiliative and proximal/attachment.

Two general factors determine just where on the molecular-molar continuum a researcher is likely to work. One is the purpose behind the research. If the interest is in the topography of facial movements in various affective states (e.g., Izard, 1989), then a molecular observational system is clearly necessary. If the interest is in the social determinants of smiling or laughing (e.g., Sroufe, Waters, & Matas, 1974), then a more molar system is probably more sensible.

The second general determinant of level of observation falls under the heading of feasibility. Whatever the researcher may wish to observe, the ultimate determinant of the observational system is what *can* be observed. Recording of molecular detail, for example, may be possible only in environments in which the observer can remain very close to the participant, or in which the behavior can be filmed or videotaped. Researchers working in other environments (which would include, of course, most natural settings) may have to settle for a more molar observational system. A further consideration, which is elaborated on shortly, concerns the reliability of the observations. Observations do not mean very much if two independent observers cannot agree on what is being observed. Sometimes the fine detail required in a molecular system may exceed the observers' capacities, forcing the researcher to move to somewhat grosser scoring categories. Sometimes the interpretations required in a molar system (Was that maternal behavior really rejection, or simply an attempt to redirect the child?) may frustrate agreement, forcing the researcher to stay closer to the level of the actual behavior (e.g., turns away from child). Whatever the specific problem, the general point is the one just made: The level of detail and interpretation in an observational system is always determined not only by what is desirable but also by what is possible.

Deciding How to Observe

Suppose that a researcher has arrived at a preschool, pen and clipboard in hand and ready to observe the interesting behaviors of preschool children. How might the researcher go about recording the data of interest?

One possibility is simply to write down, in narrative form, a description of the behavior as it occurs. Such a technique is called a **narrative record**; other often-used terms are *specimen record* and (from the pioneering work of Barker and Wright, 1951) the *stream of behavior approach.* Of course, some selectivity is necessarily involved in even the fullest narrative record. The focus is typically on just one child at a time, with other children brought in only as they interact with the target child. In observing the target child, decisions must continually be made about which behaviors are significant enough to record and which behaviors (e.g., blinks, swallows) can be ignored. Decisions must also be made about the level at which behaviors should be described—for example, did Johnny close his fingers together, make a fist, or threaten some other child? With a narrative record approach, the observer functions to some extent like a combination camera–tape

Table 4.4 An Example of a Molecular Observational System for Recording Infant Behavior During Mother–Infant Interaction

I. Type of vocalization:

1. none; 2. isolated sound; 3. grunt; 4. coo; 5. cry; 6. fuss; 7. laugh

II. Direction of visual attention:

1. Direction of Gaze: 1. toward mother's face; 2. away from mother's face; 3. following mother's face; 4. part side, nose level; 5. part side, nose down; 6. part side, nose up; 7. complete side, nose level; 8. complete side, nose down; 9. complete side, nose up.
2. Head Orientation: 1. toward, nose level; 2. toward, nose down; 3. toward, nose up; 4. part side, nose level; 5. part side, nose down; 6. part side, nose up; 7. complete side, nose level; 8. complete side, nose down; 9. complete side, nose up.
3. Left/Right Modifier of Head Position: 1. infant's left; 2. infant's right.
4. Blinks and Specific Eye Movements: 1. blink; 2. eyes crossed; 3. away and focused on specifiable object (such as chair side) that is not used by mother as part of interaction; 4. eyes shifted markedly to side relative to axis of nose.

III. Facial expression:

1. Position of Infant's Cheeks (examples only): 1. neutral, relaxed position; 2. elongated, hollow; 3. raised upward and puffed.
2. Eyebrow Position (examples only): 1. neutral resting position; 2. rounded with elevation in center; 3. flashing—rapid up and down.
3. Mouth Position (examples only): 1. neutral resting position; 2. slightly opened without tension; 3. broad smile; 4. yawn-open.
4. Eye Width: 1. neutral; 2. wide; 3. narrow; 4. closed.
5. Tongue Placement: 1. not exposed; 2. tongue exposed but not extended beyond lips; 3. tongue exposed and extended beyond lips.
6. Specific Facial Expressions: 1. cry face; 2. grimace; 3. pout; 4. wary/sober; 5. lidding; 6. yawn; 7. neutral; 8. sneeze; 9. softening; 10. brightening; 11. simple smile; 12. coo face; 13. broad smile.

IV. Body position and movement:

1. leaning forward and doubled over; 2. body turned off to one side; 3. arching; 4. leaning back; 5. slumped to one side; 6. neutral; 7. position being changed by mother; 8. moving up in vertical plane; 9. upright with head raised off cushion or neck extended and trunk elongated; 10. leaning forward with back straight.

V. Limb movements:

1. Size of Limb Movement: 1. none; 2. small; 3. medium; 4. large.
2. Number of Limbs in Movement: 1. none; 2. 1 limb; 3. 2 limbs; 4. 3 limbs; 5. 4 limbs; 6. only arms seen because of maternal position—1 moving; 7. as in 6.—2 moving.
3. Place of Movement: 1. none; 2. midline; 3. between midline and shoulders; 4. side.
4. Specific Arm and Hand Gestures: 1. eye rubbing; 2. hand to mouth; 3. swiping; 4. digit fidgets; 5. hands held together at midline; 6. all four limbs extended forward.
5. Specific Leg Gestures: 1. kicking; 2. startle.

SOURCE: From "Analysis of Face-to-Face Interaction in Infant-Adult Dyads" (pp. 43–44) by H. Als, E. Tronick, and T. B. Brazelton. In M. E. Lamb, S. J. Suomi, and G. R. Stephenson (Eds.), *Social Interaction Analysis* (pp. 33–76), 1979, Madison: The University of Wisconsin Press.

Table 4.5 An Example of a Molar Observational System for Recording Infant Behavior During Parent-Infant Interaction

	Distal/affiliative behaviors
Behavior	*Definition*
Smiling	A facial expression in which the brows were not drawn together, but the corners of the mouth were retracted and raised.
Looking	Direction of gaze toward the person concerned.
Vocalizing	Includes all directed nondistress vocalizations except giggling or laughing, which were tabulated as instances of laughing.
Proffering	Occasions on which the infant either offered, showed, or pointed out an object or toy to an adult.
	Proximal/attachment behaviors
Behavior	*Definition*
Proximity	Coded once in each 15-second unit that the infant was within 3 feet of the adult concerned.
Approach	A move from beyond to within 3 feet—that is, a move into proximity of a person.
Fussing	Any distress-type vocalization directed toward an adult.
Touching	Coded when the infant made physical contact with the body or clothing of an adult.
Reaching	The child gestures to the adult by raising and moving a hand in the direction of the person.
Seeking to be picked up	Manifested by one or more of the following: fussing, reaching to, vocalizing, or clinging to the legs of the person.

SOURCE: Adapted from "Twelve-Month-Olds and Their Parents: Interaction in a Laboratory Playroom," by M. E. Lamb, 1976, *Developmental Psychology, 12*, pp. 237–244. Copyright © 1976 by the American Psychological Association.

recorder—although a camera-recorder with a strong decision-making editor built in.

Despite the constraints just noted, a major strength of the narrative record approach is its relative completeness. More of the information about behavior is preserved than in any other method of observation. This relative completeness makes the narrative record especially useful for practitioners who require detailed information about a particular child. Narrative records are thus commonly used by teachers working in educational settings, or by clinicians compiling case studies of individual children. Such records can also be used as starting points for research programs, suggesting phenomena that can then be followed up with more systematic and more focused methods of study. Finally, narrative records need not always be just the preliminary to research; if the recording is carried out in a sufficiently skillful and systematic manner, the narrative record may serve as the basic data for a study. In such cases the narrative record provides the raw data; further coding and analysis are then necessary to winnow the behavior stream down to the units and phenomena of interest.

On the debit side, compiling narrative records can be a costly and time-consuming process. The demands on the observer may be especially great, as may be the possibilities for various forms of subjectivity and bias. The researcher may end up with a huge mass of information, only a small portion of which turns out to be of interest. Or the researcher may have clear-cut goals and hypotheses before beginning the observation, in which case the narrative record may be a very uneconomical form of data collection. In such cases, other, more focused forms of observation may make more sense.

A second general approach to observation is **time sampling** (also sometimes labeled *interval sampling*). Two features distinguish time sampling from narrative record. First, with time sampling the focus is on a relatively few specific and well-defined behaviors, not on the entire stream of ongoing behavior. The examples of molecular and molar approaches shown in Tables 4.4 and 4.5 were, as actually applied, also examples of time sampling. Here, a specific set of behaviors is precisely defined in advance, and only these behaviors are recorded. Because precise definitions already exist, there is no need to write a narrative description of the behavior; rather, some sort of checklist or coding system can be used. The second distinguishing feature is the division of the observation period into exact and usually rather brief units of time. The observer might observe for 15 seconds, turn away and record for 15 seconds, observe for another 15 seconds, record for 15 seconds, and so forth. The "sampling" part of time sampling is thus two-fold: Only a few of the ongoing behaviors are examined, and only some segments of the total observation period are included.

A third general approach can be introduced by example. In what is often referred to as one of the "classic" studies of child psychology, Dawe (1934) set out to study preschool children's quarrels. Although the perceptions of many preschool teachers might differ, quarrels turned out to be fairly infrequent, occurring at an average of 3.4 per hour. Given the infrequency of the behavior, both narrative record and time sampling would have been inefficient methods of study. Time sampling might also have been misleading; it would be quite possible for the observer to miss a quarrel if it happened to occur during one of the nonobservation periods, or to see only a portion of the quarrel if it cut across periods. The method Dawe adopted instead is labeled **event sampling**. In event sampling the target behavior itself, and not time, serves as the unit of analysis. As in time sampling, the observer begins by carefully defining the behaviors of interest. With event sampling, however, the observer simply waits until the behavior occurs and begins recording only then. The recording can take a variety of forms, ranging from a narrative description to a precoded checklist. Dawe employed a combination of predetermined categories and supplementary running notes in her research; among the categories scored were "passive behavior," "retaliative behavior," and "undirected energy." Whatever the form of recording, the focus on the target behavior as the basic unit may allow the observer to capture information (e.g., average duration of the behavior, or antecedents and consequences) that might be missed with a time-sampling approach.

Although I have presented time sampling and event sampling as mutually exclusive options, with modern technology it is possible to encompass aspects of both. Many recording devices provide running records of time, allowing the observer to record in terms of events but also to perform various time-based analyses if desired. Bakeman et al. (2005) provide a good description of the possibilities.

In general, the factors that influence choice of a recording system are the same as those that influence the molecular-molar decision: goals and feasibility. In some observational studies (e.g., clinical case reports) a narrative record is essential; in others a more focused approach such as time or event sampling is more appropriate. Whatever the goals of the study, the researcher must select a system that can be applied within the available environmental setting, does not place impossible demands on either the time or the abilities of the observers, and yields a reasonable ratio of usable data to time and effort expended. Further discussion of the plusses and minuses of different recording systems can be found in Bakeman et al. (2005); Hartmann et al. (2004); Mann, Have, Plunkett, and Meisels (1991); and Odom and Ogawa (1992).

Determining the Accuracy of Observations

This section begins with a consideration of two specific threats to the accuracy of observations. It concludes with a discussion of the thorny problem of reliability.

The behaviors recorded in an observational study may be a function of any number of antecedent or contemporary factors. One factor that we do *not* wish to have influence the behavior, however, is the mere presence of the observer. Yet the presence of the observer, and the concomitant knowledge that one is being observed, may alter behavior in various ways. Such effects fall under the general heading of *reactivity:* unintended effects of the experimental arrangements on the participant's behavior. When the context is observational research, reactive effects are often referred to as the problem of **observer influence**.

Just how important reactivity is in observational studies has long been a matter of dispute. There is evidence (e.g., Brody, Stoneman, & Wheatley, 1984; Russell, Russell, & Midwinter, 1992) that both adults and children may behave differently when they know that they are being observed; there is also evidence that under some circumstances, behavior is not at all affected by observation. Hartmann and Wood (1990) provide a helpful discussion of the factors that influence the probability of reactive effects.

Several techniques to reduce the possibility of observer influence follow from such a discussion. One possibility is to habituate the participants to the presence of the observer—that is, to introduce the observer into the setting some time before the onset of observation, and to begin recording only once participants have become accustomed to the observer and their behavior has returned to normal. This technique is sometimes referred to as "fading into the woodwork." A variant of it, which is sometimes but by no means always possible, is to have the observations made by someone who is already a natural and familiar part of the setting, such as a parent or teacher. Such use of someone who is already present in the setting to collect observations is referred to as **participant observation**. The concept of participant observation also encompasses cases in which the researchers themselves *become* a part of the setting that is to be observed. For example, a researcher of nursing homes might join the staff of a nursing home for several months, thus gaining access to the events of interest without the need to introduce a potentially biasing outside observer. We will return to the notion of participant observation in the discussion of ethnographic research in chapter 7.

Another general strategy for minimizing observer influence is to disguise the fact that observations are being made. It may be possible, for example, to film the behavior with a hidden camera, or to observe participants through a one-way mirror. Of course, it may also not be possible to do so—such techniques tend to be feasible only in certain special environments. In addition, there may be ethical as well as logistical constraints on the possibility of surreptitious observation. As we will see in chapter 9, observing people without their knowledge raises a number of ethical questions.

A second general problem in observational research is **observer bias**. Here too, the threat to observational study is merely a specific instance of a more general problem. As will be discussed more fully in chapter 5, a large body of research initiated by Robert Rosenthal (1976) indicates that the expectations researchers bring to research can sometimes bias their results, moving outcomes in the direction of what was expected or desired. In observational study the danger is that observers may see and record what they expect to occur, rather than what actually happens.

A study by Kent, O'Leary, Diament, and Dietz (1974) provides an example. The observers in this study viewed videotapes that supposedly showed the baseline and treatment phases of a program intended to reduce disruptive behavior in a classroom setting. Half of the observers were told that a decrease in disruptive behavior from baseline to treatment was predicted; the other half were told that no change was predicted. In fact, all the observers viewed the same videotape, in which no change occurred for any of the behavioral categories. When later asked to make a global rating of the effectiveness of the program, however, 9 of the 10 observers who had been led to expect a decrease in disruptive behavior reported that a decrease had in fact occurred. In contrast, 7 of the 10 observers who had been led to expect no change reported no change. It is interesting to note that there were no differences between the two groups in the actual behavioral recordings made while watching the tape. Only on the overall global rating did the expectancy manipulation have an effect.

The findings of the Kent et al. study suggest one way to reduce the probability of observer bias: Make the scoring categories as specific and objective as possible. The greater the leeway for interpretation in scoring, the greater the opportunity for observers to inject their own biases. The other general way to reduce observer bias is to keep the observer uninformed about the hypotheses of the study or the group to which the participant belongs. Such withholding of potentially biasing information is referred to as **blinding**. The rationale is straightforward: If no expectancies exist, then there is no danger of a researcher expectancy effect. Unfortunately, blinding may be difficult and in some cases impossible to achieve. Furthermore, even when blinding *is* possible, it is by no means always used.

A final set of problems revolves around the concept of reliability. As noted earlier, reliability refers to consistency of measurement. In the case of observational methods, the key issue is that of *interobserver agreement:* Can two or more independent observers arrive at the same interpretation of a behavior? Such agreement is a necessary basis for concluding that the observations are accurate. It is not a sufficient basis, however, because it is possible for two observers to arrive at the same wrong interpretation of the behavior. Two wrongs may not make a right, but they can lead to high agreement. This, too, is a specific instance of a general point: Reliability is necessary but not sufficient for validity.

There are various ways to calculate reliability (Bakeman et al., 2005; Hartmann & Wood, 1990). For some kinds of data the correlation statistic is appropriate. The higher the correlation between the recordings of two observers, the more satisfactory the reliability. Another commonly used index is percentage of agreement. Suppose that there are 20 opportunities to score the occurrence of some behavior. Agreement by two observers on 19 of the 20 instances would yield a percentage of agreement of 95%, a satisfactorily high reliability. Agreement on only 13 of the 20 instances would yield a percentage of agreement of 65%, which is not likely to be considered satisfactory. In practice, such simple percentage-of-agreement measures are often adjusted by a formula known as the *Kappa index* (Cohen, 1960). The issue to which the Kappa is directed is the possibility of chance agreement between observers, something that is especially likely when the baseline probability of a particular scoring decision is high. Suppose, to return to the example sketched earlier, that both observers record the behavior as occurring 80% of the time. If so, their scores will be in agreement a substantial proportion of the time simply from chance alone. The Kappa takes the baseline probabilities into account and thereby provides a truer picture of the degree of agreement in such cases.

Procedures useful for maximizing reliability are easy enough to state, if not always to apply. Observers should be carefully trained before the start of actual data collection. The scoring system should be as clear and as specific as possible. Pilot testing can be used both to train observers and to refine the system, with categories that prove infrequent or unscorable either dropped or transformed into more usable categories. Finally, when possible, filming or videotaping can be used to produce a permanent, replayable record of the behavior.

As the preceding suggests, it is desirable to achieve reliability as early as possible in the process of data collection. It may also be desirable, however, to continue to monitor for reliability throughout the course of a study. Research by Reid (1970; Taplin & Reid, 1973) makes this point. In the Taplin and Reid study, observers were first trained to an acceptable level of reliability. Subsequently, one group of observers was told that reliability would no longer be assessed, whereas a second group was told that periodic and unpredictable reliability checks would be made. In fact, every observer's recordings continued to be checked against predetermined criterion ratings. The results were clear: The observers who expected to have their ratings checked maintained better reliability than those who did not expect a reliability assessment. This tendency for initially reliable observers to drop in reliability when they are no longer being monitored is called **observer drift**. It is one of two ways in which observers may become less accurate over time. The other is **consensual drift**: the tendency for observers who work together to drift away from accuracy in the same way. Note that with consensual drift, in contrast to observer drift, the interobserver agreement remains high; it does so, however, because both observers are erring in the same way.

As with other problems discussed in this section, observer drift and consensual drift are specific forms of a more general threat to validity. The general threat in this case is what Campbell and Stanley (1966) label **instrumentation**: an unintended change in a measuring instrument across the course of a study. With observer or consensual drift the measuring instrument that changes is the human observer.

Thus far I have been discussing reliability as though there were a single overall index of reliability that a study either achieves or fails to achieve. In fact, in the typical case there are a number of potential reliabilities—for particular behaviors, particular aspects of a behavior, particular time periods, particular subgroups of participants, and so forth. The essential point with respect to these various possibilities is that reliability must be demonstrated at the level at which the data are to be analyzed. If, for example, the researcher wishes to examine posttest differences following some treatment, then it is necessary to demonstrate that the posttest data can be scored reliably; reliability achieved during the pretest phase is not sufficient. Similarly, if the analyses involve the total number of aggressive acts, then reliability for global ratings of aggression is not sufficient; the researcher must also show that observers can agree on specific instances of aggressive behavior.

Yarrow and Waxler (1979) provide an interesting, and somewhat less obvious, example of this same point. These authors reported various observational studies in which reliability was calculated separately for boys and girls. In some cases a behavior turned out to be more reliably scored for one sex than for the other. At least in these studies, for example, aggression could be scored more reliably for boys than for girls. Furthermore, individual differences in aggression showed sensible relations to other measures for boys—not, however, for girls. As Yarrow and Waxler note, this finding could reflect a genuine sex difference, or it might come about simply because the measures of aggression were not sufficiently reliable for girls. Again, reliability is necessary at the level at which the data are to be used.

Yarrow and Waxler (1979) also provide a good overview of the pros and cons of using a human observer as a measuring instrument. Their overview can serve as summary for the present discussion of observational methods:

> Though exceedingly practiced, the human observer, by many criteria, is a poor scientific instrument: nonstandard, not readily calibrated, and often inconsistent or unreliable. Counterbalancing these failures are the human capabilities of extraordinary sensitivity, flexibility, and precision. The challenge is to discover how to conduct disciplined observing while making full use of the discriminations of which the human observer is capable. (p. 37)

Summary

This chapter begins with a consideration of some basic principles of measurement. The variables with which we work in research are defined—operationally—by the ways in which they are measured. Such measurement always consists of translation of some global construct into some more specific, objective, and quantifiable form. This translation involves choice of particular measures from a larger pool of possibilities. Among the dimensions of choice are the aspect of the behavior that is measured (e.g., frequency or intensity), the degree to which the measurement target is overt or covert, whether the measurements are signs or samples of the construct of interest, and whether the focus is on immediate states or enduring traits. Researchers must also make decisions about methods of data collection, with choices among three general possibilities: naturalistic observation, laboratory study, and verbal report.

Several considerations are important in evaluating the quality of a measurement. The measurement should yield the desired level and range of values, avoiding both *floor effects* and *ceiling effects.* Issues of *measurement equivalence* must be addressed if distinct groups (e.g., children of different ages) are included in the research. Use of a variety of methods may be necessary to avoid the problem of *mono-operation bias.* Finally, the researcher must be able to demonstrate both the *reliability* (i.e., consistency) and the *validity* (i.e., accuracy) of the measurement operations.

Questions of reliability and validity are taken up more fully in a discussion of standardized tests. As applied to a test, the validity question is whether the test really measures what it is intended to measure. Three forms of test validity are discussed: *content validity, criterion validity,* and *construct validity.* Of these forms the most complex is construct validity, the establishment of which may involve both experimental and correlational data and the demonstration of both convergent validity and divergent validity in the correlations among measures.

The reliability question is whether the test consistently measures whatever it is that it does measure. Two forms of reliability are considered: *test-retest reliability* and *internal consistency reliability.* Also considered is a threat to validity that can arise in the absence of perfect reliability: *regression toward the mean.* Regression is the tendency for initially extreme scores to become less extreme upon retesting. Among the kinds of research in which regression can be a problem are intervention studies and certain forms of matched-group designs.

The discussion turns next to another important form of measurement: observational methods. Three general issues are examined. A first concerns the level of specificity at which behavior is recorded. A *molecular observational system* attempts to capture relatively fine-grained details of behavior; a *molar observational system* consists of more global and interpretive categories. The goals of the research are one determinant of where on the molecular-to-molar continuum a researcher will work. Another is feasibility: A particular observational system is usable only if the required observations can be made accurately.

Goals and feasibility are also relevant to the second general issue discussed: the method of recording observations. The *narrative record* provides the fullest account of ongoing behavior. More focused methods of observation include *time sampling* and *event sampling*. In both cases specific recording categories are determined in advance; observations are then made within the framework of either units of time (time sampling) or behaviors of interest (event sampling).

The last issue discussed concerns problems that can arise in observational research. *Observer influence* is a particular form of reactivity; it refers to the fact that people's behavior may be altered by the knowledge that they are being observed. Various techniques for reducing such bias are discussed. The expectancies of the observer are another possible source of bias. The best way to guard against such bias is to minimize the expectancies. Finally, the utility of observations is dependent on the demonstration of *interobserver reliability,* that is, agreement by two or more independent observers about how behavior is to be classified. Reliability should be monitored throughout the study, to guard against the phenomenon of *observer drift*. It should also be demonstrated at the level at which the data are analyzed.

Exercises

1. One point that the text stresses is that any particular theoretical construct may lend itself to a number of different operational definitions. Consider the following constructs: altruism, creativity, wisdom. For each, generate a conceptual definition and at least two operational definitions. Do the same for at least two further constructs that especially interest you.

2. Select some construct in developmental psychology that especially interests you (e.g., intelligence, creativity, temperament, self-concept—any outcome of interest). Find at least two standardized tests whose purpose is to measure individual differences in the construct. Critically compare the two measures. Which would you select for your own research, and why?

3. Imagine that you have the task of validating a new test whose purpose is to measure individual differences in creativity. What kinds of evidence would you collect?

4. The following activity assumes that you have access to at least one parent with an infant of roughly 12 months of age. If so, obtain a copy of the Lamb (1976) article discussed in the text, study the description of the observational scoring system, and then attempt to apply the observational system yourself. Note that you will have to make some adjustments of the laboratory procedure for in-the-home conditions; it should be possible, however, to replicate at least some aspects of the Lamb research in the home environment. The exercise will be most valuable if you are able to try it on several occasions with different parent-child dyads. It would also be good to pair with another member of the class and to calculate the reliability of your observations.

5

Procedure

Thus far most of our discussion has concerned the decisions that go into the planning of research. How should we go about selecting participants and assigning them to conditions? What are the best ways to compare different ages or experimental treatments? How can we select measures that are best suited for the particular questions being examined? How, in short, can we put together a study that will yield valid results?

Decisions of this sort are obviously crucial to good research; as indicated in chapter 1, however, they are not sufficient. Working out the design and the measurements brings us to the brink of research; these decisions must then be implemented by actually testing or observing research participants. In this implementation a host of challenges may arise. It is to these challenges and ways to overcome them that the present chapter is devoted.

Standardization

I begin with something that is a clearly desirable goal in most research: standardization. **Standardization** refers to the attempt to keep all aspects of the experimental procedure the same for all participants. To achieve standardization, the researcher must decide in advance exactly how each aspect of the procedure is to be handled—what the wording of the instructions will be, how and when stimuli will be presented, and so forth. Once these decisions have been made, the researcher must then make sure that the standardized procedure is in fact followed.

The basic reason for standardization was discussed in chapter 2: the need for experimental control. Anyone carrying out a study must be able to control a number of aspects of the experimental setting: the exact form of the independent variables, the way in which the dependent variables are measured, other factors in the situation. If such control is not exerted, then there is simply no way of knowing exactly what it is that the study is doing—and no way of interpreting any results that are found.

Let us briefly consider some examples of problems that might arise in the absence of adequate standardization. Imagine a study of short-term memory in grade-school children—say first- and fourth-graders. The tester presents a series of pictures of familiar objects, one at a time, to the children, and then asks the children to recall as many of the pictures as they can. Table 5.1 shows various ways in which a tester in such a study might deviate from standardization and thus bias the results. The tester might, for example, inadvertently vary the exposure time for different pictures or

Table 5.1 Examples of Possible Deviations From Standardization and Their Effects in a Developmental Study of Memory

Aspect of procedure	*Intended procedure*	*Deviation from standardization*	*Possible effects*
Wording of instructions	Tell all children that the items can be recalled in any order.	Only some children are given the instructions regarding the unimportance of order.	General: Unintended variance is introduced, and results may no longer be clearly interpretable. Specific and biasing: If the wording varies across age groups, age differences may be either obscured or artifactually produced.
Exposure time	Expose all pictures for the same length of time for all children.	Exposure time varies across pictures or across children.	General: Same as preceding. Specific and biasing: If exposure times vary across either age or condition, then false conclusions may be drawn about differences between groups.
Degree of probing	Include the same predetermined amount of probing for all children.	Some children receive more than do others.	General: Same as preceding. Specific and biasing: If the probing varies across age groups, age differences may be either obscured or artifactually produced.

different children. Or the tester might forget to tell some of the children that the items can be reported in any order, a fairly important matter when one is trying to recall a long list of items. Or the tester might probe more on the recall trials for some children than for others, thus eliciting better overall performance.

As Table 5.1 indicates, such lapses from standardization can create problems of various sorts. At the least, imperfect standardization means uncertainty: It is no longer clear exactly what a given condition consists of, and therefore no longer clear how to interpret the results. This uncertainty may cloud the descriptive information that the study would otherwise yield: How impressive is a particular level of recall, for example, if we do not know what the exposure time was or how much probing was used? It may also affect the external validity of the results. If what is found depends on some particular (unintended and unknown) mixture of procedures, then findings may not be generalizable to the contexts that the researcher hopes to generalize to.

Imperfect standardization may also threaten the internal validity of a study. Problems of internal validity arise whenever the deviations from standardization vary systematically across the groups being compared. Suppose, for example, that the tester probes more on the recall trials for older children than for younger ones (perhaps out of a belief that the correct answer is more likely to be "really in there" for older children). In such a case an apparent age difference may actually reflect a difference in procedure. We would have an incorrect conclusion concerning a cause-and-effect relation, and hence a lack of internal validity.

Standardization, then, is important. But achieving standardization might seem easy enough—all that is necessary is to make procedural decisions in advance and then act upon them. In fact, as anyone who has done much research knows, things are not always so easy. It can be difficult to anticipate all the procedural issues that can arise in the course of a study; it can also be difficult to stick faithfully to a long and complicated procedure when actually face to face with a research participant. These difficulties are compounded by the fact that participants do not always respond in the way that the researcher expects; children, in particular, are notorious for doing things for which even the most experienced tester is unprepared. The difficulties are compounded too by the fact that in most kinds of research with children it is desirable to interact with the child in a natural, spontaneous-seeming manner, the better to keep the child at ease and attentive. Reading from a script—the easiest way to achieve standardization—may undercut the naturalness that is needed to keep the child happy and responsive.

Probably the best route to satisfactory standardization is experience, both general experience in doing research and specific experience with the age groups and procedures under study. Although general experience comes only with time, the needed specific experience can be gained, at least in part, through *pilot testing:* experimenting with and practicing one's prospective procedures before beginning the actual study. Careful pilot testing can enable the researcher to iron out kinks and uncertainties in advance, thereby arriving at a well-standardized procedure before the first real participant appears for testing. Pilot testing can also help the researcher to become smooth and assured in delivering instructions and interacting with participants, thus avoiding the kind of stilted, nose-in-a-script behavior that may upset young children. And pilot testing can help the researcher to anticipate, and be ready to respond to, at least some of the problematic participant behaviors that can force departures from a standardized script. Often in research with children, the most important attribute that a tester must have is not simply the ability to follow a standardized script but the skill to adapt to the vicissitudes of the individual child in a way that neither biases the experiment nor alienates the child.

Departures From Standardization

Standardization is desirable, but it is not an unquestionable goal. Frequently, it makes sense to settle for less-than-perfect standardization. We have just seen one such case: the situation in which the child's behavior forces a departure from the standardized procedure. In this section I consider several others.

We can begin by recalling a point made in chapter 2: Total standardization, in the sense of making everything literally the same for all participants, is never possible. If we are doing individual testing, for example, we are not going to be able to test everyone at 10:00 in the morning on September 30; time of testing will necessarily vary across participants. Nor are we likely to be able to hold constant such factors as the ongoing school activities, the outside weather, the attractiveness of the day's lunch, or a number of other things that may affect the child's performance. As we saw in chapter 2, the important point is not to make everything the same for all participants; it is to ensure that any potentially biasing factors are equally distributed across the groups being compared. If we disperse such factors equally across groups, then we can avoid confounding any particular factor with the independent variables of interest.

Further departures from standardization may be forced by the attempt in developmental research to accommodate distinct age groups within the same study. Consider a study of scientific reasoning that spans the age period from preschool through college (e.g., Kaiser, McCloskey, & Proffitt, 1986). Age is one of the independent variables in such research; by the prescription just given, therefore, we should make sure that there are no systematic differences in procedure for participants of different ages. It is quite unlikely, however, that any researcher will treat 4-year-olds and 20-year-olds in precisely the same way. The immediate testing environment is likely to differ; children are typically tested in an unused room at their school, adults in a laboratory on a university campus. The wording for at least part of the procedure will probably differ; one does not recruit a 20-year-old for an experiment by inviting him to "come play a game." Even if the wording is kept the same for different ages, the pacing and tone of voice will probably not be; natural conversational style is simply not the same for 4-year-olds and 20-year-olds.

The preceding is not meant to suggest that standardization and control must fall by the wayside in studies with widely separated age groups. Standardization remains important within an age group. Furthermore, any obviously critical elements of the procedure, such as the content of the problems to be solved in a study such as that of Kaiser et al., must be kept the same for all participants; otherwise there is no point in making the age comparison. For other aspects of the procedure, however, what the researcher of different ages may opt for is a kind of functional rather than literal equivalence of procedure. The goal, in other words, may be to make sure that the procedure is *equally age-appropriate* for all of the age groups tested. A grade-school library and a university laboratory are obviously different rooms; they may, however, be equally natural and familiar settings for the groups being studied. What is equated in this case is not the room itself but the familiarity of the testing environment. A similar argument can be offered in support of age-appropriate adjustments in wording, tone of voice, feedback and praise, and so forth. In each case, what the researcher attempts to do is to devise a procedure that is appropriate within age and as comparable as possible across ages. The challenge in this attempt is to ensure that the cross-age procedural adjustments do not bias the age comparisons of interest.

A final instance of departure from strict standardization is found in **exploratory research**. As the name suggests, "exploratory research" refers to research whose goal is to explore—to break new research ground, to attack some little-studied problem in some new and mostly nonpredetermined ways. The essence of such research lies in the possibility for creativity on the part of the researcher—the possibility for following up unexpected findings, trying out various methods of study, and in general experimenting with and modifying one's procedure as one goes.

A classic example of exploratory research is found in much of Piaget's work (e.g., Piaget, 1926; Piaget & Szeminska, 1952). As we will see more fully in later chapters, an important element in Piaget's success was his ability to uncover and probe new phenomena, utilizing a flexible style of questioning and posing problems that was guided as much by the child's responses as by any preset procedure. It is precisely this sort of experimentation and change that is ruled out in a strictly standardized study.

We will see another point in the later discussions of Piaget. Testing the validity of Piaget's claims has required a large number of more tightly controlled and standardized follow-up studies. Exploratory research is uniquely suited for discovering new phenomena and generating interesting ideas; verification of these phenomena and ideas, however, may depend on the rigor and control that only a more standardized approach can bring.

Overstandardization

As we have just seen, standardization is always less than perfect. Total standardization is not literally possible; certain things will always vary across participants. No matter how standardized the researcher may intend to be, the participant's responses may force deviations from the preset procedure. The researcher of different age groups may deliberately vary certain procedural features across ages. The researcher of a new content area may deliberately forgo standardization for the possibility of exploration and discovery.

But suppose that we are dealing with a case that does not fit any of the just-named exceptions. Suppose that we have a study with a single age group, a well-worked-out procedure, good control over the situation, and an interest not in open-ended exploration but in determining as precisely as possible how certain specific variables relate. Is there any reason in this case not to aim for the highest degree of standardization possible? Or can there be such a thing, even here, as too much standardization?

Campbell and Stanley (1966) suggest that there can in fact be overstandardization. Their example concerns a study of persuasion designed to compare the effectiveness of a rational appeal and an emotional appeal. To maximize standardization, the researcher decides to tape-record a single version of each appeal, thus ensuring that all participants in a given condition receive the same stimulus. Such control might seem desirable, because we presumably will then know exactly what a given experimental condition consists of. In fact, however, the decision to hold everything constant means that we may not be able to determine exactly which aspects of a particular condition are responsible for any effects we find. Perhaps the important factor is the one that we set out to study: the emotional versus rational content of the message. Or perhaps the important factor is the sex of the speaker, or distinctive voice qualities of the particular individual who serves as speaker, or the tone or pacing with which the message is delivered. Perhaps the important factor is the combination of these particular qualities—for example, an emotional appeal delivered by a male speaker with a distinctive voice. The point is that anything that is kept the same for all participants becomes potentially part of the independent variable. A better approach might be to let these extraneous factors (sex, voice qualities, etc.) vary across participants, while holding constant the one factor in which we are really interested: the content of the message. Any effects could then be more confidently attributed to content per se.

It should be clear that overstandardization poses a threat to the external validity of research. In a study such as that just sketched, the goal is to derive at least somewhat general conclusions about the persuasive effects of an emotional or a rational message. What we do not want are effects that are dependent on idiosyncratic features of a particular speaker or particular delivery, and therefore not generalizable beyond the bounds of our study.

Overstandardization can also threaten the construct validity of a study. Construct validity requires that we arrive at a correct interpretation of any effects that are found—in the present case, that it is the emotional content of the message that is important. If characteristics of the particular speaker are in fact important, then the study lacks construct validity.

The dangers posed by overstandardization provide an argument for the importance of **replication** in scientific research. A replication, by definition, holds constant those features that are the intended focus of the research, such as the emotional or rational content of the message in the hypothetical study of persuasion. At the same time, other features of the procedure will almost certainly vary, especially if the replication is carried out by a different researcher in a different laboratory. Thus, the particular individual who delivers the persuasive appeal will probably differ, as will the tester who interacts with the participants, the room in which the testing is done, the time of year at which the study occurs, and so forth. If we find the same effects of an emotional or rational appeal despite these changes, then we can be more certain that message content really does determine the outcome. Conversely, a failure to replicate our original results would suggest that supposedly irrelevant features of the procedure, such as the particular speaker, may not be so irrelevant after all.

Some Sources of Bias

We have seen that either too little or too much standardization can cause problems in research. In this section I consider some further, somewhat more specific threats to validity. In doing so I draw from the discussions by Campbell and Stanley (1966) and Shadish et al. (2002) that were summarized in Table 2.5. I discuss four specific threats: instrumentation, selection bias, history, and reactivity.

As noted in chapter 4, *instrumentation* refers to changes in either physical instruments or human testers or observers across the course of a study. It falls under the general heading, therefore, of imperfect standardization: Some aspect of the procedure that was meant to be constant in fact changes from early to late in the study. Although the change may in some cases involve instruments in the literal sense (e.g., a stopwatch that begins to malfunction as the weather turns humid), a more common problem in most developmental research is a change in the human instrument who serves as tester or observer. Perhaps the tester becomes more skilled and assured in delivering instructions across the course of the study. Alternatively, perhaps the tester becomes bored or discouraged as the study progresses and begins to behave in an increasingly perfunctory manner. In either case, the procedure would be different for participants tested early in the project than for participants tested late.

The major threat posed by problems of instrumentation occurs when the cross-study changes are confounded with one of the independent variables being examined. Consider a study in which one of the purposes is to compare the responses of kindergarteners and second-graders to some experimental task. The researcher decides to test most or all of the kindergarteners before starting with the second-graders. There are a number of possible reasons for such a decision: Only one teacher, classroom, and set of permission slips must be dealt with at a time; the eagerness of the children within a class to "come play the game" may be less likely to dissipate; the teachers may prefer that the testing within their room be completed quickly; and so forth. The researcher who gives in to such factors of convenience, however, is allowing a potentially important source of confounding into the study. If there is any possibility of change in the tester over time, then differences between kindergarteners and second-graders will not be clearly interpretable.

The instrumentation issue applies not just to age comparisons but to comparisons between two or more experimental conditions. Often it is simplest in studies with multiple conditions to do all the data collecting for one condition before beginning another, because then only one set of materials and procedures need be mastered at a time. Again, however, the convenience is bought at the price of possible confounding. Better practice would be to distribute the testing times for each experimental condition at least roughly equally over the course of the study.

There is another possible source of bias when one experimental condition is completed before another is begun. It is a bias that is made possible by the ubiquitous parental permission slips that are required for almost all research with children these days. Imagine that a researcher sends home 30 permission slips with children from a grade-school classroom. Fifteen slips are returned promptly, and these 15 children are tested under Condition 1. A week later the other 15 slips have been returned, and these children are tested under Condition 2. In this case the threat to validity (in addition to instrumentation) is *selection bias.* Parents who are prompt to return permission slips may well differ from those who are less prompt. If so, the children of the two groups of parents may also differ in ways quite apart from the Condition 1–Condition 2 manipulation.

Still one more argument can be offered in support of not confounding age or condition with time of testing. No matter how constant testers or instruments may remain, other potentially important factors may change across the course of a study. Any developmental researcher who has ever tried to test school children the day before a holiday, or even simply on a Friday afternoon, is well aware of this phenomenon. Research occurs in the midst of numerous other events in the participants' lives, and care must be taken to ensure that these other events do not confound any comparisons being made.

The point just made provides a bridge to Campbell and Stanley's (1966) history variable. As Campbell and Stanley use the term, *history* has to do with the issue that I have just been discussing: the effect of outside-the-study events on the outcome of research. The specific reference, however, is to pretest-posttest designs in which some intervening event provides an alternative explanation, in addition to that offered by the independent variable, for any changes that occur. Let us return to the example with which the history variable was introduced in chapter 3: the effects of an educational intervention program on young children's academic skills. Imagine that a new educational TV program begins to air shortly after the start of the intervention. If so, any improvements in the children's performance between pretest and posttest may have nothing to do with the intervention; they may result instead from the TV experience.

We saw in chapter 3 the obvious solution to problems posed by history: include a no-treatment control group—that is, a group that experiences the same historical events as the experimental group but does not undergo any experimental treatment. In the case of the intervention study, we could simply test a group of children at the beginning and end of the intervention period. Note, however, that such a no-treatment control group does not solve all the problems posed by the impact of some extraexperimental event on research. In some cases a historical event may interact with an experimental treatment. Perhaps, for example, our educational intervention is effective only when it occurs in conjunction with supportive TV experience; neither intervention nor TV alone, however, is sufficient to bring about a change. In such a case a no-treatment control could not make clear what is happening. And in such a case we would have a severe constraint upon the generalizability of our results.

Reactivity and Related Problems

The next threat to validity is important enough to merit a section of its own. As noted in Table 2.5, the term *reactivity* refers to unintended effects of the experimental arrangements upon the participants' behavior—or, more simply, the fact that people may behave differently when they are being studied than when they are not being studied. In research with adults, such effects typically stem from the participants' explicit awareness that they are taking part in a psychology experiment. In research with children, such explicit knowledge is less likely; indeed, in very young children or infants it is obviously impossible. Nevertheless, the fact of being studied may alter anyone's behavior, and thus reactivity may be a problem at any age.

Research with adult participants has identified a number of specific problems that fall under the general heading of reactivity. Two problems in particular deserve mention. The first is best introduced through example. Martin Orne (1962) was interested in finding a task that could be used in his studies of hypnotic control. What he wanted was a task that was so boring that eventually any nonhypnotized participant would refuse to work further on it. One task that was tried required the participant to carry out repeated additions of pairs of random digits. A single sheet contained 224 problems, and the participant was given a stack of 5,000 sheets. Five and one-half hours later a number of participants were still working! Orne then added a further stipulation: After completing each sheet, the participant was required to tear the sheet into no fewer than 32 pieces. Despite the patently ridiculous nature of the task, some participants continued to work until the experimenter finally told them to stop.

What this study and other studies by Orne revealed was that people who are taking part in an experiment may sometimes (although certainly not always—see Berkowitz and Donnerstein, 1982, for a critique) go to great lengths to do whatever it is that they perceive the experimenter wants. Such "good-subject" behavior can include both a general willingness to follow directives and specific attempts to validate whatever hypothesis the participant believes underlies the research. In Orne's terms the participant responds to the *demand characteristics* of the experiment—that is, "the totality of cues which convey an experimental hypothesis to the subject" (Orne, 1962, p. 779). Such biased responding yields what is known as a **demand effect**.

A second source of bias has been labeled the **evaluation apprehension effect** (Rosenberg, 1965); another term, to contrast with the "good subject," is the "prideful subject" (Silverman, 1977). The reference here is to the fact that people taking part in an experiment often behave in ways that will maximize the experimenter's positive evaluation of them. They try, in short, to look good. Looking good may, of course, be compatible with doing what the experimenter apparently wants, in which case the demand factor and the evaluation factor converge on the same effect. The two are not always synonymous, however; for example, participants may try to give "sophisticated-sounding" answers that are beyond those that they believe the experimenter expects. Silverman (1977), in fact, cites evidence indicating that when the demand factor and the evaluation factor conflict, it is usually the latter that wins out.

Although they are not limited to such research, problems of evaluation apprehension are probably greatest in studies using **self-report measures**. As the term suggests, with self-report measures the data consist of people's verbal reports about themselves—their own characteristics, past experiences, typical behaviors, or whatever. It may be both tempting and easy in such research for participants to tilt their answers toward what sounds best rather than what actually is. In a study of child-rearing practices, for example, a mother may report that she never spanks her child even though in fact she does occasionally resort to spanking.

The effects discussed in this section have been studied most often with adult participants. How applicable are they in developmental research? It seems clear that if we go young enough, we do not need to worry about either demand effects or evaluation effects in anything like the sense identified in the adult literature. Indeed, researchers of infants or toddlers may bemoan the fact that their young participants show so little regard for the goals of the research and the desires of the researcher! The cooperative attitude that is part of the demand effect can be a welcome, albeit dangerous, part of the research process. Even with very young children, however, the strangeness or artificiality of an experimental situation may have reactive effects. As would be expected, infants are more likely to show anxiety or be upset when tested in a laboratory than when tested in a familiar home environment. Preschoolers may clam up or generally withdraw when questioned by a strange adult; alternatively, they may become so fascinated by the stimulus materials or the adult attention that they talk about or do almost anything except what the researcher wishes. And by grade-school age, children begin to show quite clearly some of the classic reactivity effects demonstrated in the adult literature. Anyone who has tested grade-school children is familiar with the child who continually reads the adult's face for clues as to what is expected, who answers each question with an am-I-right rise in intonation at the end, and who in general seems to be most concerned with pleasing the adult, being well thought of, or both.

How can reactivity effects be minimized? Because reactivity stems from an awareness of being studied, an obvious approach is to disguise the fact that a study is occurring. Various degrees of disguise are possible. At a very simple level it is common for researchers of young children to introduce their experimental tasks as "games" rather than "tests" or "experiments." Such a description is often reasonably accurate, may convey more to a child than would the more complicated terminology, and presumably is less likely to arouse anxiety than is reference to tests. It is also common for researchers of preschoolers to spend some time playing with their young participants in an attempt to build *rapport* prior to the start of testing. Such reassuring wording and reassuring behavior do not guarantee an absence of anxiety or resistance; still, they should make negative responses less likely.

Reactivity should be least likely in research that most closely approximates the natural setting. Consider a study of preschool children's preferences for different toys. The researcher might study this issue by taking children individually to an experimental room and there administering a toy questionnaire that asks about their preferences. In this case the measure is direct and efficient, but the possibility of reactivity (anxiety about being questioned, saying what the adult apparently wants, etc.) is maximal. Another possibility would also involve bringing the children to some special experimental room, but in this case the researcher would simply observe with which of various toys placed around the room the child chooses to play. Especially if the opportunity to play is offered in a casual and natural way (e.g., "You can play with these toys while I finish working on these papers."), reactive effects may be unlikely. A third possibility is to observe children in their classrooms; playing with toys is, after all, a major part of a preschooler's day. If the researcher can observe without being observed (through a one-way mirror, for example), then reactivity should not be an issue at all.

Finally, in some cases behaviors may be inferred from their physical effects, without the participants themselves ever being observed. The popularity of different toys might be studied, for example, by charting which toys are still untouched on the shelves at the end of the school day and which are strewn about the room. Across a longer period, toy popularity might be studied by tracing physical wear and tear; which toys are still bright and shiny at the end of the school year, and which are scruffy and broken? Here, obviously, there is no possibility of reactivity. A book by Webb, Campbell, Schwartz, and Sechrest (2000) provides a detailed discussion of the use of such "unobtrusive measures" to infer behavior patterns. For cases in which a face-to-face interview is necessary to collect the data, books by LaGreca (1990) and Greene and Hogan (2005) offer a number of helpful suggestions.

Response Sets

I turn from reactivity to another, closely related problem. The term **response set** refers to a participant's tendency to respond to a question or task in a predetermined, biased fashion that is independent of task content. By this definition, the "good-subject" behavior discussed under reactivity can be considered a kind of response set: The participant attempts to say or do what the experimenter apparently wants, rather than responding to the task itself. In this section I consider response sets that are more clearly brought to the study by the participant rather than elicited by the demands of the experimental setting. There is, however, an admittedly thin and rather arbitrary dividing line between the problems of reactivity just discussed and the response sets considered next.

FOCUS ON

Box 5.1. Archival Data

The threat of reactivity arises when people are aware that they are being studied. As the text notes, one way to combat this threat is to collect data without directly studying people—that is, to use "unobtrusive measures" of the phenomenon of interest.

Unobtrusive measures can take various forms, but undoubtedly the most important category for psychologists comes in the use of archival data. The term **archival data** refers to already collected and available sources of information about some sample of people. Archival research, then, is the analysis of such information to answer some research question or set of questions.

What sorts of archives might be informative? The book *Unobtrusive Measures* (Webb et al., 2000) gives dozens of different examples. Many involve sources of information that most of us would not think of as possible data for research; examples include tombstone inscriptions, suicide notes, letters to the editor, physicians' notes, sales records, city directories, and fluctuations in the weather. Others involve more formal and systematically collected information; examples in this category are birth, marriage, and death records. Whatever the information obtained, it may play various roles in research. In some instances the interest may be purely descriptive; this is the case, for example, for census data that help to document the current and changing nature of a society (available at www.census.gov). If you look at any developmental psychology textbook you will see numerous such findings drawn from the census or other official records (e.g., birth rates for women of different ages, infant mortality rates across different countries). In some instances the archival record may provide the independent variable for some analysis; such would be the case, for example, if we were to examine weather fluctuations as a possible contributor to changes in behavior. And in some instances the archive may furnish a dependent variable that can help us understand some phenomenon of interest. Tombstone inscriptions for males and females, for example, might be used as evidence of the status accorded the two sexes in different societies at different times.

For developmental psychologists, the most valuable archival sources are research data collected by earlier investigators and now available for use by others (Brooks-Gunn, Phelps, & Elder, 1991; Tomlinson-Keasey, 1993). Such sources include a number of national survey studies, one explicit purpose of which is the sharing of data—for example, the National Longitudinal Survey of Youth (Chase-Lansdale, Mott, Brooks-Gunn, & Phillips, 1991). They also include some of the largest and most influential longitudinal studies in the field—in particular, several life-span efforts that were initiated in the 1920s. We will see an example of a use of the latter in chapter 6 when we consider Glen Elder's work on children of the Great Depression.

Working with such information—or with archival data of any sort—is by no means straightforward. Various challenges must be overcome (Elder, Pavalko, & Clipp, 1993; Zaitzow & Fields, 2006). A basic one is that the information of interest may simply not be there; after all, these are data that were collected for purposes at least somewhat different from those for which they are now being used. In some instances it may be possible to recode the data to reflect current interests; in others the researcher may need to abandon particular questions and be guided by the available information. Questions of selectivity may arise, regarding both the kinds of data that were initially collected and those that have survived over time. Given that archival data, by definition, are somewhat old, there may be issues of cohort and historical time—how general are any conclusions that can be drawn? A final caution may have already occurred to you. If the data were collected as part of a research project—as of course is the case for the longitudinal archives just discussed—then the "unobtrusive" appeal of archival data may no longer hold, and reactivity and related problems may still need to be considered. On the other hand, it may also have occurred to you that minimizing such procedural problems is hardly the only argument for archival research. A basic argument is that archival records—especially those that preserve the data from long-term longitudinal studies—provide a wealth of information that speaks to central issues in developmental psychology and that few investigators will ever be able to gather themselves. In the words of one commentator, such archives constitute "a professional treasure" (Tomlinson-Keasey, 1993, p. 65).

In both cases we are talking about biased responses that may lead to invalid conclusions.

A concrete example can help clarify the points made in this section. Figure 5.1 shows a task that I consider more fully in chapter 12: the Piagetian conservation-of-number problem (Piaget & Szeminska, 1952). As applied to number, conservation refers to the knowledge that the number of objects in a set does not change just because its perceptual appearance changes. Figure 5.1 shows several perceptual transformations that might be used to tap this knowledge. It also shows various ways in which the conservation question might be worded.

What kinds of response sets might children bring to the conservation task? One common form of response set is *yes-saying:* the tendency to say yes whatever the question asked. Clearly, such yes-saying is a potential problem whenever a one-directional question is used, as in the first two examples in Figure 5.1. Children who are always asked "Same?" and always respond "Yes" will appear to be conservers, even though their answers may be quite unrelated to the task in front of them.

Simple yes-saying can be precluded through use of a two-directional question, as in the third example in the figure. Other problems may still arise, however. Some children have a tendency to *choose the last-named alternative*—that is, to agree with whatever happens to come last in the adult's question. If the wording of the question is always that given in the third example, the result will be consistent (but perhaps pseudo) nonconservation. Other children have a tendency to *alternate answers,* to change from one answer to another simply as a function of repeated questioning. In conservation research such alternation may apply across trials or within a trial, since the question about number is typically posed both before and after the perceptual change. Finally, with some kinds of tasks, including conservation, *positional preferences* may be a problem. Young children, for example, might always pick the row that happens to be closer to them as the one that has more.

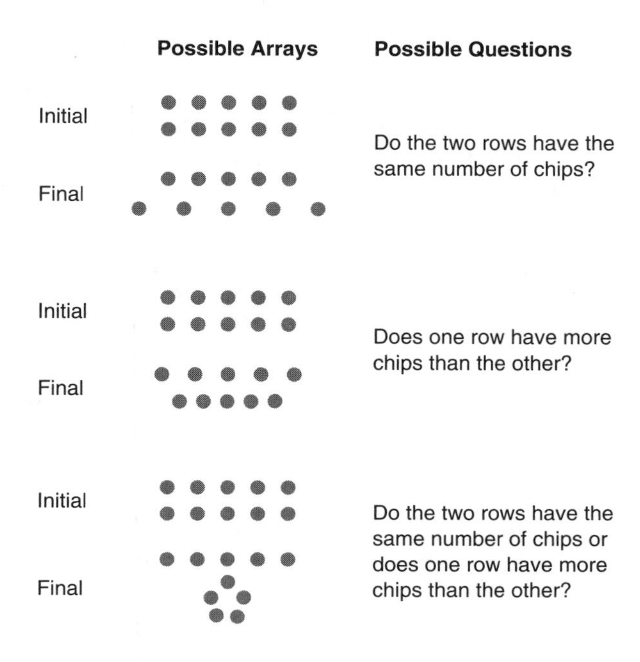

Figure 5.1 Possible perceptual arrays and conservation questions in a study of conservation of number

Several general points should be made concerning response sets. First, although I have drawn our examples from the conservation task, such problems are by no means limited to conservation, or, for that matter, to school-aged children. Whenever verbal responses are elicited, biases such as yes-saying may enter in. And as we will see in chapter 11, certain kinds of positional preference may be present even in neonates.

The second point concerns the interpretation of response sets. What does it mean if a child always answers "yes" to the conservation question no matter how the question is worded? Such a response might well be taken as evidence of an inability to conserve—the child who falls back upon such a simple response bias could hardly understand much about the phenomenon of conservation. And indeed, this interpretation may often be justified. The problem is that there is no way to know for sure. If children consistently pick the longer row as having more, then they have given a clear-cut wrong answer that is in response to the task manipulation. But children who always say "yes" are simply not responding to the task—perhaps because they do not understand the phenomenon being measured, perhaps because they are confused by the wording, perhaps because they are not motivated to think carefully, or perhaps for any of a number of other reasons. Response sets lead to no clear conclusion—except that the child has a response set.

This brings us to the final point: Response sets are clearly something that the researcher wishes to minimize. Various approaches are possible. In the case of conservation the researcher might work with maximally simple language, provide some verbal pretraining before the test itself, use incentives to try to ensure careful response, and so forth. No matter how skilled the procedure, however, response sets may in some cases be unavoidable. In such cases it is critical that the researcher at least be able to *identify* the fact that a response set has occurred, lest the participant's behavior be misinterpreted. A researcher who tests conservation by presenting a single trial can never know for certain what the child's judgment means. Presenting a number of trials of varied form provides a much firmer basis for deciding whether the child really understands conservation, really believes in nonconservation, or is showing some irrelevant response bias.

Intersubject Communication and Diffusion

An experiment by Horka and Farrow (1970) demonstrated a rather odd-looking response bias, one not fitting any of the patterns discussed in the previous section. In their study, the participants (fifth- and sixth-grade children in a public school) were required to identify a set of letters embedded as the white ground within a group of black nonsense figures. The stimuli used are shown in Figure 5.2. Half of the children, tested in the morning of a single day, were presented with the stimulus shown at the bottom of the figure; half, tested that same afternoon, were presented with the stimulus shown at the top of the figure. The children were given 4 minutes to look at the figure and say what they saw, and they received a reward of 50 cents if they were able to name the correct pattern (a sizable incentive back in 1970).

The response bias was shown by the afternoon participants. A substantial number of children tested in the afternoon indicated that they saw the word LEFT—that is, they gave the response that had been correct in the morning. This answer was approximately twice as likely as was the actual correct answer for the afternoon. Responses of LEFT were also twice as frequent in the afternoon as they had been in the morning, when LEFT was in fact the stimulus shown!

What had happened, clearly, was that some of the afternoon participants had been talking to some of the morning participants. The

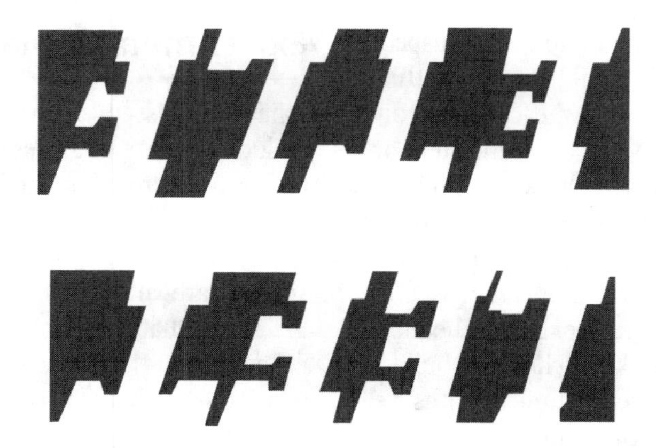

Figure 5.2 Stimuli used in the Horka and Farrow study of intersubject communication

SOURCE: From "A Methodological Note on Intersubject Communication as a Contaminating Factor in Psychological Experiments," by S. Horka and B. Farrow, 1970, *Journal of Experimental Child Psychology, 10*. Copyright © 1970 by Academic Press. Reprinted with permission.

children tested in the morning were all told the correct answer once they had finished, and they were all reminded that they could have won 50 cents for simply saying the word "left." They were also asked not to tell anyone about the study. Apparently, however, this admonition was not very effective in the face of a chance to help a friend or, perhaps, simply to appear knowledgeable about an unusual event.

The Horka and Farrow study demonstrates that another kind of response bias that participants can bring to an experiment stems from whatever it is that other participants have told them about the study. Such **intersubject communication** can have various effects. Participants may sometimes perform better than they otherwise would have, perhaps because other participants have told them the right answers, or have alerted them to some aspect of the procedure that was meant to be secret. They may sometimes perform worse, perhaps because the procedure differs for different participants (as in Horka and Farrow), or because the account they have received is too jumbled to be helpful. Certainly anyone who has ever listened to young children recount their research experiences knows that there can be a sizable gap between what happens in an experiment and the later retelling of it. Or the effects of the communication may be more general, with later participants made more or less eager, more or less apprehensive, or whatever as a function of what they have heard. Whatever the specific effect, the general import is the same: bias that may invalidate the results.

How general a problem is intersubject communication? Clearly, there are some studies in which communication is not an issue—research with babies, for example (but watch the parents!), or studies in which the participants either are unacquainted or have no opportunity for contact during the study. At the other extreme, the Horka and Farrow study includes a number of features that seem likely to maximize communication. Among these is the simple, easily verbalized, and lucrative right answer, a feature that is by no means present in every study. For these reasons, it is wise to be cautious about generalizing from Horka and Farrow's findings to research in general. Still, as Horka and Farrow argue, their procedure does bear some similarity to many learning or problem-solving studies with children, in which right answers exist and can be fairly easily transmitted from one child to another. Furthermore, the more general aspects of the study—testing in a public-school setting, with participants who know each other and have ample opportunity to communicate—are quite common in research with children. Indeed, of all the research areas in psychology, it may well be research with school-aged children in which intersubject communication is most likely to be a problem. It is prudent, therefore, for researchers who work in schools to build in safeguards against such communication, as well as techniques for detecting whether communication has in fact occurred. Brooks and Kendall (1982) provide a helpful discussion of ways to minimize intersubject communication.

One particular kind of communication effect is important enough to have earned its own label. The term **diffusion** refers to the unintended spread of a treatment effect from an experimental group to an untreated control group. Consider, for example, a preschool intervention program intended to enhance the academic readiness of children believed to be at risk for school failure. A typical approach in such research is to divide the participants into two groups: an experimental group that receives the intervention and a comparable control group that does not receive it. If both groups are drawn from the same community, however, there almost certainly will be contact and communication during the months that the treatment is in effect—the children will play together, and the parents will talk and compare notes. Any benefits of the intervention, therefore, may not be confined exclusively to the experimental group; rather, the supposedly untreated controls may also receive some new experiences and hence show some gains. It is to guard against such a dilution of the experimental-control contrast that some intervention projects set up two control groups: a *proximal control group* drawn from the same community as the experimental participants, and a *distal control group* drawn from a different community. The latter, no-contact group provides a check for the possibility of diffusion effects.

Experimenter Bias

I have discussed a variety of kinds of bias, including those that researchers bring upon themselves through procedural miscalculations. With the possible exception of the demand effect, however, the biases that have been considered are all essentially nondirectional, in the sense that they are equally likely to work for or against the researcher's hypothesis. Let us turn now to what is perhaps the most insidious form of bias: the possibility that researchers may systematically bias their research to obtain the results that they desire. Such systematic bias yields what is known as the **experimenter expectancy effect**. The possibility of such effects was discussed briefly in chapter 4 in the context of observational research; the present section takes up the issue more generally.

The pioneering work on experimenter expectancy was carried out by Robert Rosenthal (1976). In a typical Rosenthal study, a number of individuals, often undergraduate students, are recruited to serve as testers. These testers are randomly divided into two groups. As part of its training, each group is given a clear expectancy regarding the probable outcome of the experiment to be run—that is, they are told what the principal investigator expects (and perhaps hopes) to find. The groups differ, however, in that the expectancies they receive are directly opposed. One group, for example, might be told to expect good performance from their participants, whereas the other group might be told to expect poor performance. Apart from this difference in expectancy, the two groups are trained equivalently and are presumably following the same experimental procedure. Despite this equivalence, the results obtained for the two groups often diverge, and the divergence is in the direction of the induced expectancies. In short, testers find what they expect to find. Such effects have been demonstrated across a variety of topics and a variety of age groups, including children. By 2002, 479 studies of expectancy effects had been reported (Rosenthal, 2002).

Table 5.2 Possible Bases for an Experimenter Bias Effect

Unintentional bases	
1.	The experimenter may influence the subjects' behavior through unintentional paralinguistic cues—for example, variations in tone of voice.
2.	The experimenter may influence the subjects' behavior through unintentional kinesic cues—for example, changes in posture or facial expression.
3.	The experimenter may influence the subjects' behavior through unintentional verbal reinforcement.
4.	The experimenter may unintentionally misjudge the subjects' behavior.
5.	The experimenter may unintentionally misrecord the subjects' behavior.
Intentional bases	
6–10.	The experimenter may *intentionally* engage in each of the preceding practices.
11.	The experimenter may fabricate all of his or her data.

SOURCE: Adapted from "Fact, Fiction, and the Experimenter Bias Effect," by T. X. Barber and M. J. Silver, 1968, *Psychological Bulletin Monograph Supplement*, 70, pp. 1–29.

How do the researcher's expectancies exert their effects? Barber and Silver (1968) identified 11 ways in which a tester might bias the outcomes of a study. These sources of bias are listed in Table 5.2. Of greatest interest typically are the first five: those in which the bias is mediated through unintentional, and presumably often rather subtle, means. There is some evidence, across both Rosenthal's work and that of others, that each of the 11 kinds of bias can indeed play a role. It should be noted, however, that determining the exact locus for experimenter expectancy effects has often proved difficult.

There are other criticisms that have been leveled against the Rosenthal research (see Barber, 1976; Barber and Silver, 1968; and Rosenthal, 1968, for discussions). Perhaps the most important concerns the issue of external validity. In much of the research demonstrating bias effects, the testers are quite inexperienced, consisting of undergraduate students recruited solely for the purposes of the research. Each tester is given a very clear expectancy regarding the likely outcomes of the study. In some cases the testers are also promised larger payments for their services if their results support the investigator's hypothesis than if they fail to support it. Everything, in short, is geared toward maximizing the likelihood of an expectancy effect. What the studies may simply show, however, is that it is possible to do bad research. They have more general relevance—that is, they possess external validity—only to the extent that research in general embodies the flaws that are built into the expectancy studies.

Some obvious ways to minimize experimenter bias follow from the possible sources just discussed. If inexperienced testers must be used, then they clearly should be trained as carefully as possible before data collection begins. This training should emphasize standardization, since deviations from standardization provide one obvious opening for the introduction of bias. If possible, testers should not only be trained in a standardized procedure but should also be periodically *monitored* as they test, to ensure that an initial rigor is not lost in the course of the study. The use of extra payment for desired results is, needless to say, to be avoided. In addition, investigators should guard against differential use of *non*-monetary rewards—for example, responding with obvious pleasure when testers bring in desired results and obvious displeasure when they do not. Because the bias effect depends on expectancy, it is clearly desirable to prevent testers from forming clear-cut expectancies. To this end, testers are sometimes kept uninformed about the hypotheses behind a study or the status (e.g., experimental or control condition) of a particular participant. This is the technique of *blinding* that was discussed in chapter 4.

Although it is easy enough to state safeguards against experimenter bias, actually implementing these safeguards is, once again, not always so easy. In particular, blinding of testers or observers may sometimes be difficult. In some studies the principal investigator (i.e., the person who designs the study) also tests the participants, in which case blinding with regard to the study's hypotheses is obviously impossible. In other cases it may be unrealistic to think that a tester, no matter how initially naive, will not form hypotheses concerning the study's purposes and probable results. In many cases it may be impossible to blind the tester with regard to a participant's status. Testers will know, for example, when they are working with a 3-year-old and when with a 5-year-old, or when with a boy and when with a girl, and this knowledge of the participant's age or sex may bias behavior. Indeed, it is worth noting that developmental researchers are almost never blind with respect to one of their principal independent variables: the age of the participant. Finally, even in those cases when blinding *is* possible, it is by no means always employed.

Questions were raised earlier about the external validity of experimenter expectancy research. The gist of the preceding paragraph is that such research cannot be too readily dismissed as the limited product of a contrived experimental situation. There seems little doubt that experimenter expectancy effects do occur in developmental research. How often they occur, and how many false conclusions they produce, remain debatable. One goal of all researchers should be to forestall any debate in their case—that is, to design, execute, and report research in a way that minimizes the possibility of experimenter bias.

Loss of Participants

Much of this chapter has been concerned with problems that can arise when working with research participants. In some cases the problems may be so severe—and also so readily apparent—that there is no possibility of keeping the participant's data. Loss of participants (also labeled *dropout, attrition,* or *mortality*) may occur for any of a number of reasons discussed in this chapter—a critical deviation from standardization on the experimenter's part, an especially strong reactive response to the experimental arrangements on the participant's part, a persistent response set that the participant refuses to break away from. In addition, some subject groups or kinds of research may present special problems that can lead to dropout in even the best executed of studies. Babies may fall asleep or begin to cry uncontrollably partway through a study. Preschoolers may need to go to the bathroom at a critical point in the experimental procedure. In longitudinal research, participants may move away or die before the testing can be completed.

The most basic point concerning participant loss is that such loss should be minimized. Dropout from research can cause various problems. There is the practical problem of waste of time, on both the experimenter's and the participant's part. If the dropout is substantial, the researcher may end up with too small a sample to draw any conclusions. Finally—and most critically—if the dropout is at all selective, then the validity of the study may be threatened. This is the issue that was discussed at various points in chapter 3. Dropout from research often *is* selective, involving those participants who are least competent, least motivated, least willing to comply with the requests of a stranger, or whatever. Such dropout can affect the external validity or the generalizability of the results, in that experimenters may no longer be studying the population to whom they wish to generalize. If the dropout is differential for particular ages or conditions, then it may also affect the internal validity of the conclusions.

Methods useful for minimizing dropout can be inferred from the various problems and corresponding prescriptions discussed throughout this chapter. Two general pieces of advice are worth reiterating here. One is to build rapport before attempting to test children, especially toddlers or preschoolers. The second is to pilot-test the procedures as extensively as needed before beginning the actual study. Researchers who end up with procedures that simply do not work for a substantial proportion of their participants have probably neglected the pilot-test phase of the study.

In addition to minimizing the number of participants who are lost, a researcher has two further responsibilities. One responsibility is to decide, as far as possible, on objective, in-advance criteria for rejection of participants. The researcher should know at the start of the study exactly what it is that participants can do that will make their data unusable. The particular criteria will vary from study to study; they might include going to sleep or crying uncontrollably in infants, failure to pass a verbal pretest in preschoolers, or a clear recognition of some attempted experimental deception in older children or adults. The point is that it can be dangerous to collect all the data for a participant, see whether those data agree or disagree with one's hypothesis, and only then begin to worry about criteria for deciding whether the data are keepable. The danger is that researchers may be too easily tempted to keep results that fit with their expectations and discard those that do not.

The second responsibility is the one noted in chapter 2: to report clearly the criteria for rejecting participants and the number of participants rejected. Dropout is one factor that readers should take into account in evaluating a study. They cannot do so if the information is not provided.

There is one more point to be made. I have been stressing the desirability of keeping as high a proportion of one's participants as possible. Clearly, however, this advice does not mean that a screaming infant or a terrified preschooler should be forced to continue in an experiment. Two arguments can be advanced against such a practice. First, data collected from such participants are not likely to mean very much. Second and more basically, as we will see in chapter 9, a primary ethical principle in research with humans is that the participants have the right to withdraw from an experiment at any time. This right applies at least as strongly to the nonverbal infant or toddler as to any other participant. And the rights of the participant always take precedence over the preferences of the experimenter.

Summary

This chapter discusses the translation of abstract design into experimental procedure. It concerns problems that can arise in working with research participants, as well as ways to minimize the problems.

A central concept is that of *standardization:* keeping important elements of an experimental procedure the same for all participants. Standardization is the procedural counterpart to the notion of control in experimental design. If standardization is not maintained, procedures may vary from participant to participant, biases can arise, and results may no longer be clearly interpretable. The best tester is one who can combine standardization with naturalness and who is flexible enough to adapt as necessary to the individual participant.

Although standardization is desirable, some departure from standardization is both inevitable and often quite sensible. Total standardization is never possible, for some aspects of the procedure will necessarily vary across participants. In developmental research with different age groups, alterations may be necessary to make the procedure *equally age-appropriate* for the different groups. In *exploratory research* the goal is to uncover the interesting phenomena in some new research domain, and flexibility may therefore be more important than standardization. Finally, any aspect of the procedure that is held constant (such as the tester who collects the data) becomes potentially part of the independent variable, and such *overstandardization* may result in findings of limited generality. Allowing supposedly irrelevant features to vary across participants may enhance the validity of the study.

The discussion turns next to some specific threats to validity. *Instrumentation* refers to unintended changes in either physical instruments or human testers or observers across the course of a study. Such changes are of greatest concern when they are confounded with age or experimental condition, as can happen if all or most of the participants from one group are tested before the participants from other groups. Another reason to avoid confounding of condition and order of testing is the possibility of *selection bias:* Participants who are quick to volunteer (or whose parents are quick to return permission slips) may differ in various ways from those who are slower to respond. Possible effects of outside-the-study events, such as the excitement surrounding a holiday, provide a third reason to balance order of testing across groups. In pretest-posttest designs such outside-the-study events may give rise to the threat to validity labeled *history:* change produced by uncontrolled events occurring during the course of the study.

A very general threat to validity is *reactivity:* unintended and biasing effects of being studied upon the participant's behavior. Two much-studied forms of reactivity are "good-subject" behavior, in which the participant attempts to respond in ways desired by the experimenter, and "prideful-subject" behavior, in which the participant attempts to look good. The most general way to reduce reactivity is to use procedures that minimize the obviousness of the experimental manipulations and measurements. A closely related problem is that of *response sets:* participants' tendency to respond in a predetermined, biased fashion that is independent of task content. Researchers must also guard against *intersubject communication* and *diffusion,* in which participants' responses are biased by their contact with other participants.

A particularly insidious form of bias occurs when the researcher's hopes and expectations affect what is found. Such *experimenter expectancy effects* can come about in a variety of ways. Although the generality of the phenomenon has been disputed, experimenter bias remains an issue to consider and guard against in any research project. When possible, testers or observers should be blinded to minimize potentially biasing expectations. Training and monitoring of testers to ensure standardization are also important.

In some cases the problems discussed throughout the chapter are so severe that the participant must be rejected. Several points are made concerning participant loss. One is that such loss should be minimized, since it can affect both the internal validity and the external validity of the study. A second is that the researcher should decide in advance what the criteria are for rejection of participants and should convey clearly in the final report how many participants were rejected and for what reasons. A final point is that participants must not be coerced to remain in a study against their will.

Exercises

1. The exercises for chapters 11 through 14 include several hands-on activities—that is, suggestions for simple kinds of testing that you may be able to do with different subject groups. The following activity assumes that you will be able to act on some of these suggestions; it will work best if the procedure involved is a fairly challenging one (e.g., not simply handing out a questionnaire) and if you are able to try it on several different occasions. If so, evaluate your progress as a tester as you gain more experience with the instructions and stimuli and with testing in general. You may wish to tape-record the sessions for later rehearing; it might also be helpful to elicit feedback from your participants. Among the issues to think about are maintenance of standardization, naturalness and clarity of delivery, and the possibility of demand effects or experimenter bias.

2. The text cites the book *Unobtrusive Measures* (Webb et al., 2000) and describes an example of how the approach might be applied to issues in developmental psychology. Generate as many additional examples as you are able to think of. If possible, obtain a copy of *Unobtrusive Measures* and evaluate your ideas in light of the methods and examples discussed there.

3. Select one of the journals listed in Table 1.2 and examine the articles in the most recent issue from the perspective of possible experimenter expectancy effects. For each article consider two questions: (a) Is it plausible, given the issues under study and the methods used, that experimenter expectancy could affect the results; and (b) assuming a positive answer to the first question, how successful is the study in ruling out such effects?

6

Contexts for Research

Most of what has been said to this point has had to do with the *how* of research—how to construct well-designed studies, how to execute the studies with a minimum of problems and biases. But psychologists, more than most scientists, must also make decisions about the *where* of research—decisions that may in turn affect the how. Research in psychology, unlike that in some sciences, can be carried out in a variety of settings, ranging from highly controlled and artificial laboratory environments to the everyday, naturalistic world of the preschool, playground, or supermarket. Each setting has its pros and cons and its particular appropriateness for particular research questions—which, of course, is why different settings are used. These pros and cons will be the first broad topic considered in this chapter.

The second broad topic concerns the contexts within which development occurs. What are the environmental settings within which behaviors of interest develop and are expressed, and how can we best study the different settings? The discussion of this topic will be grounded in an influential contemporary model for conceptualizing the environment, the ecological systems approach of Urie Bronfenbrenner.

Lab and Field

There are various ways in which research settings can be classified. The system that I use for an initial classification is adopted from an article by Ross Parke (1979) and is summarized in Table 6.1. Parke begins with the familiar distinction between "field" settings and "lab" settings. Field research is carried out in the natural environment of the participant—for example, the playground or supermarket settings mentioned earlier. Lab research occurs in a setting specifically designed for research, a setting that may be quite different from the participant's natural environment and to which the participant is brought solely for the purpose of the research. A mobile laboratory research trailer within which changes in heart rate are recorded in response to tones that are presented through earphones would fit anyone's description of a laboratory setting.

The second factor in Parke's classification scheme concerns the independent variable–dependent variable distinction. The independent variables in a study can be manipulated in either a laboratory or a field setting. The dependent variables can be measured in either a laboratory or a field setting. The conjunction

Table 6.1 Classification of Research Settings

		Locus of dependent variable	
		Laboratory	Field
Locus of independent variable	Laboratory	1	2
	Field	3	4

SOURCE: Adapted from "International Designs" by R. D. Parke. In R. B. Cairns (Ed.), *The Analysis of Social Interactions: Methods, Issues, and Illustrations* (pp. 15–35), copyright © 1979 by Lawrence Erlbaum Associates. Reprinted with permission.

of lab and field and independent and dependent yields the four cells shown in Table 6.1.

Several qualifications are necessary before discussing the possibilities symbolized in the table. In examples like those just given, the field-lab distinction seems clear-cut. Often, however, the distinction is a good deal murkier. "Lab" settings can vary greatly in the extent to which they approximate a natural environment. "Field" settings may rapidly lose their naturalness as experimental controls and measurement procedures are imposed. For these reasons, the field-lab distinction is best thought of as a continuum rather than a dichotomy. In addition, "naturalness" is not a unitary construct; rather, there are several dimensions along which the naturalness of a setting can vary. Parke (1979) discusses three such dimensions: the general physical environment, the immediate stimulus field, and the social agents present in the situation. Because of these complexities, the use of "field" and "lab" in the discussion that follows should be recognized as a simplification, useful for making methodological points but nevertheless a distortion of a more complex reality.

Qualifications are also necessary with regard to the independent variable–dependent variable distinction. As we have seen, not all studies include a manipulated independent variable. In some cases, the nature of the variable precludes experimental manipulation. This is the case of *subject variables* discussed in chapter 2: variables such as age, sex, or race that are controlled and examined through selection of participants rather than experimental manipulation. In other cases no manipulation occurs because the interest is in examining relations among two or more dependent variables. This is the case of *correlational research* discussed in chapter 3. Thus, often the independent variable part of Table 6.1 does not apply. The dependent variable part, however, always applies: All research involves measurement of some outcome, and this measurement can occur in the lab or in the field.

A specific research example can help clarify our discussion. The example is one that we have met before: the issue of TV violence and aggression. Does exposure to violence on TV make children more aggressive? This is an interesting, much-studied, and much-debated question. Because it is a question that can be—and has been—examined via each of the four approaches shown in Table 6.1, it is a good example for present purposes. As we go, however, we also will encounter other examples, both in this chapter and in chapters to come. Indeed, a central issue to be considered is which research questions *can* be examined through each of the four approaches.

Design 1: Lab-Lab Studies

Research under this heading probably best fits most people's conceptions of an "experiment in psychology." An experimental manipulation is administered in a controlled laboratory setting, and the effects of this manipulation are assessed in the same laboratory environment. In the case of TV violence and aggression, a typical sequence might be the following. Participants are first randomly divided into two conditions, an experimental group that is to be exposed to violent TV and a control group that will receive nonviolent fare. Children from both groups are taken individually to the experimental room and there shown the particular TV segment designated for their group. Sometime shortly afterward the child is given a chance, in the same laboratory setting, to engage in aggressive behavior. Greater aggression in the experimental than the control group would be taken as evidence of the impact of TV modeling.

Within this basic paradigm a number of variations are possible. A pair of studies by Bandura, Ross, and Ross (1963) and Liebert and Baron (1972) can serve to illustrate some of the possibilities (the latter is the study introduced in chapter 4 to illustrate some principles of measurement). The TV segments presented to the children may be more or less similar to actual TV material. In some cases films have been constructed especially for the purposes of the research (Bandura et al.); in other cases episodes from commercial television have been shown (Liebert & Baron). The aggressive responses may also vary in their similarity to real-life aggression. In some cases the target of the aggression has been an inanimate object (a Bobo doll in Bandura et al.); in other cases it has been another child (Liebert & Baron). In some cases the aggressive behaviors have been physically identical to real-life aggression (hitting and kicking in Bandura et al.); in other cases they have been physically quite different (a button press in Liebert & Baron). Finally, the aggressive behaviors may vary in the ease with which they can be measured. The button-press response in the Liebert and Baron study could be automatically recorded; hits and kicks, however, typically require a human observer to make decisions about the occurrence and meaning of the behavior.

Research in a controlled laboratory setting offers several important advantages. One is the degree of control over the independent variable. In studying effects of TV on aggression, the researcher can decide exactly what sort of films will be shown to the children, exactly which children will be shown which films, and exactly what the general context will be within which the films are viewed. As we saw, it is precisely these kinds of control that are necessary if clear cause-and-effect conclusions are to be drawn. On balance, research under the lab-lab heading is most likely to maximize internal validity.

A further advantage lies in the possibility for systematic exploration of the independent variables of interest in the study of some phenomenon. Once a basic laboratory paradigm has been developed, it is possible to contrive almost endless variations of it. In the case of TV and aggression, for example, we might vary the realism of the aggressive acts, or the nature of the aggressor, or the consequences to the aggressor following the aggression, or the time period between exposure to the model and the chance to act aggressively. We can thus go beyond a general conclusion that aggressive models can increase aggression to a more exact determination of how and why the effects come about.

A final advantage of laboratory research lies at the dependent-variable end. By definition, the dependent variable is free to vary and therefore is never under the control of the researcher. Dependent variables must, however, be measured, and such measurement is generally easiest in a structured laboratory environment. It might be possible in a lab, for example, to videotape the aggressive behavior for later replaying and analysis—a luxury that is unlikely to be available on a preschool playground or in a home. It may also be possible to forgo the human observer altogether and to work with recordings that are either literally or essentially automatic. The button press in Liebert and Baron's study is one example of such automatic recording of aggression; an automated Bobo doll that registers every hit that it receives (Deur & Parke, 1970) is another. Finally, as we will see shortly, some dependent variables simply cannot be measured except in controlled laboratory settings.

The strengths of the laboratory approach are balanced by some weaknesses. Just as the strengths can be summarized by the word "control," so can the weaknesses be summarized by a single word: "artificiality." Laboratory environments may vary in the degree to which they resemble the real-life settings of interest; they are always somewhat different from these settings, however, and often they are very different. The question thus arises: Can results obtained in the laboratory be generalized to more natural settings? This is the issue of the trade-off between internal validity and external validity that was discussed in chapter 2. As we saw then, the same factors that maximize internal validity may often operate to reduce external validity.

Consider the issue of TV violence and aggression. No matter how naturally the opportunity to watch TV is presented, the viewing situation in the lab is necessarily somewhat different from that in the child's home. For one thing, the viewing is not in the home, where the child has watched most TV to this point in life. Not only the physical setting but also the social setting may differ. The child may usually watch TV in the company of siblings or peers and is now watching alone. The fact that an adult has explicitly directed the child to this particular TV segment is a further difference. The adult's presentation of the TV material may impart a kind of sanction to the content, and to the child's subsequent imitation of it, that is not found in the home. Finally, the exposure to TV in the lab is necessarily fairly brief; any generalization to the effects of long-term, hours-a-day viewing must therefore be tentative.

Limitations also exist with respect to the dependent variable. Hitting a Bobo doll is not the same thing as hitting another child; delivering a purportedly painful stimulus to an unseen child in another room is also not the same thing. At least partly for ethical reasons, laboratory measures of aggression have often involved a kind of pseudo-aggression—less clearly interpersonal than real-life aggression, and less likely to elicit negative reactions from an adult. Whether such responses are predictive of genuinely aggressive behaviors in less permissive contexts is debatable. Even when ethical factors are not an issue, capturing complex social behaviors in a laboratory setting can be difficult. As we will see repeatedly in later chapters, the laboratory analogues for behaviors of interest to the developmental psychologist are often quite distant from their real-life counterparts. In some cases the very nature of the research question may preclude laboratory study. If our specific interest, for example, is in aggression toward familiar peers on the school playground, then research in a laboratory setting is simply not one of our options.

Laboratory studies are also especially subject to the problems of reactivity and response set discussed in the last chapter. The child in a lab environment may become anxious and nonresponsive, may attempt to do whatever it is that the adult apparently wants, may become distracted by all the fancy equipment, and so forth. It is true, as we saw in chapter 5, that experimental arrangements can sometimes be disguised and reactivity minimized. It is worth noting, in fact, that the "lab" settings used in most studies with preschool and grade-school children are unused rooms in the children's own school—hardly the alien, apparatus-loaded environment conveyed by the word "laboratory." Nevertheless, the fact remains that the children are being brought to an unusual setting by an unfamiliar adult and there subjected to experiences that they would not otherwise encounter, and all of these departures from the ordinary can introduce various kinds of bias.

Rather than working through Table 6.1 in order, we will turn next to the cell that is most distinct from the one just considered: field-field studies. Once both the lab and the field approaches have been discussed, it will be easy to fill in the two combination cells in the table.

Design 4: Field-Field Studies

Our concern now is with research in which the independent variable is manipulated in the natural setting and the dependent variable is measured in the natural setting. It is true, as noted earlier, that the experimental manipulation and measurement necessarily change the "natural" setting to some extent. Still, in this case, in distinction to laboratory research, the starting point is the natural environment, and the setting remains toward the natural end of the lab-field continuum.

An example for the problem of TV and aggression is provided by a study by Feshbach and Singer (1971). The participants were preadolescent and adolescent boys living in various residential centers. For a period of 6 weeks Feshbach and Singer were able to control the TV "diets" experienced by the participants. Half of the boys were randomly assigned to a diet of violent TV shows across the 6 weeks, and half were assigned to a diet of nonviolent shows. In this case, therefore, the manipulation of TV viewing was carried out in the participants' natural environments. Effects of the TV viewing were determined from counselors' and supervisors' ratings of the boys' naturally occurring aggression across the 6-week span. Thus, the locus of the dependent variable was also in the natural setting.

The great advantage of field research is conveyed by the word "natural." What we are principally interested in with respect to TV and aggression is whether the TV that children watch at home affects the aggression that they produce at home, at school, on the playground—wherever children naturally find themselves. As we saw, laboratory research can provide only an indirect answer to this question, because laboratory research neither manipulates home TV viewing nor measures aggression in the natural setting. With field research, however, the focus is directly on the situations and behaviors of interest. This fact means that the external validity is likely to be higher than in a comparable laboratory study. If we really can exert the necessary control over the independent variable, and if we really can accurately measure the dependent variable, then the internal validity should also be high.

The disadvantages of field research are suggested by the "ifs" in the preceding sentence. Some manipulations and some measurements are difficult, if not impossible, to carry out in the natural setting. Think for a moment about the problems involved in controlling, for an extended period of time, the kinds of TV shows that large numbers of children watch. It is no surprise to learn that there are only a handful of field studies like Feshbach and Singer's or that such studies tend to be carried out with "captive" populations—for example, adolescents in boarding schools. Whether results obtained with such populations and in such settings are generalizable to the more normal home situation is debatable. Furthermore, the introduction of experimental control is in itself a major and very noticeable change in the natural environment. Few children have their TV viewing totally controlled by an adult, and the sudden imposition of such control opens the way for various reactive or confounding effects. In Feshbach and Singer's study, for example, there was some evidence that the boys subjected to the nonviolent TV diet became frustrated, and thus more aggressive, as a function of having lost their favorite shows.

It should be noted that the importance of the problems just discussed does vary to some extent across different kinds of independent variables. Some variables lend themselves more readily to natural, nonreactive manipulation in a field setting than does TV viewing. It is fairly easy, for instance, to hang mobiles over babies' cribs in a study of visual attention (e.g., Weizmann, Cohen, & Pratt, 1971) or to introduce different books into the home in a study of the effects of reading to children (e.g., Whitehurst & Lonigan, 2001). On the other hand, some variables are even harder to manipulate in the field than is TV. Sometimes practical and ethical constraints combine to make experimental manipulation essentially impossible; parental child-rearing practices are a common and important example. In other cases the interest of the researcher lies in the effects of quite specific and tightly controlled stimuli that by their very nature are somewhat artificial—for example, repeated tones in a study of auditory habituation, tachistoscopic flashes in a study of visual detection, lists of words in a study of short-term memory. It might sometimes be possible to embed such stimuli in the natural environment, but there would be little point in doing so: The environment would quickly become far from "natural," and the other factors present in the situation could well bias performance and confound the differences among participants. In such cases laboratory study is the most sensible option for the researcher's purpose.

The second general problem in field research concerns the second of the two "ifs" cited earlier: accurate measurement of the dependent variable. Consider the example of auditory habituation. The term *habituation* refers to a lessening or dropping out of an initial attentional response to a stimulus as the stimulus is repeated. Such "getting used to it" is commonly indexed by changes in heart rate as the stimulus is presented repeatedly. Clearly, heart rate is not a dependent variable that can be measured in a field setting. Even if the investigator does succeed in importing the EKG equipment into the participant's home, all the wires, electrodes, and such constitute a very obvious departure from the natural environment. The same argument applies whenever physiological responses are the dependent variable.

Nor is the argument limited to physiological measures. A number of responses of interest to the developmental psychologist are simply very difficult to elicit and measure in the natural environment. Many of the concepts studied by Piaget fall in this category. The conservation concept may well be the important component of children's thinking that Piaget believed it to be. Conservation, however, is rarely directly and clearly expressed in the child's naturally occurring behavior; rather, determining its presence or absence requires a test explicitly designed for this purpose, such as the test shown in Figure 5.1. It is true, as will be argued in chapter 12, that there is still some point in making this test as close to real life, and perhaps as nontest-like, as possible. But some explicit elicitation of the concept is required, and this elicitation necessarily shifts the study in the direction of the laboratory end of the lab-field continuum.

The point of the preceding is that some behaviors are either literally (e.g., heart rate) or figuratively (e.g., understanding of conservation) beneath the surface, and thus are hard to measure in a natural setting. But what about aggression? Aggression, after all, is a frequent, overt, observable, and inherently social behavior, and as such would seem a good candidate for measurement in a field setting. And indeed, measuring aggression in the field, as we saw, has some definite advantages over attempts to measure aggression in the laboratory. But major problems of feasibility and accuracy still remain. Two general approaches to field measurement of social behaviors can be identified (recall Table 4.3): ratings of the behavior by someone who knows the child (as in Feshbach and Singer's study), and direct observations of the behavior as it occurs. Observational techniques were discussed in chapter 4, and both ratings and observations are considered with respect to particular aspects of social development in chapter 13. The complexities involved in both sorts of measurement should become evident then. For now, I settle for simply reiterating the general point: Whatever the behavior, accurate measurement in a field setting can be very difficult.

Design 2: Lab-Field Studies

In this case the independent variable is manipulated in the laboratory, and the dependent variable is measured in the field. A study by Josephson (1987) provides an example. Second and third grade boys were randomly assigned to watch either a violent or a nonviolent TV segment in an unused room at their school. The independent variable was thus manipulated in a laboratory setting. The boys subsequently played a game of floor hockey in the school gymnasium, during which instances of physical aggression (which turned out to be frequent) were recorded. The dependent variable was thus measured in a field setting.

The strengths of a study like Josephson's are the particular strengths of lab manipulation and field measurement discussed previously. Such research affords both good control over the independent variable and ecologically valid measurement of the dependent variable. The confidence with which effects can be attributed to the experimental manipulation is thus enhanced, as is the likelihood that the effects tell us something about real-life aggression. Furthermore, the combination of lab manipulation and field measurement makes possible both a spatial and a temporal separation of independent variable and dependent variable that may be less likely in lab-lab or field-field studies. This separation may in itself increase the generalizability of the results, as well as reduce the probability of reactivity or response bias. In Josephson's study, for example, the fact that aggression was measured at a time and place different from the TV viewing may have reduced the chances that children would imitate simply because the immediate situational cues supported aggression or the adult seemed to expect imitation of the film.

The weaknesses of lab-field studies are also derivable from the discussion of lab approaches and field approaches separately. The laboratory locus of the independent variable raises the possibility of artificiality and nongeneralizability; the field locus of the dependent variable increases the problems of accurate measurement. In addition, there can sometimes be practical obstacles to combining a lab component and a field component within the same study. Despite their potential advantages, lab-field studies in fact constitute a rather small proportion of research in developmental psychology.

Design 3: Field-Lab Studies

The final possibility involves manipulation of the independent variable in a field setting and measurement of the dependent variable in a lab setting. Our example in this case is a study by Parke, Berkowitz, Leyens, West, and Sebastian (1977). The Parke et al. study was in part similar to the Feshbach and Singer (1971) study discussed earlier. Their participants were adolescent boys, half of whom were shown violent films in their dormitories across a 5-day period, and half of whom were shown nonviolent films. In this case, however, one measure of the effects of the film viewing was a later laboratory test in which the participant was given a chance to deliver shocks to an unseen peer (a measure similar to that in the Liebert and Baron, 1972, study). Thus a lab setting was used to assess the effects of manipulating TV viewing in the natural environment.

Once again, the strengths and weaknesses of the approach follow from the general points made with respect to lab and field studies. The natural setting for the independent variable is a virtue, as is the rigor with which the dependent variable can be measured. On the negative side, the attempt to impose experimental control in a natural setting may result in some loss of both naturalness and control. The measurement of the dependent variable in a lab setting may result in a precise but artificial index that has an uncertain relation to real-life aggression. And again, special difficulties may be involved in attempting to do both lab research and field research within the same study.

Overview and Evaluation

Two general themes emerge from this discussion of different settings for research. One theme concerns the trade-off among the various goals that the researcher would like to achieve. What we would like to be able to do is to carry out studies that yield clear cause-and-effect conclusions that are generalizable across a wide range of real-life situations. Doing so requires that we have exact control over our independent variables, precise measurement of our dependent variables, and a general context for the research that is similar enough to naturally occurring situations to permit generalization. It is this conjunction of goals that is difficult to achieve, because methodological decisions that work in favor of one goal often work against one of the other goals.

The second theme follows from the first. Because no single approach to studying an issue is perfect, there is no justification for adopting but a single approach. Much more informative is the use, either within a study or across studies, of a variety of different methods of attacking any particular issue. Such a multipronged research strategy, as we saw, is referred to as the method of *converging operations.* The idea behind converging operations is that the strengths of one method of study can, to at least some extent, compensate for the weaknesses of another method of study, and that conclusions based on a convergence of evidence from different methods can be held with greater certainty than conclusions based on one approach alone.

Let us apply the argument to the problem of TV and aggression. We saw that any single method of studying this issue is subject to various criticisms. If we find, however (as in fact we do), that all four of the approaches described in the preceding pages lead to the same conclusion—namely, that TV violence promotes aggression—then we can have considerably more confidence that this conclusion really is valid (Anderson et al., 2003).

The discussion thus far has been concerned with possibilities. Let us turn now to actualities. If you have done much reading in psychology, you know that the four approaches just discussed do not appear with equal frequency in reports of psychological research. Although there are some variations across topics and across journals, most studies in psychology, including developmental psychology, are laboratory-based. This is especially true with respect to the manipulation of independent variables. Measurement in field settings, such as the home or school, is found for some topics, and we will see some examples in the concluding chapters of the book. Even at the measurement end, however, laboratory procedures predominate, and completely field-based studies are rare.

Is the laboratory locus of most psychology research a problem? Various commentators have expressed concern. McCall (1977) has written that "we rarely take the time to keep our experimental hands off a behavior long enough to make systematic descriptive observations in naturalistic settings" (p. 336). Bronfenbrenner (1977) has charged that "much of contemporary developmental psychology is *the science of the strange behavior of children in strange situations with strange adults for the briefest possible periods of time*" (p. 513; italics in original). Recall also the discussion in chapter 5 of work by Orne, Silverman, and others with respect to the various artifacts and biases that can occur in lab research.

Not surprisingly, a number of authors have mounted defenses of the laboratory approach (e.g., Anderson, Lindsay, & Bushman, 1999; Berkowitz & Donnerstein, 1982; Kerlinger & Lee, 2000; Mook, 1983). These authors do not deny that external validity can sometimes be a problem in laboratory studies, nor do they contest the value of research in more naturalistic settings when such study is possible. They do, however, offer two general arguments in favor of laboratory study, one conceptual and one empirical.

The conceptual argument concerns the goals of laboratory research. The goals are not to reproduce some real-life situation down to every detail. Real life, after all, already exists, and if it were possible to explore independent variable–dependent variable links clearly in the natural environment there would be no need ever to go into the laboratory. The point of laboratory research is to rule out all the confounding factors that bedevil study in the natural setting, thus permitting a clear determination of whether some independent variable *can* influence some dependent variable. Such a demonstration, of course, does not allow us to conclude that one variable typically *does* influence the other; it is, however, a necessary step toward such a conclusion. It is true that the simplified world of the lab may often seem artificial in comparison to the complexity of real life, but this characteristic is not merely a worrisome by-product of laboratory study; rather, some degree of artificiality is essential to the logic and the success of the approach. As Kerlinger and Lee (2000, p. 580) put it, "When a research situation is deliberately contrived to exclude the many distractions of the environment, it is perhaps illogical to label a situation with a term that expresses in part the result being sought."

The empirical argument is a particular version of a general point about threats to validity. It is a point that we encountered in chapter 3 in the discussion of cross-sectional and longitudinal designs. Whatever the potential threats to validity may be in a particular form of research, it is an empirical question whether they in fact apply in specific cases. If artificiality and related threats (reactivity, demand effect, etc.) are serious problems for some line of laboratory study, then results from such study are unlikely to yield a clear and coherent picture. In particular, they are unlikely to agree with the results from other approaches to the same questions—that is, more field-based approaches. Conversely, if lab and field agree—which is the case not just for the question of TV and aggression but for a range of topics (Anderson et al., 1999)—then we have support for the validity of each. This, clearly, is again the argument of converging operations.

Before concluding this section, I should add a qualifier to one of its themes—namely, the infrequency of field-based research. Field studies do play a prominent role in the two kinds of research to be discussed in chapter 7: qualitative research and applied research. There will thus be more to say about field studies when we reach those topics. In addition, Box 6.1 discusses an influential contemporary approach to measurement in field settings: the experience sampling method.

FOCUS ON

Box 6.1. The Experience Sampling Method

An adolescent girl is sitting in her room one evening reading the history assignment for tomorrow's class. Suddenly a device attached to her waist starts to emit a beeping sound. The girl pauses momentarily in her studying, opens a nearby notebook, and records various details about her activities, thoughts, and emotions in the period immediately preceding the beep.

Our hypothetical adolescent is a participant in what is informally known as a pager or beeper study and what is more formally labeled the **experience sampling method**, or **ESM**

(Csikszentmihalyi & Schneider, 2001; Hektner & Csikszentmihalyi, 2002). In recent years ESM has emerged as a valuable complement to observational studies of the natural environment. Observational methods, as we have seen, provide an invaluable window onto naturally occurring experiences and behaviors, but they do have limitations. The presence of an observer may change how people behave. Some phenomena of interest, such as thoughts or emotions, may be difficult or impossible to observe. Finally, the scope of observations is necessarily limited, perhaps a few hours of observation in one or two different settings.

The experience sampling method is designed to provide a much fuller sampling of people's everyday experiences and behaviors. ESM is a self-report measure, and thus removes the need to introduce an observer into the setting. It differs in important respects, however, from most self-report measures. Most such measures take the form of questionnaires or interviews in which respondents attempt to remember experiences from the past or offer general conclusions about their attitudes or attributes. ESM, in contrast, provides an immediate, online record of whatever may be going on in the participants' lives. If implemented successfully (and there is evidence in support of both the validity and the reliability of the approach—Hektner & Csikszentmihalyi, 2002), ESM can provide a much more detailed, specific, and wide-ranging picture of everyday experience than any other method of measurement.

Within the basic ESM approach a number of variations are possible. The number and the spacing of the recordings (which typically are programmed to occur at random and thus unpredictable intervals) can vary greatly. The responses recorded may be largely open-ended (writing down a description of the current activity) or more closed-choice (choosing among possibilities on an answer sheet), or, of course, some mixture of the two. The responses may focus fairly narrowly and objectively on current activities or they may encompass a range of thoughts, emotions, and evaluations as well. Finally, with modern technology the notebook may be replaced by a hand-held computer, allowing the respondent to enter responses directly into the eventual data file (Christensen, Barrett, Bliss-Moreau, Lebo, & Kaschub, 2003).

To what sort of uses might such a methodology be put? The answer is a wide variety of uses across a number of different areas of psychology; in addition to developmental psychology, the disciplines for which ESM has proved informative include clinical psychology, counseling psychology, social psychology, school psychology, organizational psychology, sports psychology, marital relations, and communication research. Often the same data may be put to several uses. One strength of the approach is the fact that once a data file exists, different investigators may use it to address different questions and analyze different aspects of the participants' responses.

The lower age bound for ESM studies is about age 10, and thus the approach is not an option through much of childhood. By adolescence, however, it is, and some of the most informative ESM studies have been with adolescents. Figure 6.1 provides one example; it shows the places in which American adolescents spend their time (at least as of the mid 1980s). Table 6.2 illustrates one of the many cross-cultural applications of the approach, comparing adolescent time use across several general cultural settings. Note that the results confirm the common impression that American students spend less time in school and study than their East Asian counterparts, with correspondingly more time, however, in social interaction and organized activities.

For most of us it is difficult to recapture what it means to be an adolescent. The experience sampling method provides one of the best ways to get closer to the experience of adolescence.

(Continued)

(Continued)

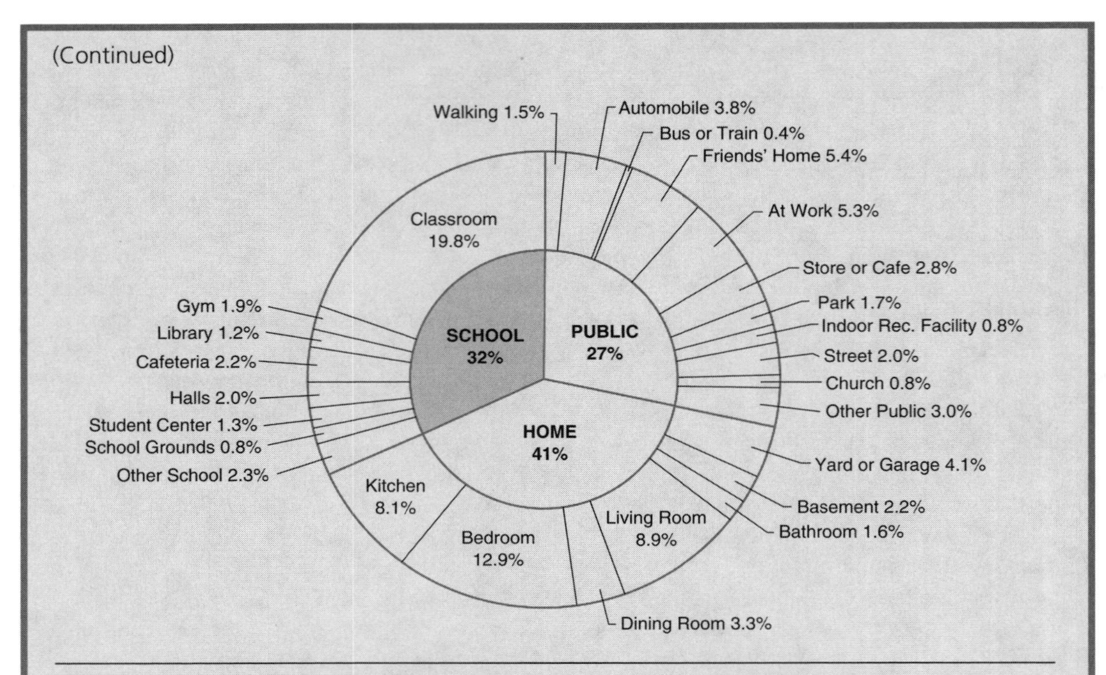

Figure 6.1 Adolescent time use as revealed by the experience sampling method

SOURCE: From *Being Adolescent* by M. Csikszentmihalyi and R. Larson. Copyright © 1984 by Basic Books, a member of Perseus Books, L.L.C. Reprinted by permission.

Table 6.2 Estimated Time Use Per Day by Adolescents in Different Cultures

	United States	*Europe*	*East Asia*
Work time			
Household labor	20-40 minutes	20-40 minutes	10-20 minutes
Paid labor	40-60 minutes	10-20 minutes	0-10 minutes
School work	3.0-4.5 hours	4.0-5.5 hours	5.5-7.5 hours
Total work time	4.0-6.0 hours	4.5-6.5 hours	6.0-8.0 hours
Free time			
Television viewing	1.5-2.5 hours	1.5-2.5 hours	1.5-2.5 hours
Social interaction	2.0-3.0 hours	insufficient data	45-60 minutes
Organized activities	40-80 minutes	30-100 minutes	0-30 minutes
Total free time	6.5-8.0 hours	5.5-7.5 hours	4.0-5.5 hours

SOURCE: Adapted from "Adolescents' Leisure Time in the United States: Partying, Sports, and the American Experiment" by R. Larson and S. Seepersad. In S. Verma & R. Larson (Eds.), *New Directions for Child and Adolescent Development: No. 99. Examining Adolescent Leisure Time Across Cultures*. Copyright © 2003 by Jossey-Bass. Reprinted with permission of John Wiley & Sons, Inc.

The work to which we turn next is also concerned with settings and context, but from a somewhat different perspective. This is the *ecological systems approach* of Urie Bronfenbrenner (1979, 1989; Bronfenbrenner & Evans, 2000; Bronfenbrenner & Morris, 1998).

Ecological Systems

The emphasis of the ecological systems approach is on the contexts within which development takes place—and on interrelations among the different contexts that are important to a child's development. The previous section, of course, also dealt with context, but in a somewhat different sense. In the discussion of settings the concern was with where *research* takes place, with a rough division into the two general settings of lab and field. By definition, however, *development* always takes place in the field—in the various natural settings (home, school, playground, etc.) where children spend their lives. The question to which Bronfenbrenner's writings have been directed is how best to conceptualize and study these various contexts or *ecological systems.*

Let us begin the exposition of Bronfenbrenner's thinking by considering the contexts sketched in the preceding paragraph—home, school, playground. Each of these contexts includes, of course, not just a physical dimension but also characteristic activities and important social agents: parents, siblings, teachers, peers. Contexts such as these constitute what is called the **microsystem** in Bronfenbrenner's formulation. The microsystem is the most proximal of the various ecological systems—the layer that is closest to the child and acts most directly on the child. Interactions with family members within the home fall under the heading of the microsystem. So too does playing with friends on the playground, talking with the teacher at school, or listening to a sermon in church.

Most research in developmental psychology focuses on the microsystem; hence relevant research examples are not difficult to find. Indeed, most of the examples discussed up to this point in the book have been forms of microsystem research. This is true, for example, for the topic used earlier in this chapter to illustrate points about settings for research: TV violence and aggression. When we study the possible contribution of TV to aggression, our interest is in the effects of direct, first-hand experience: how what children see and hear on TV affects their subsequent behavior. More generally, when we study the contribution of models—TV or otherwise—to any aspects of the child's development, our concern is again with the microsystem: the impact of direct experience with the social agents who populate the child's world.

Psychology's concern with the microsystem is, of course, quite understandable—this is the layer of the environment within which causal forces operate and the child's development takes place. The real question is why we should be concerned with anything else—that is, what other layers or systems must be added to the microsystem if we are to achieve a full picture of development?

Bronfenbrenner's answer is depicted in symbolic form in Figure 6.2. As the figure illustrates, ecological systems theory envisions three further contextual layers beyond the microsystem. The **mesosystem** refers to the system of relationships among the child's microsystems. It might include, for example, the parents' involvement in the child's school activities, or the interactions between the child's siblings and the child's friends. The **exosystem** is the term for social systems that can affect children but in which they do not participate directly. A school board that sets educational policies relevant to the child would be an element of the exosystem; so too would an employer who grants (or fails to grant) a maternity leave. Finally, the **macrosystem** refers to the culture or subculture in which the child develops. Typically, the macrosystem will itself include layers embedded within layers. For a particular child, for example, the macrosystem might include not only the general culture of the United States but also cultural features specific to growing up in a large northeastern city, and perhaps certain features specific to a particular ethnic neighborhood within the city as well.

Not shown in the figure is one more form of context. A relatively recent addition to

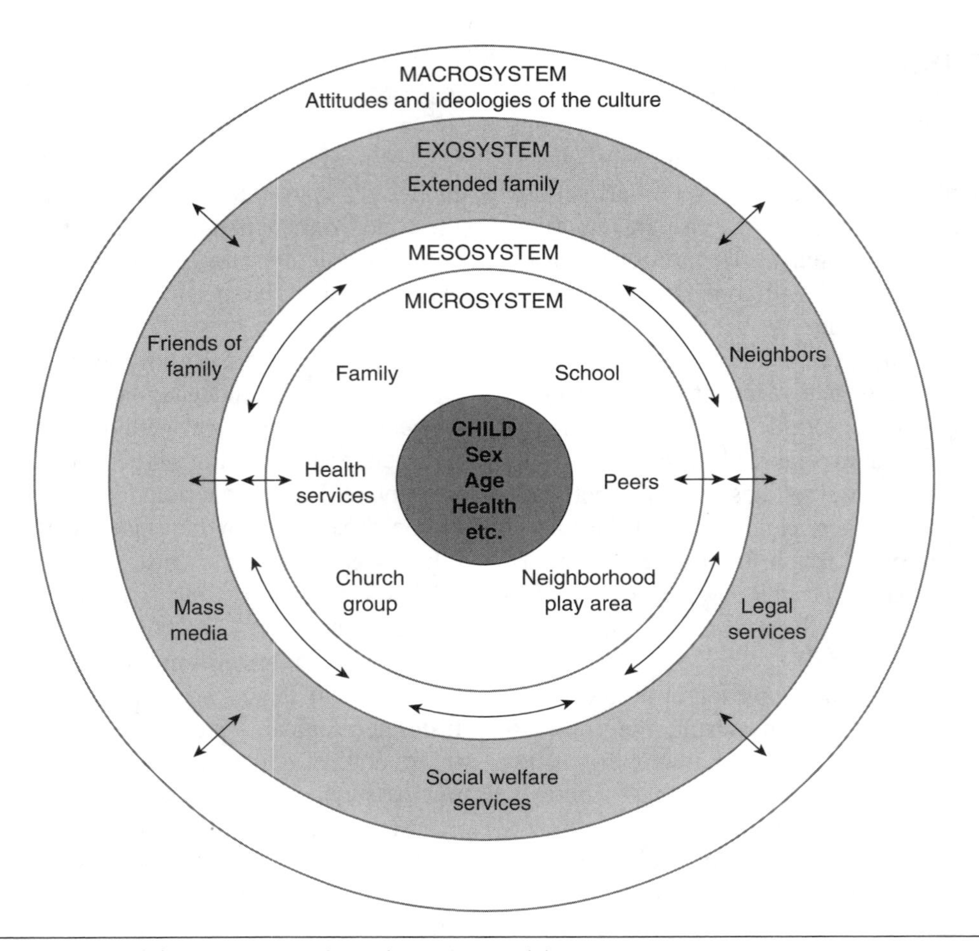

Figure 6.2 Bronfenbrenner's ecological systems model

SOURCE: From Sociocultural Risk: Dangers to Competence, by J. Garbarino. In C. B. Kopp and J. B. Krakow (Eds.), *The Child: Development in a Social Context* (pp. 630–685). Copyright ©1982 by Addison-Wesley Longman Publishing Company, Inc. Reprinted by permission.

Bronfenbrenner's model is the **chronosystem**: Bronfenbrenner's term for the passage of time as a context for studying development. The various systems and their interrelations change over time and across the development of the child, and children, of course, also change as they develop. A full understanding of contextual factors must therefore include a temporal dimension.

What are the implications of this conception of ecological systems? In particular, what are the *methodological* implications—how does ecological systems theory affect the ways in which we do and interpret research? Bronfenbrenner (1979, p. 38) expresses one major implication as follows: "In ecological research, the principal main effects are likely to be interactions." What he means by this is that an effect that emerges when we study one ecological system in isolation may prove to be qualified in some important way when we simultaneously consider a second ecological system as well. This notion of ecological systems as inherently intertwined and interacting is conveyed in the circles-within-circles representation of Figure 6.2. We may (and indeed generally do) study just one ecological system at a time—most commonly the microsystem. But the systems do not *exist* in isolation. Rather a microsystem is always embedded within a mesosystem, and both micro and meso are nested within higher and more general layers.

What sorts of interactions might a consideration of two or more ecological systems reveal? We will examine two examples here; other examples, along with a fuller presentation of Bronfenbrenner's theory and its research implications, can be found both in the sources by Bronfenbrenner cited earlier and in other writings inspired by the theory (Moen, Elder, & Luscher, 1995; Pence, 1988).

The first example anticipates some material to be covered more fully in chapter 13. One of the topics to be discussed in chapter 13 is methods of studying parental child rearing, and one of the approaches described there is an influential program of research by Baumrind (1971, 1989). Baumrind identifies several general styles of parental child rearing, one of which I focus on here. Baumrind uses the label *authoritarian* to refer to techniques of socialization that are relatively high in the use of threats and punishment and general assertion of parental power, and relatively low on the dimensions of reasoning and warmth. When summarized this way, the authoritarian style does not sound like a very promising approach to child rearing, and in fact, research suggests that it is not. In Baumrind's studies the authoritarian style was consistently associated with relatively poor child outcomes. Other investigators, using different measurement techniques and different-aged samples, have replicated this finding (e.g., Lamborn, Mounts, Steinberg, & Dornbusch, 1991).

All of the studies just cited, however, were with mostly White, predominantly middle-class samples. They were limited, therefore, to one general macrosystem. This limitation turns out to be important. Dornbusch and colleagues (Dornbusch, Ritter, Leiderman, Roberts, & Fraleigh, 1987) examined the effects of the Baumrind styles in an ethnically diverse (White, African American, Asian American, and Hispanic) sample of adolescents and their parents. Although the results were in general compatible with those of other such studies, the expected relations between parental practice and child outcome were most clearly evident for the White families; each of the other ethnic groups showed some exceptions and qualifications. Especially interesting were the Asian American families. These families ranked high on the index of authoritarian parenting, yet the Asian American adolescents were also high in school performance (the main child outcome examined). Thus, within the macrosystem of Asian American culture, authoritarian parenting apparently carried a different, and more positive, meaning from that in the Caucasian samples previously studied. Other studies have since confirmed that aspects of the authoritarian style can have beneficial effects in particular cultural groups, including Asian and African American as well as Asian American samples (Parke & Buriel, 1998).

In writing about the findings for Asian American families, Bronfenbrenner (1993, p. 39) quotes an Asian American student whose comment may capture at least part of the cultural difference. When asked what Asians thought about authoritarian parenting, the student replied, "That's how we know our parents love us."

The child-rearing example concerns an interaction of microsystem and macrosystem. The next example brings in a contribution of both the exosystem and the chronosystem. Elder and associates (Elder, 1999; Elder & Caspi, 1990) examined the effects of the Great Depression of the 1930s on children's development. To do so, they made use of archival data—that is, longitudinal data originally collected in the 1930s, which they were able to rescore and reanalyze in various ways (recall Box 5.1). The overall goal of the work was to chart aspects of what Elder refers to as the **life course**: the sequence of socially defined, age-graded roles and events that people experience throughout life, in conjunction with the historical events specific to their cohort.

The aspect of the work that we will examine focuses on four groups of boys. Half were of preschool age at the onset of the Depression, and half were in late childhood. Half in each age group had fathers who had lost their jobs, whereas for the other half the fathers remained employed.

Not surprisingly, negative effects were more marked in families in which the father had lost his job and the economic hardships were therefore greater. These effects were considerably more severe, however, for boys who were of preschool age when the Depression hit. In these families, the change in income and role relations precipitated various changes in family life (e.g., inconsistent discipline, increased tension, decreased effectiveness of the father as a role model) that had adverse consequences for the young boys' development. Boys who were older at the time of the Depression, however, were able to take on more adult-like roles and outside-the-home responsibilities and they were therefore buffered to some extent from the changes in family dynamics. It is interesting to note also that girls in general showed fewer negative effects than boys, a situation that was true regardless of their age at the onset of the Depression. What the research demonstrates, therefore, is not simply an effect of the exosystem (father's job) upon the microsystem (family relations) but an interaction of the two systems. The impact of the worsened economic circumstances varied across different children and different types of family. In addition, effects of the chronosystem are evident, both in the impact of a particular historical event (the Depression) and in the varying effects of this event depending on the developmental level of the child.

Table 6.3 summarizes the central ideas in Elder's life course model. The first two principles are clearly reflected in the two findings just discussed: the importance of historical events and the interplay between such events and the developmental level of the individual. The third principle is also illustrated by the results of the Great Depression study: The effects of the Depression on the children were mediated by its effects on their parents and the consequent changes in parent-child relations.

The applicability of the final principle may not be evident from my brief description of the research, but it is also central to Elder's conceptualization. It is a point that applies not just to life course theory but also to Bronfenbrenner's ecological model. The presentation to this

Table 6.3 Principles of Elder's Life Course Model of Development

1. The life course of individuals is embedded in and shaped by the historical times and events they experience over their lifetime.
2. The developmental impact of a succession of life transitions or events is contingent on when they occur in a person's life.
3. Lives are lived interdependently, and social-historical influences are expressed through this network of shared relationships.
4. Individuals construct their own life course through the choices and actions they take within the opportunities and constraints of history and social circumstances.

SOURCE: Adapted from *Children of the Great Depression* (pp. 304–308) by G. H. Elder, Jr., 1999, Boulder, CO: Westview Press. Copyright 1999 by Westview Press. A member of the Perseus Books Group. Reprinted by permission.

point, with its emphasis on all the forces that act upon children, may have given the impression of passive children whose development is molded by whatever environments they happen to encounter. Such is hardly the intention of the ecological approach. Indeed, Bronfenbrenner was among the first developmental psychologists to argue for a proposition that has become generally accepted: that children play an active role in their own development, to some extent creating rather than simply responding to the environments in which they live. He has written, in fact (Bronfenbrenner, 1989), that an unintended and ironic consequence of his initial writings about ecological systems was to encourage research that focused too exclusively on context without taking into account the characteristics of the child—thus "Context without Development," in place of the earlier and equally misguided "Development without Context." Clearly, we need research that encompasses both development and context.

Summary

The first section of this chapter considers the settings within which research in developmental psychology occurs. It is organized in terms of the distinction between structured laboratory settings and natural field settings. Independent variables can be manipulated and dependent variables measured in either setting, yielding four general designs.

Research under the lab-lab heading tends to maximize both control of the independent variable and accurate measurement of the dependent variable. Such research, therefore, is often high in internal validity. Furthermore, some topics in developmental psychology can be clearly studied only within the controlled environment of the laboratory. On the negative side, laboratory manipulations and measures are often quite different from the real-life situations of interest, and this artificiality raises questions of external validity. Laboratory research may also be especially subject to problems of reactivity and response bias.

The strengths and weaknesses of field research tend to be the converse of those of laboratory research. The great advantage of research under the field-field heading is the locus of both the independent variable and the dependent variable in the natural setting. Problems of artificiality and consequent lack of external validity are therefore reduced. The major limitation concerns feasibility: Control of the independent variable and measurement of the dependent variable are more difficult in the natural setting than in the laboratory, and with some variables such control and measurement are impossible.

The advantages and disadvantages of the two combined approaches (lab-field and field-lab) follow from the general points made concerning laboratory research and field research. A further advantage is the temporal and spatial separation of the independent and dependent variables, a feature that may sometimes increase validity.

Following the run-through of designs, the discussion turns to an overview and evaluation. The section stresses the value of *converging operations:* studying a topic in as many different ways as possible so that the strengths of other approaches compensate for the inevitable limitations of any one approach.

The second section of the chapter addresses the issue of the contexts for research from a somewhat different perspective. Bronfenbrenner's *ecological systems* approach identifies four interrelated layers of environmental context within which development takes place. The *microsystem* is the layer closest to the child; it comprises the social agents (e.g., parents, peers) and direct experiences that influence children's development. The *mesosystem* refers to relations among the child's microsystems—for example, parents' effects on a child's performance in school. *Exosystem* is Bronfenbrenner's term for social systems that can affect children but in which they do not participate directly; school boards, local governments, and parental workplaces are all examples. Finally, the *macrosystem* refers to the culture or subculture within which the child develops. Subsuming and potentially affecting all of these systems is the *chronosystem:* the temporal context within which development occurs.

The most general implication of Bronfenbrenner's formulation concerns the need to study environmental contexts that extend beyond the microsystem processes that are the focus of most research. Of particular interest are interactions among two or more ecological systems—for example, differential effects of a particular parental socialization practice across different cultures or subcultures. Two examples of research that demonstrate such interactions are discussed.

Exercises

1. Both of the example studies used in chapter 2 were laboratory studies. In both cases the experimental manipulation (easy vs. hard items in Dufresne and Kobasigawa, map vs. model in Cherry and Park) occurred in a laboratory setting, and in both cases the measurement of the effects of the manipulation (the memory performance) occurred in the same lab setting. Imagine that you wished, as far as possible, to carry out the same studies in a field rather than lab setting. Describe how you might proceed, and discuss the major challenges that you would face.

2. Most of us can probably reflect more readily on the various microsystem influences on our development (parents, friends, teachers, etc.) than we can about the other layers of context in Bronfenbrenner's ecological framework. Think for a moment about the role of the mesosystem, exosystem, and macrosystem in your own development. How have these higher-level contextual factors contributed to who you are? How might your developmental course have been different if the ecological context had been different?

3. As the text notes, most research in developmental psychology examines the contribution of the microsystem to development. Select some developmental outcome that interests you and discuss how you might go about studying effects of the mesosystem and the exosystem.

7

Qualitative and Applied Research

As the title indicates, this chapter considers two forms of research that are distinctive enough and important enough to merit some concentrated attention in their own right. The first half of the chapter discusses approaches that fall under the heading of qualitative research; the second half deals with applied research.

I should emphasize at the outset that many of the principles discussed in the preceding chapters remain relevant when we turn to these new topics. Good research is good research, and the same rules and the same criteria often apply whatever the approach. Still, the goals that underlie both qualitative and applied research are in part different from those that underlie the kinds of research we have considered thus far, and these differences in goals lead to some differences in how research is done and evaluated.

Qualitative Research

A reasonable way to introduce what is meant by qualitative research is by contrasting the qualitative approach with other approaches to research. The main contrast, as you might guess, is with so-called quantitative research. It is quantitative research that has been our main subject to this point in the book.

Several features distinguish quantitative research. First, as the name indicates, quantitative research is quantitative—it involves attaching numbers to the units measured and performing statistical analyses on the numbers. As we saw in chapter 4, the kinds of numbers can vary, and as we will see in chapter 8, the kinds of statistics can vary. Still, one hallmark of the quantitative approach is that there is always some sort of quantitative framework within which questions are posed and answered.

A further feature of quantitative research is an emphasis on group-level comparisons. Responses of specific individuals are seldom a concern in the quantitative approach; indeed, the attempt is often to reduce the "error variance" associated with individual differences among people. Participants within a group (that is, a particular level of an independent variable) are assumed to be the same, and the focus is on possible differences between groups.

A third characteristic of the quantitative approach is an emphasis on representativeness and generalization. This point ties into the preceding one. Individual participants in a quantitative study are not of interest in themselves; they are of interest to the extent that they tell us something about some broader population of interest. Issues of representative sampling are therefore critical.

A fourth characteristic is an emphasis on causal explanation. This emphasis on causality is reflected in the priority accorded to internal validity: the accuracy of cause-and-effect conclusions concerning how variables relate. Methodologically, it means a preference for experimental manipulation of variables when possible.

A final characteristic concerns the typical starting point for such research. The typical starting point is some basic theoretical question or set of questions that the research literature suggests is both important to answer and not yet completely answered. The goal of the research is then to provide the research community with further scientific knowledge. The emphasis is on an objective, value-free search for what is assumed to be objective reality.

All of these characteristics are evident in the two example studies introduced in chapter 2. Both began with some basic questions about memory that the investigators wished to address. Both selected samples that were intended to be representative of the populations of interest. Both created experimental contrasts that were meant to illuminate the role of some causal factor. And both—as Tables 2.1 and 2.2 indicate—applied numbers to their results and performed various group-level comparisons with these numbers.

How, then, does qualitative research differ? Let us begin with the final point concerning the starting point for research. The typical starting point for qualitative research is not the search for some aspect of objective reality that is the same for everyone. A basic assumption, in fact, is that such a search is misguided. There may (depending on one's philosophical preferences) be a constant and knowable physical world to which the natural sciences can speak. The target for the social sciences, however, is not the world itself but people's experiences of the world, and such experiences are variable and personal and self-constructed. It is the meanings with which people invest their experiences that should be our concern, according to the qualitative perspective, and these meanings may vary from person to person. The quantitative researcher's search for an invariant objective reality is thus replaced by the qualitative researcher's search for variable forms of subjective meaning.

The focus on people's interpretation of reality has implications for issues of sampling. Because people's experiences are in part personal and subjective, there is no expectation that one person or group will necessarily be representative of some other person or group. Samples tend to be studied, therefore, not for their representativeness but because they are of interest in themselves. Often (to anticipate the subject of the second part of the chapter) they are of interest because they are somehow marginalized or neglected or at risk—for example, children growing up in poverty; or adolescents with an emerging gay or lesbian sexual orientation; or women who face barriers and discrimination in the workplace. Issues of values are often an important component of qualitative research, and some (although by no means all) such research has an explicit political agenda.

If our concern is with how people interpret their worlds, it would make little sense to study them in tightly controlled and artificial laboratory environments. The quantitative researcher's emphasis on control and manipulation and causality gives way in qualitative research to a concern with context and naturalness and rich description of people's experiences. The one-way, expert-to-subject orientation of quantitative research gives way as well to a more egalitarian, interdependent relationship between researcher and research participant. The researcher needs to enter into the participant's world in order to understand it, while all the time remaining aware of the difficulties in fully doing so.

A final contrast is, of course, the one conveyed in the names for the two approaches. Qualitative research is not necessarily number-free; some qualitative projects do include numerical tallies and perhaps some statistical comparisons. Numbers are not intrinsic to the approach, however, and any quantitative component is simply a step en route to the final goal—namely, to capture how people make sense of their worlds. Rather than numbers, therefore, primary data typically take the form of words, both the participants' own words and the investigator's interpretation of those words and related behaviors. We will see some examples in the next section when we consider some particular kinds of qualitative research.

Table 7.1 provides a summary of these and other distinctions that are often evoked in attempts to contrast the qualitative and quantitative approaches. The table is a merging and reworking of several such lists of contrasts found in the literature, including Creswell (2003), Denzin and Lincoln (2000), Gliner and Morgan (2000), and McGrath and Johnson (2003). Clearly, some of the contrasts overlap, and there are others that might be added and in fact are added by some commentators. The table is intended simply as a quick and initial way to think about how qualitative research differs from quantitative.

Methods of Study

How do qualitative researchers collect their data? In part, in the same way as do quantitative researchers. There are, after all, only so many ways to gather information about what people are like. Observations of naturally occurring behavior play an important role. So do interviews with the samples of interest.

It is true that some techniques in the quantitative researcher's arsenal are unlikely to be found in qualitative studies—in particular, any that involve automatic or essentially automatic recording of behavior (e.g., physiological measures). Beyond this difference, what distinguishes qualitative from quantitative is less the specific technique of data collection than the approach taken to it. The qualitative researcher who opts for an interview approach is unlikely to use a closed-choice interview or questionnaire on which the respondent chooses among options provided by the experimenter. Much more likely is the use of an open-ended form of questioning, with the direction of the questioning determined at least in part by how the participant responds. Similarly, the qualitative researcher who decides to observe behavior is unlikely to employ a preset coding system in which boxes are checked to record the behaviors under study. More likely is some form of running record that is open enough to capture anything of interest, with, again, the possibility of change in what is recorded as the study develops.

Table 7.1 Contrasts Between Quantitative Research and Qualitative Research

Quantitative	*Qualitative*
Objective	Subjective
Single tangible reality	Multiple constructed realities
Reductionist	Holistic
Experimental	Descriptive
Deductive	Inductive
Knower and known separate	Knower and known inseparable
Context-free	Context-bound
Value-free	Not value-free

Of course, whatever the method of data collection and the nature of the data, the final step is always the attempt by some human observer to make sense of what has been found. In quantitative research, however, it seems fair to say that the goal is to minimize this interpretive aspect—to produce results that are so clear and so objective that anyone will interpret them in the same way. In qualitative research the judgments of the individual researcher play a larger, indeed intrinsic, role. It has been said that in qualitative research, "The researcher is the instrument for collecting data" (Mertens, 2005, p. 247). Note, however, that the researcher does not simply record, à la a tape-recorder or camera; the researcher also *interprets.*

The points just made relate to some of the distinctions that were introduced in chapter 5. As the preceding discussion suggests, standardization does not occupy the honored position in qualitative research that it does in quantitative. If standardization does enter in, it is often late in the data-gathering process rather than as an essential element from the start. A related point is that much of what is done under the qualitative heading corresponds to what was discussed as *exploratory* or *discovery research* in chapter 5. Quantitative researchers, of course, also do exploratory research, but such efforts do not occupy a very high proportion of most quantitative researchers' time, and they typically are simply a prelude to more systematic and standardized study. In qualitative research exploration is often the entire story and not simply the prelude.

These general points still leave unanswered the question of exactly how qualitative researchers go about doing their research. The specific methods that are used in qualitative studies can be organized in various ways. Table 7.2 shows one typology, drawn from Creswell and Maietta (2002). In what follows I elaborate on and give examples from the two approaches that have arguably been most important so far for developmental psychology: narrative research and ethnographic research. Among the

Table 7.2 Methods of Study Used in Qualitative Research

Method	*Description*	*Types of data*
Narrative	Collecting the stories of lived experiences	Interviews, observations, and documents
Phenomenology	Understanding the essence of experiences surrounding a phenomenon	Long interviews with a small number of informants
Grounded theory	Developing a theory grounded in data from the field	Interviews with 20–30 people to develop categories and detail a theory
Ethnography	Describing and interpreting a cultural and social group	Primarily observations and interviews, with additional artifacts, during an extended time in the field
Case study	Developing an in-depth analysis of a single case or multiple cases	Multiple sources, including documents, archival sources, interviews, and observations

SOURCE: Adapted from "Qualitative Research" (p. 147), by J. W. Creswell and R. C Maietta. In D. C. Miller & N. J. Salkind (Eds.), *Handbook of Research Design and Social Measurement* (6th ed., pp. 143–184), 2002, Thousand Oaks, CA: Sage Publications, Inc.

helpful treatments devoted to the other approaches—in addition to the Creswell and Maietta source—are Stake (2000) and Yin (2003) for case studies, Giorgi and Giorgi (2005) for phenomenology, and Charmaz (2005) for grounded theory.

Narrative Research

As the name suggest, **narrative research** focuses on narration—the stories people tell about their lives. In Murray's (2003, p. 95) words, "Narrative psychology is concerned with the structure, content, and function of the stories that we tell each other and ourselves in social interaction." Because even quite young children can and do tell stories, it is a method well suited for developmental issues.

The narrative approach has been employed with a variety of different populations, ranging in age from toddlerhood through old age and spanning dozens of different cultures or subgroups within Western culture. It is, in fact, one of the primary methods of study within the discipline of cultural psychology (Shweder et al., 1998).

The method for collecting narratives varies some depending on the population being studied and the goals of the researcher. Sometimes narratives may be observable in people's spontaneous interactions, sometimes they may emerge in informal conversations between researcher and participant, and sometimes they may be elicited through more formal interviews. Murray (2005) provides guidelines for collecting narratives from adult populations, and Engel (2005) adds suggestions for working with children.

Although narratives may be elicited as part of research, it is important to emphasize that they are not simply a research tool for learning about people. Rather, producing narratives is something that people do spontaneously, quite apart from the demands of research. Narratives are a basic way in which we come to understand both ourselves and those around us. They are also one of the ways in which adults socialize children.

The example I discuss, some work by Peggy Miller and associates (Cho & Miller, 2004; Miller, Cho, & Bracey, 2005; Miller, Fung, & Mintz, 1996; Miller, Wiley, Fung, & Liang, 1997), encompasses both developmental and cultural dimensions. The participants were mothers and their 2- to 4-year-old children drawn from four communities: a middle-class United States community, two working-class United States communities, and a middle-class community in Taiwan. In each setting researchers made video recordings of family interactions on several occasions and also interviewed the mothers. The goal of both the observations and interviews was to record "stories of personal experience—stories that people tell in ordinary conversation in which they relate past experiences from their own lives" (Miller et al., 1996, p. 238).

The results revealed both some commonalities and some differences across communities. In every setting, mothers did in fact tell stories, both in interactions with other adults and in conversations with their children. Stories told in the children's presence divided into three groups: stories about others for which the child was a listener, stories about the child for which the child was primarily a listener, and stories about the child for which the child was an active collaborator and co-narrator. In every setting, children did in fact participate in the storytelling, taking on more active and independent roles as they grew older. And in every setting, aspects of the mother's narrative style and emphases came to be reflected in the child's narrations.

The main difference between the Chinese sample and the United States samples concerned the didactic purpose that seemed to underlie stories about the child. Chinese mothers often used stories to teach moral lessons, focusing on the child's transgressions and the lessons to be taken away from particular episodes. Box 7.1 provides an example, one in which an older sister joins with the mother in reminding the younger child of his misbehavior.

Box 7.1. An Example of a Didactic Narrative in a Chinese Family

Mother: [Looks at child] Eh, eh, you that day with Mama . . . with older sister went to the music class. Was that fun?

Child: It was fun.

Mother: What didn't the teacher give you?

Child: Didn't, didn't give me a sticker.

Mother: Didn't give you a sticker. Then you, then what did you do?

Child: I then cried.

Sister: Cried loudly, "Waah! Waah! Waah!"

Mother: Oh, you then cried? Yeah, you constantly went: "Waah, didn't [gestures wiping eyes, makes staccato gesture of fists away from body], why didn't you give me a sticker? [whines] Why didn't you give me a sticker? [whines]," didn't you?

[Child looks up from book, gazes at mother, smiles, and looks down at book again]

Sister: [To mother] Yes, "Why didn't you give me a sticker?' [claps hand]

Mother: [To child] Sticker. [sighs] Ai, you made Momma lose face. . . . That, that, I wanted to dig my head into the ground. Right? [smiles, shakes head, smiles again]. . . .

Sister: Almost wanted to faint . . . Mommy almost began to faint.

SOURCE: From "Self-Construction Through Narrative Practices: A Chinese and American Comparison of Early Socialization," by P. J. Miller, H. Fung, and J. Mintz, 1996, *Ethos, 24*, p. 251. Copyright © 1996 by the American Anthropological Association. Adapted with permission.

Although American mothers also used stories for such teaching purposes, they did so less frequently than Chinese mothers, and this was especially true for the middle-class sample. Instead, the main concern for American mothers seemed to be to nurture their child's self-esteem. Thus stories tended to concentrate on positive behaviors and events, and when transgressions were discussed the tone was usually humorous rather than didactic.

Whereas the main China–U.S. difference lay in the content of the narratives, the main difference among the United States samples was in sheer frequency. Stories—both from adults and from children—were considerably more frequent in the working-class samples. Indeed, one group of mothers produced an average of 65 stories per 1- to 2-hour interview! What this finding suggests is that narratives, while perhaps a universal feature of self-understanding and socialization, may play especially important roles in some cultural groups. In the authors' words,

> Working-class adults participated prolifically, avidly, and artfully in personal storytelling in their homes and communities and . . . they brought children into this valued activity from an early age. The children in these families experienced home environments that were saturated with stories, and

by the time they were 3 years old, it is likely that telling stories of personal experience had become second nature to them. (Miller et al., 2005, p. 125)

Ethnographic Research

Ethnography means literally "portrait of a people." It is an approach that originated in anthropology and has gradually spread to the other social sciences. As Table 7.2 indicates, the goal of **ethnographic research** is to describe and to interpret the central practices of some cultural group.

Capturing the essence of a culture is clearly an ambitious and challenging undertaking. It is not something that can be accomplished quickly, and ethnographers typically immerse themselves for months or perhaps even years in the culture they are studying. It is also not something that can be accomplished with just one form of evidence, and ethnographers therefore use a variety of kinds of data. Observations are a part of any ethnography, and they typically take the form of what was discussed in chapter 4 as *participant observation*—that is, the ethnographer attempts as far as possible to become a participating member of the culture and not simply an on-the-sidelines observer. Interviews also are important, perhaps especially to learn about practices whose meaning may not be apparent through observation and interaction. Finally, physical records or artifacts, such as diaries or letters, may also contribute.

An ethnographic study typically proceeds through three phases (Miller, Hengst, & Wang, 2003). The first phase is developing the general questions to be examined and gaining access to the population of interest. Depending on the population—and also the fit between the researcher's background and the particular population—this first phase may be a challenging one to negotiate. The second phase is the collection of data. As we saw, this phase is likely to be a lengthy one, and the particular emphases and particular methods may evolve across the course of a study—a point that applies in general to qualitative research. Finally, the third phase is devoted to interpreting and analyzing the data.

As noted, many of the questions that are addressed in an ethnography emerge in the course of the study—something that is necessarily the case, given that the researcher is learning about a new culture. On the other hand, it is not really possible to enter into a new site with no idea of what one is looking for; rather, the researcher must begin with some sort of conceptual framework and at least a tentative set of issues to be examined. Striking the right balance between guiding assumptions and openness to experience is one of the most challenging aspects of ethnographic research. Particularly challenging is setting aside one's own cultural preconceptions when it is necessary to do so. As Miller, Hengst, and Wang (2003, p. 224) put it, ethnographers must "try not to mistake their own deeply taken-for-granted, culturally saturated understandings for those of the study participants."

Although ethnography is often associated with work in far-off lands, it can be applied to cultures that are closer to hand. For any adult, the world of children or of adolescents is in part a different culture, and thus a possible topic for ethnographic study. In recent years a number of ethnographers have taken children or adolescents as their target. The example I use is a study by Barrie Thorne (1993).

Thorne's interest was in the emergence of gender relations and gender differences during the grade-school years. To examine these issues she visited two elementary schools on a daily basis for close to a year, observing and interacting with children who ranged in grade level from kindergarten through fifth grade. As far as possible Thorne took on the role of a child herself during these visits—sitting at a child-sized desk during classroom time, joining the children in the cafeteria for lunch, entering into their games and conversations on the playground. Of course it is never possible, or sensible, for an adult to take on a child's role in total, and Thorne did not attempt to do so (she did not participate in the classroom lessons, for example). This is a point that applies to any ethnographic study of childhood. Nevertheless, the attempt was to experience the world of school from a child's rather than an adult's perspective. A further goal was to be perceived by the children not as another adult but, as far as possible, as someone more akin to a peer, and thus to be privy to conversations and interactions that normally would be restricted to other children. Through these techniques Thorne was able to document the emergence and eventual pervasiveness of gender differences and gender separation (one chapter is entitled "Boys and Girls Together . . . But Mostly Apart"), as well as to identify a number of the contributors to the differences, both in school practices and in the attitudes and behavior of children themselves.

Thorne's study is just one of a number of examples of ethnographers' attempts to enter the worlds of children or adolescents. Other examples include a study of friendships among teenage girls (Hey, 1997), an analysis of the importance of teen magazines for preteen girls (Finders, 1996), a study of interaction patterns among teenagers living in a group home (Emond, 2005), and a study of adolescents in a substance abuse treatment center (Reisinger, 2004).

Overview and Evaluation

I add three general points here, followed by some suggestions for further readings about the qualitative approach.

The first point may have occurred to you if you have already had some exposure to writings about qualitative research. Those who write about the approach are certainly in general agreement concerning its basic nature and goals. At a more specific level, however, there is often considerable divergence in the distinctions that are drawn, the points that are emphasized, and even the terminology that is used. Thus the particular summary I have presented here may map only roughly onto what you encounter in other sources.

A second point concerns the current status of the approach in developmental psychology. Opinions on this point may of course vary. My own evaluation is that the qualitative approach is gaining in prominence and importance but that it remains distinctly secondary to the quantitative approach in most of the field's mainstream writing. This situation is of course not surprising—most researchers, journal editors, and textbook authors were trained in quantitative research. The visibility of qualitative work does vary to some extent across topics and across publication outlets. Work in the Vygotsky theoretical tradition, for example, often takes a qualitative slant (e.g., Nelson, 2003) as does work in cultural psychology (e.g., Miller, Hengst, & Wang, 2003). Feminist approaches to developmental psychology are another currently active topic (e.g., Miller & Scholnick, 2000). Of the publication outlets listed in Tables 1.1 and 1.2, probably the most likely sources for qualitative research are *Human Development, Merrill-Palmer Quarterly,* and *New Directions for Child and Adolescent Development.*

The third point concerns the fit between qualitative research and quantitative research. There are authors—including some in both the qualitative and quantitative camps—who view the approaches as incompatible and who advocate forcefully for one method of study or the other. Most of us, however, regard the approaches as more complementary than contradictory—that is, as equally legitimate ways to address somewhat different questions about human development. At the least, this means that our knowledge of any topic is likely to be fullest if we are aware of both the quantitative and qualitative approaches to its study. For particular topics it may also mean the use of both quantitative and qualitative techniques within the same study or at least within an overall program of research. Both Creswell (2003) and McGrath and Johnson (2003) provide examples of how such "mixed methods" approaches can be profitably applied.

This last point about complementary approaches should sound familiar by now. Understanding human behavior is an exceedingly challenging task. It is almost always better to have multiple methods of study rather than just one.

Writings about qualitative research have proliferated in recent years, and hence there is no shortage of further sources. Among those that are most likely to prove helpful to the developmental psychologist are Camic, Rhodes, and Yardley (2003); Smith (2005); and several chapters in Greene and Hogan (2005).

Applied Research

As we have seen, one general way to distinguish quantitative and qualitative research is in terms of the goals that motivate the research. In a somewhat different sense, a difference in goals also defines the distinction to which we turn now.

Chapter 1 began with a discussion of why we do research in developmental psychology. As I argued there, we do research for two general reasons. We do research for reasons of *basic science,* to advance our understanding of human development; and we do research for purposes of *application,* to better the lives of children and other vulnerable groups. These two goals are by no means incompatible, nor do the basic methodological principles that govern good research change when we move from a basic to an applied focus. Nevertheless, applied research can present special challenges that add to the challenges and complexities of research in general. Because most of the book deals with the basic research, the purpose of the present section is to add some points with respect to research in applied developmental psychology.

What is meant by "applied research"? Fisher and Lerner (1994b, p. 4) define applied developmental psychology as "that aspect of psychology which bears upon enhancing developmental processes and preventing developmental handicaps." Lerner, Jacobs, and Wertlieb, 2003, p. 4) offer a similar definition: "the programmatic synthesis of research and applications to describe, explain, intervene, and provide preventative and enhancing uses of knowledge about human development."

A number of specific goals and related forms of research fall within the scope of applied developmental psychology. Table 7.3 summarizes these goals and gives examples of the kinds of research that relate to each; the table is derived in part from Fisher and Lerner (1994b) and in part from other discussions of the applied approach (e.g., Zigler & Finn-Stevenson, 1999). Table 7.4, which is taken from Lerner et al. (2003), provides an overview of the kinds of topics that are frequent foci for work in applied developmental psychology.

Clearly, all of the goals identified in Table 7.3 are important. In what follows I limit the discussion to two forms of applied research: research directed to socially important issues, and interventions designed either to prevent or to correct problems. Fuller discussions of both these and other kinds of applied research can be found in Fisher and Lerner (1994a), Fisher and Lerner (2005), Lerner et al. (2003), and Sigel and Renninger (2006). Note also that there are two journals devoted specifically to applied developmental psychology: *Journal of Applied Developmental Psychology* and *Applied Developmental Science.* The Society for Research in Child Development also publishes a regular *Social Policy Report* that deals with applied issues (available online at www.srcd.org/pubs.html).

Socially Important Issues

Table 7.4 conveys an idea of the kinds of topics that fall under this heading. The example on which I focus is both one of the most important and one of the most intensely studied of these

Table 7.3 Forms of Applied Research in Developmental Psychology

Category	*Goal*	*Examples*
Assessment	To identify developmentally important characteristics in an at-risk target population	Use of a neonatal screening test to measure developmental status in newborns subjected to prenatal stress. Use of the Stanford-Binet IQ test to identify possible cognitive delays in preschoolers from impoverished homes.
Intervention	To alter the environment in ways that may prevent, correct, or reduce problems of development	Project Head Start or similar programs intended to boost the academic readiness and cognitive and social competence of preschoolers from disadvantaged backgrounds. Memory training programs for elderly individuals experiencing memory difficulties.
Research on socially important issues	To provide evidence relevant to the resolution of pragmatically important questions	Research directed to the accuracy of children's testimony in cases of suspected abuse. Research directed to the determinants of teenage substance abuse.
Contribution to public policy	To use knowledge gained from research to inform policy-making decisions	Use of findings from developmental research to establish standards of quality for outside-the-home child care. Incorporation of findings from developmental research in court decisions regarding custody and visitation in cases of divorce.
Dissemination of psychological knowledge	To make available the results of research to those who might benefit from the knowledge	Production of books or pamphlets of advice for teenage parents. Consultation by psychologists in situations (e.g., school board decisions, court cases, design of nursing homes) in which their knowledge is relevant.

topics: children as witnesses in suspected cases of abuse. Research directed to this issue defines the field of **forensic developmental psychology** (Bruck & Ceci, 2004).

Some statistics can help to convey the importance of the issue. It has been estimated that at least 100,000 children testify in court cases in the United States every year (Ceci & Bruck, 1998). This figure does not include the much larger number of instances in which children provide evidence outside of court. The cases in which children testify span a range of topics, but the most frequent category among criminal trials, accounting for about 13,000 cases a year, is child sexual abuse. In most instances of alleged abuse, the child witness is also the target of the abuse. In many, the child is the only witness.

Table 7.4 Examples of Topics of Study in Applied Developmental Psychology

Adolescent pregnancy
Aggression and violence
Children's eyewitness reports
Depression
Developmental assets
Developmental psychopathology
Domestic violence and maltreatment
Early child care and education
Early childhood education
Education reform and schooling
Literacy
Marital disruption and divorce
Mass media, television, and computers
Parenting and parent education
Pediatric psychology
Poverty
Prevention science
Successful children and families

SOURCE: From "Historical and Theoretical Bases of Applied Developmental Science" (p. 12) by R. M. Lerner, D. Wertlieb, and F. Jacobs. In R. M. Lerner, F. Jacobs, and D. Wertlieb (Eds.), *Handbook of Applied Developmental Science* (Vol. 1, pp. 1–28), 2003, Thousand Oaks, CA: Sage Publications, Inc. Copyright © 2003 by Sage Publications, Inc. Adapted with permission.

Can the testimony of a young child be trusted? Should such testimony be admissible in court? As the figures just cited indicate, questions like these can be critically important to answer.

They can also be very difficult to answer. We have not yet considered how researchers typically study children's memory—that will be one of the topics discussed in chapter 12. I will tell you now, however, something you can no doubt guess, and that is that the typical study of children's memory bears little resemblance to the memory-for-abuse situation. Two general differences exist.

The first difference concerns the content for the memory. Memory studies in the basic-science literature focus on children's memory for benign content, either arbitrary material such as list of words (recall the Dufresne and Kobisagawa, 1989, example introduced in chapter 2) or personal experiences of a pleasant sort such as a family trip. In contrast, the memory at issue in cases of abuse is memory for personal experience of a very negative, often highly traumatic sort. Most basic memory studies ask about memory for one-time experiences or events; abuse may sometimes be a one-time occurrence, but it also may occur on multiple occasions across an extended period of time. In much basic memory research the child is a passive recipient of the memory materials; instances of abuse may involve the child as a participant and not merely as an onlooker or bystander. Finally, in basic memory research there is no reason for children not to report everything they remember; this, however, is not necessarily true in cases of abuse. Complex social and emotional factors may affect children's reports, such as the child's guilt about being a participant or reluctance to implicate a parent or friend. Thus what children say may not be identical to what they remember.

The second general difference concerns the method of questioning. In basic memory research the questioning is generally designed to be optimal—that is, to elicit the most accurate recall of which the child is capable. To this end, researchers use clear and straightforward questions, provide appropriate encouragement and social support, and in general do the things that research has shown help children to perform at their best. Accurate recall is certainly the goal in most real-life forensic interviews, but the questioning may be far from optimal. To begin with, children may be questioned by a range of people—parents, teachers, doctors, lawyers, police officers—none of whom is necessarily expert in the best way to question children. The sheer number of questioners introduces a potentially daunting factor not found in most laboratory research, as does the number of separate occasions on which questioning may occur—estimates are 12 occasions on the average, and the figure is much higher in some instances (Ceci & Bruck, 1998). In addition, the questioning in real-life cases is almost always delayed to some extent, and in some instances it may be very delayed; children may be questioned several years after the alleged incidents took place. Finally, the questioning that children encounter in legal cases is often far from straightforward; rather it may include various suggestions and prompts and reinforcers that lead the child in the direction of particular answers. Such shaping of the child's responses is especially likely if questioners believe that they already know the truth or if, as advocates within the court system, they have an interest in a particular outcome.

How might research speak to all these complexities? One challenge, clearly, is to discover or devise situations we can ask about that bear some similarity to the abuse situation. Two general approaches have been tried. One possibility is to create experimental settings that reproduce some elements of the real-life situations of interest. One set of investigators, for example, had the child play a Simon Says game with the experimenter, in the course of which both the child and the experimenter touched parts of each other's bodies (White, Leichtman, & Ceci, 1997). In another study the child interacted with an adult dressed in a clown suit; in the course of their play parts of the suit were put on and taken off both participants, and the adult took photos of the child (Rudy & Goodman, 1991).

Although studies such as these capture many aspects of real-life cases (e.g., removing clothing, taking photos), they do have an obvious limitation, which is that they omit a central element of the abuse situation—namely, the trauma of being abused. Obviously, researchers cannot intentionally subject children to traumatic experiences. They can, however, capitalize on instances in which such experiences have already occurred, and the second type of study falls in this category. Among the examples of painful or embarrassing experiences that children have been questioned about are going to the dentist (Peters, 1991), receiving an injection (Goodman, Hirschmann, Hepps, & Rudy, 1991), receiving a genital examination (Saywitz, Goodman, Nicholas, & Moan, 1991), and undergoing urinary catheterization (Quas, Goodman, Bidrose, Pipe, Craw, & Ablin, 1999). Although such experiences can hardly equal the trauma of abuse, they do capture some of its characteristics.

As noted, the second general challenge in this research is to simulate the types of questioning that suspected victims of abuse must undergo. This, too, has been the focus of a number of studies. A child may be questioned several times across a period of weeks, for example, or questioned by different people on different occasions. Children may be told to "keep a secret" about what happened to them during the experimental session (Bottoms, Goodman, Schwartz-Kenney, Sachsenmaier, & Thomas, 1990), or a police officer rather than a research assistant may do the questioning (Tobey & Goodman, 1992). Finally, probably the most thoroughly explored variable is the nature of the questioning, with a basic contrast between straightforward questions (e.g., "Tell me what happened.") and suggestive or leading questions (e.g., "He kissed you, didn't he?" "How many times did he kiss you?") As we saw, suggestive questioning is a part of many real-life cases; it is therefore important to determine how susceptible children are to an adult's suggestions.

Although methods rather than findings are our concern here, I will note a few general conclusions that emerge from this very active research literature (for further discussion, see Bjorklund, 2000; Eisen, Quas, & Goodman, 2002; Lamb & Thierry, 2005; Pipe, Lamb, Orbach, & Esplin, 2004). First, research verifies what we would expect: that memory improves with age and that older children typically report more of their experiences than do younger children. Second, as the delay between an event and questioning about it increases, the completeness and accuracy of recall decline, and this is especially true for young children. Third, in at least some cases, young children are more suggestible than are older children or adults—that is, they are more likely to be influenced by leading questions from an adult authority figure. On the other hand—and more positively—in many studies memory differences between younger and older children or between children and adults are not very great. Age differences, as well as memory inaccuracies in general, are most likely when specific questions are used; conversely, they are minimized by the use of free recall measures that allow children to say what happened in their own words. Finally, the memory problems that children do show are mainly errors of omission rather than of commission—that is, they are more likely to fail to report certain details than they are to introduce false information. This finding suggests that any clearly spontaneous mentions of abuse by children should be taken very seriously.

Having offered these conclusions, I will add that there remains much controversy about exactly what the research shows and what the implications are for children's legal testimony. Among the topics under active investigation is the issue of how best to question children in order to meet two goals: maximizing the accuracy of testimony and minimizing stress to the child (Goodman et al., 1992; Poole & Lamb, 1998; Poole & Lindsay, 2002). Having to testify can add to the trauma of an already traumatic situation. It is therefore important to devise procedures that protect the child from further harm.

The studies of children as witnesses illustrate several general points about the relation between basic research and applied research. First, the starting point for applied work on some topic is the relevant basic-science literature—in the present case, the literally thousands of studies of children's memory that emerged from the first century of psychology as a science. Such studies produced numerous findings (e.g., differences between recognition and recall, effects of delay, effects of retrieval cues) that can help us to understand the applied situation. Second, the basic-science research is seldom a sufficient basis for applied efforts; rather, the complexities of real-world problems may require kinds of study that go beyond what has been attempted before. This, of course, is a point I stressed in introducing the research on children as witnesses. Finally, the relation between basic and applied is a two-way one: As findings emerge from the new paradigms of applied research, they feed back on and expand our basic-science understanding—in the present case, what we know about children's memory.

Intervention

The goal of intervention research is to identify methods to correct or prevent problems in some developmentally vulnerable population. There have been thousands of such efforts, with populations ranging from newborns to octogenarians and interventions ranging from a few minutes to several years. All, of course, are intended to provide benefit to their recipients. But they are also intended to generate knowledge that can be applied more broadly to similar populations in the future—which is what makes them research and not just intervention.

The example we will consider concerns memory problems in old age. One common stereotype of old age—and one common complaint of many elderly people—is of problems with memory. As we will see in chapter 14, there are many qualifications and exceptions to this stereotype; nevertheless, memory difficulties do constitute a genuine, and distressing, reality for some elderly individuals. The issue to address now is whether interventions in the form of memory training programs can undo some of these difficulties.

The general answer to this question is yes, as reviews of the memory training literature indicate (Camp, 1998; West, 1995). Here I first present an example of a representative intervention study, after which we will consider some of the general issues raised by such efforts.

Sheikh, Hill, and Yesavage (1986) examined the possibility of memory training in a group of elderly individuals who had either volunteered for a program in memory improvement or been referred by physicians or senior centers. The form of memory at issue was an important one at any point in the life span: the ability to recall links between names and faces. An initial assessment verified that many of the participants found such recall difficult: The pretest average was only 4.13 correct name-face pairs out of a possible 12. The subsequent memory training, which was conducted in group settings and extended for 14 hours, focused on *mnemonic strategies*—that is, techniques that the participants could use to facilitate their recall performance. As will be discussed in both chapter 12 and chapter 14, a large research literature indicates that strategies play an important role in both individual and developmental differences in memory; not surprisingly, therefore, strategies are a major component in many memory training programs.

The particular emphasis in the Sheikh et al. study was on the use of *mental imagery* as a mnemonic—for example, remembering the name Whealen by imagining a whale in Mr. Whealan's rather large mouth. Training in the use of such techniques proved beneficial: The various experimental groups showed close to a 100% improvement over their pretest performance, and they also outperformed a no-training control group. Furthermore, the gains from the most effective forms of training were still in evidence on a follow-up test 6 months later.

The Sheikh et al. study illustrates many of the issues that arise both in memory training programs and in intervention projects more generally. One issue concerns *identification of participants:* To whom will the treatment be applied? Memory programs for the elderly vary along this dimension. In some cases the target group is elderly people in general, with no particular selection criteria beyond age and willingness to participate; in others the focus is on elderly people who are known to be experiencing memory difficulties; in still others the focus is more clinical, the attempt being to ameliorate some of the memory problems that are a component of pathological changes with age (e.g., Alzheimer's disease). In many other domains of intervention the target group is a more distinct subset of the general population—not preschoolers in general, for example, but only those with attentional deficits, or not adolescents in general, but only those with a history of substance abuse. It is in such instances that assessment becomes especially important: We need a measure that can accurately identify those in need of help, and perhaps also specify areas of need in a way that can guide subsequent intervention.

Once prospective participants have been identified, considerations of experimental design become important. At the core of most intervention projects is the sequence labeled in chapter 3 as the *One Group Pretest-Posttest Design:* an initial assessment, followed by the intervention, followed by the final assessment. As we saw in chapter 3, a significant pretest-posttest difference in such research is not certain evidence of the value of the treatment; this design is subject to various alternative explanations, or threats to the validity of its cause-and-effect conclusions. For this reason, many interventions (such as the Sheikh et al. study) assign a subset of the participants to an untreated control group. The control group provides an estimate of the contribution of factors other than the experimental treatment (e.g., practice effects, regression to the mean), thus allowing a clearer determination of the gains that are specific to the intervention.

As is often the case with matters of design, however, implementation of some methodologically desirable procedure may be complicated by pragmatic difficulties. Participants may balk, for example, at being assigned to a no-treatment group, thus introducing the possibility of selection bias in the formation of groups. There also may be ethical issues involved if some needy individuals are denied a potentially beneficial treatment for purposes of experimental control, an issue discussed in chapter 9 under the heading of *withholding treatment.* Because of these concerns, many intervention projects offer some form of assistance to control participants following the completion of formal data collection. In the Sheikh et al. study, for example, control participants received a 4-hour course in mnemonic training at the conclusion of the study.

Accurate assessment is important not only to identify participants at the time of the pretest but also to determine effects of the intervention at the time of the posttest. What we are interested in, clearly, is documenting improvement following the intervention—improvement above and beyond what might be attributable to factors other than the intervention (thus the need for a control group) and improvement that is not merely statistically significant but also pragmatically important. Exactly what is reasonable or desirable to expect will vary across different projects. Two general considerations, however, are important in the evaluation of many intervention efforts, including most memory training programs with the elderly.

One concerns *transfer* or *generalization:* Do the benefits of the treatment extend beyond the specific tasks, stimuli, and contexts that were involved in the intervention? In the case of a study like that of Sheikh et al., for example, is there transfer to forms of paired-associate learning other than name-face pairs, or to other kinds of memory for which imagery might be helpful, or to any sort of memory performance outside the laboratory context? (The answer, by the way, is uncertain, since none of these forms of transfer was included in the study.)

The second general issue is *maintenance:* Do the benefits of the intervention persist beyond the study itself, or do they fade away once the program is no longer in effect? Here, as we saw, the Sheikh et al. study does provide some positive data, with effects still evident 6 months later. Maintenance is by no means always found in this literature, however. Indeed, a basic methodological point is that many intervention projects—in the memory training domain and more generally—fail even to include tests of transfer or maintenance. And when such tests *are* included, a general conclusion seems safe: Immediate effects are easier to produce than longlasting effects, and narrow and specific effects are easier to produce than broad and general ones.

A final issue concerns the magnitude of the intervention. We might expect in general that more will be better (more time during which the treatment is in effect, more distinct components to the intervention, more changes in the participants' lives, etc.); and for the most part this expectation proves to be true—again, not just for memory training but for interventions of a variety of sorts with a variety of populations. Here, however, considerations of feasibility clearly enter in. Resources are always limited, and often it is simply not possible to do everything that might be desirable to do in order to bring about improvements in the people whom we are attempting to help. Furthermore, the more components we pack into an intervention, the more difficult it may be to determine which particular components are critical to its success. This is an issue in *construct validity:* arriving at the correct theoretical explanation for any effects we obtain. Ordinarily, construct validity is important for matters of theory, but here it carries pragmatic implications as well. As noted, many interventions are intended not simply to improve the lives of the participants to whom they are directed but also to furnish information that can be used to help the target population more generally. And to achieve this second goal, we need interventions that are feasible, so that they can be widely implemented, and whose elements are well understood, so that they can be accurately applied by different practitioners in different settings.

Several chapters in Fisher and Lerner (1994a) and Sigel and Renninger (2006) provide further discussion of both specific intervention programs and general issues in research on intervention. A chapter by Willis (2001) discusses issues that arise in interventions with elderly participants. Finally, two books by Zigler and associates (Zigler & Muenchow, 1992; Zigler & Styfco, 1993) are good sources for information about one of the best known and most important childhood intervention efforts: Project Head Start. Head Start, in turn, provides a natural context for consideration not just of intervention but of another major initiative under the applied heading that I have not attempted to discuss here: contributions to public policy. As a reader of these books will learn, Head Start—despite many misunderstandings, setbacks, and frustrations—remains one of developmental psychology's most successful contributions to society.

Summary

This chapter discusses two forms of research that differ in important ways from the kinds of research discussed to this point in the book.

The first section of the chapter is devoted to *qualitative research.* Qualitative research can be contrasted with the quantitative research paradigm that has dominated psychology along several dimensions. Qualitative research is not directed to a single objective reality; rather, the focus is on how people invest their experiences with meaning, and the assumption is that such subjective interpretations will vary across individuals. Because people vary, the emphasis in sampling is less on representativeness and generalization and more on identifying populations that are of interest in themselves. The goal in studying such populations is not to test theoretical predictions or to make group-level comparisons; rather it is to understand their subjective experiences, working as far as possible from the perspectives of the group members themselves. Finally, numbers and statistical comparisons are not intrinsic to the qualitative approach; the concern rather is with rich description of people's experiences.

The goals of qualitative research are reflected in the methods that are used. In general, the quantitative researcher's emphasis on control and manipulation is replaced in qualitative research by a concern with context and naturalness and exploration rather than hypothesis testing. More specifically, five general qualitative methodologies can be identified: narrative, phenomenology, grounded theory, ethnography, and case study. Of these, the text elaborates on narrative research and ethnography, giving examples of developmental study in both cases.

The first section concludes with three general points about the qualitative approach. The first point concerns the diversity in writings about the approach; the second evaluates its status in developmental psychology, concluding that it is a growing but still minority position; and the third emphasizes the complementary nature of the qualitative and quantitative perspectives and the desirability of basing general conclusions about a content area on both.

The second section of the chapter is directed to *applied research*—that is, research whose explicit goal is to prevent or correct problems in development and to optimize developmental outcomes. Two forms of applied research are discussed.

One important form of applied study is *research directed to socially important issues.* The content area used as an example is *forensic developmental psychology:* the study of children as witnesses, especially in cases of suspected abuse. Research on children as witnesses builds upon but also extends the relevant basic-science literature—a point that applies in general to applied research. In the present case two challenges must be met: devising or discovering situations that bear some similarity to instances of abuse, and determining the effects of questioning that is similar to that in real-life forensic cases. Examples of approaches to both challenges are discussed.

A second important form of applied study is *intervention research:* programs designed to correct or prevent problems. Memory training programs for the elderly serve as the example from which general points about intervention are drawn. One such point concerns the need to identify prospective participants, a task for which an accurate assessment device may be essential. Assessment is also important at the termination of the intervention to document effects; among the questions that are usually of interest are the *transfer* of effects beyond the intervention context itself and the *maintenance* of the effects over time. Finally, a goal of many intervention projects is to develop programs that can be applied more broadly; it is important, therefore, that programs be feasible and their essential elements well understood.

Exercises

1. The goal of ethnography is to enter into and learn about some cultural setting that is different from one's own. Consider your own cultural group, either broadly defined (e.g., in terms of your country or ethnicity) or narrowly defined (e.g., in terms of some college group to which you belong). Imagine an ethnographer from a very different background who wished to learn about your culture. What do you think would be the most challenging aspects to understand, and why?

2. Project Head Start is found in more than 2000 centers spread across all 50 states. Visit at least one and if possible more than one Head Start center located near you. Try to make several visits at different times so that you can see different activities. Once you have acquired some familiarity with how Head Start works, consider the task of evaluating the effects of such programs. What kinds of assessments do you think would be important to include? Locate some of the background sources cited in this chapter, and compare your ideas with the measures that have in fact been used in research on Head Start.

3. The text discusses memory training studies with the elderly as an example with which to illustrate some general points about intervention research. Select one of the topics in Table 7.4 that especially interests you and outline a possible intervention study for that content area (i.e., the sample that would be studied, the measures that would be used, the form that the intervention or interventions would take, etc.). If possible, locate a review of research on the topic in one of the sources cited in the chapter and compare your ideas with what has actually been done.

8

Statistics

We saw in the preceding chapter that numbers and statistics are not central to the kinds of study that fall under the heading of qualitative research. We saw also, however, that they *are* central to so-called quantitative research—and thus to most mainstream research in psychology. No one can hope to be a contributor to this research without some basic competence in statistical analysis. And no one can hope to understand and evaluate the research of others without some grasp of how statistics works.

Any student who has penetrated very far in the study of psychology is familiar with the points just made. Any such student also knows that mastery of statistics is not a rapid process but typically requires several courses and several depressingly thick textbooks. There will be no attempt here to compress several books' worth of information into a single chapter. The goal is more modest: to present some of the basic ideas and principles behind statistical tests, as a complement to much fuller discussions of the tests themselves. The chapter can serve as a preview of, accompaniment to, or reminder of course work in statistics.

The general direction of movement in the chapter is from the relatively simple to the more complex. We will begin with some reminders about uses of statistics, after which the familiar *t* test is introduced as an example of statistical reasoning. To a good extent, much of the rest of the chapter consists of complications of the simple case presented by the *t*—various situations in which we need other kinds of statistics, along with the possibilities available and bases for choosing among them. Because the discussion in every case remains brief and general, further sources are noted for each of the topics touched on.

Uses of Statistics

Psychologists use statistics for two purposes: to describe data and to draw inferences about the meaning of data. The first of these uses is relatively straightforward and familiar; the second is both more complicated and more challenging.

Descriptive Statistics

Let us return to one of our earlier examples. Imagine an observational study of aggression in a preschool setting. The researcher collects the (hypothetical) data shown in Table 8.1. As

Table 8.1 Number of Aggressive Responses in a Sample of Preschool Children (Hypothetical Data)

3-year-old boys	*3-year-old girls*	*4-year-old boys*	*4-year-old girls*
5	0	2	3
4	0	27	3
0	10	3	0
14	3	34	10
5	0	3	1
15	18	38	4
0	2	0	3
2	0	4	0
9	5	19	4
5	15	35	3
3	0	3	5
2	6	2	11
1	0	3	0
16	10	18	1
3	6	10	3

can be seen, there are clearly individual differences in frequency of aggression; there might also be differences associated with sex and age. But how can we move beyond this welter of numbers to determine what we really have?

The first step is to summarize the data by means of various **descriptive statistics**. Most descriptive statistics are concerned with identifying the **central tendency**, or dominant pattern of response in the sample. The most common measure of central tendency is the familiar arithmetical average or **mean**. Table 8.2 shows that the various groups in our hypothetical study do in fact show different mean levels of aggression.

In most cases the mean is the most informative descriptive statistic. It is not the only descriptive measure that can be derived, however, and in some cases the mean alone may give an incomplete picture of the results. Consider a comparison between the 3-year-old boys and the 4-year-old boys. We saw in Table 8.2 that the mean level of aggression was higher in the older boys. An examination of the raw data in Table 8.1, however, suggests that most of the aggression scores were in fact quite similar in the two groups. The higher mean for the older children resulted largely from a few very high values. Or consider a comparison between the 3-year-old boys and the 3-year-old girls. From the means in Table 8.2 we might conclude that these two groups responded in the same way. Yet the raw scores in Table 8.1 tell us that the similar means had somewhat different underlying bases.

These examples suggest the need for some further descriptive statistics beyond the mean. Two other measures of central tendency are available. One is the **median**. The median is the midpoint of the distribution, the point above which half of the scores fall and below which the remaining half fall. Consider again the comparison between 3-year-old boys and 4-year-old boys. We can see in Table 8.2 that both groups had a median score of 4. This finding suggests a basic similarity between the two distributions, a similarity that was obscured by the differences in means. In general, the median is a useful statistic whenever a distribution is *skewed*—that is, contains a few unusually large

Table 8.2 Descriptive Statistics for the Aggressive Responses Reported in Table 8.1

	Mean	*Median*	*Mode*	*Standard deviation*
3-year-old boys	5.6	4	5	5.38
3-year-old girls	5.0	3	0	5.88
4-year-old boys	13.4	4	3	13.90
4-year-old girls	3.4	3	3	3.29
Boys combined	9.5	4	3	11.09
Girls combined	4.2	3	0	4.75
3-year-olds combined	5.3	3.5	0	5.55
4-year-olds combined	8.4	3	3	11.15

or a few unusually small values. In such cases, the mean may not give a representative picture of the typical level of response.

The third measure of central tendency is the **mode**. The mode is the most common score in a particular group. It is not an often used statistic, but in certain circumstances it can be informative. Look, for example, at the scores for the 3-year-old girls in Table 8.1. We saw earlier that this group averaged 5.0 aggressive responses, essentially the same as the 3-year-old boys. The modal response for the 3-year-old girls, however, was 0 aggressive acts, which was not true for the boys. This fact might well be worth reporting.

In addition to measures of central tendency, descriptive statistics are also used to summarize the **variability** of a distribution. We need to know not only what the central tendency is but also something about the extent to which scores either cluster around or depart from this central value. The most commonly calculated measure of variability is the **variance**. To find the variance, we begin by calculating the mean for the sample. We then determine the difference between this mean and each individual score in the group. These difference, or "deviation," scores are then squared, the squares are summed, and the total is divided by $N - 1$, at which point we have the variance. The variance is thus essentially the average of the squared deviation scores—"essentially" because the denominator is $N - 1$ rather than N.[1] The greater the differences between individual scores and the mean, the larger the variance will be.

Most journal articles do not report the variance but rather the **standard deviation** as the measure of variability. The standard deviation is simply the square root of the variance. Unlike the variance, which is in squared units, the standard deviation is in the same units as the original scores and the mean; it is thus a more readily interpreted descriptive statistic. This value is presented in Table 8.2 for each of the groups in our hypothetical study. These standard deviations confirm our earlier intuitions about the degree of spread in the various groups. Note, in particular, the relatively large standard deviation for the 4-year-old boys, the group with several extremely high scores.

Inferential Statistics

Suppose that we have found the means reported in Table 8.2. It appears that aggression may vary as a function of both age and sex. But how can we decide with any certainty whether the differences we have found are genuine and not simply chance fluctuations? This is an issue in **inferential statistics**.

To explain the need for inferential statistics, we need to return to a number of (partially overlapping) distinctions that were introduced in earlier chapters. One is the distinction between true scores and errors of measurement. Any score always consists of two components: the participant's actual value on the dimension being measured and whatever errors of measurement have occurred in attempting to determine this true score. A second distinction is between primary variance and secondary or error variance. Primary variance is variance associated with the independent variables being examined; secondary variance and error variance refer to all other sources of variation in the study—that is, all the other reasons, apart from the independent variables, that scores might differ. A final distinction is between populations and samples. A population is the entire universe of observations in which the researcher is interested; a sample is the particular subset of observations actually included in the study.

What we are interested in when we compare two samples (two age groups, two sexes, two experimental conditions, etc.) is whether there is a genuine difference between the populations from which the samples were drawn. If we could somehow collect the entire population of relevant observations, rather than just a sample, and if we could somehow rule out all errors of measurement, then we would have our answer: The scores that we obtain *would* be the population values of interest. But of course we cannot do these things; samples are always partial, measurement is always imperfect, and many sources of unwanted variance always exist. It is for this reason that we need techniques for estimating, or "inferring," the likelihood that an obtained difference between samples is indicative of a true difference between populations.

Let us apply the argument to the hypothetical aggression study and the issue of sex differences in aggression. We already know that there *was* a sex difference, in the sense that boys' scores and girls' scores were not identical. We also know, however, that various errors of measurement and extraneous sources of variation have entered into this difference. Furthermore, we have observed only a minute sample of the population in which we are interested—only 60 children out of the millions of 3- and 4-year-olds attending preschool in the United States, and only a few hours of these 60 children's behavior. Perhaps if we observed these same 60 children again, we would obtain somewhat different results. Perhaps if we observed a second sample of 60, we would obtain still different results. And perhaps if we could somehow observe the entire population of interest, we would obtain yet another set of results. It is to determine the probability of these various "perhapses" that we need inferential statistics.

The preceding paragraph expresses the two typical ways of thinking about the purpose of inferential statistics. One is in terms of replicability or reliability: If we performed the same experiment over and over again, would we obtain the same results? The second is in terms of moving from sample to population: Is the difference found in the sample sufficiently large to warrant the conclusion that there is a difference in the population? However we frame the question, we must decide between two possibilities: Either our results are genuinely reflective of what is true in the population, or our results have occurred because of chance factors operating in our particular study. And however we frame the question, inferential statistics cannot tell us with certainty which of these possibilities is correct; all that statistics can do is to establish a *probability* for the two alternatives. This, in fact, is a basic point to realize about statistical inference: Conclusions are always probabilistic, not certain.

It is time for a specific example of how statistical inference works. Consider again the apparent sex difference in aggression. We want to decide whether the difference we have found reflects a genuine population difference or whether it could have occurred by chance. As noted, the t test will serve as an initial illustration of how statistical inference proceeds.

The formula for calculating a t test is shown in Figure 8.1. The logic behind the test is quite

$$t = \frac{M_1 - M_2}{\sqrt{s^2_{pooled}\left(\frac{1}{n_1} + \frac{1}{n_2}\right)}}$$

Note: M_1 = mean for Group 1
M_2 = mean for Group 2
s^2pooled = weighted average of the variances for the two groups
n_1 = sample size for Group 1
n_2 = sample size for Group 2

Figure 8.1 Formula for calculating a *t* test

straightforward. The size of the *t* statistic—and therefore the probability that the results deviate from chance—depends on three things. One is the size of the difference between the means. The larger the difference, the larger the *t* will be. A second is the variability within each of the two groups being compared. It is the variability that is represented in the denominator. The smaller the variability, the larger the *t*. Finally, the third factor is the *n*, or sample size for each group. Sample size contributes to the calculation in two ways. First, as some thought about Figure 8.1 should reveal, sample size affects variability; the larger the *n*, the smaller the denominator in the *t* formula. Second, once we have obtained a *t*, we still need to determine the probability that a *t* of such magnitude could have occurred by chance. This probability depends on both the size of the *t* and the size of the sample: The larger the *n*, the lower the probability that a particular-sized *t* could have resulted from chance variations alone.

Let us now apply the *t* test to the male-female difference in the hypothetical study. Doing so yields a *t* of 2.41. If we now consult a *t* table (available in any statistics textbook), we find that a *t* of this size or greater could occur by chance fewer than 5 times in 100. This calculation of chance probability is based upon what is known as the **null hypothesis**—that is, an assumption that there really is no difference between the groups. By convention, outcomes whose probability of occurrence through chance alone is less than 5% are regarded as statistically significant. In such cases .05 is referred to as the **alpha level** for the study—that is, the cutting point for deciding that a result did not occur by chance. Because our results meet this criterion, we can reject the null hypothesis of no difference between the sexes and conclude that boys really are more aggressive than girls.

We will return shortly to the notion of statistical significance. First, however, it is worth reiterating the reasoning behind the *t* test, for the same general rationale applies to a number of different inferential tests. As noted, this reasoning is really quite straightforward, embodying three commonsensical rules:

1. Large differences between groups are less likely to occur by chance than are small differences. Thus, many of the other mean differences summarized in Table 8.2 (e.g., 3-year-old girls versus 3-year-old boys) are too small to yield a significant *t*, and are therefore best attributed to chance.

2. Differences based on chance alone are less likely when the within-group variability in response is small than when the variability is large. If few scores deviate very much from the group mean, there is simply less opportunity for chance deviations in a particular direction to affect the mean. This factor is relevant to the comparison between the 3-year-old boys and the 4-year-old boys. Despite a fairly large mean difference, a *t* test comparing these groups falls short of significance, largely because of the large variability shown by the 4-year-olds.

3. Finally, a difference found with large samples is less likely to be attributable to chance than is the same-sized difference found with small samples. When only a

few participants are involved, one or two extreme scores may have a disproportionate effect on the mean; with larger samples such chance fluctuations are more likely to even out. This factor too is relevant to the comparison between 3-year-old boys and 4-year-old boys. Had the sample size for this comparison been 30 per group rather than 15, then the particular t obtained *would* have been significant.

Some More About Significance

It should be clear from the preceding section that the purpose of inferential tests is to establish statistical significance. It is important to be clear about exactly what is meant—and also what is *not* meant—by the term "statistically significant."

Recall first that conclusions from inferential statistics are always probabilistic. A statement that a particular mean difference is statistically significant means that a difference of this size *probably* did not occur by chance, assuming the null hypothesis of no population difference. There is, however, always the possibility of error. Two kinds of errors can occur. One consists of falsely rejecting the null hypothesis—that is, concluding that an effect exists when in fact none does. This type of error is called a **Type 1 error**. In our aggression study we would be making a Type 1 error if we concluded that boys and girls differ in aggression when in fact there is no such difference in the general population. The probability of a Type 1 error is determined by the probability level at which we reject the null hypothesis. If the probability level is .05, then we have 5 chances in 100 of making a Type 1 error. If the probability level is lower—say .01 or .001—then clearly our chances of being wrong are greatly reduced.

The second type of error consists of failing to reject the null hypothesis when there is in fact a genuine effect. This type of error is labeled a **Type 2 error**. In the aggression study we would be making a Type 2 error if 3-year-olds and 4-year-olds really do differ in aggression but we falsely conclude that they do not. The probability of a Type 2 error is more difficult to calculate than that of a Type 1 error, and no attempt is made to explain the calculation here. I will note, however, that the probabilities for the two error types vary inversely; that is, as one type of error gets less likely, the other type gets more likely. A researcher could, for example, reduce the chances of making a Type 1 error by insisting on a probability level of .001; at the same time, however, the researcher would be greatly increasing the chances of making a Type 2 error. I will note too that psychologists have generally opted for minimizing Type 1 errors. This conservativeness in drawing positive conclusions is reflected in the general convention that only results whose chance probability is less than 5% merit the label "significant."[2]

Table 8.3 summarizes the distinctions just made. Note, of course, that we never know what the reality is; if we did, then there would be no possibility of either error type.

The discussion of Type 1 and Type 2 errors brings us back to the concept of validity. Chapter

Table 8.3 Possible Outcomes of Statistical Inference

	Reality	
Decision	*Genuine effect*	*No genuine effect*
Reject null hypothesis	Correct decision	Type 1 error
Do not reject null hypothesis	Type 2 error	Correct decision

2 discussed three of the four basic forms of validity. The fourth form is **statistical conclusion validity**: the accuracy of the statistical conclusions that are drawn from the analyses of the data. Are we correct in the inferences that we make about the relations—or absence of relations—among our variables? If we can avoid both falsely concluding that a relation exists when none in fact does (the Type 1 error) and falsely concluding that there is no relation when one in fact obtains (the Type 2 error), then we have achieved statistical conclusion validity.

A finding of statistical significance allows us to say that our results are probably not due to chance. It is important to realize that the significance test is directed *only* to the possibility of chance variations. The significance test cannot rule out other possible threats to validity. It can tell us that two groups differ, but it cannot tell us why they differ.

Consider the male-female difference in our aggression study. We are interested in the possibility that this is a genuine sex difference in behavior. But a significant difference could easily come about for other reasons. Perhaps our observers expect boys to be more aggressive than girls and then slant their observations accordingly—hence a difference based on observer bias. Perhaps girls are more affected by the presence of the observer and therefore more likely to inhibit aggression while being watched—hence a difference based on differential reactivity. Perhaps we observe the girls early in the school year and the boys only later in the year when aggression has become common—hence a difference based on a confounding of group and time of measurement. The point is that any of the various threats to validity discussed throughout this book could still be operating to bias our results. A finding of statistical significance is no guarantee of overall validity. It is merely the starting point, a necessary but not sufficient basis for concluding that we really have found something.

A final caution is that a finding of significance is no guarantee of the meaningfulness of the results. "Significance" in the sense used here refers only to statistical probability, not to theoretical or practical importance. A sex difference in aggression may be genuine, in the sense that it did not occur by chance or because of some invalidity in the study. Whether the difference is large enough to mean anything, however—with respect, for example, to how preschool teachers should treat boys and girls—is a separate question. It is important to remember that the statistical significance of a difference depends not only on the size of the difference but also on the size of the sample. With a large enough sample, even a tiny difference may achieve significance. Conversely, if the sample is too small a genuine difference may fail to reach significance.

These and other limitations of significance tests have been much discussed in recent years, and various alternative approaches have been proposed, the most radical of which involve abandoning such tests altogether (Harlow, Mulaik, & Steiger, 1997; Killeen, 2005; Kline, 2004; Nickerson, 2000). Less radically, there is general agreement that two further statistics should generally be added as supplements to, if not replacements for, the significance test.

One is the presentation of confidence intervals. Let us suppose that the finding of interest is the difference between two means, such as that between boys and girls in the Table 8.2 data. As we saw, a *t* test tells us that a difference this great would occur less than 5% of the time by chance, assuming that there is no difference in the population. The specific difference found (in this case, 5.3) is referred to as a *point estimate*—our best estimate of the population difference, given the sample values. A **confidence interval** adds a range of possible outcomes to this estimate. If we are working with the .05 significance level, then the confidence interval tells us the range within which the true population value can be assumed to fall with 95% certainty. In the case of the Table 8.2 data the range is from a difference of .9 to a difference of 9.7 (Kline, 2004, gives the formula for calculating a confidence interval). A confidence interval thus provides more information than does a significance test, and it does not force us, as does a significance test, into a single accept/reject decision.

The second proposed addition is a measure of **effect size**—that is, a measure of how strong the relation is between the independent variable and the dependent variable. Significance tests, as we have seen, do not tell us this; all they tell us is that there is (probably) some greater-than-chance relation. It is possible, of course, to get some idea of the magnitude of an effect simply by looking at the means; a large difference between means obviously reflects a greater effect than a small difference. But is it possible to derive more rigorous measures of effect size?

The answer is that a number of techniques now exist for calculating effect size. Among the primary sources for descriptions of these techniques are Cohen (1977), Rosenthal (1994), and Tatsuoka (1993). Here I describe the simplest of the various procedures, developed by Cohen (1977). We will return to the notion of effect size later in the box devoted to meta-analysis.

With the Cohen approach, effect size, or *d*, is defined as the difference between two means divided by the standard deviation of the groups being compared. The approach thus takes into account the mean difference but weights it in terms of the variability in the scores. The smaller the variability, the more impressive any mean difference is. In the case of the boy-girl difference in aggression (Tables 8.1 and 8.2), calculation of effect size yields a *d* of .62, a value that is usually considered as medium in magnitude.

Seitz (1984) provides another example. The standard deviation for most IQ tests is 15 points. A mean IQ difference between groups of 12 points (as is reported to exist, for example, between college freshmen and people with PhDs) would therefore yield an effect size of .8. If the mean difference were 7 points, the effect size would be .5; if the difference were 3 points, the effect size would be .2.

One way to interpret what these various values mean is to plot the extent to which scores for the two populations overlap. Curves for the three situations just described are shown in Figure 8.2. Note that the degree of overlap decreases as the effect size increases. As Seitz notes, the middle pair of curves may be especially informative. A mean difference of 7 points—or, more generally, a difference of one-half standard deviation—may not seem very large. What such a difference means, however, is that 70% of one population will score above the mean of the other population.

Measures of effect size can provide useful information that is not directly given by most inferential tests. The APA *Publication Manual* (APA, 2001) recommends their use, and so do the editorial policies of most journals in the field. Thus far, however, articles that provide effect sizes remain in the minority—a conclusion that applies also to the use of confidence intervals (Fidler, Thomason, Cumming, Finch, & Leeman, 2004; Kirk, 2001). This situation almost certainly will change in the next generation of research reports, and my advice is to watch for—and perhaps contribute to—the change.

Choosing a Statistical Test

To many students "statistics" connotes memorization of formulas and endless hours of painful calculations. In fact, most professional researchers remember few if any formulas, and most spend little time in calculations. There is simply no need; the formulas can be looked up in texts or programmed into a computer, and the calculations can be performed on a calculator or a computer (or by a student assistant!). What is much more important is to know which kinds of statistical analysis are appropriate and informative for which kinds of data. Many

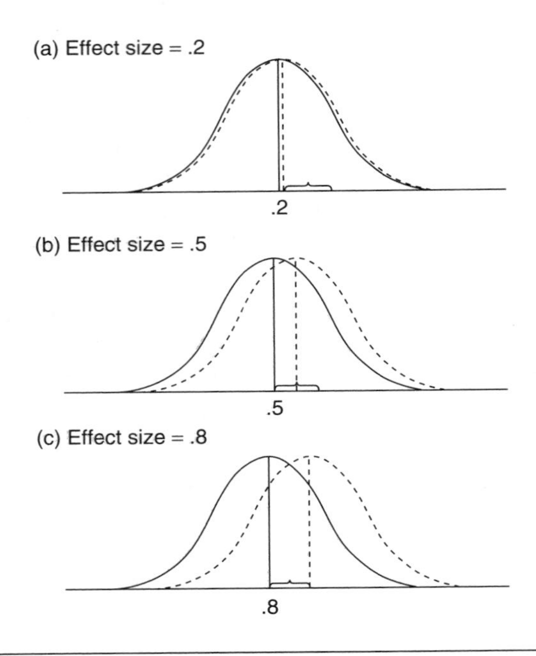

Figure 8.2 Population differences corresponding to different effect sizes

factors go into determining which statistic is best to use. In this section we will consider three such factors: the level of measurement at which the dependent variable has been assessed, the distribution of scores shown by the dependent variable, and the design of the study.

Level of Measurement

The notion of levels or scales of measurement was introduced in chapter 4. Recall that measurement theorists identify four levels of measurement: *nominal,* or the assignment of qualitative labels to the observations; *ordinal,* or the rank ordering of the observations along some quantitative dimension; *interval,* or the placement of the observations along a quantitative dimension whose points are not only rank ordered but equally spaced; and *ratio,* or the placement of the observations along a quantitative dimension with not just equal spacing but a true zero point.

The level of measurement is one factor that determines which inferential test is appropriate to use. Some tests, including the *t* test, require that the measurements constitute either an interval or a ratio scale. One reason for this requirement should be evident from a consideration of the formula in Figure 8.1. To calculate a *t* test we must perform various arithmetical operations on the numbers that we have obtained—adding them together and then dividing to get the mean, subtracting each number from the mean to get a deviation score, and so forth. These operations make sense only if we can assume that the numbers we are dealing with are accurate reflections of the quantities involved and not merely labels or ordinal rankings. The frequency scores in Table 8.1 meet this criterion, and the *t* test is therefore appropriate for these data. A *t* test would not be appropriate, however, if our data came from the kind of rating scale described earlier. We could not, for example, add together a rating of 5 ("very aggressive") and a rating of 1 ("very unaggressive") to get an average of 3 ("moderately aggressive"). (I will note some qualifications to this statement shortly. Also, you should be aware that not all measurement theorists and statisticians agree on the proper relation between scales of measurement and statistics—see Cliff, 1993; Michell, 1986.)

Distribution of Scores

Some inferential tests make assumptions about the distribution of scores that go into the test. In particular, so-called **parametric tests** are dependent on certain assumptions about how the data are distributed. This, in fact, is the meaning of "parametric": that the test depends on the validity of certain assumptions about "parameters" of the population from which the sample is drawn. The *t* test discussed earlier is one example of a parametric test; the analysis of variance (ANOVA) that I consider in the next section is another example.

More specifically, two assumptions about distribution underlie the use of most parametric tests. One assumption is that the scores are normally distributed. The second is that the variances of the groups being compared are equal. The second assumption is less pervasive than the first, but it does apply to many often-used parametric tests, including the *t* test and the ANOVA.

I have already discussed the concept of variance. Let us consider for a moment the requirement of a **normal distribution**. The (a) portion of Figure 8.3 illustrates a normal distribution. "Normal" refers to the classic bell-shaped curve, a distribution in which the mean, median, and mode are synonymous, and in which the scores slope gradually away from this center point on each side. The curves in parts (b) and (c), in contrast, represent distinctly nonnormal distributions.

There is a relation between level of measurement and distribution. Scores obtained from a nominal or ordinal scale cannot be normally distributed. With a nominal scale there is no quantitative significance to the measures and thus no question of their distribution along some quantitative dimension; all that is possible is a frequency count of the number of cases in each category. With an ordinal scale there is no way to know the distance between scores and hence no way to determine their true distribution. Furthermore, in a perfectly ordered scale (i.e., one with no ties) there can be just one instance at each level of the scale; thus, theoretically, the distribution from an ordinal scale must always be a flat one. A ratio or interval scale, therefore, is a necessary basis for a normal distribution. It is not a sufficient basis, because the scores might still look like those in

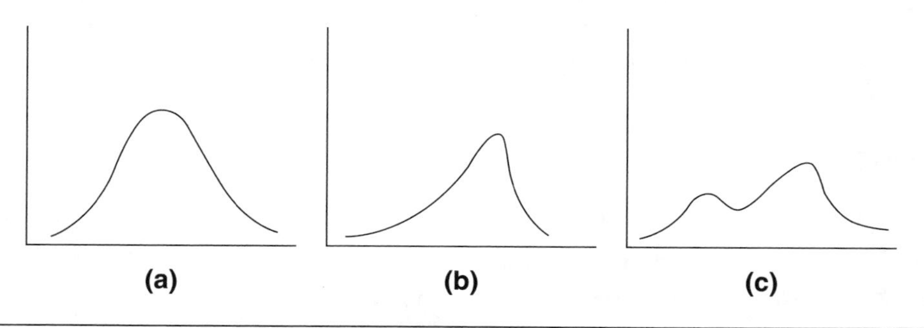

Figure 8.3 Examples of normal and nonnormal distributions

Logic of the chi square test: Determine the extent to which the observed frequency in each cell deviates from the expected frequency assuming no difference between groups.

Observed Frequencies:	Cars and Trucks	Construction Toys	Dolls and Furniture	Art Supplies
Boys	12	10	1	5
Girls	1	4	12	11
Expected Frequencies:				
Boys	6.5	7	6.5	8
Girls	6.5	7	6.5	8

$$\text{Formula: } \chi^2 = \Sigma \frac{(\text{observed frequency} - \text{expected frequency})^2}{\text{expected frequency}}$$

$\chi^2 = 23.42$, probability $< .01$

Conclusion: There is a significant difference in toy preference between boys and girls.

Figure 8.4 Illustration of the chi square test

Figures 8.3 (b) or (c). But it is only if the scores are from the right sort of scale that the possibility of a normal distribution even arises.

I have been discussing assumptions that underlie the use of parametric tests such as the *t* test and ANOVA. Let us consider for a moment the alternative to such parametric tests, after which several further points can be made concerning choice of statistic.

As might be guessed, the alternatives to parametric tests are called **nonparametric tests**. Figure 8.4 illustrates a widely used nonparametric test, the chi square. The hypothetical data in the figure came from the toy-preference study sketched earlier; the hypothetical finding is that toy preference varies significantly as a function of sex.[3] The chi square is intended for nominal data such as those in the figure, data for which the *t* test would not be appropriate. Nonparametric tests exist for data at each of the four levels of measurement: nominal, ordinal, interval, and ratio. The approach is thus more generally applicable than is the parametric approach. Furthermore, nonparametric statistics do not depend on the assumptions about distribution that underlie parametric statistics; thus, nonparametric tests may be applicable to interval or ratio data that fail to satisfy parametric assumptions. (Among the texts devoted to nonparametric statistics are Gibbons, 1993; Marascuilo & McSweeney, 1977; and Siegel & Castellan, 1988.)

How do researchers decide between parametric tests and nonparametric tests? As just indicated, in some cases there is no decision to make, for the only possible test is a nonparametric one. In other cases a decision is necessary, and here several concepts become relevant. I discuss two: power and robustness.

The term **power** refers to the probability that an inferential test will reject the null hypothesis when the null hypothesis should in fact be rejected. The more powerful the test, the more likely it is to detect a genuine difference and thus lead to a correct rejection of the null hypothesis. This concept should sound familiar, for power is simply another way of talking about the probability of Type 2 error. The more powerful a test, the lower the probability of making a Type 2 error.

In some cases, parametric tests are more powerful than comparable nonparametric tests. Essentially, this greater power comes about because a parametric test uses more information about the data than does a comparable nonparametric test. Many nonparametric tests, for example, are limited to ordinal properties of the data—in particular, the relative rank orders of scores in the samples being compared. A *t* test, in contrast, is able to make use of the actual scores and the absolute differences between them; it can therefore sometimes reveal differences not revealed by a nonparametric test. It should be added that the difference in power is often small, and that the difference is found primarily when the samples are large. Furthermore, the difference is not inevitable. There are many situations in which parametric and nonparametric statistics are equally powerful. If the assumptions that underlie the parametric test are seriously violated, then the nonparametric alternative may actually be more powerful.

The point about assumptions leads to the concept of robustness. **Robustness** refers to the impunity with which the assumptions that underlie a test can be violated. A robust test is relatively insensitive to the violation of its assumptions—that is, it tends to yield accurate conclusions about significance even when the assumptions are not met. It turns out that both the *t* test and the ANOVA are fairly robust tests. It is for this reason that such tests are so common in the literature even for data that do not fit the criteria that I have been discussing—data that come from rating scales, for example, or that depart markedly from a normal distribution, or that reflect unequal variances in the groups being compared. This robustness does *not* mean that the researcher can automatically apply a parametric test to any sort of data; it does mean, though, that the possibility of a parametric test should not be too quickly abandoned just because some assumption of the test is violated. It may be worthwhile, rather, to seek some expert advice with respect to whether parametric statistics might nevertheless be applicable to the data.

Design of the Study

I have discussed two determinants of the choice of statistic: level of measurement and distribution of scores. A third important consideration is the design of the study.

Various aspects of the design are relevant. One factor is the number of levels of the independent variable. In our example of aggression in the preschool, this factor was kept simple: two different ages and two different sexes. It was easy enough then to run *t* tests comparing the two levels of each variable. Suppose, however, that we complicate matters by adding more levels. Because it is a bit difficult to imagine doing so for sex, we will make the additions for age. Suppose that instead of two age groups, we have six. What happens then to our *t* test?

The most obvious thing that happens is that we need many more tests. With six different age groups there are 15 possible pairwise comparisons. We therefore need to run 15 separate *t* tests to determine what we have found. Running, and reporting, 15 *t* tests is obviously rather unwieldy. The more important objection, however, concerns the level of significance for the tests. We want this level to remain fixed at whatever cutoff point for significance we have decided upon—for example, the traditional .05 level. With multiple tests, however, the interpretation of significance becomes very difficult. If we run 15 tests, each at a .05 level, then we have a .54 probability that at least one of these tests will be significant by chance alone.[4] How then can we interpret any statistically significant result?

The problem is actually more complicated still. The .54 probability rests on the assumption that the 15 *t* tests are independent of each other. Often, however, multiple *t* tests in the same study are not independent; rather, they are related, in the sense that the same data enter into a number of different comparisons. This, in fact, would be the case in the age comparisons sketched earlier; each of the six age groups would contribute data to 5 of the 15 tests. When this sort of interdependency among tests exists, it becomes impossible to determine the exact probability level for any one test. The researcher may calculate a *t* and report significance at the .05 level; the actual probability level, however, might be quite different from .05.

There is a further problem with multiple *t* tests. Suppose now that we complicate our study not by adding more levels of an independent variable but by adding more independent variables. In addition to studying age and sex as determinants of aggression, we might examine possible effects of classroom structure, of indoor versus outdoor locus for the observations, of presentation of a violent cartoon to half the children prior to the observation period, and so forth. Clearly, the more independent variables we add, the more possible *t* tests there will be. But there is another problem beyond simply the proliferation of tests. Whenever multiple independent variables are studied, it is possible that the effects of one variable will depend on the level of the other variables. It is possible, in short, that there will be an *interaction* of variables. Such interaction effects are important to identify, but they can be difficult to uncover through *t* tests alone.

The most popular alternative to multiple *t* tests is the *analysis of variance* or *ANOVA*. The ANOVA is essentially an extension of the *t* test to situations involving more than two means. The method of calculation is different from, and more complicated than, that of the *t*, and no attempt will be made to describe the calculation here. The underlying logic, however, is identical: We test for significance by determining the extent to which the primary variance associated with the groups being compared exceeds the secondary and error variance within groups. The statistic that the test yields is called an *F*, and the *F*, like the *t*, can be checked for significance in standard tables found in any statistics textbook.

Let us consider how the ANOVA would be applied to the aggression study. Because we have two independent variables, age and sex, the analysis is referred to as a *two-way ANOVA*. To make the superiority of ANOVA to *t* clearer, we will assume that there are in fact six levels of the age variable, rather than just the two represented in Table 8.1. Application of the ANOVA will then yield *F*s for each of the independent variables, or *main effects*, of the study. If the *F* for sex is significant, then our work with regard to this variable is complete; because there are only two levels of sex, we can simply look at the means to determine where the effect comes from. A significant main effect for age would be more complicated. In this case the *F* would be based on a simultaneous comparison of all six age groups, and significance would mean that at least one of the possible pairwise comparisons among these groups is significant. We would then need follow-up tests to determine which specific comparison or comparisons are significant. These follow-up tests are similar to a *t* test but typically more conservative, and they are run only after the overall *F* has indicated significance.

The ANOVA will also yield a third *F*: that for the interaction between age and sex. In general, an ANOVA will yield interaction *F*s for all of the possible combinations of independent variables in a study. If there are three independent variables (hence a three-way ANOVA), for example, then the ANOVA will give four interaction *F*s: one for each of the three two-way combinations of variables and one for the three-way combination. As with a significant main effect, significant interactions can be followed up by specific tests to determine the exact basis for the effect.

One final aspect of design is relevant to the choice of statistic. Thus far the focus has been on between-subject designs—that is, cases in which each participant contributes data to only one of the conditions or groups being compared. But we saw in chapter 3 that many independent variables can be examined via within-subject designs, in which each participant contributes data to each experimental condition. What happens to our statistics when each participant is represented in each condition?

The answer is simple enough: We merely switch from the between-subject tests that have been discussed thus far to comparable within-subject tests. There are, in fact, analogous within-subject tests for each of the between-subject procedures that has been considered. There is a within-subject *t* test, for example, as well as a within-subject or repeated-measures analysis of variance. There are also nonparametric tests that are appropriate for within-subject data (for example, a test called the McNemar test of change, which constitutes a kind of repeated-measures chi square). The logic of such tests is similar to that of between-subject tests; in most within-subject tests, however, actual difference scores (e.g., a participant's performance in Condition 1 minus the performance in Condition 2) are the grist for the analyses. Because difference scores are the focus, the tests are appropriate not only for literal repeated-measures designs but also for cases in which each participant in one condition has been matched with a comparable participant in another condition.

One further point can be made concerning within-subject statistics. It is a reiteration of a point made in chapter 3 in the discussion of the relative merits of within-subject and between-subject designs. As we saw then, within-subject tests are often more powerful than comparable between-subject tests. This greater power comes about because of the reduction in secondary variance associated with participants. If the same participants appear in each condition, then there is less possibility for preexisting differences among participants to contribute unwanted variance to the group comparisons. This greater power provides one possible basis for deciding between a within-subject approach and a between-subject approach.

Correlations

Thus far the emphasis has been on tests whose purpose is to identify differences among groups. Not all statistical tests fit this model. Imagine a study in which we collect the data shown in Table 8.4. We are interested in determining whether there is a relation between IQ and performance on standardized achievement tests. How can we proceed?

The statistic that is appropriate for the data in Table 8.4 is the **correlation statistic**. A correlation statistic is a measure of the relation between two variables. As we saw in chapter 3,

Table 8.4 IQ and Achievement Test Scores for a Sample of Fifth-Grade Children

Participant	*IQ*	*Achievement test*
1	82	22
2	85	18
3	90	43
4	92	28
5	95	23
6	99	24
7	101	48
8	102	30
9	104	56
10	107	35
11	108	38
12	112	46
13	114	27
14	116	54
15	124	50
16	140	60

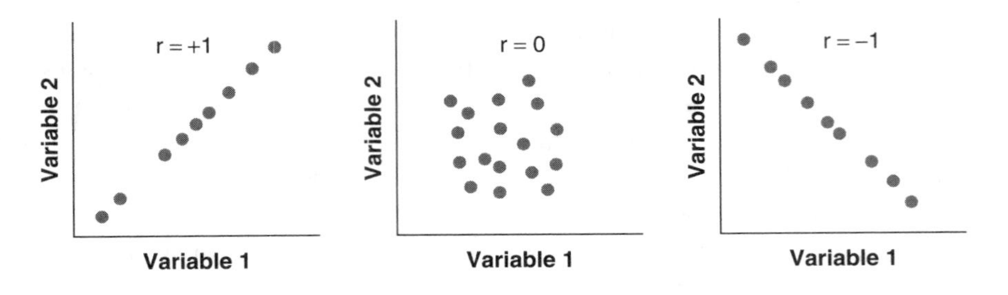

Figure 8.5 Scatter plots illlustrating different degrees of correlation

correlations have a possible range of +1 to −1. A correlation of +1 indicates a perfect positive relation between variables, a correlation of 0 indicates no relation at all, and a correlation of −1 indicates a perfect negative relation. These possibilities are shown pictorially in the **scatter plots** in Figure 8.5. Correlations between 0 and 1 or 0 and −1 indicate varying degrees of positive or negative relation, with the strength of the relation increasing as the value approaches either 1 or −1.

What about the data in Table 8.4? To determine the correlation for these data, we must first decide which correlation statistic to use, for there are in fact several different methods of calculating correlations. Just as with inferential tests, which method is appropriate depends on certain assumptions about the nature of the data. The two most often-used tests are the Pearson product-moment correlation and the Spearman rank-order correlation. The Pearson is a parametric test that shares assumptions common to all parametric tests—namely, that the measures come from an interval or ratio scale and that the scores are normally distributed.[5] The Spearman is a nonparametric test that is based solely on the ordinal properties of the data and hence is more widely applicable than the Pearson. Both tests, it should be noted, are dependent on another important assumption: that the relation between variables is a *linear* one. If the relation is something other than linear (for example, a curvilinear relation in which scores on one variable first rise and then fall in conjunction with scores on the other), then a standard correlation test would not be appropriate.

Because it is the simpler of the two tests to illustrate, I will apply the Spearman to the data in Table 8.4. The formula for the Spearman, as well as its application to the illustrative data, is shown in Figure 8.6. As can be seen from a consideration of the formula, the Spearman is a measure of the extent to which the paired scores in the two distributions share the same rank order. If the agreement in rank is perfect, then there will be no deviation scores, the solution on the right side of the equation will be 0, and the correlation will be +1. The more frequent and marked the deviations in rank, the further the correlation will depart from +1. In our sample data the correlation between IQ and achievement scores is .70, indicating a fairly strong but far from perfect positive relation. It is worth noting that application of the Pearson product-moment correlation to these data yields a very similar value: .71. For most data sets, in fact, the Spearman and the Pearson give quite similar values.

What do we know when we have established the correlation between two variables? A correlation, like a mean or median, is a descriptive statistic, a description in this case not of central tendency but of the association between variables. Before it can be interpreted, it must be tested for statistical significance. The null

Logic of the Spearman rank-order correlation: Determine the extent to which the paired scores in the two distributions share the same rank order.

Rank Order			
IQ	Achievement Test	IQ	Achievement Test
1	2	9	15
2	1	10	8
3	10	11	9
4	6	12	11
5	3	13	5
6	4	14	14
7	12	15	13
8	7	16	16

Formula: $\text{rho} = 1 - \dfrac{6 \sum D^2}{N(N^2 - 1)}$

Note D = deviation score, i.e., the absolute difference between ranks

rho = .70, probability < .01

N = number of pairs

Conclusion: There is a significant positive correlation between IQ and achievement test score.

Figure 8.6 Illustration of the Spearman rank-order correlation

hypothesis for such tests is that of a 0 correlation between variables; the question then is whether the obtained correlation deviates significantly from 0. Answering this question is easy enough, for statistics texts provide tables in which the probability level for correlations of any magnitude can be checked directly (many computer programs also indicate probability levels). Both the size of the correlation and the size of the sample affect the determination of significance; as either increases, the probability of significance also increases. Reference to such a table tells us that a correlation of .70 with a sample size of 16 (i.e., 16 *pairs* of scores) is significant at the .01 level; thus, there *is* a relation between IQ and achievement.

Significance is important, but it is only part of the story. Recall that we are interested not just in the existence of a relation but also in the strength of the relation. The usual way to interpret the strength of a correlation is in terms of goodness of prediction: Given that we know the participant's score on one variable, how well can we predict the score on the other variable? If the correlation is 0, then the relation between the variables is random, and knowing one score adds nothing to our ability to predict the other. As the correlation departs from 0, its predictive power increases, the extremes being the perfect correlations of +1 or −1.

Another (equivalent) way to think about correlations is in terms of variance accounted for. When we use scores on one measure to predict scores on a second measure, we are "accounting," in a statistical-predictive sense, for a certain proportion of the variance on the second measure. The higher a correlation, the more variance accounted for. We can be more precise than this, however. When the correlation statistic is the Pearson *r*, then the proportion of variance accounted for is equivalent to r^2. Thus, the correlation of .71 between IQ and achievement means that variations in one measure account for 50% of the variations in the other.

This last statement should have a rather sobering sound. A correlation of .71 seems high, yet it still leaves half of the variance unaccounted for. As the correlation drops toward 0, the proportion of variance accounted for also drops, and quite rapidly. A correlation of .5 accounts for 25% of the variance. A correlation of .3 accounts for only 9% of the variance.

This discussion raises again the distinction between significance and meaningfulness. A correlation may be statistically significant but also so small that it has little theoretical or practical meaning. Such statistically significant but trivial correlations are especially likely when the sample size is large. With a sample of 50, a correlation of .27 achieves significance at the .05 level. With a sample of 100, a correlation of .19 is significant.

In addition to sample size, another important variable to consider in evaluating correlations is the range of scores on the two measures. Two sorts of problems with range can arise. Probably the more common is the problem introduced in chapter 4: restriction of range. Restriction of range occurs when scores on one variable are so close together that the differences among them do not relate to variations on other measures. Suppose that we decided to limit the sample for our IQ-achievement comparison to children attending "gifted" or "enrichment" classes. There is typically an IQ cutoff for such classes; let us say in this case that the criterion is an IQ of 130 or above. The decision to focus only on the very high IQs means that we have greatly restricted the range of variation for one of our measures; instead of a range of perhaps 60 to 70 IQ points, our actual range might be about 20 points. When all the IQ scores are so closely bunched, it is unlikely that the small differences among them will show much relation to anything else, including differences in achievement.

It is also possible to have a range that is too broad rather than too narrow. Suppose now that we sample IQ in 20-point jumps, beginning with a child whose IQ is 40, moving next to one whose IQ is 60, and continuing until we reach an eighth child whose IQ is 180. With such enormous variability it is likely that IQ will correlate significantly and substantially with almost any other psychological measure that we might collect from our sample. It is doubtful, however, that the size of such correlations would mean very much.

Whether the problem is too little range or too much range, the underlying issue is one of external validity. For a correlation to mean anything, it must be generalizable beyond the immediate sample on which it was calculated to some broader population of interest. The sample must therefore be representative—in both central tendency and range of variation—of the larger population. If the sample is not representative, then correlations obtained with it cannot have much external validity.

Alternatives to and Extensions of ANOVA

In this section we return to the issue of comparisons among groups. It seems safe to say that analysis of variance has been the most widely used statistical technique for making such comparisons for at least the past 50 years. Despite its popularity, however, ANOVA is not always the best statistical choice, even in cases in which the assumptions behind the test can be met. In this section I briefly discuss two of the alternatives to ANOVA.

Planned Comparisons

I noted earlier the problems of multiple t tests and the consequent superiority of analysis of variance as an overall test for the presence or absence of significant effects. Overall tests are not always superior, however. ANOVA is most clearly appropriate when the investigator does not have clear-cut hypotheses to test but is interested instead in identifying any interesting results that may emerge from the research. In some studies, however, there *are* definite hypotheses, and the main goal of the statistical analysis is to provide a clear answer with respect to each hypothesis. In these cases ANOVA may be an inefficient form of analysis, for it will include comparisons that are not of interest and yield less power for the comparisons that the researcher cares about.

Consider an example (adapted from an example in Hays, 1981). We are interested in the effects of training on response to Kohlberg's moral dilemmas (see chapter 13). We decide to examine two forms of training in children: exposure to an adult model who demonstrates advanced forms of moral reasoning and peer group discussions in which moral issues can be debated. We are also curious about the possible combined effects of these two experiences, and hence we include a third training condition in which children receive both modeling and peer group interaction. We know that to evaluate the effects of our training we will need a control group; thus, a fourth condition consists of just the pretest and posttest with no experimental intervention. We have some concern, however, that mere exposure to a model or a peer group may affect response, quite apart from the moral content on which we wish to focus. Hence we include two more control conditions: one in which children see an adult model talk about something other than morality and one in which children engage in peer discussions of some topic other than morality.

In a study like this we are not really interested in a main effect of condition, an effect that could result from a significant difference between *any* of the possible pairs of means. Our interests, rather, are both more specific and more limited, in that we know in advance that there are only certain comparisons between means that are important to examine. We will probably want to know, for example, whether each experimental condition differs significantly from its appropriate control condition, as well as whether the three experimental conditions differ among themselves. These comparisons make sense. Other comparisons make considerably less sense—for example, the peer group experimental condition versus the adult control condition. An overall *F* test lumps all of these comparisons together. We could, of course, begin with the *F* and then go on to specific follow-up tests of the comparisons of interest. There is always the danger, however, that the main effect *F* will not be significant, in which case we really have no warrant for follow-ups. Furthermore, the tests that follow a significant ANOVA are conservative and hence low in power, which means that we run the risk of failing to detect some genuine and important effect.

An alternative to ANOVA in such a situation is to use *planned comparisons.* With planned comparisons we specify in advance which of the means will be compared, and *only* these comparisons are carried out. In our hypothetical training study, for example, we could draw the comparisons between treatments that we wish to draw while ignoring other comparisons of less interest. The techniques for making such specific, in-advance comparisons are beyond the scope of this book; descriptions can be found in most statistics textbooks (e.g., Hays, 1981). There is, of course, some potential loss of information entailed in such a selective approach. If we really know what we want to look for, however, the loss may be minimal. And because in-advance tests are more powerful than post hoc tests, we have a better chance of gaining clear answers to the questions of interest.

It is important to stress that planned comparisons do imply planning and selectivity; that is, we cannot look at *every* possible contrast. How many contrasts can be examined has been a subject of dispute among statisticians. Keppel (1991) provides a helpful discussion of the different viewpoints, as well as a set of techniques for adjusting the probability level in cases in which multiple and overlapping comparisons are made.

Multiple Regression

In the words of Kerlinger and Lee (2000), multiple regression "is a method for studying the effects and the magnitudes of the effects of more than one independent variable on one dependent variable using principles of correlation and regression" (p. 755). In slightly different words, what multiple regression does is to tell us how two or more independent variables, or "predictors," relate to a single dependent variable, or "criterion." We might do a study, for example, in which we examine performance on a laboratory problem-solving task as a function of IQ, SES, nature of the task instructions, and time allowed to reach a solution. Applying multiple regression to our data will give us an estimate of the contribution, both singly and in combination, of each of these four predictor variables to variations in the criterion variable.

There is a close relation between regression and the correlation statistic discussed earlier in the chapter. In both cases, we use knowledge of a participant's values on one variable to predict scores on another variable. As the inclusion of IQ and SES in the example suggests, a further similarity is that regression is applicable to variables that, from a design point of view, are correlational, in that they are merely measured rather than experimentally manipulated.

There is also a close similarity between regression and analysis of variance. Both procedures have the same goals: to determine the effects, both singly and in interaction, of a set of independent variables upon some dependent variable. The study sketched earlier as an example of multiple regression could also be analyzed via a four-way ANOVA, with four main effect *F*s and a large number of interaction *F*s as well. This conclusion is, in fact, a quite general one: Most problems that lend themselves to multiple regression could also be analyzed with analysis of variance. As many authors have pointed out, however, regression is the more general of the two concepts, because ANOVA constitutes a special case under the general heading of techniques of regression. Thus, *any* problem that can be analyzed as a multifactor ANOVA can also be analyzed with multiple regression.

Given the wide applicability of multiple regression, what factors might lead us to prefer regression to ANOVA as a method of analyzing some data set? I will note first that such decisions are often a matter of personal preference, for in many cases the two approaches are equally appropriate and informative. There are some ways, however, in which multiple regression does have advantages. I note here three of the arguments in favor of regression.

1. Regression is especially appropriate when an independent variable is continuous—that is, encompasses a large number of different values rather than just a few discrete levels. IQ is an example of a continuous variable. If we included IQ in our problem-solving study, we would probably obtain a range of scores spanning some 60 or 70 IQ points. There is no way with ANOVA to capture all this variation; the best that we are likely to do is a gross classification system such as "high," "medium," and "low." With regression, however, there is no loss of information, for the variable is treated continuously, and every distinct score enters into the analysis.

2. Regression is especially appropriate when two or more of the independent variables are correlated. This point ties into the earlier point about the suitability of regression for correlational designs. In a completely experimental design there is no correlation among the independent variables; rather, one aspect of experimental control is to ensure that the levels of one variable (e.g., age) are equally distributed across the levels of another variable (e.g., experimental condition). In a

correlational design the levels of the variables are measured rather than experimentally assigned, and there is therefore no guarantee that such independence will hold. In the problem-solving study, for example, IQ and SES might well turn out to be correlated. Regression provides a means (similar to the partial correlation technique discussed earlier) to control for one variable when examining the effects of another—to determine, for example, whether SES makes a contribution even when IQ is statistically held constant.

3. Regression is especially well suited to the issue of effect size discussed earlier in the chapter—that is, determining the magnitude of the effects that the independent variables have upon the dependent variable. The basic statistic that multiple regression yields is R^2: the proportion of variance in the dependent variable that is accounted for by the set of independent variables under study. R^2, then, tells us how well our predictors "work"—how well we can predict variations in the criterion given the particular set of independent variables that we have selected. Although doing so is by no means simple or straightforward, it is also possible with regression to derive separate estimates of the contribution of each of the individual predictor variables to the criterion, both the size of the contribution and the form of the relation between predictor and criterion.

Among the textbooks devoted to multiple regression are Cohen, Cohen, West, and Aiken (2003) and Pedhazur (1997). Among the general methods texts that provide good summaries are Kerlinger and Lee (2000) and Pedhazur and Schmelkin (1991).

FOCUS ON

Box 8.1. Meta-Analysis

Not all uses of statistics are in the context of individual studies. In recent years statistical analyses have increasingly been brought to the task of research review—that is, attempts to synthesize bodies of related research. Such quantitative approaches to research synthesis are referred to as **meta-analysis**.

In introducing meta-analysis it is helpful to contrast it with the alternative approach to synthesizing research. The alternative is a traditional narrative review (which is what you have encountered in any textbook you have ever read). In a narrative review the author identifies the general topic to be considered, as well as the specific questions and theoretical positions that will be the focus of the review. The subsequent review then includes descriptions of particular studies and illustrative findings, as well as general conclusions about the variables of interest. Typically, the review concludes with the author's overall assessment of what is known and what still needs to be done.

Meta-analyses retain the goals of the traditional narrative review: to identify questions of interest, review relevant research, and draw accurate conclusions. What they add is a set of quantitative rules and criteria whose purpose is to bring more objectivity and more quantitative precision to the task of research synthesis. These rules parallel the statistical procedures that guide analysis of individual studies.

(Continued)

(Continued)

Suppose that our topic is intervention programs designed to reduce levels of aggression. One basic question in any individual study of this issue is whether there is a statistically significant effect of the program being examined. This is also an issue in a meta-analysis of the literature as a whole. In the case of a meta-analysis, however, conclusions will be based on dozens of research studies rather than just one, and on perhaps thousands of participants rather than just a few dozen. In comparison to a single study, there is much greater likelihood that the sample values are representative of those in the population as a whole. There is also considerably greater statistical power, and therefore a greater chance of detecting genuine effects. (I should add—although the explanation is beyond our scope here—that the greater power is not simply a function of having more participants. Cohn and Becker, 2003, explain how it comes about.)

We saw earlier that a finding of significance is of limited value without a corresponding measure of effect size. A second basic goal of meta-analysis is to determine the effect size for any significant relations that are found. Again, however, meta-analysis allows us to move beyond a single value based on a single sample to an average of dozens or hundreds of effect size values. We can be much more confident, therefore, that we really have identified the strength of the relation between independent variable and dependent variable—in the example, the effectiveness of intervention programs in reducing aggression.

Although there are various ways to calculate effect size, the measures group into two general "families" (Rosenthal & DiMatteo, 2001). One family is based on differences between the means being compared. Earlier I discussed one such measure: Cohen's *d*, which is the difference between means divided by their standard deviation. The other family includes effect size measures derived from correlational relations. Among these is the Pearson *r* statistic, which is appropriate when the data fit parametric assumptions and the relation in question is a linear one. The Pearson *r*, as we saw, is a direct measure of the strength of a relation; thus it is one of the most immediate and accessible measures of effect size. If desired, mean-difference measures can be converted to *r* or vice versa.

In addition to establishing the existence and the size of an effect, a third goal of most studies is to explore factors that might influence the nature or extent of any effects that are identified. This is also a goal of meta-analysis. The usual starting point is the discovery that there is variation in the magnitude of effect sizes across studies. Such a discovery suggests the operation of *moderator variables*—that is, variations among studies that are affecting the results. Further analyses then test for the effects of possible moderator variables. In the case of intervention programs and aggression, moderator variables might include the intensity or duration of the program, or the age of the child, or the form of the aggression being studied, or the country in which the data are collected. They might also include such across-study variables as the date of publication or the experience of the principal investigator. Note that these last two factors are not variables that could be examined in an individual study; it is easy, however, to incorporate them in a meta-analysis.

Meta-analysis seems, then (at least to its proponents), to retain all the virtues of a traditional literature review and to add several further ones as well. What of possible criticisms? There *have* been criticisms, and the most common objections can be summarized in three phrases (Sharpe, 1997): "apples and oranges," "garbage in, garbage out," and the "file drawer problem."

The first phrase refers to the criticism that meta-analyses group together studies that are too disparate to be considered as studies of the same thing. In the aggression example, this might mean a lumping together of intervention programs that in fact differ in important ways. The result would be what is always the result when the independent variable is inadequately specified: uncertainty about exactly what has been found.

The "garbage in, garbage out" criticism is perhaps self-explanatory. It refers to what is seen as a kind of uncritical egalitarianism in meta-analyses—a tendency to weight bad studies equally with good in drawing conclusions.

Finally, the "file drawer problem" refers to the fact that most meta-analyses are limited to the published work on a particular topic. It seems reasonable to predict (and for some topics there is supporting evidence) that published studies are more likely to report significant effects than unpublished studies—that is one reason that the former have been published and the latter have not. Thus meta-analyses may overestimate the certainty or the size of particular effects.

Proponents of meta-analysis offer two general responses to these criticisms (Rosenthal & DiMatteo, 2001; Schmidt & Hunter, 2003). One is to note that the criticisms are not unique to meta-analysis; traditional narrative reviews may also group studies inappropriately, for example, and they also rely primarily on the published literature. The second response is that the problems are not intrinsic to meta-analysis. That is, a well-designed meta-analysis will distinguish studies that need to be distinguished, and a well-designed meta-analysis will code studies for quality and weight good studies more heavily than bad. How successfully these goals are achieved in particular cases can still be debated, of course; still, the general argument seems reasonable.

To date, probably the common applications of meta-analysis in developmental psychology have concerned the study of sex differences in development. We will consequently return to meta-analysis in chapter 13.

Some General Points

I conclude the chapter with a few general pieces of advice about how to approach the statistical part of a study.

The first piece of advice was offered originally in chapter 1: Plan ahead. It should be clear from the discussion throughout the chapter that there may sometimes be aspects of the data analysis that could not have been completely anticipated. Researchers may not know, for example, what the distribution of scores will be until they have actually collected the data. Unanticipated but interesting findings may emerge that require statistical tests that could not have been foreseen at the start. As far as possible, however, such after-the-fact statistical decisions should be minimized. The researcher should know at the outset both the major questions that the analyses will address and the particular statistical tests that will be used to address them. Such planning can help to guard against the tendency to engage in post hoc "data snooping," in which multiple tests are run in an attempt to ferret out whatever looks as though it might be significant in the data. Such planning can also help to ensure that there will in fact be appropriate tests for the data and issues of interest. And it can help to ensure that the most powerful and informative of the potentially available tests will in fact be utilized.

A second piece of advice is to be careful. The conclusions that can be drawn from a study are completely dependent on the accuracy of the statistical calculations. An error in calculation can lead to an error in conclusions, in some cases a major error. Calculation errors, moreover, tend to be less detectable than other mistakes that a researcher might make; readers of a report may be able to spot other methodological flaws, but they generally have to take the calculations on faith. It is critical, therefore, that every calculation be checked very carefully. Ideally, each calculation should be done independently by at least two people. It may also be useful to compare two or more methods of performing the same calculation (e.g., hand calculator versus computer).

In addition to the general need for care, there may sometimes be statistical outcomes that are so unlikely that they cry out for rechecking. In particular, suspicions should arise whenever there is a marked discrepancy between the descriptive statistics of the study and the inferential conclusions about significance. Descriptive and inferential are based on the same data, and they should always show a reasonable correspondence. Finding that a tiny difference is statistically significant, or that a very large difference is not significant, should be a stimulus for rechecking the calculations.

One more point can be made about checking. Investigators should guard against the tendency to engage in differential checking, in which outcomes that contradict their expectations receive close scrutiny while those that are supportive go unchecked. Natural though this tendency is, it provides another opening for possible experimenter expectancy effects: If only the negative errors get corrected, the result may be a pseudoconfirmation of the researcher's hypotheses. Better practice, obviously, is to check and correct everything.

A last piece of advice is to seek help. So many sources of help are available—textbooks, computer programs, statistical experts—that it is foolhardy for nonexperts to attempt to navigate through tricky statistical waters on their own. Expert help can be sought at any point in the research process, beginning with the initial planning of measurements and design and continuing through the write-up for publication.

It may be helpful to say a bit more about written sources of help. I have already mentioned some of the best general textbooks in psychological statistics, as well as more specialized sources for particular topics. There are also some statistical issues and corresponding procedures that are especially prominent in certain kinds of developmental psychology research. This chapter has not attempted to talk about statistics that are specific to developmental, for two reasons: Most of the statistical issues with which developmental psychologists grapple are not in fact specific to their field but are common to psychology as a whole, and those issues that *are* at all specific are too complex to go into here. Note, however, that there are several helpful (although sometimes difficult) sources for discussions that are oriented specifically to statistics in developmental research. These sources include Applebaum and McCall (1983), Gottman (1995), Hartmann and Pelzel (2005), Hertzog and Nesselroade (2003), and several chapters in Teti (2005).

Summary

Psychologists use statistics for two related purposes. *Descriptive statistics* summarize data; they provide a kind of first-level description of what a study has found. Of interest are both the *central tendency* of response, typically indexed by the arithmetical *mean,* and the *variability* in response, typically measured by the *variance* or *standard deviation.*

Inferential statistics go beyond description to the determination of *statistical significance.* At issue is whether the obtained results deviate significantly from what could have occurred by chance, with chance based upon the *null hypothesis* of no difference between the groups being compared. The *t* test is described as a specific example of an inferential test. As in most inferential tests, significance in the *t* test depends on three factors: the size of the difference between the groups, the amount of variability within each group, and the size of the sample.

It is stressed both that conclusions from inferential tests are always probabilistic rather than certain and that two kinds of errors can occur: *Type 1 errors,* in which there is an incorrect rejection of a true null hypothesis; and *Type 2 errors,* in which there is a failure to reject a false null hypothesis. It is also stressed that statistical significance is concerned only with ruling out chance variations as an explanation for the results, and that a finding of significance does not guarantee either the validity of the study or the meaningfulness of the results. Two supplements to significance tests are described. A *confidence interval* provides a range of values within which the population value can be assumed to fall with a certain probability. Measures of *effect size* provide estimates of the magnitude of effect associated with an independent variable.

The discussion turns next to the issue of which kinds of statistical analysis are appropriate for which kinds of data. Three factors are important to the choice of test. One is the level of measurement: whether the measurements are made on a *nominal, ordinal, interval,* or *ratio* scale. A second is the distribution of scores—in particular, whether the scores are normally distributed. So-called *parametric tests,* such as the ANOVA or *t* test, are dependent on certain assumptions about how the data are distributed. They are sometimes more powerful than *nonparametric tests*—that is, more likely to detect a true effect. At the same time, parametric tests are dependent on more assumptions than are nonparametric tests and are therefore less widely applicable. The third factor is the design of the study: how many different independent variables are included, how many different levels of each variable there are, and whether the comparisons are made within-subject or between-subject.

In addition to inferential tests of group differences, another major use of statistics is to establish the correlation between variables. *Correlation statistics* measure the degree of linear relation between variables. The statistical significance of a correlation depends on both the size of the correlation and the size of the sample. The meaningfulness of a correlation depends on its size: The closer the value is to 1, the better our ability to predict one score from knowledge of the other.

Flexible and informative though ANOVA is, there are some situations in which other statistics may be more appropriate. In studies with clear hypotheses, *planned comparisons* may be preferable to an overall ANOVA. Such planned, in-advance tests can provide greater power for specific comparisons of interest. A general alternative to a multifactor ANOVA is *multiple regression,* a broad system of analysis that in fact encompasses ANOVA as a special case. Although the two analyses are often interchangeable, there are situations in which regression may be preferable to ANOVA—most notably, when the independent variables are continuous rather than discrete.

The chapter concludes with several general points about statistics. These points include three general pieces of advice: to plan ahead, to be careful in one's calculations, and to seek expert help whenever necessary.

Exercises

1. Researcher A compares 10 experimental participants and 10 control participants and reports a significant difference between the groups at $p = .04$. Researcher B compares 100 experimental and 100 control participants and reports a significant difference between groups at $p = .04$. For which of these findings, if either, would you be more confident that the difference really is greater than chance? Which, if either, is likely to reflect a stronger difference between the conditions? Justify your conclusions.

2. Imagine a study with two independent variables, A and B, each of which has two levels—hence a 2×2 design. The dependent variable, C, is measured on a scale that can range from 0 to 50. For each of the following outcomes, draw a figure that illustrates one form that the result might take, and then say in words what the outcome would mean.

   (a) significant main effects of A and B with no interaction

   (b) a significant interaction of A and B with no main effects

   (c) a significant main effect of A and a significant interaction of A and B

3. Imagine the following set of correlations between Variable A and Variable B:
   $r = .80, p < .01$
   $r = -.50, p < .05$
   $r = .05, p > .10$

For each correlation:

(a) Say in words what the correlation means.

(b) Draw a scatter plot that shows the relation between the variables.

4. As you have seen in this chapter—and will see further whenever you read a Results section—statistical symbols play a role whenever statistics are discussed. Correspondingly, a review of the various abbreviations for statistical terms can be a helpful way to test one's understanding of what the terms and concepts mean. For each of the symbols listed below, explain as clearly as you can what the symbol stands for.

p r F
t N SD

Notes

1. The formula given is the one for calculating the variance of a sample. For the variance of a population the denominator is N rather than $N - 1$.
2. Among the critics of the all-powerful .05 cutting point are Rosnow and Rosenthal (1989), who suggest that "Surely, God loves the .06 nearly as much as the .05" (p. 1277).
3. The formula given in the figure is called the *definitional formula* for chi square. There is also a *calculational formula:* a mathematically equivalent but computationally simpler method of calculating the value. Many other statistics show a comparable division between definitional and calculational forms.
4. Perhaps the simplest way to see where this probability comes from is to ask what the chances are of *not* obtaining a significant result by chance alone. With a single test the probability of avoiding such an error is .95. With two independent tests the probability is the product of the two individual probabilities, or $.95^2$. With 15 independent tests the probability is $.95^{15}$, or .46. The probability that we will obtain at least one significant result by chance is therefore $1 - .46$.
5. Actually, r itself, as a descriptive statistic, is not parametric; determination of its statistical significance, however, depends on parametric assumptions.

9

Ethics

Chapter 8 stressed that questions of statistics should not be left until the concluding phase of a study. The same point applies even more strongly to questions of ethics. Research must always meet two criteria: scientific merit and ethical soundness. If there are serious doubts about either criterion, then the research should not be done.

For the first 50 or so years of psychology as a science, decisions about ethics were left largely to the conscience of the individual investigator. This is no longer the case. Recent years have seen the development of multiple sets of guidelines concerning ethical principles in research with human subjects. There are also multiple layers of protection afforded the participant in research today. When the participants are children, the guidelines and the layers of protection become even more extensive.

I begin this chapter by reviewing the various guidelines under which researchers in developmental psychology operate. This introductory section outlines the steps that must be followed in determining whether a research project is ethical, as well as the issues that must be considered in this determination. We will then move on to a fuller consideration of the basic rights possessed by every person who participates in research. Three such rights are discussed: the right to informed consent prior to participating in research, the right to freedom from harm as a result of research, and the right to confidentiality with regard to the information obtained in research. Fuller discussions of many of the points made can be found in a number of sources, including Fisher (2003), Hill (2005), Sieber (1992), and Stanley, Sieber, and Melton (1996). An article by Fisher et al. (2002) considers ethical issues that can arise in the study of minority children and adolescents. A book by Fisher and Tryon (1990) provides an extended consideration of ethical issues in research in applied areas of developmental psychology.

A cautionary point is in order before the review of guidelines and procedures. Adherence to established rules is a necessary step in determining the ethics of a research project. It is not a sufficient step, however. Guidelines are of necessity always somewhat general, and researchers must still think carefully about how the guidelines apply to their special cases. Codes of ethics exist to aid the decision-making process, not to replace it. Furthermore, no matter how many levels of approval have been obtained in the course of a project, it is still the researcher who retains final responsibility for the ethics of the research.

Guidelines and Procedures

Imagine that a researcher has devised a study that she plans to carry out with grade-school children as participants. She is comfortable with her proposed research from a scientific point of view but wishes also to be sure that all of her projected procedures are ethically acceptable. In this section we consider two questions: To what sources can a researcher turn for guidance about ethics? And through what channels *must* a researcher work before she can begin collecting data?

Guidelines for ethical conduct in research are available from a variety of sources. (Fuller accounts of the historical development of these standards can be found in Cooke, 1982, and Rheingold, 1982.) At the most general level, both the Nuremberg Code (Trials of War Criminals . . . , 1949) and the Declaration of Helsinki (World Medical Association, 1964) set forth international standards for the involvement and treatment of human subjects in research. The United States government also provides detailed standards for the use of human participants, primarily in the form of rules supplied by the various agencies that dispense federal research grants. The Web site for the Office for Human Research Protections (www.hhs.gov/ohrp) is a helpful source with regard to these rules. Finally, various professional societies whose members carry out research have developed their own sets of ethical standards. The American Psychological Association, for example, has published Ethical Principles of Psychologists and Code of Conduct (2002), and the Canadian Psychological Association (2001) has issued the Canadian Code of Ethics for Psychologists. The Society for Research in Child Development, the major interdisciplinary organization for researchers of child development, has also published guidelines for research with children (Committee for Ethical Conduct in Child Development Research, 1990). Note that all three codes are available online (www.apa.org/ethics; www.cpa.ca/publications; www.srcd.org/ethicalstandards.html).

The various publications just described are not, of course, independent of each other. The points made are similar and frequently identical, and the authors of one document often draw explicitly from the content of another. Because its guidelines are probably the most relevant for the developmental researcher, the ethical code of the Society for Research in Child Development is reprinted in Box 9.1.

Box 9.1. Ethical Standards for Research With Children (Society for Research in Child Development, 1990)

Principle 1. Non-harmful procedures: The investigator should use no research operation that may harm the child either physically or psychologically. The investigator is also obligated at all times to use the least stressful research operation whenever possible. Psychological harm in particular instances may be difficult to define: nevertheless its definition and means for reducing or eliminating it remain the responsibility of the investigator. When the investigator is in doubt about the possible harmful effects of the research operations, consultation should be sought from others. When harm seems inevitable, the investigator is obligated to find other means of obtaining the information or to abandon the research. Instances may, nevertheless, arise in which exposing the child to stressful conditions may be necessary if diagnostic or therapeutic benefits to the child are associated with the research. In such instances careful deliberation by an Institutional Review Board should be sought.

Principle 2. Informed consent: Before seeking consent or assent from the child, the investigator should inform the child of all features of the research that may affect his or her willingness to participate and should answer the child's questions in terms appropriate to the child's comprehension. The investigator should respect the child's freedom to choose to participate in the research or not by giving the child the opportunity to give or not give assent to participation as well as to choose to discontinue participation at any time. Assent means that the child shows some form of agreement to participate without necessarily comprehending the full significance of the research necessary to give informed consent. Investigators working with infants should take special effort to explain the research procedures to the parents and be especially sensitive to any indicators of discomfort in the infant.

In spite of the paramount importance of obtaining consent, instances can arise in which consent or any kind of contact with the participant would make the research impossible to carry out. Nonintrusive field research is a common example. Conceivably, such research can be carried out ethically if it is conducted in public places, participants' anonymity is totally protected, and there are no foreseeable negative consequences to the participant. However, judgments on whether such research is ethical in particular circumstances should be made in consultation with an Institutional Review Board.

Principle 3. Parental consent: The informed consent of parents, legal guardians or those who act in loco parentis (e.g., teachers, superintendents of institutions) similarly should be obtained, preferably in writing. Informed consent requires that parents or other responsible adults be informed of all the features of the research that may affect their willingness to allow the child to participate. This information should include the profession and institution affiliation of the investigator. Not only should the right of the responsible adults to refuse consent be respected, but they should be informed that they may refuse to participate without incurring any penalty to them or to the child.

Principle 4. Additional consent: The informed consent of any persons, such as school teachers for example, whose interaction with the child is the subject of the study should also be obtained. As with the child and parents or guardians informed consent requires that the persons interacting with the child during the study be informed of all features of the research which may affect their willingness to participate. All questions posed by such persons should be answered and the persons should be free to choose to participate or not, and to discontinue participation at any time.

Principle 5. Incentives: Incentives to participate in a research project must be fair and must not unduly exceed the range of incentives that the child normally experiences. Whatever incentives are used, the investigator should always keep in mind that the greater the possible effects of the investigation on the child, the greater is the obligation to protect the child's welfare and freedom.

Principle 6. Deception: Although full disclosure of information during the procedure of obtaining consent is the ethical ideal, a particular study may necessitate withholding certain information or deception. Whenever withholding information or deception is judged to be essential to the conduct of the study, the investigator should satisfy research colleagues that such judgment is correct. If withholding information or deception is practiced, and there is reason to believe that the research participants will be negatively affected by it, adequate measures should be taken after the study to ensure the participant's understanding of the reasons for the deception. Investigators whose research is dependent upon deception should make an effort to employ deception methods that have no known negative effects on the child or the child's family.

Principle 7. Anonymity: To gain access to institutional records, the investigator should obtain permission from responsible authorities in charge of records. Anonymity of the information should be preserved and no information used other than that for which permission was obtained. It is the investigator's responsibility to ensure that responsible authorities do, in fact, have the confidence of the participant and that they bear some degree of responsibility in giving such permission.

(Continued)

(Continued)

Principle 8. Mutual responsibilities: From the beginning of each research investigation, there should be clear agreement between the investigator and the parents, guardians or those who act in loco parentis, and the child, when appropriate, that defines the responsibilities of each. The investigator has the obligation to honor all promises and commitments of the agreement.

Principle 9. Jeopardy: When, in the course of research, information comes to the investigator's attention that may jeopardize the child's well-being, the investigator has a responsibility to discuss the information with the parents or guardians and with those expert in the field in order that they may arrange the necessary assistance for the child.

Principle 10. Unforeseen consequences: When research procedures result in undesirable consequences for the participant that were previously unforeseen, the investigator should immediately employ appropriate measures to correct these consequences, and should redesign the procedures if they are to be included in subsequent studies.

Principle 11. Confidentiality: The investigator should keep in confidence all information obtained about research participants. The participants' identities should be concealed in written and verbal reports of the results, as well as in informal discussion with students and colleagues. When a possibility exists that others may gain access to such information, this possibility, together with the plans for protecting confidentiality, should be explained to the participants as part of the procedure of obtaining informed consent.

Principle 12. Informing participants: Immediately after the data are collected, the investigator should clarify for the research participant any misconceptions that may have arisen. The investigator also recognizes a duty to report general findings to participants in terms appropriate to their understanding. Where scientific or humane values justify withholding information, every effort should be made so that withholding the information has no damaging consequences for the participant.

Principle 13. Reporting results: Because the investigator's words may carry unintended weight with parents and children, caution should be exercised in reporting results, making evaluative statements, or giving advice.

Principle 14. Implications of findings: Investigators should be mindful of the social, political and human implications of their research and should be especially careful in the presentation of findings from the research. This principle, however, in no way denies investigators the right to pursue any area of research or the right to observe proper standards of scientific reporting.

SOURCE: From "Ethical Standards for Research with Children," Society for Research in Child Development *Directory of Members,* copyright © 1996. Reprinted by permission of the SRCD.

Let us return now to our hypothetical researcher. We assume that she is familiar with the principles set forth in the documents just listed. At the local level these principles are embodied in the form of an organization called an **institutional review board**. It is this board through which the researcher will work.

Institutional review boards are a reflection of one of the basic principles of research ethics: the principle of **independent review**. The idea behind independent review is that the researcher who designs a study has too great a personal investment in the research to be a trustworthy judge of its ethical soundness. What is needed is an independent evaluation of ethics by competent people who are not themselves directly involved in the research. To this end, every university at which research with human participants is conducted maintains an institutional review board whose purpose is to judge the ethics of proposed research projects. Such boards consist of a number of members drawn from diverse backgrounds. Among the conditions set forth in the federal regulation that established such boards were the following: "The members could not have a direct interest in the research under review, could not all be employees or agents of the institution, and could not be members of a single professional group" (Cooke, 1982, p. 167).

Each institutional review board publishes its own set of ethical guidelines, guidelines that are drawn closely from the various federal and professional sources described previously. Each board also provides standard forms on which researchers can submit proposed research projects. Typically, these forms begin by asking for a general statement of the purposes of the research and of the procedures to be followed. If the board deems it necessary, copies of experimental protocols, questionnaires, and so forth may also be requested. In addition to such general information, the form also solicits information about a number of specific points of ethical concern. These points (which are discussed more fully shortly) include the following: What are the risks, if any, to the participants from participating in the research? What are the benefits, if any? How will participants be recruited for the study, and what inducements, if any, will be used to secure their cooperation? What exactly will participants be told about the research before their agreeing to participate? Will it be clear to the participants that participation in the study is voluntary and that they can withdraw at any time? Where appropriate, will participants be debriefed with regard to any deception and told the true purpose of the study at the conclusion of testing? And how will the confidentiality of the results be preserved?

Approval from an institutional review board allows the researcher to submit a grant for funding or, if the research is unfunded, to seek out participants and begin testing. In the case of our hypothetical grade-school study, the researcher will probably need to work through a public-school system. The exact procedure for doing so varies from one community to another, although there are some common elements. In some school districts there is a central office through which all research proposals must be channeled, and a proposal can be forwarded to a school only after it has been approved at this initial level. Once the proposal reaches the school it will need to gain the approval of the principal. In some schools the principal is the sole arbiter; in other schools the principal consults with the teachers whose classrooms are involved and grants approval only if the teachers agree. Once the proposal reaches the classroom, it will need to be conveyed to the parents, usually via a letter describing the study and requesting permission to include their children. Only children whose parents return a signed permission form may be tested. Finally, the child must be given some idea of what the research involves and must be offered the choice of whether to participate.

As this description should make clear, there are numerous layers of protection between the researcher's initial idea for a study and the eventual testing of children. The primary purpose of these layers is, of course, to ensure that the rights of the child are safeguarded. From the researcher's point of view, the multiple safeguards have a couple other implications as well. First, it is important not only to think carefully about ethics when planning a study but also to communicate persuasively about the ethical soundness of one's ideas when presenting the research to others. Research can proceed only after many people have been convinced of the value and the ethical propriety of the project. Second, it is important to allow sufficient time for the various layers of approval to be obtained. Depending on the ease with which the different channels can be negotiated, weeks or even months may intervene between the planning of a study and the eventual start of data collection.

Rights of the Research Participant

Informed Consent

Participation in research must be voluntary. No one can be studied without his or her knowledge, no one can be forced to be a subject in a study, and no one can be forced to continue in a study. These statements sum up the doctrine of **informed consent**: Participants must know in advance what a research project involves, and they must give their explicit agreement to be included.[1]

The principle of informed consent is easy enough to state, but implementing it can give rise to a host of complexities. It is these complexities on which I concentrate here. I will begin by considering more fully the two components of the phrase, "informed" and "consent." I then move on to some special issues that arise when the participants are children.

As "informed" implies, simple agreement to participate is not sufficient; rather, the agreement must come after the participant has been fairly informed about what the research will entail. The principal way that prospective participants learn about research is via an "informed consent form"; it is also this form that the participant signs to indicate willingness to participate. The general goal in devising a consent form is to convey in nontechnical, comprehensible language all of the information about the study that a participant needs in order to make a fair decision about whether to take part. All institutional review boards pay close attention to the consent form that the researcher plans to use, and all review boards have specific and detailed criteria for what should be included on the form. Box 9.2 reproduces in slightly abridged form the guidelines provided by the University of Florida's review board, guidelines that are modeled closely after those of the U.S. government's Office for Human Research Protections.

Box 9.2. Guidelines for Informed Consent (University of Florida Institutional Review Board)

Identify yourself and your connection to the University of Florida.
Provide a statement that the study involves research.
State the scientific purpose of this research.
Explain exactly what the participant will be asked to do.
Note the amount of time the participant can expect to allow for participation.
Indicate where and when the procedure will take place.
Disclose alternative procedures or treatments that may be advantageous to the participants.
State who will administer the procedure.
Describe any anticipated risks or discomforts (if there are no risks, indicate this).
Describe potential benefits to participant or others (if there are no direct benefits, indicate this).
Explain how and to what extent the participant's identity will be protected.
Indicate whether or not compensation will be awarded.
Make a statement that the participant has (when a minor is involved state that the parent *and* child have) the right to withdraw consent at any time without consequence.
State that the participant does not have to answer any question that s/he does not wish to answer, if applicable.
If audio or video recordings are to be made, indicate who will have access to the tapes and what the disposition of the tapes will be.
Indicate with whom (if anyone) the results (group or individual) will be shared.
Offer to answer questions and provide a means of contacting the principal investigator.

SOURCE: Adapted and used with permission.

Despite the availability of explicit guidelines, consent forms may sometimes fail to inform fully. Various problems can arise. Researchers may have difficulty meeting the criterion of "language that is reasonably understandable" to participants (American Psychological Association, 2002, p. 1065). Most researchers have limited experience in writing for nonprofessionals, and many are not very successful when they have to attempt the task. A survey of 284 consent forms approved by five review boards found that most were written at a 12th-grade reading level or above. Fewer than 10% were worded at a 10th-grade or lower level (Goldstein, Frasier, Curtis, Reid, & Kreher, 1996). Similar results had emerged in earlier examinations of the issue (Ogloff & Otto, 1991; Tannenbaum & Cooke, 1977). The point that these data make is obvious: Consent forms cannot do a very good job of informing if they are written at levels that are not appropriate for many of the people reading them.

Consent forms also cannot do a full job of informing if some information about the research is deliberately withheld. In some projects the nature of the research requires that participants not know in advance about some of the purposes or some of the procedures of the study. Sometimes the withholding of information takes the form of **incomplete disclosure**: simply not telling participants about some aspect of the study. Recall, for example, the Horka and Farrow (1970) study of intersubject communication described in chapter 5. If such a study is to be valid, the children cannot be told that the true focus of the research is communication among the participants, nor can they be told that the right answer is to be switched from one test session to the next. In other kinds of research the absence of full disclosure takes the form of active **deception**: not simply withholding information, but instead deliberately misleading participants about some aspect of the research. The Liebert and Baron (1972) study described in chapter 4 provides an obvious example. In this study there was no game-playing child in the room next door, yet the measure of aggression depended on the participants' believing that they were either helping or hurting another child. A number of research procedures in developmental psychology require either deception or (more commonly) some degree of incomplete disclosure. It will be a useful exercise, as we move through different research paradigms in the concluding chapters of the book, to try to identify procedures that might require either incomplete disclosure or deception.

As might be expected, incomplete disclosure is generally regarded as less problematical from an ethical point of view than is active deception. Nevertheless, both incomplete disclosure and deception do involve some violation of the principle of informed consent, in that participants are not given full information about the study prior to agreeing to participate. When can such procedures be justified?

From the various guidelines on ethics cited earlier, a number of rules can be abstracted. Such practices should be used only if the topic is an important one and if there is no other way to gather the necessary information. The practices should be used only after the researcher has convinced an independent review board that such procedures are in fact justified in the study. Although participants cannot be told everything in advance about the research, they should be told as much as possible, short, of course, of anything that might bias their later responses. Among the facts that *must* be conveyed is any information about potential risks from participating in the research. Once the testing is completed, participants should be fully informed about any deception that has occurred, and they should be given the opportunity to refuse the use of their data if they wish (the latter is a recent addition to the APA guidelines). Finally and most critically, both the review board and the researcher must be convinced that the use of incomplete disclosure or deception will not result in any harm.

As always, it is easier to state general principles than to handle every specific application.

I return to these matters, especially deception, in the next section when I discuss various kinds of research that may be harmful to participants.

Let us turn now to the second component in the phrase "informed consent," the concept of consent. To be meaningful, consent must involve more than simply signing a consent form; consent must be *voluntary*. The notion of voluntary does not mean that people must seek out researchers simply for the chance to be studied, nor does it mean that researchers must forswear all attempts at persuading prospective participants to take part. It does mean, however, that participants should not be lured into a study through unfair inducements or coerced into participating through the greater power of the experimenter. It means also that participants should not be forced to continue in a study if they indicate clearly that they wish to withdraw.

Issues of coercion are generally greatest when there is a clear imbalance of power between participant and experimenter. Such is the case, of course, when children are the participants. It may also be the case for the relation between the experimenter and the children's parents, especially if the experimenter is associated with (or perceived to be associated with) some institution that has control over the child, such as a hospital or school. In such cases parents may not realize that they have the right to refuse permission for their child's participation. Whatever the specific situation, researchers must strike a careful balance between the goal of eliciting maximum participation and the goal of ensuring that the participation is truly voluntary. Researchers are generally convinced of the scientific importance and ethical propriety of what they do; they may also be convinced that their research will yield social benefits, either directly to the participants involved or more generally to people who are like the participants. Under these circumstances, it is perfectly proper for the researcher to make a concerted effort to recruit and maintain a sufficiently large and representative sample. It is important, however, that this effort stop well short of misrepresentation or coercion, either in recruiting participants in the first place or in persuading them to continue once the study is under way.

Consent Procedures With Children

It is time now to consider how informed consent works when the participants are children. Children are not competent, from either a legal or a psychological point of view, to give informed consent for their own participation in research. Consequently, permission must be sought elsewhere. In the great majority of cases it is the child's parent or parents who give the permission. Typically, permission from one parent is sufficient, although a review board may occasionally require that both parents give their agreement. In rare cases a review board may decide that parental permission is not necessary, just as it may occasionally decide that informed consent is not necessary in research with mature participants. Even here, however, the researcher may be required to obtain permission from the child's school; indeed, such permission may be a condition for waiving parental approval. There are also cases in which the child's parents are not available—for example, when the child is institutionalized or living with a legal guardian. The principle, however, remains the same: Permission must be obtained from someone who is legally responsible for the child and who is competent to evaluate what the research will require of the child.

The requirement that parental permission be obtained in almost all research with children is a fairly recent development. Another recent development is a greatly increased concern with the *child's* right to decide whether to participate. In contrast to the written procedures that are typical for adult consent, the consent process with children is usually oral: The tester provides a brief oral description of the research, and the child is then asked whether he or she is willing to participate. The child's agreement is referred to as "assent," to distinguish it from the legally effective "consent" that an adult can give.

Securing assent from child participants can be a delicate business. The approach that is appropriate depends on both the age of the child and the nature of the study. Although systematic data do not exist, it seems safe to conclude that researchers typically tell their child participants considerably less in advance than they do their adult participants. This discrepancy can be justified in terms of the child's lowered capacity to understand the purpose of research; the fact that various adults have already agreed to a particular project is a further justification. Nevertheless, if the assent is to mean anything, the child must be given *some* basis for deciding whether to come with the experimenter. Except with very young children, it probably makes sense to attempt to convey the following pieces of information: a general idea of what will be done ("play some remembering games"), where the research will take place ("in Mrs. Smith's office"), how many people will be involved ("just you and me"), how long the session will be ("take about 20 minutes"), whether other children have done or will do the same thing ("lots of kids from the class will be doing it"), whether a reward will be offered ("get a little prize at the end"), and finally, an opportunity to say yes or no ("Would you like to come?"). As the age of the sample increases, the description can be moved closer and closer to the kind of informed consent that would be used with adults (see also Tymchuk, 1991).

Do the typical assent procedures succeed in conveying the information they should convey? Several recent studies have explored what children understand about the assent process and about their rights as research participants more generally (Abramovitch, Freedman, Henry, & Van Brunschot, 1995; Abramovitch, Freedman, Thoden, & Nikolich, 1991; Bruzzese & Fisher, 2003; Hurley & Underwood, 2002; Nannis, 1991). These studies suggest a mixed picture with regard to children's understanding. On the positive side, most children (at least in these studies) were able to understand the purpose of the studies in which they participated, although understanding was less complete for the one study in which deception was involved (Hurley & Underwood). Most also realized that they could terminate their participation at any time if they wished. Many, however, were less clear about their right to refuse participation in the first place, even though this right had been spelled out in the assent procedure. Many also were unclear about the confidentiality of their responses, believing that their parents would find out how they responded. Again, the misunderstanding occurred despite the fact that the assent procedures had indicated explicitly that the results would be confidential.

As the authors of these studies note, children are accustomed to being told what to do by adults, especially in school settings (which is where most studies with school-aged children take place). They are also accustomed to their parents being informed of what they do in school. It may be a challenge, therefore, for researchers to convince children that they have the right to refuse an adult's request to participate in a study, and also the right to have their responses held in confidence from parents and teachers if they do participate.

Although I have focused the discussion on children, children are not the only population whose participation in research poses special ethical challenges. Many of the same issues of informed consent and freedom from coercion apply to other vulnerable populations, such as adults with mental illness or elderly victims of Alzheimer's.

Freedom From Harm

Undoubtedly the most basic right that participants possess is the right not to be harmed in research. Correspondingly, the greatest responsibility that the researcher has is to ensure that no harm comes to any participant.

How might a participant be harmed in research? In medically oriented projects the possibility of physical harm may sometimes arise. A new medical treatment, for example, may carry risks as well as expected benefits, and there may be no way to rule out all such risks before the research is conducted. It is in just such cases that informed consent is especially important, as is prior review by an independent review board. Indeed, it was to monitor medical research that institutional review boards were first set up. Their application to psychological research came later.

In psychological research there is generally not the slightest chance of physical harm. The concern in this case lies in the possibility of psychological harm. Psychological harm is a more nebulous concept than physical harm, and as such has been the subject of numerous debates, both in general discussions of research ethics and in evaluations of specific research projects. I begin with a few general points and then discuss kinds of research that could conceivably result in psychological harm.

The notion of psychological harm does *not* mean that nothing unpleasant can happen to the participant in the course of the study. Participants may be bored by a repetitive task and wish they were elsewhere, may find parts of the procedure frustrating or confusing, may fail on some task on which they would very much like to succeed, and so forth. It is, of course, desirable to minimize such experiences, especially if boredom, frustration, or whatever is not the focus of the study. But the mere fact that the participant wishes things were otherwise does not make a study harmful. Two questions must be asked. Does the research expose the participants to experiences that are significantly different from those that they routinely encounter anyway? And is there any possibility that the effects will be permanent and not just transitory?

Consider a study of problem solving in grade-school children. The children are given a variety of problems to solve, some children are more successful than others, and all children fail on at least some problems. As described, the research experience sounds quite similar to what children routinely encounter every day in school, and there is no reason to think that any negative effects from the failures will persist beyond the study. Suppose, however, that one goal of the research is to determine the effects of anxiety on problem solving. To heighten anxiety, the experimenter tells half the children that the problem-solving battery is a test of their ability to succeed in school during the coming year. In this case failures on the tasks may well lead to strong negative reactions. And in this case the negative reactions may well persist beyond the study itself.

This example suggests one class of experimental manipulations that may result in psychological harm: manipulations that create negative feelings or a negative self-image in the participant. Induced failure experiences, especially on tasks that are labeled as important, fall under this heading. So do manipulations that induce the participant to engage in some unethical or forbidden behavior. Cheating is an obvious example. In a cheating study, children (or at least *some* children) are enticed into doing something that they know is against the rules and will lead to censure if discovered. Even if the child is left to believe that cheating was undetected, feelings of anxiety or guilt may result. If the child is not somehow reassured, these feelings may well persist.

The last sentence suggests a possible solution to the problem. Why not use the poststudy period to dispel whatever negative feelings may have arisen in the course of the study? In the problem-solving study, for example, the experimenter could explain to the child that the test does not really predict school performance and that all children get some problems wrong. In the cheating study the experimenter could emphasize that no harm was done by the cheating and that many other children behaved in the same way. Such after-the-fact clarification of the true nature of a study is referred to as **debriefing**. The goals of such debriefing efforts are two: to make clear the actual purpose of the experimental manipulations, and to leave the participants feeling as positive as possible about their behavior.

Debriefing is a recommended procedure whenever important information is initially withheld from the participant. If negative feelings are induced, then debriefing becomes even more important. With children, however, debriefing can be difficult, and it may sometimes create more problems than it solves. Consider the cheating example again. It is not clear that the child will really feel better if the experimenter ends the study by explaining that cheating was being studied and that the child's cheating was in fact known about all along. After all, the child *has* misbehaved and now no longer even has the solace of having gotten away with it. Nor is it clear how fully young children can understand and benefit from explanations of the true purpose of research. The experimenter who attempts to explain the rationale behind a cheating study to a 5-year-old may succeed only in confusing and possibly alarming the child. Perhaps adults can see through any mirror! Under such circumstances, it may be better to forgo the debriefing altogether.

Debriefing has another possible unwanted outcome. By definition, debriefing occurs only after the experimenter has somehow misled the participant. In many cases "misled" is a euphemism, for what has happened is that the experimenter has lied to the participant. The debriefing then spells out exactly what the lie was. It may be that the main message that the participant takes away from the debriefing is that researchers (or psychologists or adults in general) are not to be trusted. Children in particular may be susceptible to negative reactions when they learn that an adult has deceived them.

The preceding paragraph has a kind of "shoot the messenger" quality to it. Clearly, the prime culprit here is not the debriefing but the original act of deception. The obvious way to avoid problems attendant on lying to participants is not to lie in the first place. Some issues, however—such as cheating—are very difficult to study without the use of deception. If all the criteria discussed in the preceding section can be met, then most researchers would agree that the use of deception can sometimes be justified (not all researchers, however; cf. Baumrind, 1985). It is in these cases that the researcher and the institutional review board must decide whether debriefing would mitigate or add to the problems posed by the deception.

Thus far I have been discussing cases in which a researcher deliberately brings about some sort of undesirable outcome in the participant, such as a heightened level of anxiety or a feeling of guilt. Such cases violate the basic ethical principle known as the *principle of nonmaleficence:* Do not intentionally inflict harm on another (Thompson, 1990). The final ethically problematical area I consider is in a sense the converse: the case in which a researcher may deliberately *fail* to bring about some desirable outcome. The basic ethical principle at issue here is the *principle of beneficence:* When possible, remove existing harm and provide benefits to others. In research this issue is usually discussed under the heading of **withholding treatment**—that is, failing to make some potentially beneficial experimental treatment available to someone in need.

Issues of withholding treatment first arose in medical research. Suppose that we have developed a new drug therapy that we expect to be effective in combating a particular form of cancer. To test our hypothesis, we need to apply our therapy to a sample of people who are suffering from the disease. We also need some sort of untreated control group against which to evaluate the effects of the treatment. In some cases already existing data about the natural, untreated course of the illness may be sufficient for a control. In some cases, however, the only way to get a clear comparison may be by testing two initially equivalent groups: an experimental group to whom we apply the treatment and a control group from whom we withhold the treatment. We then would have a clear test of the effects of our treatment. But how can we justify withholding a potentially beneficial treatment from people who need it?

Consider a somewhat analogous psychological example. The last 40 years have seen the development of numerous preschool intervention programs designed to enhance young children's ability to succeed in school. Typically, these programs are oriented to children who for various reasons (e.g., low socioeconomic status) are perceived as being at risk for school failure. Determining how well any particular program works requires two groups of children: an experimental group that receives the program and a comparable control group that does not receive the program. We already know, however, that without some intervention many of the control children will fail in school. How can we justify withholding our program from them?

There are various justifications that can be offered for the deliberate withholding of an experimental treatment. Note first that we do not *know* that our treatment will work; if we did, there would be no need to carry out an experimental test. It is only if we can test our ideas experimentally that we can verify our treatment's value, a verification that may then prove beneficial for much larger numbers of people in the future. Note also that researchers seldom have the resources to offer a treatment to all people, or even a very large proportion of people, who might benefit from it. Given this limitation, the sensible course is to adopt an approach that can simultaneously benefit as many as possible while also providing a scientific test of the treatment's worth. Whenever possible, untreated controls should be offered the treatment, or at least some form of help, at the conclusion of the experimental phase of the project. Finally, it may sometimes be possible to draw a contrast not between an experimental group and untreated controls but between two or more different experimental treatments. In this way all participants can potentially derive some benefit from the research.

This section has only touched on the difficult concept of psychological harm. An article by Ross Thompson (1990) provides a fuller treatment, including an analysis of how susceptibility to different forms of harm may vary with the developmental level of the child.

Confidentiality

A final basic right of the participant is the right to **confidentiality**. Information obtained in research must be confined to certain well-defined scientific uses that should be clear to the participant at the time of informed consent. Such information must not be made generally available in a way that could ever embarrass or harm the participant.

Issues of confidentiality arise at two points in the research process. One is at the time of data collection and storage. Clearly, if participants' names appear on the data sheets, there is always the possibility that unauthorized persons might see the sheets and thereby learn something that should not be divulged. The obvious solution to this possibility is to record and store data in terms of code numbers rather than names. If there might ever be a need to link data and name, a separate list of number-name translations can be stored somewhere apart from the data. Information might also be transmitted to unauthorized persons if researchers talk about their participants by name or in ways that make a participant's identity apparent. The obvious solution to this possibility is for researchers not to indulge in such talk.

The second point at which the issue of confidentiality arises is at the time of publication. The goal of publication, of course, is to share what was done and found with the broadest possible audience. In this sharing, however, it is critical that the anonymity of individual participants be preserved. In most psychology research reports there is no difficulty in preserving anonymity: Large samples of participants are tested, and all analyses are in terms of group averages and group differences. Occasionally, however, cases arise in which individual participants might be identifiable—in case-study reports, for example, or in studies with small and distinctive samples. In such cases various strategies (e.g., use of pseudonyms, omission of geographical information) may be necessary to ensure confidentiality.

As this discussion may suggest, in most research projects there is no particular difficulty in maintaining confidentiality. All that the researcher need do is to avoid the carelessness of loose talk or indiscreetly exposed data sheets. I turn next to two situations that pose somewhat more complex problems of confidentiality.

Researchers of child development are often asked to provide information about the performance of particular children. Teachers may wish to know how a particular child from their classroom responded to some cognitive test. Parents may want to know what the researcher's measures have revealed about the psychological development of their child. Such requests can pose a delicate problem. On the one hand, it is only through the good graces of the parent or teacher that the researcher has been allowed to test the child at all. Refusal to share information may seem unreasonable and may lead to diminished cooperation in the future. On the other hand, the child, who is the subject of the research, should presumably have the same right to confidentiality as any other participant. In some kinds of research (e.g., children's perceptions of their parents' child rearing) it is essential that children be certain that their responses will be held in confidence.

Several general rules can help guide response to teacher or parent requests for feedback. A first rule is to be clear from the start about what information will or will not be shared. If no specific information about the child's performance will be provided, then this fact should be explicit on the permission form that the parent signs. A second rule is that student testers should be especially careful about providing feedback; the teacher or parent who persists in asking for specific information should be referred to the more experienced principal investigator. A third rule is to be positive, within reasonable bounds of honesty, when making evaluative statements. It is generally possible to find something good to say about a child's performance, although it is of course important to avoid conveying a misleadingly positive picture. A final rule is to be clear about the diagnostic limitations of one's measures. Most studies in developmental psychology are not set up to diagnose individual children, and most research measures in developmental psychology do not provide information about specific children that is useful in any clear-cut diagnostic or predictive sense. If teachers or parents are clear about these limitations, then requests for specific feedback may not arise. Note that the child participants themselves also have a right to be told something about what can and cannot be concluded from their responses.

Although most research measures do not yield deep insight into a particular child, exceptions do occur, and these exceptions provide a second possible threat to confidentiality. Suppose that the research uncovers evidence of some serious problem in the child's development, a problem that seems so serious that it should be called to the attention of those responsible for the child's welfare. A battery of cognitive tests may reveal an alarmingly low score, so low that the researcher feels that school authorities should be informed. A set of personality measures may elicit bizarre responses suggestive of some deep-seated personality disorder. A study may uncover, quite inadvertently, what seems to be evidence of child abuse. As in all research, the information in these cases has been gathered in confidence. What then is the researcher's responsibility?

The Society for Research in Child Development code of ethics, reprinted in Box 9.1, provides a general answer. Principle 9 of the code states that "the investigator has a responsibility to discuss the information with the parents or guardians and those expert in the field in order that they may arrange the necessary assistance for the child." This statement, general though it is, makes two important points. The first is that confidentiality is not an absolutely inviolable principle. The ultimate goal remains the welfare of the child, and in some cases securing the participant's welfare may require a violation of confidentiality. The second point is that the researcher should not act alone in making the decision to violate confidentiality. It is perhaps unnecessary to note what a serious undertaking it is to tell parents that their child may be psychologically disturbed or to inform legal authorities that parents may be abusing their child. The costs of being wrong, either in acting or in failing to act, are very great. The researcher should obtain as much expert advice as possible before deciding how to proceed.

Summary

This chapter discusses the ethical principles that guide research in developmental psychology. It begins by reviewing the various sets of guidelines under which researchers in developmental psychology operate. In recent years numerous codes of ethical conduct have been developed to which researchers can turn when evaluating the ethics of their own research. The code of standards of the Society for Research in Child Development is reprinted as part of the chapter.

At a local level, ethical principles are embodied in an organization called the *institutional review board.* All research with human participants must be approved in advance by the institutional review board. The requirement of such approval is a reflection of a basic ethical principle: the need for *independent review* in determining the ethics of a project. In research with children, further layers of protection are provided by the requirement that parents and in some cases school authorities give permission for the child's participation. The discussion stresses, however, that the approval of others is a necessary but not sufficient step in the determination of ethics. Researchers are always ultimately responsible for the ethical standards of their own research.

The review of guidelines and procedures is followed by a consideration of the basic rights of participants in research. One right is to give *informed consent* prior to participating in a study. Participants must be told in advance what a research project entails, and they must agree freely to take part in it. Various obstacles to fully informed consent are discussed. In some cases researchers may fail to convey the purposes and demands of their research in language that is appropriate for the participant. In some cases researchers may deliberately not reveal certain aspects of their research, either through *incomplete disclosure* of information or through active *deception.* Although some topics may require such secrecy, it is still important that participants be told as much as possible, including a fair account of any risks involved. When children are the participants, special problems of consent arise. With child participants the parent or some responsible adult must be informed of the study and must give permission for the child's participation. Children should also receive an age-appropriate description of the study and should give their assent before participating.

A second basic right that research participants possess is *freedom from harm.* In psychology research, the concern is generally with psychological harm, a difficult-to-define and much debated concept. Two questions are important when assessing the likelihood that an experimental manipulation will prove harmful: Are the experimental experiences significantly different from those that participants routinely encounter anyway? Are the effects of the experiences likely to persist beyond the experiment itself? Among the kinds of manipulations that may prove harmful are those that induce a negative self-image, those that create unpleasant feelings (e.g., hostility, anxiety), and those that induce the participant to engage in undesirable behavior. Whenever such manipulations are used, a period of *debriefing* and reassurance is generally appropriate at the conclusion of testing. Another ethically problematical kind of research involves *withholding treatment:* Some desirable treatment is withheld from people who might benefit from it. Whenever possible in such research, the untreated control group should receive some benefit from participation, such as an alternative form of treatment or later administration of the experimental treatment.

The final right that the chapter discusses is the right to **confidentiality**. Data obtained in research are confidential and must never be made available in a way that could harm the participant. It is important, therefore, that the anonymity of individual participants be safeguarded. Requests from parents or teachers for information pose one threat to confidentiality, and suggestions for responding to such requests are therefore offered. A more difficult challenge arises when the research uncovers evidence of some serious problem in a child's development (e.g., evidence of child abuse). Confidentiality is not an inviolable principle in such cases, for the well-being of the child remains the most important consideration.

Exercises

1. Construct (in at least outline form) a simple study on some topic in child psychology that interests you. Imagine that you are submitting your study to an institutional review board and draft the parental permission letter that you would include with your submission. If possible, solicit feedback on the letter from faculty members in your program.

2. Submissions to an institutional review board must include documentation of the informed consent process. If the research is with children, the submission must include a *child assent script*—that is, an indication of the information that will be given to the children prior to their being asked whether they want to participate. Suppose that you had the task of writing the child assent script for the Dufresne and Kobisagawa study described at the beginning of chapter 2. Draft scripts for the youngest (first grade) and oldest (fifth grade) participants in the study. Imagine that preschool children (3- and 4-year-olds) had been included, and draft a script for this age group as well.

3. In an article entitled "Vulnerability in Research: A Developmental Perspective on Research Risk," Ross Thompson provides three vignettes of types of research in developmental psychology that pose ethical questions. Read the Thompson vignettes (the source is *Child Development,* 1990, *61,* 1–16) and offer your own evaluation of the ethics of the three kinds of research. Respond from two perspectives: as a member of an institutional review board and as a parent who is being asked to allow his or her child to participate in the research.

Note

1. One exception to this statement is provided by some forms of observational research, especially those that occur in the natural setting. Imagine an observational study of purchase decisions in a supermarket, or of traffic patterns at a busy intersection, or of spacing or crowding in a movie theater. It is unlikely that any review board would insist on informed consent from the persons who contribute data in such instances. Several justifications can be offered for the absence of consent: The behaviors are naturally occurring ones that are unaffected by the decision to study them; the behaviors are innocuous and in no way revealing or embarrassing; and the participants are anonymous and certain always to remain anonymous.

10

Writing

Chapter 1 outlined eight steps involved in doing good research. We have now reached the last of these steps: communicating one's work to others.

Scientists communicate about research in two ways. One way is through oral presentation. This is the method of communication used at professional conferences; it is also, of course, the familiar medium of classroom instruction. No student who sits through a dozen lectures a week needs to be told that oral presentations can vary greatly in their clarity and success.

The second method is written communication, usually in the form of publication in one of the many specialized research journals in the field. It is this method on which we will concentrate in this chapter. The basic focus, therefore, is on how to write reports for publication in an APA (American Psychological Association) journal. Many of the points made, however, have a broader application as well. Thus, much that is said is also relevant to preparing oral presentations, as well as to writing papers other than psychology research reports.

As always, there are a number of further sources that can be consulted. An indispensable source of information about writing in psychology is the *Publication Manual of the American Psychological Association* (APA, 2001). The *Manual* provides both general guidelines for effective writing and specific rules for the way that various parts of a manuscript (e.g., headings, footnotes, references) should be handled. Throughout this chapter there are frequent references to the *Manual.* Advice about not only research reports but also other kinds of writing in psychology (e.g., literature reviews, research proposals, posters) is available in several recent books directed to psychology students (Parrott, 1999; Rosnow & Rosnow, 2006; Sternberg, 2000, 2003). Two chapters by Bem (1995, 2004) provide helpful advice about how to prepare literature reviews and empirical reports. Among the many good general books on how to write, Strunk and White's (2000) *The Elements of Style* remains a classic. Finally, no reader with a sense of humor should miss Harlow's (1962) "Fundamental Principles for Preparing Psychology Journal Articles."

The organization of this chapter is as follows. The chapter begins with some general suggestions of how to write effectively, both in psychology papers and elsewhere. It then considers the various sections of a psychology research report, moving in order through Introduction, Method, Results, and Discussion.

The chapter concludes with a focus on common writing problems of two sorts: violations of APA conventions and violations of English grammar.

Some General Points

There are two basic issues to be resolved in writing for publication in psychology. One issue is what to say; the other is how to say it.

This book has already said quite a bit about what kinds of information should be included in psychology journal articles. I have talked, for example, about what readers should be told about participants, about methods of data collection, and about analyses. More generally, having a clear conception of the important components of research—the major goal of the first nine chapters—is necessary for deciding what is important to say when reporting research. More will be said about questions of "what" when I discuss the different sections of a research report shortly.

Questions of "how" are divisible into two sorts. A number of points of style are essentially matters of convention—that is, agreed-upon rules for handling particular aspects of writing. Some examples were mentioned earlier. There is, for instance, no single "right" way to do footnotes or references, and any reader has undoubtedly encountered a number of different approaches across different books or articles. Precisely because various possibilities exist, however, it is important for members of a discipline to agree on a common approach. Thus, in psychology there *is* a right way to handle footnotes, references, and dozens of other stylistic conventions, and this way is spelled out in the APA *Manual*. In the words of the *Manual* (2001), such rules "spare readers a distracting variety of forms throughout a work and permit readers to give full attention to content" (p. xxiii).

No attempt will be made in this chapter to summarize all of the stylistic rules presented in the *Manual*, rules that occupy the bulk of the *Manual's* 412 pages of text. What I will attempt are two things. One is to highlight a small set of rules that seem to give students special difficulty. The other is to emphasize the importance of having and working from the *Manual* when writing a paper in psychology. A good rule of thumb is to assume that *any* stylistic issue that might arise will be covered somewhere in the *Manual*. And indeed it probably will be, starting with what kinds of spacing for which to set the margins and then moving on from there.[1]

Adhering to the APA stylistic conventions is the relatively easy part of the "how" question. The more difficult part is not unique to psychology: It is the general question of how to write clear and readable prose. Needless to say, this question will not be answered in a brief chapter like the present one (wouldst that things were so easy!). What I can attempt to achieve are some more modest goals. In this section several general points about writing are made. Later, a number of specific stylistic points that seem to give students special difficulty will be noted. This latter section is a counterpart to the discussion of specific APA conventions.

I will begin by emphasizing the importance of writing in a clear and grammatical fashion. That such a seemingly obvious point needs to be emphasized at all is a reflection of two natural but nevertheless mistaken beliefs. One is the belief that the really interesting and important work of the researcher is done when the study is completed and the researcher knows what has been found. Communicating one's findings to others is merely a bothersome addendum to the real business of doing research. The second is the belief that all that really matters in this communication is that all the important content be included; "style" is a nicety that can be left to English classes.

What both of these beliefs ignore is the basic point stated in chapter 1: Science is a matter of shared information. For a research finding to mean anything, it must be made accessible to the scientific community as a whole. Furthermore, it must be made accessible in a way that heightens the probability that it will be attended to and assimilated. The written presentation must therefore be interesting, or the busy scientist may quickly turn elsewhere. It must be understandable, with all important details presented in a coherent fashion, or the reader will have no basis for evaluating the contribution. And it must be persuasive, with well-motivated arguments offered for the importance of the problem, the soundness of the methods, and the validity of the conclusions. It must, in short, be well written. The better the writer, the more likely it is that the work will have an impact on the field.

An even more obviously pragmatic justification can be offered for the importance of writing well. Before an article can be made available to the scientific community, it must be accepted for publication in a professional journal. Most journals are selective in what they publish, and the best journals have rejection rates of up to 90%. A poorly written article is simply much less likely ever to see the light of day than a well-written one. Busy editors and reviewers may be unwilling to make the effort to penetrate the poor writing to get to underlying content, and may be unable to find the content if they do make the effort. Furthermore, because the purpose of a research report is to communicate, the quality of the writing is a quite legitimate part of the evaluation process. It is an illusion to think that style and content can, or should, be separated in the evaluation of a written work.

Recognizing the importance of good writing is easy enough; achieving good writing is much harder. As noted, I settle in this chapter for offering a few general suggestions.

One suggestion is to read psychology before writing psychology. A basic difficulty that many students have in writing psychology research reports is simply that they have read few such reports themselves. Writing in psychology is not fundamentally different from other sorts of writing, as will be stressed shortly. Nevertheless, there is a kind of feeling for what is appropriate in a research report—for how things are said, for what should be included and what left out—that can be gained only by exposure to a number of real-life examples. Such exposure is not sufficient to ensure success, but it may well be necessary. There is no reason, of course, to seek models indiscriminately—there are plenty of bad examples in even the best journals. What makes more sense is to enlist some guidance in finding especially good examples and then learn from them.

The second suggestion is to seek simplicity in writing. The danger in a paragraph like the preceding one is that it may reinforce the notion that scientific writing is an abstruse business that is somehow basically different from other kinds of writing. In particular, scientific writing is *difficult*, packed with arcane technical terms, long and complex sentences, densely reasoned arguments, and so forth. It is true that scientific writing requires a kind of formality of discourse that may not be necessary in other kinds of writing. It is true too that technical terms exist in any science and are often preferable to less precise everyday language. It is not true, however, that scientific writing should *aim* for difficulty; rather just the reverse is true. There will be difficulties aplenty in the content being conveyed; the goal of the writing should be to help the reader surmount the difficulties and arrive at understanding. Thus, in general the short, simple word is preferable to the long, complex one; the familiar word to the obscure one; and the short, simple sentence to the long, convoluted one. The aim, after all, is to communicate, not to impress the reader with one's sophistication.

The third suggestion is to seek variety in writing. Simple sentences may be desirable, but an unbroken string of simple sentences quickly begins to pall. The shorter word is not always the best one, and in any case an occasional long word or long sentence can impart a kind of rhythm to the prose that enhances readability. Recall that one goal of effective writing is to interest the reader. The writer who consistently uses the same vocabulary, the same sentence structure, and the same paragraph structure may succeed in being clear but is unlikely to be very interesting. Clarity of expression and grace of style are not incompatible, and both should be sought.

The fourth suggestion is to seek economy in writing. All readers have limits on their time and patience, and all journals have definite limits on their available space. Articles whose length is out of proportion to their content invite a negative response. If submitted to a journal they may be rejected outright; at the least, reviewers are likely to demand substantial reductions in length. As they write, authors should continually ask themselves two questions: Is this information necessary to include? Have I expressed it in the most efficient manner possible? Or as Sternberg (2003) puts it, authors must always be ready "to recognize redundancy, repetition, reiteration, rehashing, restating, and duplication!" (p. 68).

The advice to be economical does not mean that scientific writing should be telegraphic. Brevity achieves nothing if important content is lost or if the terseness of the writing makes the paper unreadable. The primary goal remains communication, not space saving. A good rule is this: When in doubt, include. It is generally easier, for both author and reader, to pare down an unnecessarily long draft than to try to make sense of a bewilderingly short one.

The final suggestion is to be careful in writing. All of us at times write less well than we might, simply because we do not take the time to think sufficiently before writing or to reread after writing. At a basic level, there is absolutely no excuse for not *proofreading* a paper before submitting it. A paper replete with typographical errors is not only an insult to the reader but also a sure stimulus for negative reactions. At a higher level, it is important to reread papers for ideas as well as for grammar and spelling. Many papers contain blatant misstatements, contradictions, inconsistencies, and so forth that obviously would never have survived if their authors had simply taken the time to reread what they had written. Not everyone can be a master stylist; everyone, however, can be careful, and thereby can avoid the most obvious mistakes in writing.

Sections of an Article

In this section I consider the major parts into which psychology research reports are typically divided. This discussion is meant as a highlighting of important issues rather than as an exhaustive consideration of every question that might arise. The APA *Manual* provides further discussion of all the points considered here.

Introduction

The Introduction serves to orient the reader to the study. The kind of orientation that is appropriate depends to some extent on the particular study, and rigid guidelines about what should be said and when it should be said must therefore not be taken too seriously. In general, however, the Introduction should attempt to answer several questions. What is the problem under investigation? Why is this an interesting problem to study? What has previous research on this problem shown? What are the limitations in previous research that make further study necessary? What exactly is the gain in knowledge that the new study is intended to produce? And how (in a general sense) will this gain in knowledge be achieved?

The typical direction of movement in the Introduction is from the general to the specific. Thus, an Introduction might start with a statement of the general area of study (e.g., Piagetian training studies) and move from there to the specific question being examined in the new research (e.g., Can preschool children be trained with a modeling approach?). The first part of the Introduction is usually devoted to a review of relevant past research; the last part is more likely to point specifically to the new study. As noted, however, there are no fixed rules about sequence; the important point is to answer the questions sketched in the preceding paragraph, and to do so in a clear, readable, and persuasive fashion.

In reviewing past studies, the key word is "relevant." An exhaustive literature review is not appropriate for a research report. Not only is there inadequate space for such a review, but the readers to whom the report is directed can be assumed to be professionals who are already familiar with the general background for the research. In the case of a Piagetian training study, for example, there is no need to spend paragraphs detailing the conservation phenomenon or citing sources that document its validity; this material can be assumed to be familiar to anyone who would seek out such a report. Nor is there any need to try to review every past attempt to train conservation; there are far too many such studies, and most will be only tangentially related to any new effort. What *does* make sense, if possible, is to cite one or a few good review articles that summarize what past research has shown. And what is absolutely essential is to discuss any previous studies that have direct bearing on one's own research. Thus, in the training study example, any previous work with modeling would have to be discussed; so too would any studies with preschool children. Only in the context of a full and accurate consideration of what has gone before can the potential contribution of a new study be evaluated.

The concluding paragraphs of the Introduction typically constitute a bridge between the background material with which the article opens and the new research that occupies the remaining pages. It is here that any important new aspects of the research can be spelled out for the reader. Here too is the place to present specific hypotheses, if the study in fact fits the hypothesis-testing mode. Even if the author does not want to pose specific hypotheses, it is usually helpful to summarize the major questions (e.g., "Can 3-year-olds be successfully trained?") with which the study is concerned. Finally, the Introduction often concludes with an overview of the design and procedures that will be used to examine the questions under study. Details of procedure are the province of the Method and therefore not appropriate here; some orientation to what is to come, however, is often helpful.

Method

The Method section tells what was done in the study. The overall Method is usually divided into several subsections. At the least, divisions into *Participants* and *Procedure* are typical. Depending on the study, separate sections for *Materials, Apparatus, Design,* and *Scoring* may also be appropriate. Alternatively, information relevant to these matters may be incorporated under a general *Procedure* section. The decision as to how many subsections to include depends on both the complexity of the study (e.g., Was complicated apparatus used?) and the preferences of the researcher.

I have already talked about the kinds of information that should be conveyed under the heading of Participants. Recall that this information should include not only obvious demographic data about the sample (e.g., number, age, sex) but also details about methods of selection, proportion agreeing to participate, and proportion remaining in the study to its completion. In the remainder of this section I concentrate on the other parts of the Method.

It is often said (in the APA *Manual,* for instance) that a Method section should be detailed enough to permit an experienced investigator to replicate the study. The validity of this statement depends on what is meant by "replicate." If the term is taken to mean "perform the same study in every detail," then few Method sections would suffice; except perhaps in the simplest of studies, there is simply not space to spell out every aspect of the procedure. If the term is taken instead to mean "perform the same *kind* of study, with all *important* aspects of the procedure kept the same," then the goal of replicability becomes more reasonable. The challenge then is to determine which aspects of the procedure are important, and therefore require description in the Method.

There are various bases for deciding that a procedural detail is *not* important enough to warrant inclusion. In any study there are a number of procedural minutiae (e.g., Did the participant sit to the left or the right of the experimenter?) that are so unlikely to affect the results that they can safely be omitted. In studies with verbal instructions it is seldom necessary to reproduce the instructions verbatim; paraphrasing what was said is usually sufficient. In studies with multiple tasks of a similar nature it is sometimes sufficient to describe one or two tasks that can then serve as examples of the rest. And just as in the Introduction, some kinds of information can be assumed to be already available to any professional reader. If a standard piece of apparatus was used, for example, then the name and perhaps the model number should be sufficient; readers who are not already familiar with the apparatus can easily look it up. Similarly, an often-used test need not be described in detail; a published source, coupled perhaps with a brief indication of contents, should be enough.

The important details of the Method are those that the reader must have in order to understand and to evaluate the study. The information about participants discussed earlier falls in this category. So do various aspects of the experimental procedure, aspects that should be derivable from the discussion throughout this book of the important components of research. How were participants assigned to experimental conditions? What exactly did the conditions consist of, and how were they conveyed to the participants? (Note that verbatim quotation *may* sometimes be justified here.) What were the dependent variables, and how were they measured? In what order were the various events of the study presented? How many testers and how many observers were involved? Were the testers and observers blinded? Was reliability assessed, and if so, what was it? The particular information that needs to be conveyed will vary to some extent across studies. The replicability criterion is a helpful guide in thinking about what needs to be said. The really critical criterion, however, is the *evaluation* one: Has the reader been given enough information to evaluate the validity of the method?

Results

The Results section tells what was found in the study. It includes both descriptive statistics—for example, mean levels of performance in the different experimental conditions—and inferential statistics—for example, tests of significance for the differences between the means. It is essential that both kinds of statistics be provided. As we saw in chapter 8, descriptive statistics are hard to interpret without corresponding inferential tests. By the same token, an F test or a p level is of limited value without information about the means on which it is based. The reader needs to know both the statistical tests and the data that went into the tests.

Just as with the Introduction and the Method, the researcher must decide what information is important to include in the Results and what information can be omitted. If the earlier sections of the paper have done their job, they will have set up definite expectations with regard to exactly what questions are to be examined in the Results. Some answer should be given to each of these questions, even if in some cases the answer consists of a single sentence indicating that no significant effects were found. Furthermore, there is no law stating that every analysis in the Results must be clearly anticipated by what has gone before; unexpected or incidental findings of interest may emerge, and if so they should be included. On the other hand, there is also no law stating that *every* conceivable analysis of the data must find its way into the Results. Researchers should beware of cluttering their Results section with peripheral analyses that add little to the study.

The Results section is usually the hardest part of an article to read. Authors can do various things to make their readers' task easier. One is to adopt a logical order of presentation. A common practice is to begin with the major analyses and move only later to more secondary findings. If specific hypotheses were presented in the Introduction, then the results can be organized in terms of the hypotheses, with outcomes relevant to each hypothesis discussed in turn. If several dependent variables were assessed, then it may make sense to present results for each dependent variable separately. If different kinds of statistics were used (e.g., ANOVAs and correlations), then the Results might be organized in terms of type of analysis. Again, the particular approach that will work best depends on the particular study. The important point is to adopt a logical order, and to be sure that the logic is evident to the reader.

Authors should also provide summary statements that help the reader make sense of the data. A common misconception is that a Results section must consist solely of numbers and statistical tests. It is true that the purpose of the Results section is to present data, whereas the purpose of the Discussion section is to interpret data. It is *not* true, however, that statements about the meaning of the data are forbidden in the Results. Indeed, quite the reverse is the case. An unbroken string of *t*s, *F*s, and correlations can be both difficult to decipher and extremely frustrating to read. Much better practice is to spend a sentence or two, either in preview before the tests or in summary afterward, to say in plain English what the numbers mean.

It is important to be clear about what is meant here. What is permissible in the Results is a kind of first-order, close-to-the-data interpretation. Thus, a passage like the following might be appropriate:

> There was a significant interaction of age and experimental condition, $F(1, 96) = 8.95$, $p < .01$ (see Table 1 for relevant means). The training did produce significant effects, but only at age 5; 3-year-olds showed essentially no progress.

It would *not* be appropriate to go on to say, "The greater susceptibility of the older children is compatible with Piaget's emphasis on the importance of operational structures. . . ." This sort of second-order interpretive statement should be saved for the Discussion.

Authors can also help their readers through judicious use of tables and figures. The word "judicious" is important: Tables and figures are expensive to set in print, and journal editors tend therefore to frown upon their overuse. Whether a table or figure is justified depends on how easily the information in question could be conveyed in the text. If only a small number of means must be presented, then there is no need for a table, since the numbers can be easily given in the text. If 15 or 20 means are involved, however, then a table is much easier for the reader to process than a string of numbers in the text. In general, tables are preferable to figures for conveying exact values and for handling large sets of data (such as 15 or 20 means). Figures are especially useful for illustrating trends or interactions.

Several further rules govern the use of tables and figures. A table or figure should not repeat data given in the text or in another table or figure. If the same information appears in two places, one of the sources should be dropped. A table or figure should be largely self-explanatory—that is, comprehensible without reference to a detailed explanation in the text. Finally, although the table or figure may be self-explanatory, discussion of what it shows is still quite appropriate. Tables and figures supplement the text; they do not replace it.

For the most part, tables should be used to present descriptive, rather than inferential, statistics. Tables can sometimes be used for inferential statistics—for example, a table might summarize an ANOVA or regression printout. Such uses are justified, however, only when the

Table 10.1 An Example of a Table of Means (Hypothetical Data)

	Recognition		*Recall*		*Total*	
Age	*M*	*SD*	*M*	*SD*	*M*	*SD*
25	24.00	2.25	19.50	4.00	21.75	5.15
50	23.50	3.15	16.25	6.80	19.88	6.20
75	21.00	4.40	10.25	8.33	15.63	9.11
Total	22.83	3.96	15.33	7.56		

Table 10.2 An Example of a Table of Correlations (Hypothetical Data)

	Age	*IQ*	*Impulsivity*	*Problem solving*
Age		.07	−.35*	.46**
IQ	–	–	−.26	.53**
Impulsivity			–	−.32*
Problem solving				–

*$p < .05$. **$p < .01$.

analysis is both unusually complex (and therefore difficult to summarize in the text) and of central importance to the study. In most cases, it is the descriptive information that is more important to convey to the reader.

Probably the most commonly reported descriptive statistics are means and correlations. Tables 10.1 and 10.2 provide examples of what tables to report such information might look like. Note that a table of means should include a measure of variability (the standard deviations) along with the means. Note also that the table should be as informative as possible without becoming overwhelmingly large. In the example, the table provides the descriptive information for three effects from the study: the interaction of age and type of memory (the six cell means) and the main effects of age and type of memory (the marginal means).

The array in Table 10.2 is known as a *correlation matrix*. In the example all possible correlations are presented; in some instances only the significant values are included. Note that there is no need to give the values that fall along the diagonal (because these are the correlations of each measure with itself) or those that fall below the diagonal (because these are redundant with those above the diagonal).

Discussion

The Discussion section is the author's opportunity to draw everything together and to state the conclusions that the reader should take away. This section should have close links to both the Introduction and the Results. The Introduction lays out the questions that the research is intended to answer; the Discussion summarizes the answers that the researcher believes have emerged. The Results presents the data that the study has generated; the Discussion interprets these data and works always within their constraints. Whatever the author may have expected or hoped would occur, the data are the ultimate determinants of what the Discussion has to say.

For many authors, the Discussion is the hardest part of the paper to write (especially when the results are not all that was hoped for!). I offer here a few "don'ts"—practices to avoid when writing a Discussion section (and to watch for when reading one).

The Discussion should not simply reiterate the results already presented in the Results section. Some summarizing of findings is fine as a prelude to their interpretation. It is the job of the Results section, however, to make clear what the basic data are. The job of the Discussion is to explain the data.

With rare exceptions, findings not previously presented should not be introduced in the Discussion. The place to introduce new data is the Results.

As with other sections of the article, the Discussion should follow a clear and logical sequence. Some students write Discussions in what has been called a "striptease" manner, beginning with peripheral points and only gradually working up to their main conclusions. Usually, it is most helpful to the reader to have the major conclusions stated early in the Discussion. Elaborations, qualifications, and subsidiary points can be worked in later.

The Discussion need not devote equal space to every hypothesis or every finding. Some weighting in terms of importance is one of the author's responsibilities. Too much selectivity, however, can result in a distortion of the study's results. It may be misleading, for example, to focus on the one finding that supported a particular hypothesis, while ignoring three others that failed to support it.

It is quite appropriate for a Discussion section to venture beyond the immediate data into various speculations about the more far-reaching implications of the research. Some authors, however, venture so far afield that they lose sight of the study that they are supposedly discussing. Such wandering may sometimes be a method of compensating for results that were not as strong as had been hoped for. In any case, every point made in the Discussion should have a clear link to the research being reported.

One common method of concluding a Discussion section is with suggestions for future research. This practice is fine, as long as the suggestions are (a) specific, (b) sensible, and (c) concisely stated. Merely saying that more research is needed, however, does not tell the reader anything.

Another common method of concluding a Discussion is with a consideration of any weaknesses in the study that the author believes may have affected the results. This practice too is fine; indeed, some discussion of possible shortcomings can be seen as a basic form of honesty that the author owes the reader. It is important, however, not to dwell on the shortcomings or to become overly defensive or overly apologetic. For if the Discussion section is devoted mainly to negatives, the obvious question in the reader's mind will be "Why should I take this research seriously?" And the obvious question in an editor's mind will be "Why should this research be published?"

Other Sections

Although the Introduction, Method, Results, and Discussion constitute the main body, they do not exhaust the distinct parts of a research report. Table 10.3 provides a more comprehensive account of the different sections into which research reports are divided. The sections are listed in the order in which they would be typed in a submission to a journal. Again, a fuller discussion of what is appropriate for each section can be found in the APA *Manual.*

Some Specific Stylistic Points

APA Conventions

As noted, the discussion here is quite selective, focusing on a few rules that are often violated in student papers.[2]

Table 10.3 Sections of a Research Report

Section	*Description*
Title Page	Title should be brief (suggested length: 10 to 12 words), clear, and informative. Avoid unnecessary phrases like "A Study of." Title page also includes running head and author affiliation and byline.
Abstract	Summarizes the study in 100 to 150 words (depending on the journal). Should include information about the problem being studied, method of study, major results, and major conclusions.
Introduction	Conveys to the reader the problem under investigation, the reasons for interest in this problem, and the status of current knowledge based on a review of relevant past research. Concludes with an overview of the present study, including specific goals and specific hypotheses.
Method	Summarizes how the study was carried out, including information about participants, apparatus or materials, and experimental procedures. Should be detailed enough to permit evaluation.
Results	Presents the major data obtained in the research, as well as the statistical analyses of the data.
Discussion	Interprets the results of the study. Should place the data in the theoretical context provided by the Introduction and should attempt to answer the specific questions or evaluate the specific hypotheses presented in the Introduction.
References	Presents the sources for any references that were cited in the text. References are listed in alphabetical order.
Appendix	Presents relatively lengthy material (e.g., mathematical proofs, lists of stimulus materials) that would be difficult or inappropriate to convey in the text. Use should be minimized.
Author note	Presents identifying information about the authors and acknowledges sources of support.
Footnotes	Presents relatively brief material that would be difficult or inappropriate to convey in the body of the text. Use should be minimized.
Tables	Summarizes bodies of data that can be more efficiently conveyed in a table than in the text.
Figure captions	Describes the contents of each figure included in the article.
Figures	Summarizes bodies of data for which a graphical presentation is efficient and informative.

Table 10.4 Levels of Heading in APA Journals

Type of heading	*Example*
CENTERED UPPERCASE HEADING (Level 5)	EXPERIMENT 1: AN INTERVIEW VALIDATION STUDY
Centered, Uppercase and Lowercase Heading (Level 1)	External Validation
Centered, Italicized, Uppercase and Lowercase Heading (Level 2)	*Method*
Flush Left, Italicized, Uppercase and Lowercase Heading (Level 3)	*Participants*
Indented, italicized, lowercase paragraph heading ending with a period. (Level 4)	*The nonclinical group.*

Headings. APA style permits five levels of heading. These levels are shown in Table 10.4. When only one heading is needed (usually in very short articles), Level 1, the centered uppercase and lowercase heading, is used. When two headings are needed, Levels 1 and 3 are used. When three levels of heading are needed, the indented paragraph heading (Level 4) is added. Use of four or five headings is rare, being confined mainly to reports of multiple studies or to lengthy reviews or theoretical articles.

Measurements. APA policy is to express measurements in metric units—thus millimeters rather than inches, meters rather than yards, and so forth. Measurements expressed in nonmetric units are acceptable if accompanied by the metric equivalent—for example, "The rod was 3 ft (.91 m) in length."

Numbers in the text. The APA rule is that the numbers zero through nine are generally expressed in words, whereas numbers of two digits or more are generally expressed in figures. Thus, "There were eight experimenters and 120 participants." There are, however, a number of exceptions to this rule. Perhaps the most commonly occurring exceptions are that units of measurement must be accompanied by figures, not words (thus "4 years old," "6 weeks," etc.), and that a figure cannot be used to start a sentence (thus "One hundred and twenty participants were included"). Other exceptions are spelled out in the APA *Manual.* In my experience, treatment of numbers is probably the most common violation of APA style in student papers; thus careful consultation of the *Manual* is especially recommended whenever numbers are involved.

Statistics in the text. The author must always make clear which statistical tests were used on which aspects of the data. Thus, "Correct responses on the posttest were analyzed via a two-way analysis of variance, with age and condition as the independent variables." Sources do not need to be given for statistical tests, unless the test is an unusual one or its use is in some way controversial. Summaries of the results of inferential tests should include the name of the statistic, the degrees of freedom, the value of the statistic, and the probability level—for example, "There was a significant effect of age, $F(1, 96) = 7.90, p < .01$." Note that the summary of an inferential test is set off from the text with commas, not parentheses. Summaries of the corresponding descriptive statistics should include the appropriate measures of central tendency (usually means), along with a measure of variability (usually standard deviations)—for example, "Older children *(M* = 6.88, *SD* = 2.45) performed better than younger children *(M* = 4.90, *SD* = 2.26).

References in the text. The APA form for references in the text is the one used throughout this book: the name of the author and the date of publication, enclosed in parentheses—for example, "(Smith, 1992)." When the author's name is part of the text, only the date appears in parentheses—for example, "Smith (1992) reported." When a work has two authors, both names are always cited; in the text they are connected by "and," and within parentheses they are connected by an ampersand (&). When a work has three to five authors, all names are cited on first mention; on subsequent mentions the citation consists of the name of the first author followed by "et al." When a work has six or more authors, the "et al." is used even on first citation. When a parenthesis contains multiple citations by the same author, the references are listed chronologically; when a parenthesis contains multiple citations by *different* authors, the references are listed *alphabetically.* Thus, "(Smith, 1984, 1989, 1992)" and "(Brown, 1979; Jones, 1974; Smith, 1992)." Note that multiple references by the same author are separated by a comma, whereas those by different authors are separated by a semicolon.

Reference list. The reference for every source cited in the text must be given in the References at the end of the article. *Only* those sources cited in the text are included in the References. Authors must be certain that information provided in the References is accurate, so that interested readers can track down sources if they wish. Authors must also follow APA rules for References, rules that occupy a full 67 pages in the *Manual*![3]

General Matters of English

The discussion here is even more selective (and idiosyncratic). What follows is a potpourri of points of English usage that are frequently violated in student papers (and elsewhere).

Affect-effect. Both *affect* and *effect* can be used as either nouns or verbs. In their most common uses, however, *affect* is a verb meaning "to influence," and *effect* is a noun meaning "outcome" or "result." Thus, "Did the manipulation affect performance? The posttest showed a clear effect."

Among-between. Although exceptions exist, *between* should generally be limited to statements involving two elements, with *among* used when three or more elements are involved. Thus, "There were no differences among the three conditions," not "There were no differences between the three conditions."

Contractions. Don't (i.e., do not) use contractions.

Data. *Data* is a plural noun. Thus, "The data were clear," not "The data was clear." Similarly, *criteria* and *phenomena* are plural nouns and therefore require plural verbs. The singular forms are *datum* (rarely used), *criterion,* and *phenomenon.*

e.g.-i.e. The abbreviation *e.g.* means "for example"; the abbreviation *i.e.* means "that is." Thus, "Numerous theorists (e.g., Brown, 1979; Smith, 1992) have claimed" and "What seemed critical was the experimental manipulation (i.e., the presence or absence of reward)." The common mistake is to use *i.e.* where *e.g.* is appropriate. (Note that APA policy is to restrict the use of *e.g.* and *i.e.* to material within parentheses.)

Fewer-less. *Fewer* refers to number when countable objects are involved; *less* refers to amount or degree along some continuous, noncountable dimension. Thus, "The control participants gave fewer correct responses" and "The control participants showed less success." The common mistake is to use *less* in place of *fewer,* as in "The control participants gave less correct responses."

Hyphens. Decisions about whether to hyphenate a compound word or compound phrase can be difficult, and the APA *Manual* provides five pages of guidelines on the question. Included among the guidelines is a list of prefixes that do *not* require a hyphen in APA style; among the prefixes in this category are *inter, mid, multi, non, post, pre,* and *semi.*

It's-its. *It's* is a contraction meaning "it is" or "it has"; *its* is a possessive adjective meaning "belonging to it." The mistake here is to use *it's* as a possessive, apparently out of the belief that possessives should have apostrophes in them.

Only. Although exceptions exist, the word *only* should usually come immediately before the word that it modifies. Thus, "The failure manipulation produced visible distress in only two children," not "The failure manipulation only produced visible distress in two children."

Pronouns. Various issues related to the use of pronouns can be problematic. The APA *Manual* indicates that the use of the first person singular pronoun is now acceptable in scientific writing, and is even preferable to an exclusive reliance on the third person and passive voice. Students should be aware, however, that the use of *I*, although more common than formerly, is still fairly rare, being confined mainly to senior (or perhaps very confident) researchers and theoretically oriented papers. The use of *I* in a routine research report is likely to be jarring to many readers. The use of *we* in multiple-author reports, however, is fairly common (note, though, that *we* should never be used in the editorial sense if there is only one author).

The *Manual* also encourages the use of nonsexist terms, including nonsexist pronouns, and it provides an extended discussion of how to achieve not only nonsexist wording but also unbiased language more generally. Students should read this section carefully and should refer to it when writing. It is important, however, that the search for nonsexist language does not impair the grammaticality or readability of the paper. Thus, sentences of the following sort should be avoided: "The child was told to push his or her button as soon as he or she saw the flash appear on his or her screen." In many cases (including this one), the pronoun problem can be solved by replacing the singular *child* with the plural *children.* Note also that "it" should never be used when referring to a child, even a very young infant.

A final point about pronouns is a matter of basic English: Pronouns should agree in number with their antecedents. Thus, "Each participant filled out his or her response sheet," not "Each participant filled out their response sheet." Note, in particular, that if you refer to "the child" you are talking about one child and cannot go on to use a plural pronoun. In my experience, constructions of the "child . . . they" sort are easily *the* most common grammatical error in student papers.

Since and while. The *Manual* recommends that *since* and *while* be used only in their temporal senses. Thus, *since* should be used to refer to the passage of time and not as a synonym for *because,* and *while* should be used to refer to the simultaneity of two events and not as a synonym for *although* or *whereas.* Many guides to writing style, however, disagree with one or both of these admonitions (see, for example, Bernstein, 1971; Follett, 1966; Fowler, 1996; Strunk & White, 2000), and for this reason I have not followed them in this book. Nevertheless, students should be aware that some reviewers or editors may frown on the uses of *since* and *while* that are modeled here.

Tense. A journal article reports events that have already occurred. For the most part, therefore, the past tense is appropriate—thus, "Smith reported," "Participants were told," "The analysis revealed." The present tense can be used when discussing results that are literally there before the reader—for example, "The table shows." The present tense is also appropriate when making statements that are essentially timeless—"The theory states," "Young children are often impulsive," "Conservation is an important ability."

This. An occasional use of the indefinite *this*—that is, *this* without a specific referent immediately following—is acceptable. Be careful not to overdo it, however. Thus, in general, write "this point should be clear by now," not "this should be clear by now."

Where. The word *where* denotes spatial location, and it should be reserved for this use. Thus, do not write "Children heard a story where the critical information was presented." Replace the *where* with *in which.*

Summary

The final step in a research project is communicating one's work to others. This chapter discusses how to write research reports in psychology.

The chapter begins by emphasizing the importance of writing well, noting that the clarity and persuasiveness with which research is presented have a direct bearing on the impact that the research is likely to have. A well-written article is more likely to be published than is a poorly written one, and it is more likely to be attended to and assimilated after publication.

A number of suggestions are offered for improving the effectiveness of one's writing. A first suggestion is to read good examples of writing in psychology, drawing from such models ideas about both what to say and how to say it. A second suggestion is to seek simplicity in writing, for jargonese and needless complexity are obstacles to clear communication. A third suggestion is to aim for variety in writing, the goal being to produce prose that is not only clear but also interesting and readable. A fourth suggestion is to be economical in writing, because both journal space and reader patience can be quickly exhausted by irrelevant detail or unnecessary wordiness. The final suggestion is to be careful when writing, because many obvious errors can be avoided if the author takes the time to be sure that every part of the manuscript is exactly as intended.

The middle part of the chapter discusses the major sections into which psychology research reports are divided. The Introduction presents the issues and goals that underlie the study; it sets the research in the context of relevant past work and prepares the reader for the specific study to follow. The Method describes the participants included in the study and the procedures used to collect the data; it should be detailed enough to permit critical evaluation of the research. The Results section summarizes the data produced by the study; it presents both descriptive statistics (e.g., mean levels of response) and inferential statistics (e.g., tests of significance for the differences between means), along with summary statements of what the statistics mean. The Discussion interprets the results of the study; it answers questions posed in the Introduction, evaluates any specific hypotheses that were offered, and summarizes the author's assessment of what the research has and has not accomplished.

The concluding section of the chapter discusses specific stylistic points that often prove troublesome for students. Included are both matters of APA style (e.g., rules for headings and references) and matters of English grammar—i.e., don't write passages like this one, which would be likely, even if they contained less errors (and there are five here), to effect the reader negatively.

Exercises

1. With your instructor's approval, exchange drafts of one of your papers with a fellow student from the course. Prepare detailed critiques of each other's papers, evaluating content, general writing style, and adherence to APA rules.

2. Chapter 2 introduced two research examples that recur at various points throughout the text: the Dufresne and Kobasigawa study of memory and study time in grade-school

children, and the Park and Cherry study of spatial memory in young and elderly adults. Using both the descriptions in the text and your own inventiveness (since you will have to create details not provided to you), write a Method section for one of these studies and a Results section for the other. Compare your drafts with the comparable sections in the original articles.

3. Select any two of the tables in the text that present data (e.g., Table 2.1, Table 2.2, Table 8.2) and prepare APA-style figures that present the same information. Select any two of the figures in the text that present data and prepare APA-style tables that present the same information.

Notes

1. A helpful supplement to the *Manual* is a Web site maintained by APA: www.apastyle.org. The Web site summarizes the most important changes to the most recent edition of the *Manual,* gives the answers to frequently asked questions, provides updates with regard to evolving stylistic issues (e.g., how to do electronic references), and gives information about further sources of help with APA style (e.g., an electronic "APA-Style Helper").

2. I should note that some instructors may follow APA style selectively, preferring that some aspects of papers for their courses adhere to a different set of stylistic guidelines. Obviously, in such cases the instructor's preferences take precedence over APA rules.

3. Students are not the only ones who find the APA rules for references challenging. A survey of editors of psychology journals (Brewer, Scherzer, Van Raalte, Petitpas, & Andersen, 2001) revealed that references were the most common source of deviations from APA style in manuscripts submitted for publication.

11

Infancy

In the final four chapters we turn from general principles of research to specific research topics in developmental psychology. The organization is partly chronological and partly topical. The section begins with a chapter on infancy and concludes with a chapter on aging. In between are chapters devoted to cognitive development and social development. Untidy though this chronological-topical division may seem, it is in fact typical of the field. Some developmental psychologists identify themselves primarily in terms of age group studied, and others identify themselves primarily in terms of topic. Similarly, some methodological issues are peculiar to the study of particular age groups, and others are more closely linked to particular topics.

The discussion of infancy is divided into two broad sections. We will begin by considering some of the general issues involved in doing research with babies. For the most part these issues consist of specific applications of points discussed in the opening chapters of the book. As we will see, however, some methodological challenges are especially acute for the researcher who decides to study infants.

The second section is devoted to particular kinds of infancy research. The coverage of topics is necessarily selective, as indeed it is throughout the next four chapters. It should be possible, however, to highlight some of the most important, and methodologically interesting, kinds of infant study.

General Issues

Sampling

We begin with one of the basic steps in research identified in chapter 1: getting research participants. It was suggested in chapter 1 that this problem tends to be especially great for developmental psychologists. I can add now that the researcher of infancy can probably lay claim to the greatest woes with respect to subject recruitment. Infants cannot volunteer for experiments, as can older children or adults. Nor can they be readily recruited from some institutional setting, such as a grade school, introductory psychology class, or nursing home. How, then, do researchers find babies to study?

The answer is: in a variety of ways, with the particular way dependent on both the age of the infant and the resources of the researcher. When newborns are the targets, an institutional source *may* be available, if the babies can be tested before being taken home from the hospital. Note, however, that such early hospital testing has become less feasible than it used to be, given the decline in length of hospital stay for both newborn and mother. When older babies are the target group, recruitment may be possible through the rolls of pediatricians or well-baby clinics. Babies of any age may be solicited through newspaper advertisements describing the research and inviting interested parents to call. Or the direction of contact may be reversed: Some researchers keep track of birth announcements and then call parents as the baby reaches the appropriate age for the study. Whatever the initial approach, the willingness of the parent to allow the baby to be studied remains the sine qua non.

As the preceding suggests, finding infants to study may be a time-consuming and expensive process. The real issue, however, is the representativeness of the sample obtained. I noted in chapter 2 that truly random sampling from a target population is a desirable but rarely attained goal. Some deviation from randomness is the norm in research. At least within the span of childhood, however, such deviations are probably greater for infancy than for any other age group. Babies who are brought to well-baby clinics are a particular subset of the population of babies as a whole; babies who are brought to pediatricians are a different subset. Parents who respond positively to phone calls soliciting participation are a particular subset of parents; those who respond to newspaper ads are another, undoubtedly smaller, subset. For all of these reasons, babies who find their way into research are often a distinctly nonrandom sample of the larger population in which the researcher is interested.

A second issue concerns how many babies make it through to the end of the study. Dropout in infant research is substantial. Indeed, the dropout rate is almost certainly greater in studies of infancy than in any other sort of research with human participants, figures of up to 50% to 60% being not uncommon. The most frequently cited reason for dropout is fussiness; drowsiness or going to sleep is also a common problem. To the extent that babies who drop out of research differ from those who remain in, then an initially nonrandom sample will become even more nonrandom.

Is the nonrandom nature of most infant samples a serious problem? Not necessarily. The fact that a sample is nonrandom does not automatically mean that findings obtained with it cannot be generalized to some larger group. This is the point that was made in chapter 2 in the general discussion of sampling. As noted then, the issue in selecting a sample is not randomness per se; it is whether any deviations from randomness are at all likely to bias the results. With many kinds of research it is simply not plausible that the nonrandom nature of the sample could make any difference. This argument is probably more applicable to infancy and to the kinds of issues studied with infants than it is to older, more diversified age groups. And it can be applied not only to initial sampling but also to the problem of dropout. All babies are at times fussy and sleepy. Some babies happen to be fussy or sleepy during the experiment and therefore drop out; other babies happen not to be and therefore remain in. But there is no reason to think that the two groups are systematically different.

Although this kind of plausibility argument may be generally valid, its application to particular cases still requires scrutiny. Particularly when the initial selection is unusual or the dropout substantial, questions of external validity may be quite legitimate. In addition, there are some data to question the assumption that dropout in infant research is random and therefore nonbiasing. For example, Lewis and Johnson (1971) tested 3- and 6-month-old infants on a series of measures of visual attention. They compared responses on a subset of the measures for two groups of babies: 22 infants who completed the entire series of tests, and 15 infants who provided data on the target measures but dropped out before the testing was completed. Infants who remained available for the entire series showed greater ability to distinguish between simple and complex stimuli than did infants who eventually dropped out. Lewis and Johnson concluded from these results that "our implicit assumption about the composition of those infants who are excluded from data analysis may be wrong. . . . The elimination of infants, especially in the large numbers, may result in serious biasing of the obtained data" (p. 1055). (See also Richardson and McCluskey, 1983.)

The Importance of State

The infant is a difficult experimental subject, often too fussy or too sleepy to make it through an experiment. This observation has general significance, for it highlights the importance of **state** in attempting to make sense of infant behavior. How a baby responds to the environment depends very much on the baby's immediate state of arousal. Is the baby awake and alert? Drowsy and about to drift off to sleep? Hungry and therefore irritable? Screaming with rage? Or perhaps deep in peaceful sleep?

As this run-through of possibilities suggests, the categories of "fussy" and "sleepy" that have been used thus far are rather gross indices of state. Psychologists have in fact devised more refined classification schemes. One common scheme is shown in Table 11.1.

The importance of state is not limited to infants, of course. Any of us is more likely to be responsive to an environmental stimulus when wide awake than when sound asleep, and any of us may behave differently when hunger pangs are gnawing than when they are not. For various reasons, however, state looms larger in the study of infants than it does in the study of older children or adults. States tend to be more labile in infancy, and shifts are frequent from one state to another. As this lability suggests, infants are less able to control their state than are older children or adults. One can assume that a 10-year-old who is awake and alert at the start of an experiment will still be reasonably awake and alert 20 minutes later. No such assumption can be made for the infant. Finally, the proportion of time spent in the various states changes with development. Most obviously (although not only), the proportion of time awake and alert increases dramatically with age. The newborn sleeps an average of 16 to 18 hours a day (Davis, Parker, & Montgomery, 2004). Only about 10% of the newborn's time is spent in the state of alert inactivity (Berg & Berg, 1987).

The most general implication of state is the one noted earlier: Understanding infant behavior requires knowledge of the infant's state. In some studies, state is included as one of the variables whose effects are being determined. Researchers have looked, for example, for differences in response to tactile stimuli during quiet and active sleep (e.g., Rose, Schmidt, & Bridger, 1978), or at heart-rate responses to auditory stimuli in sleeping and awake infants (e.g., Berg, Berg, & Graham, 1971). In some studies, state has served as a dependent variable, the goal being to identify factors that affect state. There are studies, for example, of the effects of experimental alteration of sleep cycles on the baby's subsequent state (e.g., Anders & Roffwarg, 1973) or of the influence of different kinds of stimulation on maintenance or change of state (e.g., Jahromi, Putnam, & Stifter, 2004). Most commonly in infant research, however, the goal has not been to study state but to *control* for it. What the researcher seeks, in other words, is to test all babies in the same state, whatever state is deemed most appropriate for the issue being examined. In most cases the goal has been to work with babies while they are quiet and alert, because this is the state most conducive to perceptual and cognitive functioning.

Researchers of infancy resort to various stratagems in the attempt to get babies while they are in the desired quiet and alert state. It is common practice to ask parents to bring babies in for testing at a time of day when the babies are usually awake and happy. Various

Table 11.1 States of Arousal in Infants

State	*Description*
Regular sleep	During regular sleep babies lie quite still with their eyes closed and unmoving. Their respiration is even, and their skin is pale.
Irregular sleep	During irregular sleep babies' muscular response to pressure is stronger than in regular sleep. Babies jerk, startle, and grimace spontaneously. The eyes are clearly closed but sometimes they move. The skin may be flushed, and breathing is irregular.
Periodic sleep	Periodic sleep is a combination of regular and irregular sleep: it consists of bursts of rapid breathing, jerks, and startles, followed by spells of quiet.
Drowsiness	During drowsy states babies are moderately active. Their eyes open and close intermittently, and they look glazed. Respiration is regular but more rapid than in regular sleep.
Alert inactivity	Alert, inactive babies are awake. Their eyes are open and shining, and they look at their surrounding with interest. Babies' bodies are relatively still, and their respiration is rather fast and irregular.
Waking activity	During waking activity babies are awake but their eyes focus less often, and they have spurts of vigorous activity. During these random spurts they move their legs and arms and twist their torsos. The length and intensity of these activity spurts vary.
Crying	During crying states infants engage in vigorous activity. Their skin is flushed and they cry (although without tears yet).

SOURCE: From *Understanding Infancy* (pp. 38, 40), by E. Willemsen, 1979, San Francisco: W. H. Freeman & Co. Willemsen's descriptions are adapted from Wolff (1966).

manipulations are also possible in the experimental setting. The position of the infant can have an effect on state. It has been found, for example, that very young infants are typically more alert when upright than when lying down (Korner & Thoman, 1970). It has long been known that change in position can serve to wake up a drowsy baby or quiet a fussy one. Pacifiers have also been used to calm fussy babies; some evidence indicates, however, that pacifiers may also affect other aspects of the baby's response, and thus researchers must be careful about their use (Field, 1982).

However ingenious, the researcher still remains at the mercy of the infant. This dependence on the whims of the baby has some definite methodological implications. Testing sessions must be kept fairly short, lest the infants move out of the desired state. Multiple experimental sessions may therefore be necessary, or perhaps adoption of a between-subject rather than within-subject design. If within-subject testing *is* used, counterbalancing of tasks or conditions is even more important than usual. An assumption that order is irrelevant is dubious at any age; with the rapidly changing infant, however, such an assumption is out of the question.

The most serious methodological implication concerns the selectivity involved in research on infancy. This selectivity operates at two levels. As we have already seen, there is selectivity with respect to who stays in the study and who drops out. Only babies who can maintain the desired state end up as participants in research. There is also selectivity with regard to which aspects of these eventual participants' behavior are considered. Studies tend to take babies at their very best—at those times when the babies are happy, awake, and alert. Especially for young infants, such times do not occupy a very large proportion of the baby's day. The focus in research is typically on the infant's *optimal* performance, a quite legitimate and important question. It is important to remember, however, that the infant's *typical* performance may fall well short of this optimum.

Response Measures

Coping with state is one of the two great challenges of infant research. The other is finding responses from which the baby's capacities can be inferred.

Consider the most challenging case of all: the newborn infant. The newborn is not totally helpless; rather he or she comes into the world equipped with various adaptive reflexes—that is, wired-in, automatic responses to particular kinds of stimuli. The newborn does not come equipped, however, with much in the way of skilled, voluntary motor behavior. Nor, of course, does the newborn possess language. The latter deficiency persists throughout infancy; indeed, the term "infant" means "without language." It is worth reflecting for a moment about the extent to which our assessments of older children and adults depend on the use of language. With infants, however, all of the verbally based techniques that we routinely use to study older participants are ruled out. How, then, can we figure out what a baby is thinking or experiencing?

To a good extent, the history of research on infancy is a history of the gradual discovery of more and more responses from which the infant's experience of the world can be inferred. Much of the remainder of this chapter discusses what these responses are. The present section is intended simply as a preview of some of the most informative and generally applicable response measures in infancy. Although the responses that are considered might at first seem diverse, they do share several features: All place minimal response demands on the infant, all can therefore be emitted by even a very young infant, all can be accurately measured, and all lend themselves (at least sometimes) to clear interpretation.

One common measure is *visual fixation.* Even newborns can, and do, look at objects. Furthermore, even newborns can exercise some selectivity in what they look at. As we will see shortly, the fact that infants tend to look more at some stimuli than at others has proved invaluable in the study of perceptual abilities and perceptual preferences. Looking responses also play a role in the study of the other two domains considered in this chapter: early cognitive development and early social development.

A second frequently utilized response is *sucking.* Sucking, too, is a response that is available from birth, and it too is a response in which all infants engage. Most infants, in fact, spend more time in nonnutritive sucking than they do in sucking for food. What is commonly examined in studies that measure sucking are changes in sucking in response to environmental change around the baby. For example, does the baby stop sucking if a novel stimulus suddenly appears in his or her field of vision?

Physiological responses constitute a third important class of dependent measures. Here, obviously, the requirement of an overt response disappears completely. With modern technology a wide range of physiological responses can be assessed. The most popular measure in research on infancy has been heart rate. Heart rate has several desirable qualities from the point of view of a researcher. Unlike some physiological responses, heart rate can be recorded in response to any sort of stimulus in any modality. The physiological system that mediates heart rate is relatively mature at birth, making heart rate a usable measure even in a newborn (indeed, even in a third-trimester fetus). Finally, changes in heart rate are directional—that is, the rate either speeds up or slows down. As we will see, considerable evidence indicates that heart-rate deceleration has a different meaning than does heart-rate acceleration. Changes in heart rate can thus provide information that may not be available when a unidirectional response is measured.

The three responses just considered by no means exhaust the infant behaviors that are of interest to developmental researchers. Nor, of course, does this brief introduction make clear how fixation, sucking, or heart rate is actually used in research. I consider response measures more fully in the discussion of various kinds of infancy research.

Age Comparisons

The general issue of age comparisons was discussed in chapter 3. The focus now is on age comparisons that involve infants. Drawing valid age comparisons is in some ways easier and in some ways more difficult when infants are the target. The ease or difficulty depends on both the particular ages and the particular response measures examined.

Consider a comparison of different ages within the span of infancy. Such studies tend to be primarily cross-sectional rather than longitudinal, just as does developmental research in general. In research with infants, however, some of the problems of cross-sectional designs discussed in chapter 3 are greatly reduced. In particular, the possibility of cohort effects—that is, effects stemming from differences in generation rather than differences in age—is much less than when older samples are compared. With rare exceptions, researchers can assume that 4-, 8-, and 12-month-old babies belong to the same generation and that any age differences that may be found are therefore not confounded with differences in cohort. It is, of course, still important to determine why the age differences appear; one major class of explanations, however, can be ruled out.

Longitudinal research also tends to be more clearly interpretable when the time span is confined to infancy. Recall that a major problem in longitudinal designs is the confounding of age and time of measurement. This confounding holds whatever the age of the sample, but it is much less likely to be a factor in studies of infants. It is implausible, for example, that a baby will behave differently at 12 months than at 6 months simply because one assessment is made in March 2006 and the other in September 2005. Both the short time span and the nature of the infant make it unlikely that historical-cultural change can account for a change in behavior.

Other possible problems in longitudinal research may also be reduced when the focus is on infants. Awareness of being repeatedly studied, for example, is less likely to be a biasing factor for infants than for older children or adults. Finally, interpretive issues aside, longitudinal research is simply more feasible with infants than with older samples. As we saw, a major obstacle to longitudinal study is the fact that the researcher must wait for the participants to age across some developmentally interesting span—for 10-year-olds to turn into 18-year-olds, 50-year-olds into 70-year-olds, and so forth. The researcher of infancy, however, can chart dramatic developmental changes within the span of a few months.

There is a downside, however, to the point just made. The rapidity of change in infancy raises the issue of measurement equivalence—that is, the question whether a measure means the same thing at different ages. The heart-rate response discussed earlier provides an example. There is ample evidence to indicate that newborns tend to respond to most new stimuli with heart-rate acceleration, whereas older infants tend to respond with heart-rate deceleration (Berg & Berg, 1987). In older participants, heart-rate acceleration is associated with fear or defense, and heart-rate deceleration is associated with attention. Some investigators, therefore, have interpreted this change in infancy as evidence of a shift from a primarily defensive to a primarily attentive response to new stimulation. Others have argued, however, that the shift may be unrelated to defense-attention, but may result instead from further maturation of the physiological system that controls heart rate (Porges, 1979). Thus, the fact that the same response can be measured at different ages does not guarantee that the response has the same meaning.

Further discussions of the issues considered here are available in various sources, including Holmes and Teti (2005) and Lamb, Bornstein, and Teti (2002). Our discussion turns next from general issues to particular kinds of infancy research. The remainder of the chapter considers three topics that have been popular foci for research in infancy: infant perception, infant cognitive development, and infant social development.

Infant Perception

Two general issues have been of interest in studies of infant perception. One is the question of *perceptual ability.* How well developed are the baby's perceptual abilities at birth, and what are the developmental changes in these abilities across the first 2 years? Numerous specific questions fall within this broad domain. How clearly can a newborn see an object, and what kinds of visual discriminations can a newborn make or not make? What is the newborn's threshold for detecting a sound, and what differences in sound are or are not discriminable? Does a newborn or young infant possess perceptual constancy, and if not, when does constancy develop? Any question that can be asked of perception in general can be asked from the developmental point of view, starting with infancy.

The second general issue is that of *perceptual preference.* Given the abilities that infants have, toward what sorts of stimuli do they direct attention? What do babies find interesting to look at, listen to, or touch? And why do babies show the preferences that they do?

Several general points about research on perception are worth making before the discussion of specific methods. First, the majority of studies of infant perception are cross-sectional rather than longitudinal. Although this emphasis is potentially biasing, it seems doubtful, for reasons discussed in the previous section, that the cross-sectional focus has led to incorrect conclusions about changes with age. Furthermore, most studies of infant perception have been concerned with commonalities of development rather than individual differences among children. An ability like size constancy, for example, is eventually present in all children, and the typical research interest has been in when and how it develops, not in individual differences in the speed or quality of development. As we saw in chapter 3, a prime motivation for longitudinal research is to examine the consistency of individual differences over time. Because individual differences are seldom a concern in work on infant perception, a longitudinal design may add little.[1]

A second point has to do with the lab-field continuum discussed in chapter 6. The great majority of studies of infant perception are carried out in laboratory rather than field settings. Furthermore, the settings that are used are definitely toward the lab end of the continuum—artificial, highly controlled, distinct from "real life" in various ways. This is not, of course, because perception does not occur in the natural environment; perception is the most pervasive of psychological processes. Studying perception, however, typically requires exact knowledge of what the stimulus is and precise measurement of rather subtle responses to it, and these criteria can usually be met only in the laboratory.

A final point concerns the two obstacles to infancy research discussed earlier: the baby's shifting state and limited response repertoire. There is no aspect of infancy research for which these problems are more pressing than the study of perception. Researchers of perception typically attempt to work with babies in the quiet and alert state that is optimal for most kinds of perceptual functioning. This attempt embodies all of the practical difficulties and selectivity discussed earlier. The problem of response measures arises not only because of the infant's limited repertoire but also because of the nature of perception: an internal, experiential phenomenon that is not necessarily linked to overt, measurable behavior. Finding response measures from which the baby's perceptual experience can be inferred is thus especially challenging.

Preference Method

The **preference method** is a technique devised by Robert Fantz (1961) to study visual perception in infancy. The specific question that it is designed to answer is the question of visual discrimination: Given any two visual stimuli, can a baby tell them apart? This question, clearly, is central to any assessment of babies' visual capacities.

An early version of Fantz's apparatus is shown in Figure 11.1. The infant is positioned in the looking chamber so that he or she can look at either of two stimuli suspended above. The stimuli are just far enough apart so that they cannot be fixated simultaneously; rather, a slight head turn is necessary to direct the eyes to one or the other. This head turn is a response that even a newborn can make. Apart from the head turn, the only behavior required of the infant is fixation—that is, looking at one or the other stimulus. The dependent measure is how long the infant looks at each stimulus.

In a typical study with the Fantz procedure, the two stimuli that are being compared are presented for a number of trials. A particular stimulus appears on the left side half of the time and on the right side half of the time. The question of interest is whether, over trials, the infant looks significantly longer at one stimulus than at the other. If the infant does look longer at one stimulus, the infant is said to show a preference for that stimulus. And if the infant shows a preference, the conclusion drawn is that the infant must be able to discriminate between the stimuli. The reasoning is straightforward: Only if the infant can see a difference could a preference possibly emerge. If the stimuli look the same, then there is no basis for looking systematically longer at one stimulus than at the other.

The preference method has been applied to a wide range of visual stimuli. It has been used, for example, to determine whether infants can discriminate between a patterned stimulus and a comparable nonpatterned stimulus (e.g., a bull's eye and a plain circle), between a colored stimulus and a comparable noncolored one, between two-dimensional and three-dimensional versions of an object, between a novel stimulus and a familiar one, and between their mothers' faces and those of strangers. Note that the method provides information not only about discrimination but also about the second general question introduced earlier: perceptual preference. Suppose we find (as we do) that a newborn looks longer at a patterned stimulus than at a nonpatterned one. This finding tells us something about the baby's ability to discriminate patterning in stimulation. It also tells us something about the baby's preferences: Not only can the baby see patterning, but the baby *prefers* patterning.

The strengths of the preference method lie in its minimal response demands and wide scope of application. The method also has some limitations. Specifying the exact basis for a discrimination can sometimes be difficult. If babies show a preference, we know that they can see *some* difference between the stimuli,

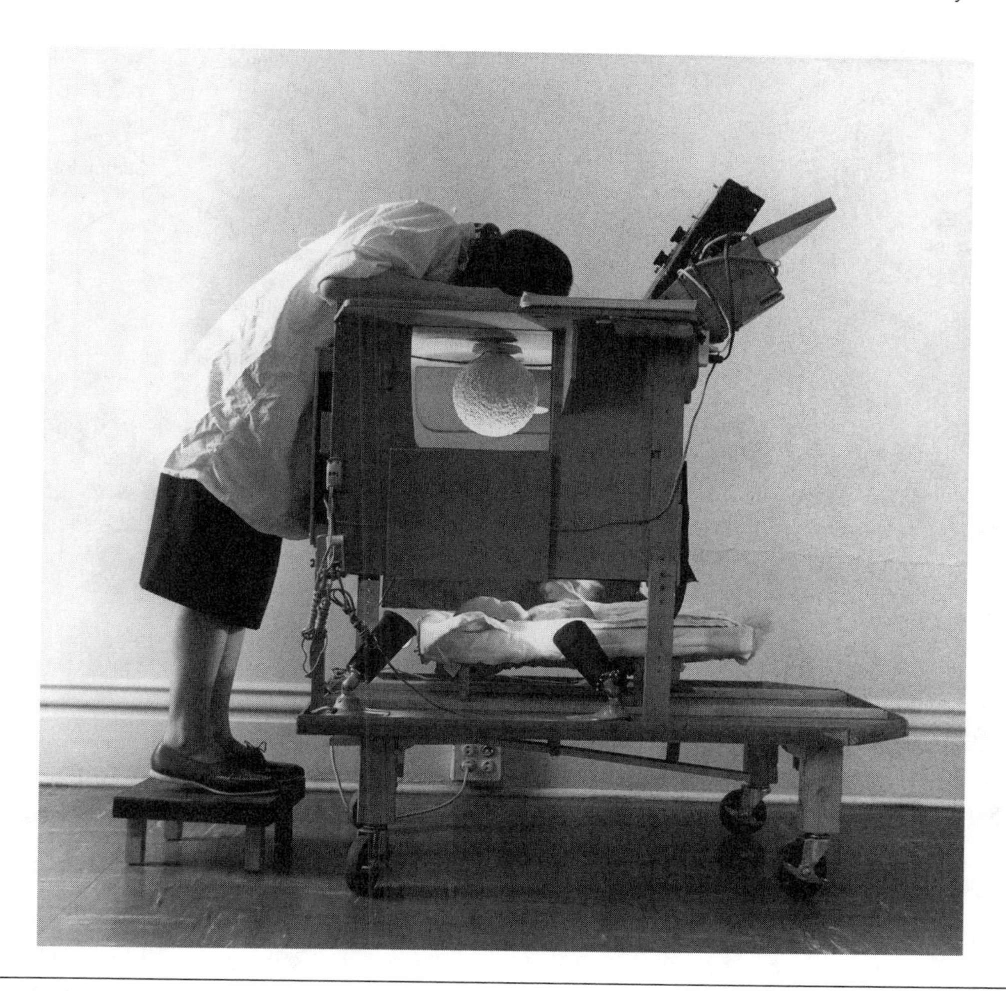

Figure 11.1 Apparatus used in the Fantz preference method to study infants' visual abilities

SOURCE: From "The Origin of Form Perception," by R. L. Fantz, 1961, *Scientific American, 204*, p. 66. Copyright © 1961 by Scientific American, Inc. Reprinted with permission. Photograph by David Linton.

but we do not necessarily know what information was used to make the discrimination. One way around this problem is to attempt to design stimuli that differ in only one critical respect, such as a two-dimensional circle and an otherwise identical three-dimensional sphere. If only a single difference exists, then presumably infants must be using this difference if they show a preference.

Another possibility is to record not only at which stimulus the infant looks but also exactly where on the stimulus the infant's eyes fixate. With modern infrared photography, it is possible to obtain very exact records of eye movements and eye fixations, and this information can help to pin down the basis for a discrimination. Such **eye-movement recording** constitutes, in fact, a valuable tool in its own right for studying early visual development, quite apart from its application to the preference method. Figure 11.2 shows a schematic representation of a typical system for recording eye movements.

The main limitation of the preference method is that negative results cannot be interpreted.

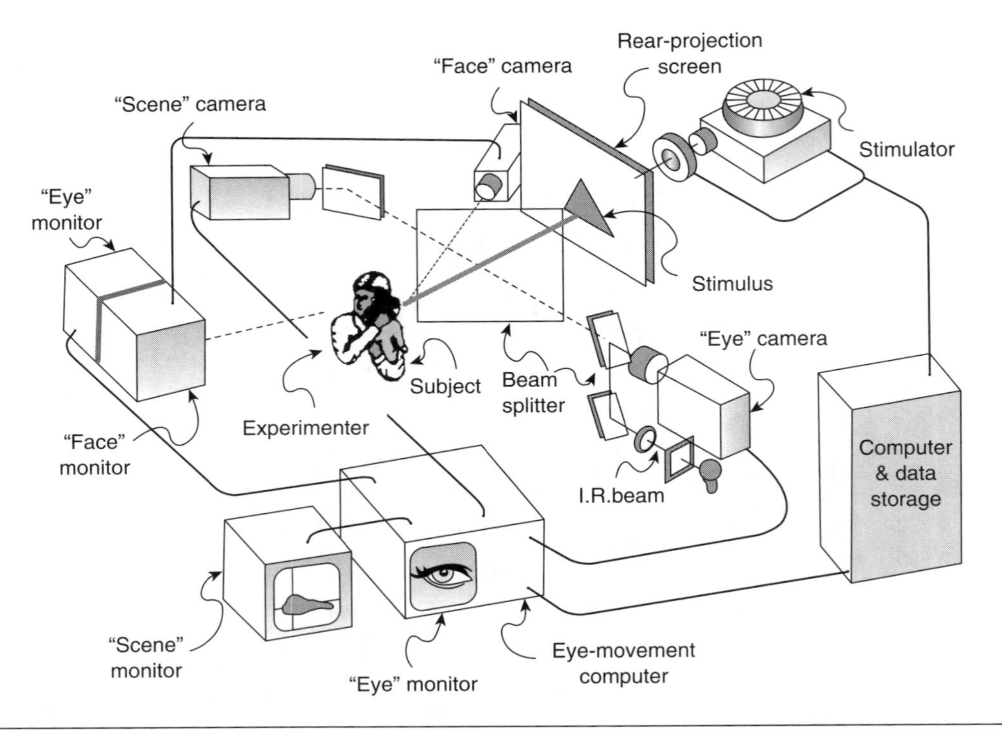

Figure 11.2 A schematic representation of an eye-movement recording system for use with infants

SOURCE: From "Infants' Scanning of Geometric Forms Varying in Size," by L. Hainline and E. Lemerise, 1982, Journal of *Experimental Child Psychology, 33*, p. 241. Copyright © 1982. Reprinted by permission of Elsevier.

Suppose that the infant does not show a preference. This result could come about for various reasons. In the simplest case, the infant shows a response bias—that is, looks only or mainly to one of the two sides. Many young infants do show such positional preferences, generally favoring the right side (Acredolo & Hake, 1982). Because the stimuli are balanced for position across trials, a consistent positional bias will result in equal fixation on each stimulus, and thus no preference. This case is simple in the sense that the response bias is readily detectable and no false conclusions need be drawn about the infant's ability or lack of ability. Because of the bias, however, the question of whether the baby can discriminate remains unresolved.

In a more complicated case, the infant does not show a response bias yet distributes attention equally to the two stimuli. This result could occur because the infant cannot discriminate between the stimuli; as argued previously, preference is possible only if there is discrimination. Such a negative conclusion cannot be drawn with certainty, however, because there is another possibility: Perhaps the infant can discriminate perfectly well but simply does not care which stimulus he or she looks at. As Flavell (1985) puts it, "preference logically implies discrimination but discrimination certainly does not logically imply preference" (p. 170). And this is the basic ambiguity in the preference method: Negative results may mean absence of discrimination, or they may simply mean absence of preference.

Although the preference method is specific to the study of vision, babies' naturally occurring preferences have proved informative in the study of other perceptual modalities as well. The fact that babies grimace to some smells and respond positively to others provides evidence of their ability to discriminate different odors (e.g., Steiner, 1979), just as their acceptance of some foods and rejection of others tells us that they can discriminate various tastes (e.g., Rosenstein & Oster, 1988).

Habituation-Dishabituation

Imagine the following sequence of events. You enter a new room and are at first very aware of a loudly ticking clock. You sit and read for a few minutes and soon no longer notice the clock at all. The clock then stops ticking, and you immediately react to the change.

This sequence illustrates three related and important psychological processes. When a new stimulus appears, many organisms, including humans, show what is called an **orienting response**. The orienting response is actually a complex of related responses whose purpose is to maximize attention to a new event. These responses typically include a momentary cessation of any ongoing behavior, orienting of receptors toward the new stimulus, and a number of characteristic physiological changes (e.g., heart-rate deceleration). This set of behaviors is simple but very adaptive, for it represents an automatic way to pay attention to what is new in the environment.

Consider next what happens when a stimulus that elicits the orienting response continues to occur, as does the ticking of the clock in the example. Eventually, the orienting response will diminish and perhaps drop out altogether. This dropping out of the orienting response to a repeated stimulus is referred to as **habituation**. Habituation, too, is an adaptive process: Once a stimulus has become familiar, there is no longer need to pay close attention to it.

Consider finally what happens when there is a change in a stimulus to which the organism has habituated, such as the disappearance of the clock's ticking. This change in the accustomed stimulation will evoke a new response of attention from the organism. The renewal of attention when a habituated stimulus changes is referred to as **dishabituation**. Dishabituation is also adaptive: When a stimulus changes, there is again need to orient to what is new.

As might be guessed from their inclusion in this chapter, all three of the phenomena just described are found in infancy. Probably the most common dependent variable in infant research has been change in heart rate: Cardiac deceleration has been measured as an index of orienting, and reduced change or no change as an index of habituation. The other two response measures introduced earlier have also been used. When sucking is measured, the baby sucks on a specially wired pacifier; interruption of sucking is considered as evidence of orienting, and continued, uninterrupted sucking as evidence of habituation. When visual fixation is measured, the interest is in total fixation time on the stimulus; initial high attention indicates orienting, and a drop-off in fixation indicates habituation.

How can these processes be used to study infant perception? At the simplest level, the orienting response can be used to answer the question of detection: Can the infant detect the stimulus at all? Clearly, if the stimulus is below the threshold for a particular perceptual modality, there is no possibility of orienting to it. Although this point is obvious, the behaviors that can be elicited as signs of detection are often rather subtle. In particular, physiological responses may sometimes tell us that an infant has detected a stimulus when there is no overt response to the stimulus at all.

Suppose that we are interested not just in detection of a single stimulus but in discrimination between two stimuli. Here the habituation-dishabituation part of the sequence becomes relevant. Let us say that we wish to know whether the infant can discriminate between two auditory stimuli, say the speech sounds "pa" and "ba." What we could do is present the "pa" sound repeatedly until the infant habituates to it. At the time that the next "pa" would normally appear, we substitute a "ba." If the infant dishabituates to the "ba," we know that the infant can discriminate between the sounds.

This example suggests an obvious strength of the habituation approach in comparison to the preference method: its wider scope of applicability. The preference method cannot really be applied to auditory perception; it is difficult to present two sounds simultaneously and see which one the infant prefers to listen to. It is easy, however, to present a single sound until response to it disappears and then to present a second sound that differs in some critical way from the first. This sort of contrast is possible with any of the senses, and habituation has in fact been used in the study of every perceptual modality.

A further strength of the approach is that it lends itself—assuming sufficient ingenuity on the part of the investigator—to the study of a number of issues that go beyond simple discrimination or nondiscrimination. We will see some examples in the next section in the discussion of infant cognitive development. Here I describe an application to one of the classic issues in perceptual development: the question of perceptual constancy.

Figure 11.3 shows a procedure designed to test size constancy in newborn babies (Slater, Mattock, & Brown, 1990). The stimuli were two cubes, one twice as large as the other. The baby first received a series of habituation trials on which one of the cubes was presented at varying distances. For example, the smaller cube might be presented at a distance of 23 cm, then at 53 cm, then at 38 cm, and so forth. On the subsequent test trials both cubes were presented simultaneously, with the larger cube twice as far away as the smaller one. The bottom portion of the figure, a photograph taken from the baby's perspective, shows the resulting, equal-sized retinal images. The question was whether the infants would look longer at one of the cubes.

The answer was yes: All the infants looked longer at the cube they had not seen before. They apparently had habituated to the original cube and thus now found it less interesting than the new one. Note, though, that they could do so only if they could perceive the constant size in the initial phase, despite the change in the distances, and only if they could perceive a difference between the two cubes in the second phase, despite the identical retinal images. The conclusion, therefore, was that even the newborn possesses some degree of size constancy.

Conditioning

Habituation can be regarded as a kind of learning—learning not to respond to a stimulus that is already familiar. In this section we consider another basic form of learning: learning to repeat a response that pays off in reinforcement. Learning that is supported by the fact that it results in reinforcement is referred to as *operant conditioning*.

Operant conditioning has been used in several ways to study infant perception. The simplest way is similar to the habituation-dishabituation paradigm just discussed. Consider an application to the issue of speech perception. We begin by reinforcing the infant for sucking on a pacifier, with the rate of reinforcement tied to the rate of sucking. The faster the infant sucks, the more frequent the reinforcement. The reinforcement consists of a speech sound, say the sound "ba." At first this reinforcement is quite effective, and the infant sucks rapidly. Eventually, however, the "ba" begins to lose its fascination, and the rate of sucking drops off. This drop-off is a reflection of *satiation:* the decrement in a reinforcer's value as a function of repetition of the reinforcer. When the sucking response has reached a low level, we change reinforcers, substituting a "pa" for the "ba." If the rate of sucking increases significantly, we know the infant can discriminate between "pa" and "ba."

A second sort of conditioning approach, labeled **conditioned head turning**, makes use of the fact that infants can learn to attach a response to a particular stimulus. Suppose that we sound a tone to the baby's right. At first the baby is likely to orient toward the sound; if the tone is presented repeatedly, however, the orienting response will eventually diminish.

(a)

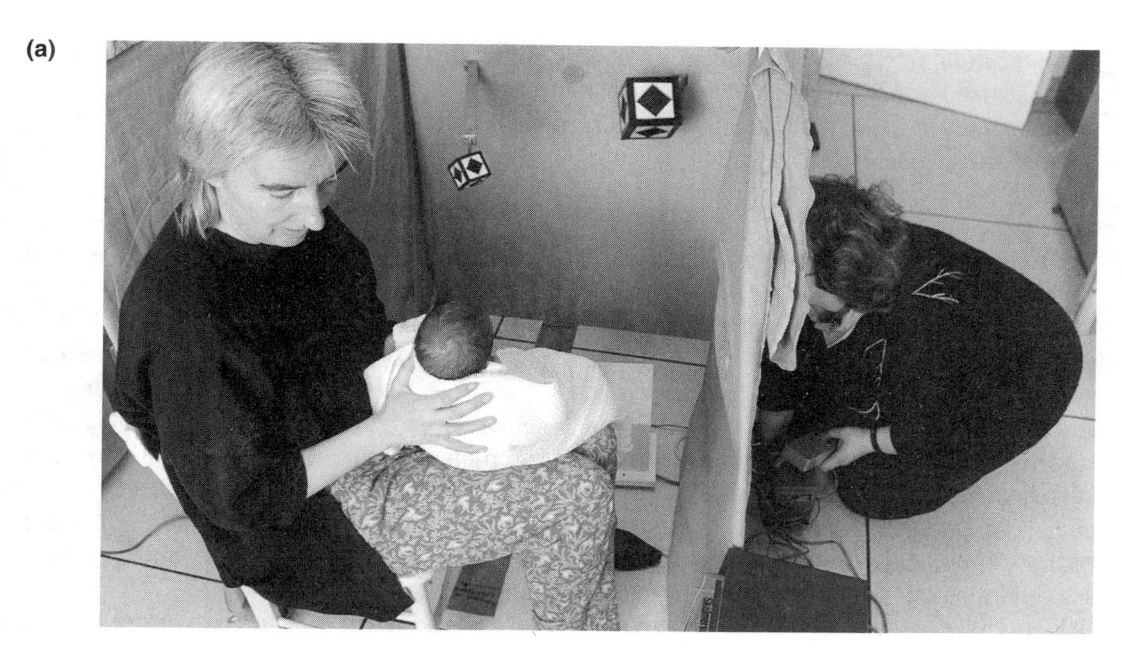

(b)

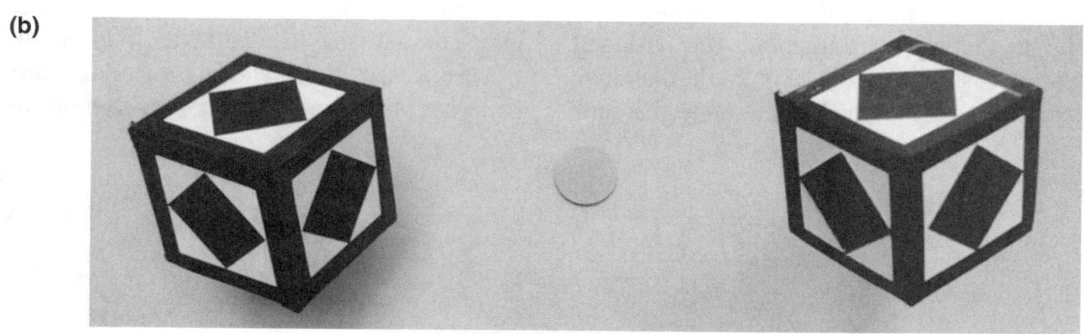

Figure 11.3 Procedure for testing size constancy in newborn babies. The top (a) shows the experimental arrangement, and the bottom (b) shows the stimuli for the test trial.

SOURCE: From "Size Constancy at Birth: Newborn Infants' Responses to Retinal and Real Size," by A. Slater, A. Mattock, and E. Brown, 1990, *Journal of Experimental Child Psychology, 49*, pp. 317, 318. Copyright © 1990. Reprinted by permission of Elsevier.

But what happens if we add a reinforcer? Suppose that every time the infant turns in response to the tone an interesting visual stimulus appears, perhaps an array of flashing lights or an animated toy. In this case the head-turn response will remain strong. And in this case the tone will become a signal, or *discriminative stimulus,* for the response: As soon as the infant hears the tone, the infant will turn to the right in anticipation of the reinforcement. Once this contingency has been established, the discriminative stimulus can be altered in various ways to probe the baby's auditory capacities. We might change the intensity or frequency of the tone to test for the baby's auditory threshold. Or we might present two different stimuli, only one of which is associated with the reinforcement. Head-turns in response to one but not the other stimulus would be evidence that the baby can discriminate between the stimuli.

A final conditioning-based approach makes use of the phenomenon of generalization.

Generalization refers to the tendency for a response that has been learned to one stimulus to be emitted also to other, similar stimuli. Because the degree of generalization depends on the similarity between the original stimulus and the new stimulus, we can use generalization to draw inferences about how similar two stimuli look or sound.

This method has been used, for example, in the study of size constancy (Bower, 1966). The rationale and general approach are similar to that described for the habituation approach to studying constancy: establish a response to one stimulus and then test for response to related stimuli. In this case, however, we first condition the baby to respond to a stimulus of a particular size, and then test for generalization to stimuli that are either the same size or the same retinal image. The specific prediction is therefore opposite to that with habituation: The strongest response should occur to stimuli that look the same as the original, not to those that look different. This method, too, suggests that some degree of perceptual constancy is present early in life.

Visual Cliff

The final procedure to be considered is more limited in scope than the techniques discussed so far. The **visual cliff** (developed by Walk & Gibson, 1961) is directed to a single question: Can babies perceive depth? The particular form of depth perception at issue is perception of a drop-off or "cliff." For example, if the baby leans over the edge of a bed to peer down at the floor, does the baby perceive the drop-off? This is a question of both theoretical and practical importance.

The visual cliff is pictured in Figure 11.4. As the figure illustrates, the cliff consists of a large glass-covered table divided by a center board. In the typical method of testing, a checkerboard pattern is placed under the glass on each side of

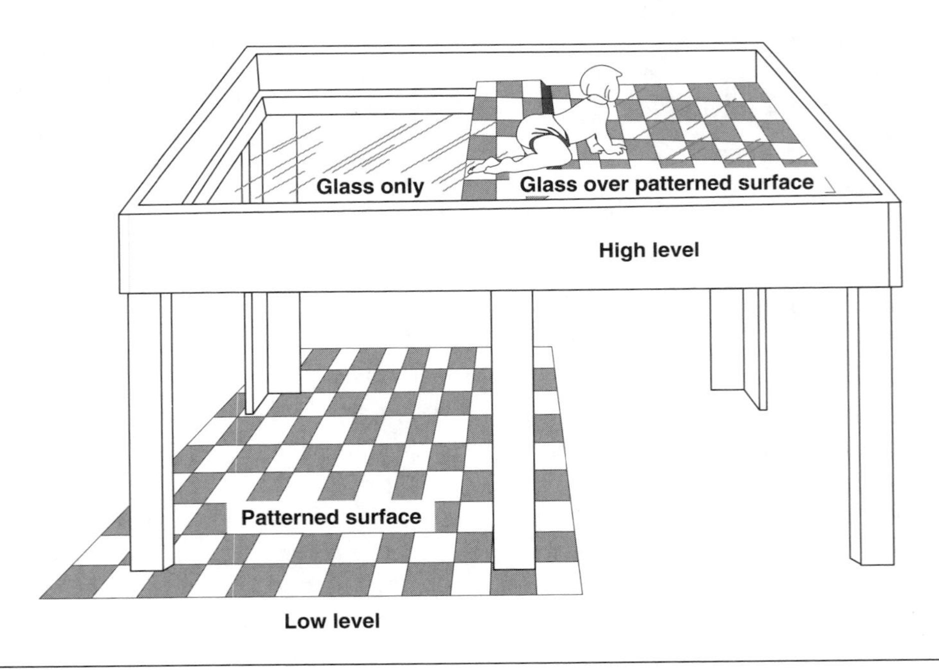

Figure 11.4 Visual-cliff apparatus for testing depth perception

SOURCE: From "A Comparative and Analytic Study of Visual Depth Perception," by R. D. Walk and E. J. Gibson, 1961, *Psychological Monographs,* 75(15, Whole No. 519), p. 8. Copyright © 1961 by the American Psychological Association.

the center board. On one side the pattern is directly under the glass, thus creating the perception of a solid surface emanating from the center board. This side is referred to as the "shallow side." On the other side the checkerboard pattern is several feet below the glass, thus creating the perception of a drop-off or cliff. This side is referred to as the "deep side." Testing typically begins with the infant placed in crawling position on the center board. The mother then goes to each side and attempts to coax the infant to crawl across the glass to her. The interest is in possible differential response: Will the infant crawl across the shallow side but refuse to venture onto the deep side? If so, the inference drawn is that the infant can perceive the depth.

Tests of this sort have revealed that infants as young as 6 months old will often (although by no means always) succeed in avoiding the deep side, thus demonstrating some ability to perceive depth (Bertenthal & Campos, 1990). Infants younger than 6 months cannot be tested with the methods described, because these methods depend on ability to crawl. Thus, the classic visual-cliff test cannot tell us how early depth perception is present. An alternative testing method developed for younger infants consists of simply placing an infant face down on each side of the cliff and measuring the reaction. Crying and changes in heart rate are the variables usually measured. Researchers using this approach have found that infants as young as 2 months old do respond differentially to the two sides, thus showing some ability to discriminate between the deep and shallow surfaces. They do not show a clear-cut fear reaction to the deep side, however, and thus their ability to perceive depth remains uncertain (Bertenthal & Campos, 1990).

The visual cliff is not the only procedure that has been devised to test depth perception in infants. Among the other measures that have been utilized are the baby's reaching behaviors in response to objects of different distances (e.g., Field, 1977) and avoidant behaviors in response to rapidly approaching, or "looming," objects (e.g., Yonas, 1981). More generally, this survey of methods for studying infant perception has barely skimmed the surface of this large and rapidly expanding literature. Fuller discussions can be found in Kellman and Arterberry (2006) and Saffran, Werker, and Werner (2006).

Infant Cognitive Development

For years, Piaget's work (Piaget, 1951, 1952, 1954) dominated the study of infant cognitive development, accounting for both the field's major theory and a high proportion of its empirical work. Such is no longer the case—various limitations in Piaget's studies have become evident, and new issues and new procedures have moved to the fore. Nevertheless, the Piagetian approach remains an important contributor to what we know about infant cognition, and our discussion consequently begins with Piaget.

The Piagetian Approach

Piaget's studies of infancy are far too extensive to be exhaustively summarized here. What I aim for instead are two goals. One is to convey some of the general aspects of the Piagetian approach to studying the infant. The other is to describe the procedures used to study one of the most important achievements of infancy—the object concept.

Piaget's conclusions about infant development are based on his study of his own three infants during the first 2 to 3 years of their lives. His approach is an extension of one of the historically earliest methods in child psychology, the baby biography. In a **baby biography**, a scientist-parent makes extensive observations of the development of his or her own child, observations that then serve as the database for drawing some more general conclusions about human development. Such studies became fairly popular enterprises during the last half of the 19th century. Prior to Piaget's work, the most famous such effort was Charles Darwin's (1877) study of his infant son.

In Piaget's case the studies of infancy went well beyond a simple compilation of interesting-looking behaviors. From the start the work was theoretically guided, with a consistent attempt to answer basic philosophical questions, such as the origins of concepts of space, time, and causality. I consider one such concept, object permanence, shortly. The work also included more than simple observation of naturally occurring behavior. Naturalistic observation *was* used—many hundreds of hours, in fact, of quite painstaking observation. But such observations were continually supplemented by small-scale experimentation. If, for example, Piaget was interested one day in his daughter's response to obstacles, he would not necessarily wait until an obstacle happened to come along in her path. Instead, he might interpose a pillow between daughter and favorite toy, and then record how she responded to this challenge.

Piaget's observations are summarized in hundreds of "protocols" of infant behavior, and it is impossible to convey the flavor of the work without quoting some of these protocols. Box 11.1 includes a sampling of protocols dealing with the infant's gradual mastery of principles of causality. Note the interweaving of naturally occurring and experimentally elicited behavior. Note also the interweaving of straight description and more interpretive commentary.

In trying to abstract general features of the Piagetian approach, it is helpful to contrast Piaget's studies with the work on infant perception discussed in the previous section. There are a number of differences. Piaget's studies were carried out in the child's natural environment; research on perception, as we saw, has been almost totally confined to laboratory settings. Piaget's studies made essentially no use of special apparatus; the child interacted with the objects and people of the everyday environment, and the only recording instrument was the father. Studies of perception, in contrast, have relied heavily on complicated apparatus such as eye-movement cameras and EKG machines. Piaget's conclusions depended on observations by a human observer; in perception research, automatic recording has been the norm. Finally, Piaget's work was longitudinal, whereas studies of infant perception have been primarily cross-sectional.

Why so many differences in approach? The differences are probably partly a function of the content being studied and partly a result of certain idiosyncrasies in Piaget's style of research. As we saw in the previous section, the issues of interest in the study of infant perception do not lend themselves to observational assessment in the natural environment. Through what naturally occurring and observable behaviors could we determine whether infants perceive a receding object as constant or changing in size? Behaviors indicative of cognitive development, in contrast, often *are* discernible in the child's natural dealings with the environment, especially during infancy. The infant's explorations of causality described in the protocols of Box 11.1 are one example. The work on object concept that we will consider shortly is another. Although the distinction is an admittedly shaky one (as is the perception-cognition contrast in general), it can be argued that cognitive functioning, especially in infancy, is more readily visible than is perceptual functioning.

The second basis for the difference lies in Piaget's preferred approach to research. As we will see more fully in the next chapter, many of the characteristics of Piaget's infancy research are also found in his studies of older children. These characteristics include a preference for flexible probing over tight standardization, for simple and familiar materials over complicated apparatus, and for interpretive analyses of individual protocols over standard statistical tests. There are also some characteristics of Piaget's infancy research that follow necessarily from the circumstances under which he was working.

Box 11.1. Examples of Piagetian Protocols Concerned With the Infant's Understanding of Causality

Obs. 128

At 0;3 (12) that is to say, several days after he revealed his capacity to grasp objects seen, Laurent is confronted by a rattle hanging from his bassinet top; a watch chain hangs from the rattle. . . . From the point of view of the relationships between the chain and the rattle the result of the experiment is wholly negative: Laurent does not pull the chain by himself and when I place it in his hands and he happens to shake it and hears the noise, he waves his hand but drops the chain. On the other hand, he seems immediately to establish a connection between the movements of his hand and those of the rattle, for having shaken his hand by chance and heard the sound of the rattle he waves his empty hand again, while looking at the rattle, and even waves it harder and harder. . . .

Obs. 134

At 0;7 (7) Laurent looks at me very attentively when I drum with my finger tips on a tin box of 15 x 20 centimeters. The box lies on a cushion before him and is just two centimeters beyond his reach. On the other hand, as soon as I pause in my game I place my hand five centimeters from his, while he watches, and leave it there motionless. So long as I drum Laurent smiles delightedly but when I pause he looks for a moment at my hand, then proceeds very rapidly to examine the box and then, while looking at it, claps his hands, waves goodbye with both hands, shakes his head, arches upward, etc. In short, he uses the whole collection of his usual magico-phenomenalistic procedures. With regard to my hand, placed before his eyes, he grasps it for a moment twice in succession, shakes it, strikes it, etc. But he does not lead it back to the box, although that would be easy, nor does he try to discover a specific procedure to set its activity in motion.

Obs. 157

At 1;6 (8) Jacqueline sits on a bed beside her mother. I am at the foot of the bed on the side opposite Jacqueline, and she neither sees me nor knows I am in the room. I brandish over the bed a cane to which a brush is attached at one end and I swing the whole thing. Jacqueline is very much interested in this sight: she says "cane, cane" and examines the swinging most attentively. At a certain moment she stops looking at the end of the cane and obviously tries to understand. Then she tries to perceive the other end of the cane and to do so, leans in front of her mother, then behind her, until she has seen me. She expresses no surprise, as though she knew I was the cause of the sight. . . .

SOURCE: From *The Construction of Reality in the Child* (pp. 261, 276, 334), by J. Piaget, 1954, New York: Basic Books.

With a sample of three, there hardly *could* be much in the way of standard statistical tests. In-the-home recording sets definite constraints on the apparatus that can be used, and in any case many of the technological tools that are mainstays of modern research had not been invented when Piaget began his work 75 years ago.

Piaget's approach to studying the infant has both strengths and weaknesses. The weaknesses are perhaps more obvious. The sample is both very small and distinctly nonrandom. Despite the total reliance on observational data, Piaget made no attempt to demonstrate any sort of interobserver reliability. Even with a sample of three, there are opportunities for standardization and control that are missed. The method of presenting the results often blurs the distinction between data and conclusions. And Piaget's emphasis on overt motor behavior in diagnosing infant development may lead to an underestimation of the infant's ability. We will return to this criticism shortly.

Many of the strengths of Piaget's studies were noted earlier. The discussion of exploratory research in chapter 5 cited Piaget's work as a prime example of the value of a flexible, discovery-oriented approach to studying the child. The discussion of longitudinal research in chapter 3 cited the Piaget infancy studies as an example of the value of an intensive, case-study approach to the development of individual children. There is probably no better example in the field of the value of naturalistic observation in the hands of a skilled observer. Most generally, the fact that Piaget's work continues to influence the study of infant cognition more than 60 years after its original publication is ample testimony to its viability.

Let us turn now to the work on the object concept. The term **object concept** refers to the knowledge that objects have a permanent existence that is independent of our perceptual contact with them. It is the knowledge, thus, that an object does not cease to exist simply because at the moment we cannot see it, hear it, feel it, or whatever. It is hard to imagine a more basic piece of knowledge than this. Yet Piaget's studies conclude that the object concept is not present at birth, that it develops only gradually across the first 2 years, and that there is a definite sequence of stages through which the infant passes in mastering the concept. This set of claims is of considerable theoretical and empirical importance, a fact that accounts for the attention that the concept has received in follow-up research. For our purposes it is the methodological challenge that is of interest. How does Piaget—or anyone—determine what an infant knows or does not know about objects?

As noted, Piaget's observations are reported in the form of protocols—some 66 protocols, many of them with multiple observations, in the case of the object concept. A small sampling of these protocols is quoted in Box 11.2. These sample protocols illustrate some of the points made across the next several paragraphs.

Piaget's studies of the object concept are centered on the infant's response to the disappearance of objects. Such disappearances are, of course, a common occurrence in the baby's natural interactions with objects. A toy falls from the high chair or crib. Mother or father walks out of the room. The baby turns around and can no longer see bottle or mother. In addition to such naturally occurring events, it is easy for an experimenter to contrive particular kinds of disappearance. Papa Piaget, for example, can drop a handkerchief over a toy, or hide a small toy in his hand and move it from one hiding place to another. We can see here one of the general characteristics of Piaget's approach noted earlier: the mixture of naturally occurring and experimentally elicited observations.

The basic question when the object disappears is whether the infant realizes that it still exists. Because a baby cannot tell us in words, we must select some observable behavior from which we can infer the infant's knowledge. Piaget looks primarily at various kinds of search behavior. Does the baby attempt to track the object with her eyes as it moves outside her field of vision? Does she lean over and gaze at the floor when a toy drops? Does she reach out and remove a cover that has been dropped over a toy? Does she turn immediately to the spot where some plaything was left, even though she has not looked at it for several minutes? There are many behaviors by which the infant can demonstrate that she knows, or does not know, that an object still exists. It is characteristic of Piaget's approach that he looks at a wide range of relevant behaviors. It is also characteristic that many of the behaviors that are stressed,

Box 11.2. Examples of Piagetian Protocols Concerned With the Infant's Understanding of the Object Concept

Obs. 2

. . . Jacqueline, as early as 0;2 (27) follows her mother with her eyes, and when her mother leaves the visual field, continues to look in the same direction until the picture reappears.

Same observation with Laurent at 0;2 (1). I look at him through the hood of his bassinet and from time to time I appear at a more or less constant point; Laurent then watches that point when I am out of his sight and obviously expects to see me reappear.

Obs. 28

At 0;7 (28) Jacqueline tries to grasp a celluloid duck on top of her quilt. She almost catches it, shakes herself, and the duck slides down beside her. It falls very close to her hand but behind a fold in the sheet. Jacqueline's eyes have followed the movement, she has even followed it with her outstretched hand. But as soon as the duck has disappeared—nothing more! It does not occur to her to search behind the fold of the sheet, which would be very easy to do. . . .

I then take the duck from its hiding-place and place it near her hand three times. All three times she tries to grasp it, but when she is about to touch it I replace it very obviously under the sheet. Jacqueline immediately withdraws her hand and gives up. The second and third times I make her grasp the duck through the sheet and she shakes it for a brief moment but it does not occur to her to raise the cloth.

Obs. 40

At 0;10 (18) Jacqueline is seated on a mattress without anything to disturb or distract her (no coverlets, etc.). I take her parrot from her hands and hide it twice in succession under the mattress, on her left, in A. Both times Jacqueline looks for the object immediately and grabs it. Then I take it from her hands and move it very slowly before her eyes to the corresponding place on her right, under the mattress, in B. Jacqueline watches this movement very attentively, but at the moment when the parrot disappears in B she turns to her left and looks where it was before, in A.

During the next four attempts I hide the parrot in B every time without having first placed it in A. Every time Jacqueline watches me attentively. Nevertheless each time she immediately tries to rediscover the object in A. . . .

Obs. 55

At 1;8 (8) Jacqueline is sitting on a green rug and playing with a potato which interests her very much (it is a new object for her). She says "po-terre" and amuses herself by putting it into an empty box and taking it out again. For several days she has been enthusiastic about this game.

I. I then take the potato and put it in the box while Jacqueline watches. Then I place the box under the rug and turn it upside down thus leaving the object hidden by the rug without letting the child see my maneuver, and I bring out the empty box. I say to Jacqueline, who has not stopped looking at the rug and who has realized that I was doing something under it: "Give papa the potato." She searches for the object in the box, looks at me, again looks at the box minutely, looks at the rug, etc., but it does not occur to her to raise the rug in order to find the potato underneath. . . .

SOURCE: From *The Construction of Reality in the Child* (pp. 8, 39–40, 56, 75–76), by J. Piaget, 1954, New York: Basic Books.

especially at the more advanced levels, involve some active motor response on the baby's part.

Just as search can occur in many ways, so can an object disappear in many ways. Did the infant's own actions make the object disappear, or was some external source responsible? Is the object completely gone, or are parts of it still visible? If the object has vanished completely, are there nevertheless auditory or tactile cues to its presence? Is more than one hiding place involved, and, if so, can the child keep track not only of the fact that the object still exists but of *where* it exists? We can see in these diverse questions another characteristic of Piaget's approach: the emphasis on the use of many different kinds of problems in studying the development of any concept. This methodological emphasis has a theoretical corollary: the belief that important cognitive acquisitions, such as the object concept, do not emerge full-blown but rather are gradually mastered through a sequence of stages. It is to chart these stages that Piaget looks at many problems and many behaviors.

Post-Piaget Research

Piaget's ingenious studies have left no shortage of issues for later researchers to explore. I begin with research directly inspired by Piaget and devoted to testing claims from his work. I then consider some recent research that departs more radically from the Piagetian model.

As you might expect, there are a number of differences between contemporary work on infant cognition and Piaget's studies. The more recent work (almost necessarily) has looked at larger and more representative samples than did Piaget. Along with the larger samples has come an emphasis on standardization and experimental control that is seldom found in Piaget's studies. Laboratory settings have become the norm for research on infant cognition; in-the-home study has become rare, and truly naturalistic study rarer still. The laboratory locus has made possible the use of various technological tools (e.g., videotaping) that were not available to Piaget. In general, more modern research has attempted to shed many of the idiosyncrasies of Piaget's original work while still maintaining the focus on basic and developmentally interesting forms of knowledge.

At a more specific level, the most common impetus for follow-up work has been concern with the adequacy of Piaget's assessment techniques. Do Piaget's observations and simple experiments really give an accurate picture of what the infant understands or does not understand about the world? Or are there instances in which Piaget may be miscalculating the baby's true abilities? In particular, are there instances in which Piaget is underestimating the baby's ability, perhaps especially because of the emphasis that he places on overt motor behaviors? It seems possible, for example, that an infant knows perfectly well that an object still exists but is simply unable to engage in the kind of active search behaviors that Piaget often requires.

How else might we study what infants know about objects? The most informative approach has been the **violation-of-expectation method**: see how the baby responds if some physical law, in this case the principle of object permanence, is apparently violated. Suppose, for example, that an object magically disappears or that one solid object seemingly passes through another solid object. *You* would certainly find such an event puzzling or surprising. Perhaps the young infant will also—and will thus demonstrate some knowledge of object permanence even in the absence of search behavior.

A study by Baillargeon (1987) provides an example. Baillargeon's research made use of the habituation-dishabituation procedure described earlier in the chapter. The infants (3- and 4-month-olds) first viewed the event shown in the top part of Figure 11.5—that is, a screen that rotated, like a drawbridge, through a 180-degree arc. Although this event was initially interesting, eventually the babies' looking times dropped off,

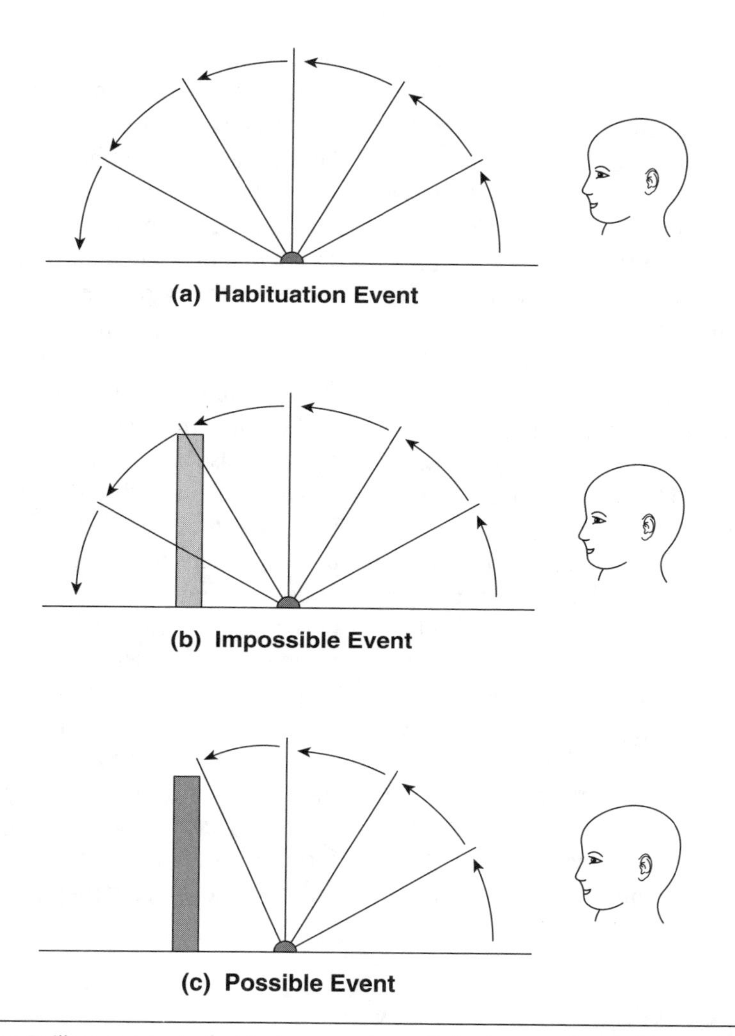

Figure 11.5 The Baillargeon test of object permanence. Infants were first habituated to event (a), after which response was measured to either event (b) or event (c).

SOURCE: Adapted from "Object Permanence in 3.5- and 4.5-Month-Old Infants," by R. Baillargeon, 1987, *Developmental Psychology, 23*, p. 656. Copyright © 1987 by the American Psychological Association. Used with permission.

reflecting the fact that they had habituated to the repeated stimulus. At this point a wooden box was placed directly in the path of the screen (see the bottom part of the figure). As the figure indicates, the box was visible at the start of a trial, but it disappeared from view when the screen reached its full height.

Two experimental conditions were contrasted. In the *Possible Event* condition, the screen rotated to the point at which it reached the box and then stopped—as of course it should, given the existence of a solid object in its path. In the *Impossible Event* condition, the screen rotated to the point of contact with the box and then (thanks to a hidden platform that dropped the box out of the way) kept right on going! The question was whether the infants would respond differentially to the two events—in particular, would they dishabituate more strongly to one than to the other? The answer is that they did, and in pretty much the way that you or I might. Interest remained low when the screen stopped at the point of reaching the box; looking times rose significantly, however, when the screen appeared to pass through the area occupied by the box. The most plausible explanation of this finding is that the infants realized that the box must still exist, and thus were surprised when another object passed through it.

The drawbridge methodology was the forerunner of a number of ingenious procedures through which researchers have explored what young infants know about object permanence. Baillargeon's own research program has gone on to examine infants' understanding of several different ways in which an object can disappear from view (Baillargeon, 2004). Some studies examine response to *occlusion events,* that is, cases in which one object blocks the view of another. The drawbridge study is one of several examples in this category. Others have explored understanding of *containment events* (one object disappears inside another) and *covering events* (a cover is lowered over an object). In each case, an apparent violation of object permanence is engineered—for example, an object disappears into a container but when the cover is removed the container is empty. And in each case, infants only a few months old respond with increased interest to the apparent violation. Such studies, I should add, do not demonstrate total competence in young babies, for there are often limitations to be overcome and developmental advances to be made. But they do suggest earlier understanding than was evident in Piaget's work.

As noted, the violation of expectation paradigm has also been applied to a number of forms of knowledge not included in Piaget's studies. Figures 11.6 and 11.7 show two examples. The procedure depicted in Figure 11.6 tested infants' understanding of principles of gravity and support, pitting response to possible events against response to impossible ones (the ball floating magically in air). Six-month-olds looked longer at the inconsistent than the consistent outcome, thus suggesting some understanding of the principles in question. Figure 11.7 shows a test of infants' understanding of very simple forms of addition and subtraction. Five-month-olds reacted more strongly to the impossible

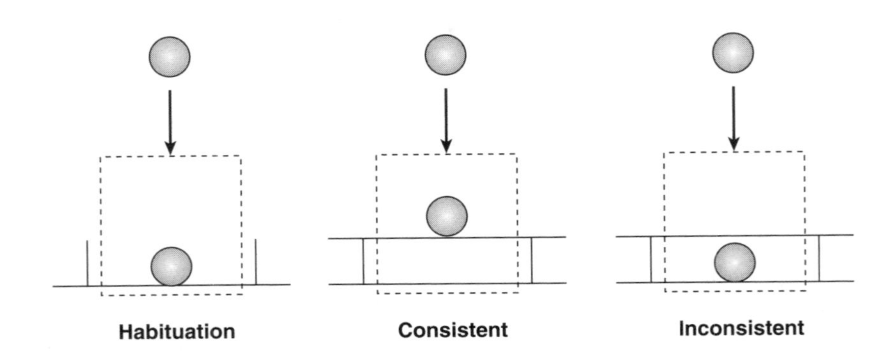

Figure 11.6 Consistent and inconsistent outcomes in Spelke et al.'s study of infants' understanding of object movement and gravity. Infants were first habituated to the event shown on the left: a ball that dropped behind a screen and was revealed, when the screen was removed, to be lying on the floor. A table was then placed in the ball's path, the ball was again dropped behind the screen, and removal of the screen revealed the ball resting either on the table (consistent outcome) or on the floor (inconsistent outcome).

SOURCE: From "Origins of Knowledge," by E. Spelke, K. Breinlinger, J. Macomber, and K. Jacobson, 1992, *Psychological Review, 99*, pp. 611 and 621.

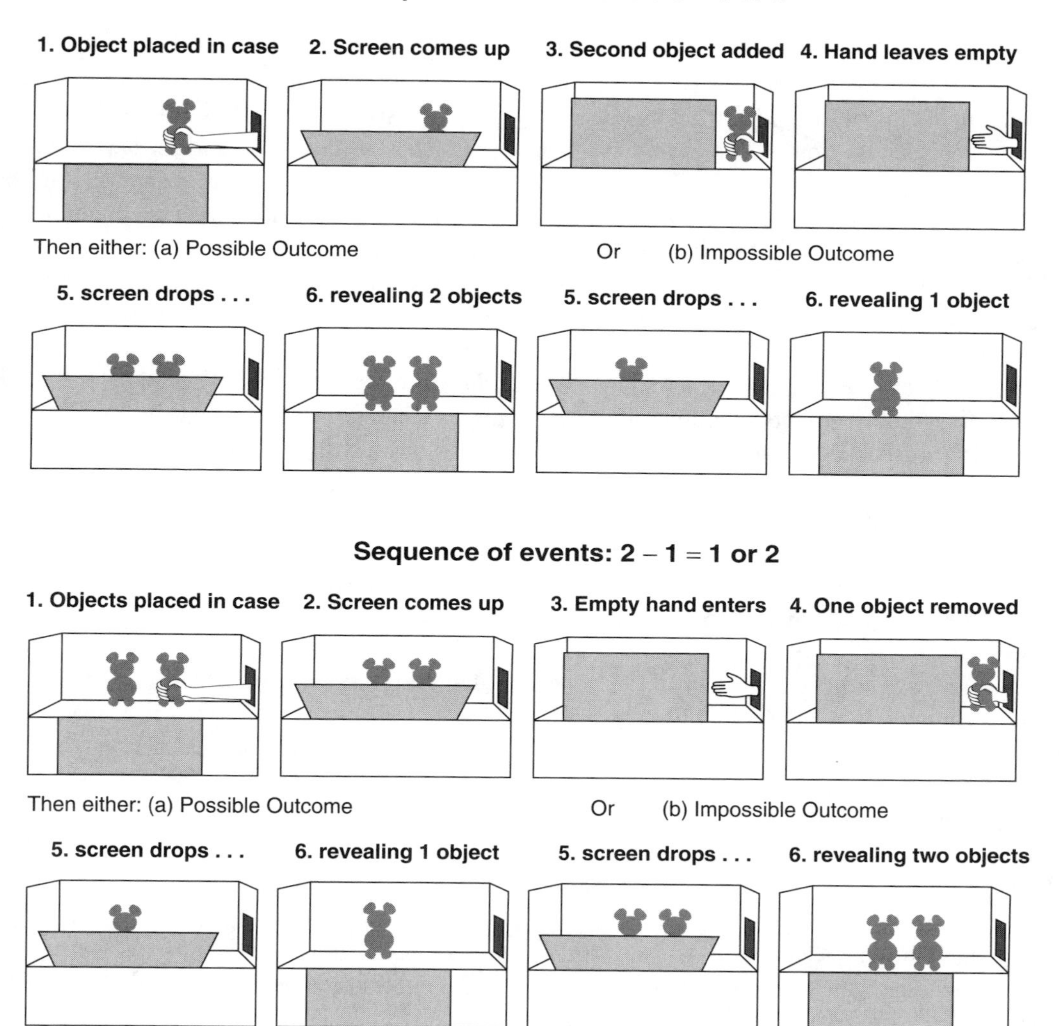

Figure 11.7 Possible and impossible outcomes in Wynn's study of infants' arithmetical competence

SOURCE: From "Addition and Subtraction by Human Infants," by K. Wynn, 1992, *Nature, 358*, p. 749. Copyright © 1992 by Macmillan Magazines Ltd. Reprinted by permission of Nature Publishing Group.

than to the possible outcomes, again suggesting some basic early understanding. The question of what infants understand about number has become an especially active focus for contemporary research (see, for example, Clearfield & Westfahl, 2006; McCrink & Wynn, 2004; Xu, Spelke, & Goddard, 2005).

Having noted the positive findings from this research, I should add that not all researchers of infancy have been persuaded that infants' knowledge is really *as* great as some investigators have claimed (Bogartz, Shinskey, & Schilling, 2000; Cohen & Cashon, 2006; Haith & Benson, 1998; Hood, 2004). It is difficult to design experiments of this sort to rule out all possible explanations for the results other than the one being tested. It is also difficult to decide exactly how much knowledge must be credited to the infant to explain the increases in looking behavior (the only dependent variable measured in the great majority of these studies.) Everyone, I should emphasize, agrees that infants understand more about the world than indicated by Piaget's studies. The challenging nature of the infant, however, has made it hard to agree on exactly how much more.

The kinds of research just discussed hardly exhaust the study of infants' cognitive abilities. Some further aspects of infant cognition will be considered in chapter 12, in the context of the general topics under which they fall.

Infant Social Development

Like perception and cognition, social development in infancy is a large topic. I begin with some general points about how the infant's social world is typically studied. I then focus on what have probably been the two most popular topics for research on infant social development: the concepts of *attachment* and *temperament.*

It is again instructive to do some contrasting of the new topic with topics already covered in this chapter. There are both similarities and differences. On the one hand, the researcher of infant social development faces many of the same problems as does the researcher of infant perception or cognition. In particular, the infant remains a nonverbal, often recalcitrant research participant, difficult to recruit for research in the first place and difficult to keep in a study once things are under way. On the other hand, many of the "problem behaviors" that bedevil the researcher of perception or cognition are no longer problems, for they constitute the responses of interest. Does the infant start to cry when the mother attempts to separate from him or her in the experimental setting? Does the infant direct attention only to the mother, refusing to respond to stimuli proffered by a strange adult? These behaviors, so vexing in some studies, may be precisely what the researcher of social development seeks to discover.

There are other differences as well. As we saw, studies of infant perception, as well as most post-Piaget studies of infant cognition, have been conducted primarily in laboratory settings. Research on social development includes a much higher proportion of studies carried out in the natural setting—that is, the infant's home. Even when a laboratory setting *is* used, the lab environment tends to approximate the natural setting much more closely than is true in other kinds of infant research. Toys and picture books are more likely to populate the experimental environment than are video cameras or EKG machines. It is true, of course, that the testing environment is still a strange room rather than the familiar home, and many infants are clearly aware of this. Indeed, one well-established finding from research on infant social development is that various distress reactions of interest—for example, crying upon separation from the mother, upset upon being confronted by a stranger—are more marked in the lab than in the home.

Two other differences between work on social development and work on perception and cognition can be noted. The first is implied by the term "social." What we are interested in now is not the infant in isolation but the infant in interaction with other people. The most commonly studied other person has been the infant's mother. In recent years the father has also been accorded a prominent role in some research programs. Interactions with other children are sometimes studied, especially once the infant has achieved the status of "toddler" (at about 18 to 24 months). And response to such familiar people is often contrasted with response to strangers—that is, anyone whom the infant has not encountered before.

The second difference concerns measuring instruments. Work on perception has relied mainly on automatic recording of responses. Although Piaget's studies were observational, more recent research on infant cognition has also moved in the direction of automatic recording—for example, eye-movement and heart-rate measures. Studies of social development have relied heavily on observational assessment by a human observer. This reliance on observational assessment follows necessarily from the goal in research on social development: to identify naturally occurring, socially meaningful units of response—"smiles," "cries," "seeks contact," "resists." Technology can help in this endeavor; the human observer, however, remains a necessary component.

Attachment

I turn now to research on **attachment**. Attachment is a broad construct that encompasses many phenomena in the social development of the infant. The core reference is to what is sometimes called "the first love relationship": the strong emotional bond that forms between the infant and the parents. What we are interested in is how this bond comes about, and also why some babies develop less satisfactory attachments than do others.

From a methodological point of view, the study of attachment presents several challenges. A first question concerns the behaviors from which we can determine the presence or absence of attachment. "Attachment," like "object concept," "depth perception," or any other developmental outcome of interest, is not an immediate behavioral given; rather it is a higher-level construct that must be inferred from a variety of relevant behaviors. Across the last 30 or so years there has been a steady expansion in the range of behaviors that are considered relevant to the assessment of attachment. The earliest systematic studies of the construct focused primarily on *separation distress*—that is, the infant's tendency to be upset when separated from the attachment object. Does the baby cry, for example, when the mother walks out of the room or when left alone in the crib at night? Separation distress has a good deal of intuitive appeal as a measure of attachment; we would expect a baby who has formed an emotional bond to another person to be more likely to protest separation than a baby who has not yet formed an attachment. Furthermore, separation distress shows the developmental course that we would expect of a measure of attachment: It is absent in the early months of life but typically emerges, often in quite strong form, some time between 6 and 12 months.

More recent research retains separation distress as one useful measure but adds a number of other behaviors as well. These additional behaviors include several more positive ways by which an infant can convey that an attachment to the parent is being formed. Does the baby smile or babble more readily to the mother than to other people? Such *differential responsiveness* is not a full-blown attachment, but it is usually considered as a phase in the movement toward full attachment. Is the baby more secure in the mother's presence, better able to venture out and explore new things or to interact with new people? The ability to use the mother as a *secure base* was first emphasized in Harlow's (1958) work with infant monkeys; it turns out to be important in human infants also. Does the baby brighten when the mother enters the room and engage in *greeting or contact-eliciting behaviors?* In general, does the baby seem to enjoy the mother's presence and do various things either to keep her near when she is already present or to bring her back when she is gone? There are a variety of attachment behaviors, and they vary across situations, developmental levels, and individual children; all, however, have this general quality of pleasure and security in the presence of the attachment object.

Deciding which behaviors to study is just one step in the research process. Another important step is deciding how to get evidence with respect to these behaviors. How can we find out whether a baby protests separation from the mother, reacts more positively to familiar people than to strangers, uses the mother as a secure base, or does whatever else it is that we have decided is relevant to attachment? The answer is that three general approaches can be taken (recall Table 4.3): We can ask someone who knows the child, typically the parents (the *parental report* approach); we can observe the behaviors in the home setting (the *naturalistic observation* approach); or we can set up a laboratory situation to elicit the behaviors of interest (the *structured laboratory* approach). The last of these approaches has dominated the study of attachment, and I consequently begin with it.

It is the work of Mary Ainsworth and colleagues that has given the field its most influential measure of attachment. Ainsworth, Blehar, Waters, and Wall (1978) devised a procedure called the **Strange Situation**. There are three participants in the Strange Situation procedure: mother, stranger (an adult female), and baby. The procedure takes place in a room in a university building that is unfamiliar to both baby and mother; hence it qualifies as "lab" rather than natural setting. The environment is designed, however, to have a comfortable, even playroom-like quality to it. Thus, there are chairs and magazines for the adults, bright pictures on the wall, and toys aplenty for the baby to play with. There is a further feature as well: a one-way mirror through which the room can be viewed surreptitiously. The mother is aware of the mirror and the observer behind it; the baby, however, is not. The baby's behavior, therefore, should be unaffected by the presence of an observer.

The Strange Situation test consists of eight ordered episodes, summarized in Table 11.2. Throughout the eight episodes the mother's behavior is largely (although not totally) controlled. The baby's behavior, however, is not controlled, and it is the baby who is the focus of the study. As the descriptions in the table suggest, the eight episodes are designed to elicit a variety of attachment behaviors. Thus, the baby is observed in interaction with the mother, both before and after being separated from her, and both alone and with a stranger present. The baby is also observed with a stranger, and similarities and differences in response to mother and stranger can be noted. The sequence includes both separation episodes and reunion episodes; when the mother departs, the baby is sometimes left alone and sometimes left with the stranger. In general, the procedure is an attempt to capture, within a span of about 20 minutes, a high proportion of the situations and behaviors that have been used to study attachment.

Ainsworth et al.'s measurements come from direct observations of the infant's behavior. Three different levels of scoring are used. At the most molecular level, specific, discrete behaviors are recorded. Examples of categories at this level include crying, smiling, vocalization, and locomotion. The second level concerns interactive behavior. There are six categories of interactive behavior: proximity and contact seeking, contact maintaining, resistance, avoidance, search, and distance interaction. Scoring at this second level requires more interpretation than does scoring at the first (e.g., "resistance" versus "crying"); a further difference is that the categories at the second level are explicitly social in nature. Finally, the third level of scoring involves a qualitative classification of the overall attachment relation between infant and mother. In most research, three general categories are used. Of these three attachment patterns, the most adaptive is Type B (*securely attached*), which is the label used for a secure, satisfactory form of attachment. The Type B infant is clearly happiest in the mother's presence and is upset when she leaves; the infant is not devastated by the mother's absence, however, and is in general able to adapt well to the cumulative stresses of a strange environment, a strange person, and various comings and goings of the mother. The other two forms of attachment are less secure and satisfactory. The Type A (*avoidant*) infant shows little distress at separation from the

Table 11.2 Summary of Episodes of the Strange Situation

Number of episode	*Persons present*	*Duration*	*Brief description of action*
1	Mother, baby, & observer	30 s	Observer introduces mother and baby to experimental room, then leaves.
2	Mother & baby	3 min	Mother is nonparticipant while baby explores; if necessary, play is stimulated after 2 minutes.
3	Stranger, mother, & baby	3 min	Stranger enters. First minute: stranger silent. Second minute: Stranger converses with mother. Third minute: Stranger approaches baby. After 3 minutes mother leaves unobtrusively.
4	Stranger & baby	3 min or less[a]	First separation episode. Stranger's behavior is geared to that of baby.
5	Mother & baby	3 min or more[b]	First reunion episode. Mother greets and/or comforts baby, then tries to settle him again in play. Mother then leaves, saying "bye-bye."
6	Baby alone	3 min or less[a]	Second separation episode.
7	Stranger & baby	3 min or less[a]	Continuation of second separation episode. Stranger enters and gears her behavior to that of baby.
8	Mother & baby	3 min	Second reunion episode. Mother enters, greets baby, then picks him up. Meanwhile stranger leaves unobtrusively.

SOURCE: From *Patterns of Attachment* (p. 37), by M. D. S. Ainsworth, M. C. Blehar, E. Waters, & S. Wall, 1978, Hillsdale, NJ: Lawrence Erlbaum.

a. Episode is curtailed if the baby is unduly distressed.

b. Episode is prolonged if more time is required for the baby to become reinvolved in play.

mother and little joy at reunion with her; this infant also demonstrates relatively little differentiation between behaviors directed to the mother and those directed to the stranger. The Type C (*resistant*) infant has a more upset and angry look. This baby may be strongly distressed during the separation, and may either actively resist the mother during reunion or else respond in an ambivalent fashion, perhaps both clinging to her and pushing her away.

In recent years some investigators (e.g., Main & Solomon, 1991) have identified a fourth attachment classification, Type D, or *disorganized.* Type D is another form of insecure attachment; it is characterized by a mixture of avoidant and resistant behaviors, with the infant often appearing confused or apprehensive upon reunion with the caregiver.

As this description should suggest, the Strange Situation elicits a wealth of behaviors relevant to attachment. A major strength of structured laboratory assessment is this rich behavioral yield; the approach is both more economical than naturalistic observation and more direct than an interview. On the negative side, the focus on behavior in the natural setting, a strength of the other two approaches, is lost with the Strange Situation. There are also questions about its appropriateness for certain populations, a point to which I return.

The major alternative to the Strange Situation is a procedure called the **Attachment Q-Set**, developed by Everett Waters and associates (Waters, 1995; Waters & Deane, 1985). The Attachment Q-Set, or AQS, is an example of a general methodology for assessing individual differences that is labeled the *q-sort approach.* Its application to attachment is typical of the approach in general. The AQS consists of 90 items that describe infant behaviors or characteristics that are relevant to an assessment of the attachment relation. Table 11.3 presents a sampling of the items. Note that in some cases (e.g., the first several items listed) a positive response is indicative of a satisfactory attachment, whereas in other cases (e.g., the last two items) it is the infrequency of the characteristic that is desirable. As with any q-sort, the AQS involves a sequential assignment of items to categories. The person judging the child first divides the 90 items into three roughly equal categories: descriptive of the child, not descriptive of the child, and neither/cannot judge. Following this initial division, the judge further subdivides each category into three more differentiated categories, to each of which 10 items are assigned. The final result is thus a 9-level ranking of the infant's characteristics, ranging from the 10 items that are most typical to the 10 that are least typical. It is the overall picture provided by this rich array of information that constitutes the child's attachment assessment.

I noted that the alternatives to laboratory study are either home observation or parental interview. The AQS lends itself to either method of gathering data. In some studies with the AQS, trained observers visit the home for several hours and make observations of mother-infant interaction; in other cases the mothers themselves provide the AQS assignments. Independent AQS ratings by mothers and observers typically correlate positively, although far from perfectly (Waters & Deane, 1985). AQS assessments also correlate positively with attachment classifications derived from the Strange Situation; the relation is stronger, however, when trained observers rather than mothers provide the assessments (van IJzendoorn, Vereijken, Bakermanns-Kraneburg, & Riksen-Walraven, 2004).

The purpose of both the Strange Situation and the Attachment Q-Set is to assess individual differences in attachment. What more might we want to discover once we know that such differences exist—that is, what further kinds of research are interesting to do beyond the basic assessment? Two broad questions are clearly important.

One concerns origins—where do the differences come from? Researchers of attachment have probed the origins question in various ways. Perhaps the most common approach, beginning with some of Ainsworth's own research (Ainsworth et al., 1978), has been to attempt to identify parental *child-rearing practices* that underlie the different attachment patterns. Child-rearing research is both one of the most important and one of the most methodologically challenging topics in the field, and we will therefore consider it at length in chapter 13. Another possibility is *cross-cultural* study—explorations of the extent

Table 11.3 Examples of Items From the Attachment Q-Set

Item number	*Description*
1	Child readily shares with mother or lets her hold things if she asks to.
11	Child often hugs or cuddles against mother without her asking or inviting him to do so.
21	Child keeps track of mother's location when he plays around the house. Calls to her now and then; notices her go from room to room.
71	If held in mother's arms, child stops crying and quickly recovers after being frightened or upset.
73	Child has a cuddly toy or security blanket that he carries around, takes to bed, or holds when upset.
76	When given a choice, child would rather play with toys than with adults.
79	Child easily becomes angry at mother.
81	Child cries as a way of getting mother to do what he wants.

SOURCE: From "The Attachment Q-Set (Version 3.0)," by E. Waters, 1995, *Monographs of the Society for Research in Child Development, 60* (2–3, Serial No. 244), pp. 236–246.

to which patterns of attachment vary across different cultural settings, especially settings in which infants' experiences are somewhat different from those in the Western cultures that are most often studied. Some on-the-average cultural differences in attachment patterns *are* sometimes found, although their interpretation remains controversial (van IJzendoorn & Sagi, 1999). Finally, in addition to searching for the experiential origins of attachment, researchers have looked for possible biological bases. The most often studied biological contributor has been the construct that we will consider next: inborn differences in *temperament.* Studies show that temperament *does* relate to attachment, although here too there are controversies with respect to how strong the relation is and why it occurs (Kagan, 1998; Vaughn & Bost, 1999).

In addition to learning about origins, we clearly want to know the *consequences* of the early attachment relation. Does the quality of attachment during infancy have implications for the child's later development? To answer this question, we need a longitudinal design, in which the same children are followed from infancy to some point in later childhood. A number of such studies now exist, and they have revealed a variety of links between early attachment and later social and cognitive functioning. It has been found, for example, that children who were securely attached in infancy are on the average more socially competent and have better peer relations than is the case for children whose attachments were less secure (Bohlin, Hagekull, & Rydell, 2000; Sroufe, 1983). Similarly, later problem solving and performance in school can be partly predicted from knowledge of the early attachment relation (Jacobsen & Hofmann, 1997). Indeed, some relations between attachment and later

FOCUS ON

Box 11.3. Attachment and Infant Day Care

One of the most striking changes in U.S. society in recent decades has been the increasing proportion of families in which both parents are part of the workforce. A corresponding change has been a substantial increase in the number of infants and toddlers who spend time in some form of outside-the-home child care.

This societal change has raised some important questions to which developmental research can speak. What are the consequences of outside-the-home care in the first year or two of life? A variety of specific issues can be and have been examined. The most prominent, however—and certainly the most hotly debated—has concerned possible effects on the baby's attachment to the parents. Are babies who spend time in infant day care at a heightened risk of developing insecure attachment?

The brief answer to this question is that there is no agreement on an answer. Probably the majority of researchers who have examined the issue would say no, at least based on current evidence. This is the usual conclusion drawn from the largest study to date of the issue, a longitudinal, multisite study of more than 1300 families funded by the National Institute of Child Health and Human Development (NICHD, 1997, 2003, 2004; online at secc.rti.org). The results of the NICHD study are complex and span many outcomes, but the examinations of attachment typically reveal no differences between day-care and home-reared infants. In the words of one summary of this research: "Child care by itself constitutes neither a risk nor a benefit for the development of the infant-mother attachment relationship" (NICHD, 1997, p. 887). The same conclusion was reached in a major review of the issue: "It now seems clear that most infant-mother attachments are not affected by regular nonmaternal care" (Lamb, 1998, p. 92).

On the other hand, the word "most" in the quoted summary indicates why controversy remains. Some studies *have* reported higher levels of insecure attachment in infants with day-care experience; thus, not all commentators are in agreement with the no-negative-effects conclusion. Jay Belsky, in particular, has argued that extensive early day care carries some (although certainly not inevitable) risks for children's development (Belsky, 2001, 2002).

The fact that controversy remains after decades of research is a clear indication that the issue of day care and attachment is difficult to study, and it is these methodological challenges that are our concern here. The challenges divide into two general sorts.

One issue concerns the assessment of attachment. Most studies of day care and attachment, like most studies of attachment in general, have used the Strange Situation. Critics have questioned, however, whether the Strange Situation is as appropriate for day-care infants as it is for the mostly home-reared babies for whom it was developed (Clarke-Stewart, 1989). A baby in day care, after all, experiences frequent separations from the parents in a setting outside the home. Perhaps the Strange Situation is not stressful enough for such babies to activate the attachment behaviors that it is intended to measure, which would mean that the procedure is simply not doing what it is supposed to do. (I will note that the same argument has been applied to the cross-cultural research mentioned in the text.)

That the Strange Situation may at times not be an optimal or sufficient measure is generally agreed. Conclusions about day care, however, do not seem to be an artifact of this one procedure. Although the point is difficult to establish with certainty, there is some evidence that the Strange Situation works equivalently for day-care and home-reared samples (Belsky & Braungart, 1991). In addition, the same general conclusions about day care and attachment—that is, mostly no relation but with occasional exceptions—emerge on other measures, such as the Attachment Q-Set discussed in the text (Belsky & Rovine, 1990).

Questions of assessment fall on the dependent variable side of research. A second set of methodological challenges revolve around the independent variable. The basic problem here is that studies of day care and attachment are always "natural experiments" rather than true experiments. That is, no one randomly assigns some infants to be in day care and some not, let alone controls all of the other aspects of infant experience that may be important (the timing or length of day care experience, the quality of day care, the quality of parenting, etc.). Rather, investigators must take all these events as they naturally occur. The best that researchers can do is to select and match samples as carefully as possible in an attempt to rule out confounds, as well as employ various statistical procedures to try to pull apart the independent contributions of all the potentially important factors.

There is one reasonably clear conclusion that does emerge from this work, and it is that early day-care experience is not an inevitable or an across-the-board risk factor. Rather, negative effects are found only in conjunction with other factors—in particular, relatively low quality of care and relatively insensitive parenting. As Belsky (2001) notes, this conclusion fits well with Bronfenbrenner's statement (quoted in chapter 6) about the complexity of outcomes in contextual research—namely, that in such research the principal main effects are likely to be interactions.

development have even been shown into adolescence (Schneider, Atkinson, & Tardif, 2001). As is always true with predictions from infancy, the links between early attachment and later measures are far from perfect, and the bases for the links that do emerge are not always clear (Thompson, 2000). Nevertheless, the evidence is compatible with what many theorists have claimed: that a satisfactory early attachment starts development on the optimal path.

Temperament

The pioneering work on childhood temperament was carried out by Thomas, Chess, and Birch as part of the New York Longitudinal Study (Thomas & Chess, 1977; Thomas, Chess, & Birch, 1968). We begin, therefore, with a passage from these authors that summarizes their use of the term "temperament":

> Temperament may best be viewed as a general term referring to the *how* of behavior. It differs from ability, which is concerned with the *what* and *how well* of behaving, and from motivation, which accounts for *why* a person does what he is doing. Temperament, by contrast, concerns the *way* in which an individual behaves. . . . Temperament can be equated to the term *behavioral style*. (Thomas & Chess, 1977, p. 9)

Temperament is a fairly recent subject for explicit scientific study. Many of the ideas behind such study are not new, however, for they reflect beliefs that have long been held by many parents. Children—even children within the same family—often seem to have distinctly different ways of approaching the world. Some children are bursting with energy and always on the go; other children are more placid and happier with quiet activities. Some children maintain attention for lengthy periods and persist in activities until they are completed; other children are more easily distracted and more likely to flit from one thing to another. These behavioral styles may generalize across a variety of situations and behaviors. They may also persist across time, appearing first in infancy and remaining characteristic of the child as he or she grows older. And they may appear so early in infancy that they appear to be at least partly biological, and not environmental, in origin.

As noted, Thomas, Chess, and Birch initiated the systematic study of temperament, and their work has served as a starting point for later research and theory. Thomas et al. identified nine dimensions of temperament, each of which is scored on a 3-point scale. These nine dimensions are listed and briefly described in Box 11.4.

Other conceptualizations of temperament typically include some version of many of these dimensions, although they may add other emphases as well. Buss and Plomin (1984), for example, stress the dimensions of emotionality, activity, and sociability. Rothbart (1986) emphasizes reactivity and self-regulation, with a further distinction between positive and negative emotionality. Kagan and associates (e.g., Kagan, Reznick, & Gibbons, 1989) emphasize the early emergence of an inhibited behavior pattern.

How do Thomas et al. obtain their evidence about temperament? The answer—not only for

Box 11.4. Dimensions of Temperament Identified in the New York Longitudinal Study

1. Activity Level: the motor component present in a given child's functioning and the diurnal proportion of active and inactive periods. Protocol data on motility during bathing, eating, playing, dressing and handling, as well as information concerning the sleep-wake cycle, reaching, crawling and walking, are used in scoring this category.
2. Rhythmicity (Regularity): the predictability and/or unpredictability in time of any function. It can be analyzed in relation to the sleep-wake cycle, hunger, feeding pattern and elimination schedule.
3. Approach or Withdrawal: the nature of the initial response to a new stimulus, be it a new food, new toy or new person. Approach responses are positive, whether displayed by mood expression (smiling, verbalizations, etc.) or motor activity (swallowing a new food, reaching for a new toy, active play, etc.). Withdrawal reactions are negative, whether displayed by mood expression (crying, fussing, grimacing, verbalizations, etc.) or motor activity (moving away, spitting new food out, pushing new toy away, etc.).
4. Adaptability: responses to new or altered situations. One is not concerned with the nature of the initial responses, but with the ease with which they are modified in desired directions.
5. Threshold of Responsiveness: the intensity level of stimulation that is necessary to evoke a discernible response, irrespective of the specific form that the response may take, or the sensory modality affected. The behaviors utilized are those concerning reactions to sensory stimuli, environmental objects, and social contacts.
6. Intensity of Reaction: the energy level of response, irrespective of its quality or direction.
7. Quality of Mood: the amount of pleasant, joyful and friendly behavior, as contrasted with unpleasant, crying and unfriendly behavior.
8. Distractibility: the effectiveness of extraneous environmental stimuli in interfering with or in altering the direction of the ongoing behavior.
9. Attention Span and Persistence: two categories which are related. Attention span concerns the length of time a particular activity is pursued by the child. Persistence refers to the continuation of an activity in the face of obstacles to the maintenance of the activity direction.

Thomas et al. but for the majority of temperament studies—is that researchers learn about child temperament by asking the parents about the child's temperament. In the New York Longitudinal Study the main data came from interviews with the children's parents. The interviews concerned a variety of common situations in which infant temperament might become apparent. The parents were asked, for example, about the baby's sleeping and eating patterns, about typical behavior when meeting a new person or being taken to the doctor, about responses to being washed, changed, or dressed. To increase the accuracy of the reports, an attempt was made to tie the questions to concrete situations and recent behaviors. An attempt was also made to elicit a description of the child's behavior, as opposed to the parent's interpretation of the behavior (e.g., "He spit the cereal out" rather than "He hated it").

Interviews are one method of eliciting parental reports. The other general approach is to administer a questionnaire. With a questionnaire the oral response of the interview is replaced by a written response. This written response may take different forms; simplest and most common is for the participants to indicate their extent of agreement with some statement by marking a point along an ordered scale. Table 11.4 shows examples of items from a questionnaire directed to infancy and the Thomas et al. dimensions. Questionnaires are probably the most common approach to assessing infant temperament (Slabach, Morrow, & Wachs, 1991).

Whether the format is interview or questionnaire, the data from such measures remain verbal reports about behavior rather than direct observations of behavior. Even if the validity of the parental-report approach is acknowledged, it is clearly desirable to complement such measures with a direct observational assessment of the behaviors from which temperament is inferred. Table 11.5 provides a sampling of items from one such measure: the Laboratory Temperament Assessment Battery, or LAB-TAB, developed by Goldsmith and Rothbart (1991, 1992). As can be seen, the LAB-TAB presents a variety of episodes (half of which are summarized in the table) to which infants can react—and from which differences in dimensions such as activity level, fearfulness, and persistence can be measured.

We turn now from instruments to the kinds of research issues to which the instruments are directed. The two most general issues in the study of temperament are the same two general issues that have inspired research on attachment: the questions of *origins* and *consequences.* Where does temperament come from, and what implications do early differences in temperament have for the child's later development?

Discussions of the origins of temperament typically focus on the role of biology. Much of the interest in measures of temperament has stemmed from the possibility that such measures might be tapping genetically based differences among children, differences that predate parental socialization efforts but also have a definite impact on eventual socialization.

Various lines of evidence suggest that temperament is in part genetically based. The fact that individual differences emerge so early, prior to much chance for socialization to operate, is one source of evidence. In the New York Longitudinal Study, differences were apparent by 2 months, the youngest age sampled. Other research has documented individual differences, including differences in temperament-like qualities, during the neonatal period (Wachs, Pollitt, Cueto, & Jacoby, 2004). Clearly, the earlier in development a difference emerges, the more plausible a genetic contribution becomes.

The relation between genes and temperament has also been a topic within the field of study known as *behavior genetics.* I discuss the methods of behavior genetics more fully in chapter 12 in the context of the heredity-environment controversy with respect to individual differences in IQ. Here I will note simply

Table 11.4 Examples of Items From the Infant Temperament Questionnaire

Almost never	*Rarely*	*Variable usually does not*	*Variable usually does*	*Frequently*	*Almost always*
1	*2*	*3*	*4*	*5*	*6*

1. The infant eats about the same amount of solid food (within 1 oz.) from day to day.	almost never	1 2 3 4 5 6	almost always
2. The infant is fussy on waking up and going to sleep (frowns, cries).	almost never	1 2 3 4 5 6	almost always
3. The infant plays with a toy for under a minute and then looks for another toy or activity.	almost never	1 2 3 4 5 6	almost always
4. The infant sits still while watching TV or other nearby activity.	almost never	1 2 3 4 5 6	almost always
5. The infant accepts right away any change in place or position of feeding or person giving it.	almost never	1 2 3 4 5 6	almost always
6. The infant accepts nail cutting without protest.	almost never	1 2 3 4 5 6	almost always
7. The infant's hunger cry can be stopped for over a minute by picking up, pacifier, putting on bib, etc.	almost never	1 2 3 4 5 6	almost always
. . .			
91. The infant's first reaction to any new procedure (first haircut, new medicine, etc.) is objection.	almost never	1 2 3 4 5 6	almost always
92. The infant acts the same when the diaper is wet as when it is dry. (no reaction)	almost never	1 2 3 4 5 6	almost always
93. The infant is fussy or cries during the physical examination by the doctor.	almost never	1 2 3 4 5 6	almost always
94. The infant accepts changes in solid food feedings (type, amount, timing) within 1 or 2 tries.	almost never	1 2 3 4 5 6	almost always
95. The infant moves much and for several minutes or more when playing by self. (kicking, waving arms and bouncing)	almost never	1 2 3 4 5 6	almost always

SOURCE: From "Revision of the Infant Temperament Questionnaire," by W. B. Carey and S. C. McDevitt. Reproduced with permission from Pediatrics, 61, pp. 735–739, copyright (c) 1978, by the American Academy of Pediatrics.

Table 11.5 Examples of Items From the Laboratory Temperament Assessment Battery

Fearfulness episodes
Mechanical toy dog races across table toward child.
Male stranger approaches and picks up child.
Anger proneness episodes
Gentle arm restraint by parent while playing with toy.
Attractive toy placed behind Plexiglass barrier.
Pleasure episodes
Reaction to sound and light display in nonsocial setting.
Modified peek-a-boo game with mother's face appearing behind various doors.
Interest/Persistence episodes
Task orientation while playing with blocks.
Attention to repeated presentation of photographic slides.
Activity episodes
Activity while in corral filled with large rubber balls.
Locomotor activity during free play.

SOURCE: From "Contemporary Instruments for Assessing Early Temperament by Questionnaire and in the Laboratory" (p. 264), by H. H. Goldsmith and M. K. Rothbart. In J. Strehan and A. Angleitner (Eds.), *Explorations in temperament: International perspectives on theory and measurement* (pp. 249–272), New York: Plenum Press. Reprinted with permission of Springer Science and Business Media.

that the methodology from behavior genetics that has been most important in the study of temperament is the *twin study* approach—that is, comparisons of the two types of human twins, identicals and fraternals. The logic of the approach is straightforward: If genes are important for some aspect of development, then identical twins (who share 100% of their genes) should be more similar than fraternal twins (who share only 50% of their genes). This in fact is the finding for temperament: A substantial research literature is consistent in reporting greater similarity in temperament between members of identical twin pairs than between members of fraternal twin pairs (e.g., Goldsmith, Buss, & Lemery, 1997). This greater similarity is evident as early as 3 months and remains true throughout childhood, it holds for a variety of different temperamental dimensions, and it is found across a range of different instruments for measuring temperament.

A further point about the twin studies will serve as a transition to the second general issue in the study of temperament. The twin studies make clear that genes may be important to temperament but that they are not the sole explanation for the individual differences that are found. Identical twins may be similar in temperament, but they are far from identical—and any differences between such twins must be the result of the environment. Thus, temperament is not completely set by the genes; the environment makes a contribution as well.

The conclusion that the environment is important emerges also from another major form of temperament research: studies of the *stability* of temperament as children develop. Do early individual differences in temperament remain consistent as children grow up? To answer this question we need a longitudinal approach, in which the same children are studied repeatedly over time. As the title of the Thomas et al. project reveals, longitudinal analysis has been a part of temperament research from its inception. In the New York Longitudinal Study the initial interview procedure covered the span of infancy; through various other procedures—parent and teacher questionnaires, observations, interviews with the children themselves—many of the participants were eventually followed until adolescence.

The most general conclusion to emerge from this research—a conclusion that is confirmed by more recent longitudinal studies (e.g., Rothbart, Derryberry, & Hershey, 2000)—was that temperament shows some, but far from perfect, stability as children develop. Thus early temperament *does* have consequences for later development; there is a tendency for the shy child to remain shy, the irritable child irritable, and so forth. But this is simply a tendency, not a certainty, and development, as is always the case, depends on the interplay of biology and experience. If, for example, an initially irritable and difficult-to-manage infant is met with patient and sensitive parenting, then the early difficulties need not be predictive of later problems.

Temperament has been a much discussed concept in recent years, and further sources are therefore plentiful (Rothbart & Bates, 2006; Sanson, Hemphill, & Smart, 2002; Wachs & Bates, 2001). Because the present coverage has emphasized the measurement of temperament in infancy, I should note that these sources include a variety of instruments designed to assess temperament in later childhood and even adulthood (e.g., McClowry, 1995; Rothbart, Ahadi, Hersey, & Fisher, 2001; Windle & Lerner, 1986). The same point applies, in fact, to the other construct discussed in this section of the chapter: attachment. Although attachment, even more than temperament, has traditionally been a topic of infancy, in recent years its study too has expanded to childhood and even adult forms (e.g., Main, 2000; van IJzendoorn, 1995).

Summary

This chapter begins with a discussion of some of the challenges involved in doing research with infants. Three general difficulties are discussed. The first is finding and retaining babies. Infants can be difficult to recruit for research, and infants have the highest dropout rate of any group of research participants. The second difficulty concerns the influence of state on the infant's behavior. The infant's state of arousal is an important determinant of how the infant responds to the environment; furthermore, infants are often in states (e.g., sleepy, distressed) that preclude optimal responsiveness. The final difficulty concerns response measures. Infants are nonverbal, and young infants in particular have limited motor skills; the result is that considerable ingenuity is necessary to find responses from which the infant's experience of the world can be inferred. Three response systems that have proved especially informative are introduced: *visual fixation*, *sucking*, and *physiological responses*.

The discussion turns next to the issue of age comparisons. Many of the problems associated with both cross-sectional and longitudinal designs are reduced when research is confined to the time span of infancy. On the other hand, the rapidity of change during infancy does raise the issue of *measurement equivalence:* finding responses that are psychologically equivalent at the different ages.

The first specific topic that the chapter considers is infant perception. Research on infant perception has concentrated on responses that can be precisely, often automatically, recorded in laboratory settings; a further virtue of the laboratory locus is that it allows precise control of the stimulus. Four methods of studying perception are discussed. The *preference method* is directed to the question of visual discrimination;

two visual stimuli are presented, and differential fixation on the stimuli is taken as evidence that the infant can discriminate between them. The *habituation-dishabituation* technique is broader in scope, in that it can be applied to any perceptual modality. Response is first habituated to one stimulus; reemergence of the response to a second stimulus constitutes evidence of discrimination. *Conditioning* has been used in various ways to study infant perception, including satiation, in which the recovery of reinforcer effectiveness signals discrimination; conditioned head turning; and generalization. Finally, the *visual cliff* is used to study a single issue of considerable theoretical interest: depth perception in infancy.

Research on infant cognition has been dominated by the work of Piaget. Piaget's studies differ from studies of infant perception in several ways, most notably in the Piagetian emphases on naturalistic observation and on flexible, discovery-oriented probing of the child's abilities. Among the many basic forms of knowledge studied by Piaget is the *object concept:* the knowledge that objects have a permanent existence that is independent of immediate perceptual contact. Piaget's methods of studying the object concept center on the infant's search for vanished objects; a variety of forms of search and a variety of kinds of disappearance are examined.

The object concept has also been the most popular topic for post-Piaget studies of infant cognition. This more recent work is more tightly controlled and standardized than were Piaget's studies, with more use of laboratory settings and automated recordings of response. A question of particular interest has concerned the effects of less motorically demanding methods of assessment on the child's performance. An especially informative methodology has been the *violation of expectancy paradigm:* measurement of looking time in response to apparent violations of physical laws.

Research on social development has been less bound to laboratory settings than has most work on perception or cognition. The interest now is in the infant's interactions with other people, most commonly the mother. Of particular interest has been the development of *attachment,* an emotional bond with the caregiver that is expressed through a variety of behaviors. Information about attachment can be obtained through *parental interview, naturalistic observation,* or *structured laboratory assessment.* The *Attachment Q-Set* is presented as an example of the first two approaches, and the *Strange Situation* as an example of the third. Studies using these procedures have revealed important individual differences in the quality of children's attachments. Further studies have examined both the origins of these differences and their consequences for children's later development.

The chapter concludes with a discussion of *temperament.* "Temperament" refers to behavioral style—to individual differences along dimensions such as activity level, approach/withdrawal, and distractibility. Temperament is most often measured through parental reports, although some programs of research incorporate observational measures as well. As with attachment, a basic research question has concerned the origin of individual differences, with a special focus on the possibility of a genetic contribution. Also of interest has been the stability of temperament over time, an issue addressed through longitudinal study.

Exercises

1. A goal of modern researchers of infant cognition has been to devise measurement procedures that are less motorically demanding than the methods used by Piaget. The

Baillargeon studies of object permanence, described in the text, are one major example. Imagine that you wished to study infants' understanding of causality, using procedures similar to those of Baillargeon. What kinds of experiments might you devise?

2. The following exercise assumes that you have access to at least one infant between about 4 and 8 months of age. If so, devise and administer several object concept tasks appropriate for this age range (you will need to consult Piaget's *The Construction of Reality in the Child*, 1954, for specific procedures). In addition to the standard Piaget approach, contrive an apparent violation of the object concept, as in the Baillargeon studies, and compare results from the two procedures.

3. Table 11.3 reproduces 8 of the 90 items that make up the Attachment Q-Set. Imagine that you had the task of creating the items for such a measure. What other infant behaviors or characteristics can you think of that might be relevant to an assessment of attachment? Once you have generated your list, obtain a copy of the Attachment Q-Set (the publication source is given in the text) and compare your ideas with the actual test items.

4. The following exercise assumes that you know at least one and preferably several parents of infants whom you can interview. Obtain a copy of one of the parental-report measures of temperament discussed in the chapter, either the Infant Temperament Questionnaire summarized in Table 11.4 or one of the other instruments cited in the review by Slabach et al. If you have several parents with whom to work, you may want to compare two of the measures. Administer the instrument in the standardized way, and once you have done so, ask the parents for their impression of how well it captures characteristics of the infant.

Note

1. Some research on the development of depth perception provides an exception to this statement. Researchers have been interested in the possibility that the onset of a fear response on the visual cliff (an apparatus described later in the chapter) may be tied to locomotor experience. One approach to this question has been longitudinal: studies of the same sample across several weeks, with the goal of determining whether babies who crawl relatively early or relatively often are most likely to show the fear response. The usual finding is that they are (Bertenthal, Campos, & Barrett, 1984).

12

Cognitive Development

The topics into which the field of developmental psychology can be divided are as many and as diverse as the topical divisions for psychology as a whole. The organizational scheme that we will follow in the next two chapters—a division into "cognitive" and "social"—is perhaps the broadest and most general cut that can be made. This scheme does not encompass every possible topic in the field, nor is the borderline between cognitive and social always clear. Nevertheless, the division is a typical and generally useful one.

The coverage of cognitive development in the present chapter is divided into five sections. The chapter begins with discussion of two of the most influential general approaches to the study of children's intelligence: the Piagetian approach and the intelligence test or IQ approach. The remaining sections are devoted to three of the most active topics for current research in cognitive development: memory, theory of mind, and conceptual development.

The age period for most of the research considered in this chapter is from 2 to 16—that is, postinfancy childhood. I discussed the Piagetian approach to the study of infant cognition in chapter 11, and we will briefly consider some other forms of infancy research at various points in this chapter. We will also return to several of these approaches in chapter 14 in the discussion of research on aging.

The Piagetian Approach

Piaget's Studies

We saw in chapter 11 that Piaget's work dominated the study of infant cognition for years but is no longer as influential as was once the case. The same can be said of the Piagetian approach to later childhood. This work is even harder to summarize briefly, for it constitutes a much larger literature than the infancy studies—some 25 books by Piaget and associates, as well as literally thousands of related studies by others. The goal of this section is to highlight some central themes and important research examples. Fuller discussions can be found in Flavell (1963), Ginsburg and Opper (1988), Voyat (1982), and Miller (1982).

I begin with an example. Box 12.1 presents two protocols from Piaget and Szeminska's (1952) *The Child's Conception of Number.* The concept under examination is **conservation**: the realization that the quantitative properties of an object or collection of objects are not changed by a change in perceptual appearance. The specific form of conservation at issue is conservation of

Box 12.1. Examples of Responses to Piaget's Conservation of Number Task

Hoc (4; 3): 'Look, imagine that these are bottles in a cafe. You are the waiter, and you have to take some glasses out of the cupboard. Each bottle must have a glass' He put one glass opposite each bottle and ignored the other glasses. 'Is there the same number?—*Yes.*—(The bottles were then grouped together.) Is there the same number of glasses and bottles?—*No.*—Where are there more?—*There are more glasses.*' The bottles were put back, one opposite each glass, and the glasses were then grouped together. 'Is there the same number of glasses and bottles?—*No.*—Where are there more?—*More bottles.*—Why are there more bottles?—*Just because.*'

Boq (4; 7): 'Put as many sweets here as there are there. Those (6) are for Roger. You are to take as many as he has.—(He made a compact row of about ten, which was shorter than the model.)—Are they the same?—*Not yet* (adding some).—And now?—*Yes.*—Why?—*Because they're like that* (indicating the length).—(The 6 in the model were then spread out.) Who has more?—*Roger.*—Why?—*Because they go right up to there.*—What can we do to make them the same?—*Put some more* (adding 1).—(The 6 were then closed up and his were spread out.)—*Now I've got more.*'

SOURCE: From *The Child's Conception of Number* (pp. 44, 75), by J. Piaget and A. Szeminska. New York: Humanities.

number: the realization that number is invariant in the face of an irrelevant perceptual change. As the protocols reveal, young children do not at first understand conservation; rather, they tend to judge quantities in terms of immediate perceptual appearance. Thus, to 4-year-old Boq it is obvious that the longer of two rows must contain more sweets.

We can see immediately several similarities between Piaget's studies of infancy and his approach to later periods of development. He again focuses on basic, epistemologically central kinds of knowledge. Just as the object concept represents a major achievement in the infant's mastery of the sensorimotor world, so are principles of conservation central to the older child's capacity for more advanced forms of thought. Object concept and conservation show a more specific similarity as well. Both represent important *invariants:* aspects of the world that stay the same even though other, more obvious aspects are changing. Throughout his research career Piaget was interested in the invariants that the child comes to understand at different points in development. The phrase "comes to understand" reflects still another similarity: Both object concept and conservation, basic though they seem, are not always present; rather, they must be developed in the course of childhood. Certainly one of the reasons for interest in Piaget's work has always lain in his ability to surprise us with respect to what children, at least for a while, do *not* know.

In addition to these similarities in content, the comparison of object concept and conservation reveals some more general similarities in approach to research. Once again Piaget eschews a highly standardized approach in favor of a flexible, discovery-oriented method of probing the child's knowledge. And once again Piaget reports his results mainly in terms of individual protocols rather than group means and statistical tests.

There are also some important differences between the studies of infancy and the studies of later childhood. One obvious difference concerns sample size. As we saw, the sample for the infancy work was limited to Piaget's own three children. The samples for the work on later childhood are considerably larger and more representative. Beyond this rather general statement it is difficult to say much, for Piaget seldom provided precise information about sample size or composition (although one book, *The Early Growth of Logic in the Child,* does report a total sample of 2,159!). Piaget's failure to describe the samples that he studied is just one of the sins of scientific reporting that he routinely committed. Nevertheless, it is safe to say that his samples for later childhood are much larger than those for infancy.

Some further differences follow from this difference in samples. Piaget's studies of infancy were longitudinal. With the exception of some work on long-term memory (Piaget & Inhelder, 1973), the research on later childhood is all cross-sectional. Similarly, the studies of infancy were within-subject, in the sense that the phenomena of interest were examined in all three babies and interrelations in development were probed for each child. The work on later childhood is, apparently, just about all between-subject. The "apparently" stems from the fact that Piaget often failed to make clear whether a particular conclusion was based on within-subject or between-subject comparisons. Generally, however, the latter appears to be the case.

A final difference concerns the locus for the observations. The studies of infancy were carried out in the home and focused largely on naturally occurring situations and naturally occurring behaviors. The studies of older children have concentrated mainly on elicited responses to tasks presented in some laboratory context. As the protocols from the *Number* book suggest, the procedure may still be more game-like than test-like, and the interchanges between adult and child may resemble spontaneous conversations more than school-like inquisitions. The fact remains, however, that the measurement is of task-elicited behavior in experimentally contrived situations, and not of spontaneously occurring cognitive activities. We will return later to this issue.

It is time to add some more illustrations of Piagetian tasks to the conservation-of-number example. I will note first that Piaget studied conservation in many quantitative domains in addition to number. Indeed, most of the books devoted to cognition in early and middle childhood include tests of conservation. There are studies of conservation of mass, weight, and volume; of length, area, and distance; of time, speed, and movement. All embody the same general approach: Two stimuli are shown to be equal on some quantitative dimension, one of the stimuli is then perceptually transformed so that the quantities no longer *look* equal, and the child is asked whether the quantities are now the same or different. All also show the same developmental progression from perceptually based nonconservation to logically based conservation.

Conservation is just one of dozens of basic logical or physical concepts that Piaget and his coworkers have studied. I settle here for briefly describing two other important examples. One is **class inclusion**: the principle that a subclass cannot be larger than the superordinate class that contains it. Class inclusion is the knowledge, for example, that there can never be more poppies than flowers, or more ducks than birds. Both of these problems were in fact included in Piaget's studies of classification (Inhelder & Piaget, 1964; Piaget & Szeminska, 1952). Box 12.2 presents a third example. The stimuli for the task are a set of wooden beads, most of which are brown but two of which are white. The response of 6-year-old Bis indicates that class inclusion, like conservation, is another basic concept that is not at first present but rather must develop.

In addition to classes, Piaget's research includes a focus on the child's understanding of relations. Of particular interest is the relational concept of **transitivity**. Transitivity is embodied in reasoning of the following sort: If A is equal to B and B equal to C on some quantitative dimension, then A must be equal to C. Or if A is greater than B and B is greater than C, then

Box 12.2. Example of Response to Piaget's Class Inclusion Task

Bis (6; 8): 'Are there more wooden beads or more brown beads?—*More brown ones, because there are two white ones.*—Are the white ones made of wood?—*Yes.*—And the brown ones?—*Yes.*—Then are there more brown ones or more wooden ones?—*More brown ones.*—What colour would a necklace made of the wooden beads be?—*Brown and white* (thus showing that Bis clearly understood the problem).—And what colour would a necklace made with the brown beads be?—*Brown.*—Then which would be longer, the one made with the wooden beads or the one made with the brown beads?—*The one with the brown beads.*—Draw the necklace for me. (Bis drew a series of black rings for the necklace of brown beads, and a series of black rings plus two white rings for the necklace of wooden beads.)—Good. Now which will be longer, the one with the brown beads or the one with the wooden beads?—*The one with the brown beads.*' Thus, in spite of having clearly understood, and having correctly drawn the data of the problem, Bis was unable to solve it by including the class of brown beads in the class of wooden beads!

SOURCE: From *The Child's Conception of Number* (p. 164), by J. Piaget and A. Szeminska. New York: Humanities. Copyright ©1952 by Humanities Press.

A must be greater than C. Such reasoning has been studied most often with respect to length and weight, the usual stimuli being sticks of different length or clay balls of different weight. Whatever the specific quantity involved, the approach is the same: demonstration of the A-B and B-C relations, followed by a request to judge A and C. Note that the quantitative relation between A and C is not perceptually apparent; hence the child must use the information in the initial two comparisons to deduce the correct answer. According to Piaget's studies it is not until about age 8 or 9 that children are capable of such logical reasoning.

The tasks described thus far are directed to thought in middle childhood, or what is labeled the concrete-operational period in Piaget's theory. In a book entitled *The Growth of Logical Thinking From Childhood to Adolescence*, Inhelder and Piaget (1958) examined a more advanced form of thinking, which they found to emerge only around adolescence. This final period in Piaget's theory is the period of *formal operations*. The essence of formal operations is the capacity for hypothetical-deductive reasoning, the ability to go beyond immediate reality to work systematically and logically within the realm of the possible. The prototype of such reasoning is scientific problem solving; Inhelder and Piaget's tasks for studying formal operations in fact consisted mainly of problems drawn from the physical sciences. One of these problems, the pendulum task, is the source for the protocols reprinted in Box 12.3. The child's job is to determine what factor or factors influence the frequency of oscillation of a simple pendulum. Solution requires identifying each of the potentially important variables (weight, length of string, force of push, etc.), systematically testing out each variable while holding the other variables constant, and finally drawing logical conclusions from the overall pattern of results. The table presents examples of both the failure of the younger child and the success of the older child on such problems.

We will consider one final example of a Piagetian task. Not all of the studies fit the logical or physical mold of the work described so far. Piaget's first book, *The Language and Thought of the Child* (1926), was directed to an important component in the child's understanding of the social world: the ability to take

Box 12.3. Examples of Responses to the Inhelder and Piaget Formal-Operational Pendulum Task

PER (10; 7) is a remarkable case of a failure to separate variables: he varies simultaneously the weight and the impetus; then the weight, the impetus, and the length; then the impetus, the weight, and the elevation, etc., and first concludes: *"It's by changing the weight and the push, certainly not the string."*—"How do you know that the string has nothing to do with it?"—*"Because it's the same string."*— He has not varied its length in the last several trials; previously he had varied it simultaneously with the impetus, thus complicating the account of the experiment.—"But does the rate of speed change?"—*"That depends, sometimes it's the same. . . . Yes, not much. . . . It also depends on the height that you put it at [the string]. When you let go low down, there isn't much speed."* He then draws the conclusion that all four factors operate: *"It's in changing the weight, the push, etc. With the short string, it goes faster,"* but also *"by changing the weight, by giving a stronger push,"* and *"for height, you can put it higher or lower."*—"How can you prove that?"—*"You have to try to give it a push, to lower or raise the string, to change the height and the weight."* [He wants to vary all factors simultaneously].

EME (15; 1), after having selected 100 grams with a long string and a medium length string, then 20 grams with a long and a short string, and finally 200 grams with a long and a short, concludes: *"It's the length of the string that makes it go faster or slower; the weight doesn't play any role."* She discounts likewise the height of the drop and the force of her push.

SOURCE: From *The Growth of Logical Thinking from Childhood to Adolescence* (pp. 71, 75), by B. Inhelder and J. Piaget. New York: Basic Books. Copyright © 1958 by Basic Books, Inc.

someone else's point of view. Can the child figure out what someone else sees at the moment, or thinks, or feels, or wishes? In particular, can the child make such judgments when the other's perspective is different from the child's own? Such *perspective taking* is critical to social understanding and social interaction. Perspective taking has a converse in **egocentrism**: the inability to break away from one's own perspective to take the perspective of others. It should come as no surprise, in light of the cognitive deficits discussed so far, to learn that Piaget finds that young children are often egocentric. Figure 12.1 shows one basis for this conclusion, the "three mountains" test of visual or spatial perspective taking. After walking around the display, the child is seated on one side of the model; the child's task is then to describe what would be seen by a doll placed at various locations around the display. To 4- or 5-year-olds the answer is clear: The doll sees whatever they see.

I return to several of the tasks just described when I discuss follow-up work. Before doing so, however, it is worth noting a few more points about the general Piagetian approach to research. The interview procedure illustrated in the various protocols in this chapter is referred to as the **clinical method** of testing. Piaget (1929) adopted the term "clinical method" because of the similarity of his approach to that of a skilled clinician attempting to diagnose and treat emotional problems. In both cases the essence of the approach is flexibility: the freedom for the investigator to deviate from preset procedure to probe the individual participant's response in a variety of nonpredetermined ways.

Piaget's fullest discussion of the clinical method is found in one of his early books, *The*

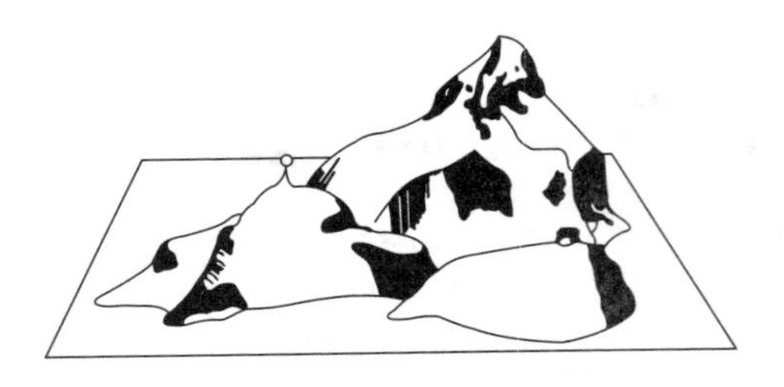

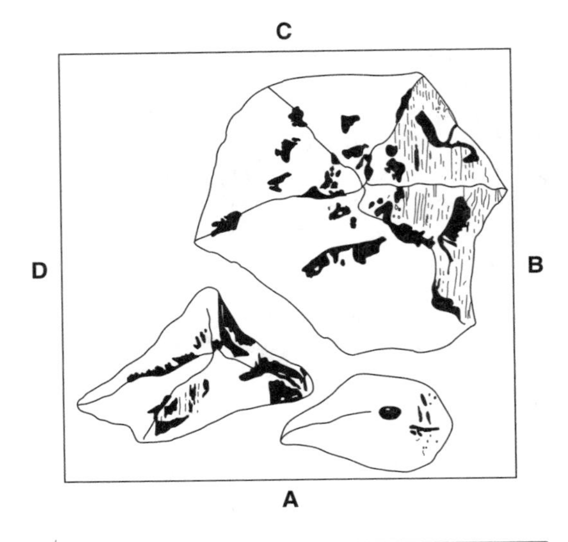

Figure 12.1 Three-mountains task for assessing visual perspective taking

SOURCE: From *The Child's Conception of Space* (p. 211), by J. Piaget and B. Inhelder, 1956, London: Routledge & Kegan Paul. Copyright 1956 by Routledge & Kegan Paul. Reprinted by permission of Taylor & Francis.

Child's Conception of the World. The following passage is worth quoting in full:

> It is our opinion that in child psychology as in pathological psychology, at least a year of daily practice is necessary before passing beyond the inevitable fumbling stage of the beginner. It is so hard not to talk too much when questioning a child, especially for a pedagogue! It is so hard not to be suggestive! And above all, it is so hard to find the middle course between systematisation due to preconceived ideas and incoherence due to the absence of any directing hypothesis! The good experimenter must, in fact, unite two often incompatible qualities; he must know how to observe, that is to say, to let the child talk freely, without ever checking or side-tracking his utterance, and at the same time he must constantly be alert for something definitive, at every moment he must have some working hypothesis, some theory, true or false, which he is seeking to check. To appreciate the real difficulty of the clinical method one must have taught it. When students begin they either suggest to the child all they hope to find, or they suggest nothing at all, because they are not on the look-out for anything, in which case, to be sure, they will never find anything. (Piaget, 1929, pp. 7–8)

The clinical method is an important component in Piaget's research, but it would not mean very much unless there were interesting concepts to which it could be applied. And here we have reached perhaps the greatest strength of Piagetian research: the incredible range of interesting forms of knowledge and insightful procedures for probing this knowledge. This richness can be only hinted at in the brief sampling of tasks and findings included here. Book after Piagetian book is full of novel and informative methods for studying how children think—methods that have set the mold for a substantial proportion of later research on cognitive development. I noted in chapter 1 that technical skill in executing research must always be joined with good ideas about what is interesting to study and how to go about studying it. It is doubtful that anyone else in the history of developmental psychology has had as many good ideas as Piaget.

Issues and Follow-Up Studies

I turn immediately from praise to problems. Piaget's studies have elicited an enormous amount of follow-up research, much of it motivated by perceived deficiencies in the original

Piagetian research. Although this follow-up work has addressed a variety of specific issues, I settle for discussing just one of them here. It is the same basic question that was discussed with regard to the infancy work: that of *assessment.* Do Piaget's procedures really give an accurate picture of the child's abilities?

As with the infancy studies, the main concern has been that Piaget may have underestimated what the young child really knows. Although various possible sources of misdiagnosis have been identified, probably the most common criticism concerns the heavy verbal emphasis in many Piagetian tasks. This emphasis should be clear from the protocols quoted earlier in the chapter. Consider the conservation task. The purpose of this task is to assess the child's understanding of the logic underlying conservation. This logic, however, is not directly observable; rather, its discovery depends on the use of language, both in the questions that are asked of the child and in the responses that the child must make. It seems quite possible that a young "nonconserver" does not really believe in nonconservation but is simply confused by the use of words like "same," "more," "less," and "number." Perhaps, for example, the child thinks that "more" refers to length of row rather than number of objects.

Various approaches have been taken in response to this possibility. Some investigators have used verbal pretests in an attempt to ensure that the child understands the words that are used on the conservation test (e.g., Miller, 1977). Others have gone beyond pretesting to attempt verbal pretraining—that is, teaching the relevant words to the child prior to the test (e.g., Gruen, 1965). And some have attempted to do away with the potentially confusing language altogether by devising "nonverbal" procedures for assessing Piagetian concepts—for example, having the child select a row of candies to eat or a glass of juice to drink (e.g., Miller, 1976a). Such procedures are seldom literally nonverbal; they are simplified linguistically, however, and they do avoid such potentially troublesome words as "same" or "more."

Not all diagnostic revisions have been directed to language. Another concern has been with the general context within which conservation is typically assessed. This context is in fact a rather strange one, and it includes a number of features that may bias the child toward a nonconservation answer. Among these features are the explicit focus on quantity, the seemingly arbitrary nature of the transformation (why is the adult spreading out the candies?), and the presentation of the identical conservation question twice within a short period, a repetition that may suggest to young children that they should change their answers. Perhaps if the context could be made more natural and familiar, the child would be less likely to look like a nonconserver.

This possibility has been tested in various ways. Rose and Blank (1974) examined the effects of the usual two-questions format by simply omitting the initial pretransformation question for half their participants and asking only the final conservation question. McGarrigle and Donaldson (1974) replaced the intentional transformation by an adult experimenter with an apparently accidental transformation by a "naughty" teddy bear, the expectation being that children might find the latter more familiar and less imbued with magical quantity-changing properties. Light, Buckingham, and Robbins (1979) performed a similar manipulation but with an incidental rather than accidental change—that is, a transformation that was not directed solely to the issue of conservation but occurred naturally in the course of an ongoing game.

So far nothing has been said about results from such modified-assessment studies. Three general conclusions seem tenable (for further discussion, see Chandler & Chapman, 1991; Miller, 1976b; Siegal, 1991). The first is that Piaget's methods do in fact result in some underestimation of the young child's abilities, for children often perform better on modified tests than on standard Piagetian tests. The second is that the underestimation is probably not great, and that phenomena such as nonconservation are by no means totally explicable on the basis of verbal confusions or contextual biases. The third is that the development of concepts such as conservation and perspective taking is more extended and multifaceted than Piaget imagined, with a number of earlier levels and precursor skills not tapped by Piaget's own procedures. This conclusion comes not only from the studies discussed but also from programs of research whose explicit focus has been on simpler and developmentally earlier skills than those examined in Piaget's research. Notable in this regard is the work of John Flavell on perspective taking (Flavell, 1992) and Rochel Gelman on concepts of number (Gelman, 1991).

The Intelligence Test Approach

The Nature of IQ Tests

The intelligence test, or IQ approach, is in many ways quite different from the Piagetian approach. Because the emphasis so far has been on Piaget, it is instructive to begin by noting what some of the differences are.

Piaget's interest was always in commonalities of development—that is, ways in which all children are alike as they grow. A concept like conservation of number, for example, is eventually mastered by all normal children. More broadly, a stage like concrete operations is eventually attained by virtually every child. What individual differences there are seem to lie in the rate of development, and such differences were never of interest to Piaget. In contrast, the whole point of IQ tests is to identify individual differences among children. Such tests, moreover, measure not only differences but *ordered* differences—we say that one child is "higher" or "lower" in intelligence than another, or that a particular child is "above" or "below" average in intelligence. There is an evaluative component that is impossible to escape from in using IQ tests. The fact that such tests force us to make value judgments about children is one reason that their use has always been so controversial.

From the start Piaget's work was theoretically guided, the goal being to answer basic epistemological questions about the child's understanding of domains such as number, space, time, and causality. Whatever practical applications the work has had (e.g., influences on school curriculum) have come later, and Piaget himself was never a major contributor to such applications. IQ tests, in contrast, have been pragmatically oriented from the start. The first successful IQ test, an instrument designed by Binet and Simon in Paris in 1905, was constructed for the very pragmatic purpose of predicting how well children would do in school. Indeed, the ability to predict school performance was an explicit criterion in the selection of items for the test. IQ tests ever since have had similar practical groundings and practical applications.

A final difference concerns the quantitative emphasis in IQ tests. IQ tests are directed to questions of how much and not to questions of how. What the tests yield is a number that tells how much intelligence a particular child has. Some tests yield several numbers, corresponding to different kinds of intelligence; the approach, however, remains basically quantitative. All that is of concern in scoring an IQ test is how many right answers the child gives, and all that is done with the right answers is to add them together to get an overall point total. The focus is thus on the products of cognitive activity, and not the underlying processes from which these products came. For Piagetians, in contrast, the interest is always more in processes than in products. The attempt in Piagetian research is to move beyond the child's right or wrong answer to identify the qualitative nature of the underlying thought system and the qualitative changes that the system undergoes as the child develops. The Piagetian focus is thus more on the how than on the how much.

What has been said so far about IQ tests has a rather negative sound to it. Such tests are more quantitative than qualitative, more concerned with products than underlying processes, more pragmatic than theoretical in their origins and construction. Nor are these the only criticisms that can be lodged against IQ tests; I have not even mentioned the common complaint that such tests are biased against certain groups. Given these various problems, the obvious question becomes: Why should anyone take IQ tests seriously? What is the evidence that such tests are really measuring intelligence? What, in short, is the evidence for the *validity* of the tests?

The general issue of test validity was discussed in chapter 4. Recall that the validity of a test is generally established through demonstrating that the test correlates with other measures to which it ought to relate, "ought to" either for purposes of pragmatic prediction or for reasons of theoretical cogency. The validation of IQ tests has always rested upon such correlational power. The first such test, that of Binet and Simon, was deemed successful because it was able to differentiate among children who were likely to do well in school and those who were likely to do poorly in school. Ever since the original Binet and Simon test, correlations with school performance or academic achievement tests have been a major validity index for IQ tests designed for children. Typically, such correlations are in the neighborhood of .5 to .6, a figure that indicates a moderately strong but certainly far from perfect relation. The correlational power of IQ is not limited to academic contexts, however. IQ also correlates with occupational status in adulthood and with performance on a wide range of learning and cognitive measures (Jensen, 1981). It is this ability to predict (albeit imperfectly) to so many contexts that clearly require intelligence that constitutes the validity argument for IQ tests as measures of intelligence.

A Sampling of Tests

A number of tests purport to measure intelligence, and they vary along several dimensions. Some provide a single overall score as a measure of global or general intelligence. Probably the best-known test of general intelligence is the Stanford-Binet (Roid, 2003), the direct historical descendant of the original Binet and Simon test. Other tests are more specific in focus. An often-used test with children, for example, is the Peabody Picture Vocabulary Test (Dunn & Dunn, 2007), which furnishes a measure of receptive oral vocabulary. The various Wechsler tests (Wechsler, 1997, 2002, 2003) all provide measures of both verbal IQ and performance IQ; summed together, the Verbal and Performance scales yield an overall IQ.

Tests also vary in the age group for whom they are intended. The three Wechsler tests are designed for three different age groups: the Wechsler Preschool and Primary Scale of Intelligence, or WPPSI, is intended for ages 4 to 6; the Wechsler Intelligence Scale for Children, or WISC, is intended for ages 6 to 16; and the Wechsler Adult Intelligence Scale, or WAIS, is intended for adults. The Stanford-Binet is broader in range; although it is most often used within the span of childhood, the test can be given at any age from 2 through adulthood. Finally, although infancy is the only age group excluded from the Wechsler and Stanford-Binet, there are tests that are specifically designed to measure development in infancy—for example, the Bayley Scales of Infant Development (Bayley, 2005).

So far little has been said about the content of IQ tests. Table 12.1 shows examples of the types of items that appear on one of the major tests of childhood IQ: the Wechsler Intelligence Scale for Children or WISC. The other leading childhood test, the Stanford-Binet, is in most respects similar to the WISC. Both tests are composed of a number of subtests directed to different kinds of abilities—11 subtests on the WISC and 10 on the

Stanford-Binet. On both tests the subtests can be grouped into larger scales: a Verbal and a Performance scale on the WISC; scales of Fluid Reasoning, Quantitative Reasoning, Visual-Spatial Processing, Knowledge, and Working Memory on the Stanford-Binet. The kinds of cognitive abilities that are stressed are similar on the two instruments. Verbal skills are important in both tests. Memory is also important, both memory for meaningful material and rote memory for unrelated items. Arithmetical ability is the focus of a number of subtests. So too is reasoning ability. And so is the child's store of real-world factual knowledge, a store that is tapped most explicitly by items such as the Information subtest of the WISC. In general, IQ tests for children are oriented to the kinds of skills that are needed for success in school—vocabulary, memory, arithmetic, problem solving. It is not surprising, therefore, that performance on such tests correlates with performance in school.

The Stanford-Binet and the WISC share other similarities as well. In both tests, a child's IQ is a function of how fast the child is developing in comparison with other children of the

Table 12.1 Simulated Items Similar to Those in the Wechsler Intelligence Scale for Children—Fourth Edition

Subtest	*Items*
Information	How many wings does a bird have? How many nickels make a dime? What is pepper?
Arithmetic	Sam had three pieces of candy, and Joe gave him four more. How many pieces of candy did Sam have altogether? If two buttons cost $.15, what will be the cost of a dozen buttons?
Vocabulary	What is a -----? or What does ----- mean? Hammer Protect Epidemic
Subtest	*Performance scale*
Object assembly	Put the pieces together to make a familiar object. 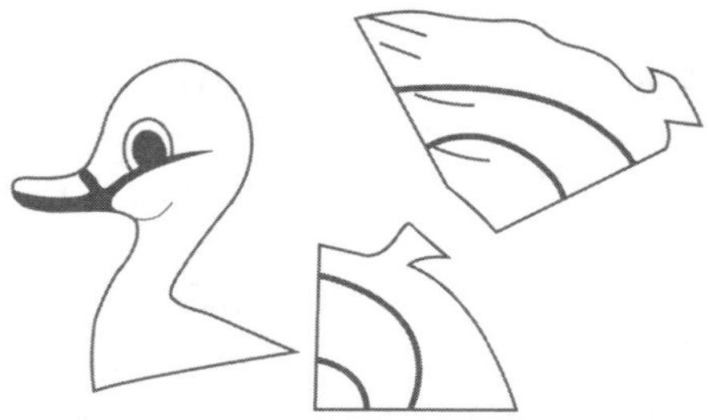

SOURCE: *Simulated Items similar to those in the Wechsler Intelligence Scale for Children – Fourth Edition.* Copyright © 1997 by Harcourt Assessment, Inc. Reproduced with permission. All rights reserved.

same age. The original formula for calculating Stanford-Binet IQs expressed this conception directly: Intelligence Quotient equals Mental Age (as determined by how far in the test the child was able to go) divided by Chronological Age times 100. For various reasons this formula is no longer used; instead, IQ is based on the deviation between the child's score and the average score for his or her age group. The logic, however, remains the same: Children who are developing faster than average have above-average IQs; children who are developing more slowly than average have below-average IQs. Childhood IQ is thus a measure of rate of development. It is also an inherently *relative* measure. There is no absolute metric for measuring a child's intelligence, as there is, for example, for measuring physical characteristics such as height or weight. Instead, IQ is always a matter of how the child compares with other children.

A final set of similarities between the Stanford-Binet and the WISC concerns the method of administration. There are two central emphases in the administration of any IQ test. One is the need for *standardization.* As was just stressed, IQ tests are relative measures, a child's IQ being a function of how that child's performance compares with that of other children. The only way that a score is interpretable is if the test is administered and scored in the same way for all children. The second emphasis is on the need to establish and maintain *rapport.* An IQ score is supposed to be a measure of the child's optimal performance, and this optimum can be achieved only if the child remains at ease and is motivated to respond carefully. The best testers are the ones who can successfully combine these two goals, maintaining the necessary standardization while at the same time using their clinical skills to elicit the best performance that the child can give.

Issues and Research Paradigms

Many of the issues that have always surrounded IQ tests have concerned their pragmatic uses—for example, tracking children in school on the basis of IQ scores. The concentration here is on more theoretically oriented questions about the development of intelligence. Two such questions have provoked much research and much controversy: the issue of the stability of IQ and the issue of the determinants of differences in IQ.

The consideration of the stability issue can be brief, for most of the relevant points were made in chapter 3 in the discussion of longitudinal designs. Studying stability in fact requires a longitudinal approach, because our interest is in the relation between a child's performance early in life and that same child's performance later in life. The specific form of stability at issue is the stability of individual differences. Do children maintain their relative standing on IQ tests as they develop, those who are high remaining high and those who are low remaining low, or can changes occur? Typically, this question has been examined through correlations between first test and second; the higher the correlation, the greater the stability. Because IQ tests for children are designed to yield the same mean IQ at each age, it is also possible to look at the constancy of the IQ value itself. IQ, after all, *is* relative standing, and thus constancy of relative standing implies constancy of IQ. We can ask, for example, whether a child with an IQ of 90 at age 4 will still have an IQ of 90 at age 6 or 10 or 20.

IQ has been a popular topic for longitudinal study for close to 80 years now. Such studies can and often do encounter all of the problems of longitudinal research that were discussed in chapter 3. Participants who are both able and willing to be tested repeatedly may not be a representative sample of the population as a whole, a bias that limits the generalizability of the results. Dropout in the course of the study may be selective, and, if so, the sample will become even more biased. Repeated administrations of the same test can result in practice effects, thus inflating later scores relative to early ones. And questions of measurement equivalence may arise if the study spans distinct age groups that require different IQ tests (e.g., infants and older children).

Several conclusions from the stability studies can be noted. A first set of conclusions concerns predictions from infancy. Except for extremely low scores, performance on traditional infant tests such as the Bayley is generally not predictive of later IQ (Lipsitt, 1992). This finding has long been evident, and it has led to the conclusion that there is a *discontinuity* in the nature of intellectual skills from infancy to later childhood—that is, what we mean by "intelligence" is simply not the same in the sensorimotor, preverbal infant as it is in the older child or adult.

There is undoubtedly some truth to this discontinuity argument. Nevertheless, recent research suggests that there is also one important qualification to it. A number of investigators have found that *response to novelty* in infancy *does* predict later IQ—not perfectly, by any means, but with typical correlations of .35 to .40 (Kaveck, 2004; McCall & Mash, 1994). Thus, babies who are especially interested in and responsive to novelty are the ones who, on the average, grow up to have the highest IQs. This finding, in turn, has led to the creation of a new instrument for assessing infant intelligence that is built around response to novelty: The Fagan Test of Infant Intelligence (Fagan & Detterman, 1992; Fagan & Shepherd, 1986).

What about predictions from childhood IQ tests? Once we move beyond infancy, scores do begin to correlate significantly from one age period to another; the stability, however, is far from perfect. In general, correlations—and thus similarity in IQ—are higher the closer together the ages being compared; they are also higher the older the child is at the time of initial testing. The latter statement is equivalent to saying that there is increased stability of IQ with increasing age.

The question of where differences in IQ come from has been hotly debated ever since the first IQ tests were developed. Part of the reason for the debate lies in the difficulty of getting clear evidence on the question. There are two possible sources for differences in IQ: the different genes with which people are born, or the different environments in which they grow up. It is easy enough to imagine a well-designed scientific study that would disentangle these two factors; all that need be done is to hold one factor constant while systematically varying the other. It is equally easy to see that such studies are impossible to do. The result is that we must fall back upon less satisfactory sorts of evidence. Two kinds of evidence have been prominent in the heredity-environment debate: studies of twins and studies of adopted children.

We considered twin studies briefly in chapter 11 in the discussion of temperament. Twin studies capitalize on the fact that there are two kinds of twins. Monozygotic or identical twins come from the same egg and are thus genetically identical; dizygotic or fraternal twins come from different eggs and thus have only a 50% average genetic overlap—the same as ordinary siblings. We have, then, a naturally occurring experiment with variation in the genetic variable. If genes are important for IQ, identical twins should be more similar in IQ than are fraternal twins. And this, in fact, is the finding. Reported correlations in IQ for identical twin pairs are in the .80s; correlations for fraternal twin pairs are typically in the .50s or .60s (McGue, Bouchard, Iacono, & Lykken, 1993; Plomin, 1990).

There is an obvious criticism of this genetic interpretation of the twin data. Perhaps environments are on the average more similar for identical twins than for fraternal twins. Identical twins, after all, look and in some ways act more alike than fraternal twins, and they may elicit a more similar treatment from their environments. There is an obvious answer to this obvious criticism: Study identical twins who have been separated early in life and brought up in different environments. Such twins are not easy to come by—there are only a handful of such studies, and none has a very large sample size. Furthermore, no one separates twins for the purposes of scientific study; separations occur for a variety of reasons under a variety of circumstances, and this lack of control hampers clear interpretation. Nevertheless, the data that emerge from such studies appear strongly supportive of a genetic model (Bouchard, 1997; Segal, 1999). Reported correlations in IQ for identical twins reared apart average around .75—only slightly lower than those for nonseparated identical twins, and higher than those for fraternal twins brought up in the same home.

Studies of adopted children constitute a larger literature than studies of twins. The starting point for such studies is the finding that parents and children typically correlate about .5 in IQ. Children tend, therefore, to resemble their parents; the problem is to figure out why. Each parent contributes 50% of the child's genes; thus there is a genetic basis for the correlation. But each parent also contributes a major part of the child's environment; thus there is also an environmental basis.

Studies of adopted children offer the possibility of pulling apart these two contributors. What we can look at are two sets of correlations. One is the correlation between an adopted child's IQ and the IQs of the adoptive parents. In this case the environmental basis remains; the genetic contribution, however, is ruled out. The other is the correlation between an adopted child's IQ and the IQ of the biological parents. In this case the environmental contribution (apart from the prenatal and perhaps early postbirth environment) is ruled out; the genetic basis, however, remains.

Two main findings emerge from the studies of adopted children (Turkheimer, 1991; van IJzendoorn, Juffer, & Poelhuis, 2005). One is that the child's IQ correlates more highly with those of the biological parents than with those of the adoptive parents. This is evidence in support of the importance of genetic factors. The second finding is that the mean IQ for samples of adopted children is typically above average—and typically above the mean for their biological parents. Because correlation is a measure of relative standing, a mean difference of this sort can come about even though the two sets of scores are fairly highly correlated. Part of the parent-child difference can be attributed to regression to the mean, a phenomenon that also applies to cross-generation comparisons (i.e., parents with below-average IQs tend to have children whose IQs are higher than their own). Part of it, however, is almost certainly a reflection of the above-average nature of adoptive homes. Such homes tend to be privileged in various ways, and they apparently boost the IQs of children who grow up in them. Thus, the studies of adopted children provide evidence for both genetic and environmental effects.

It should be noted that the adopted child studies, like the twin studies, do have some limitations. I mention two here. One is the possibility of *selective placement*—that is, the tendency of adoption agencies to do some matching of the adoptive home with characteristics of the biological parents. To the extent that such selective placement occurs, interpretation of the parent-child correlations becomes very difficult. A second possible problem is *restriction of range* among adoptive homes. As the preceding paragraph noted, such homes are not a random subset of the population of homes in general; rather, they tend to be above average in various ways. This also means, however, that they tend to be fairly homogeneous, and such homogeneity is a problem in a correlational study. The lower the variation in a variable (in this case, characteristics of adoptive homes, including adoptive parents' IQs), the less likely it is that that variable will correlate significantly with other variables. This factor sets an important qualification on the low correlation between adopted child and adoptive parents.

Memory

The remaining sections of the chapter shift from general approaches to specific topics in the study of cognitive development. The first topic is one of venerable standing. Memory has been a focus of study in psychology since the earliest days of psychology as a science, and children's memory has long been of interest to both parents and psychologists. I begin with memory in infancy, then move on to work with older children.

Memory in Infancy

Memory was not one of the topics in chapter 11's discussion of research on infancy. Nevertheless, many of the procedures considered there, whatever their primary focus, necessarily tell us something about infant memory as well. Piaget's work, for example, shows infants remembering things and acting accordingly from very early in life. Similarly, many phenomena in the development of attachment, such as the infant's preference for mother over stranger, clearly imply the operation of memory.

The main procedures for the explicit study of infant memory have come from two other approaches discussed in chapter 11: the habituation-dishabituation paradigm and the Fantz preference method. Memory is an intrinsic part of the habituation-dishabituation procedure. The only way that an infant can habituate to a stimulus is if the infant can store information about the stimulus over time and recognize it as familiar when it is encountered again. If there were no memory for past events, there could be no habituation. Similarly, the only way that an infant can show dishabituation to a new stimulus is if the infant remembers the original stimulus and realizes that the new stimulus is in some way different. Habituation and dishabituation can both be demonstrated in the newborn; thus, we know that some memory capacity is present from birth.

Memory is not an intrinsic part of the preference method, but it can easily be built into the procedure. All that need be done is to make familiarity the critical dimension along which the two stimuli differ. Suppose, for example, that we run a series of trials on which one of the two stimuli is always a triangle; the other stimulus, however, changes from trial to trial. The question is whether, over trials, the infant begins to look longer at the relatively novel or the relatively familiar stimulus. Infants older than about 2 months in fact show a preference for novelty; infants younger than 2 months may prefer familiarity (e.g., Wetherford & Cohen, 1973). The point for now is that either kind of preference—for the novel or for the familiar—implies the operation of memory.

In addition simply to demonstrating the presence of memory, the paradigms just discussed can be used to probe the nature of the early memory system. For example, we can examine the duration of the infant's memory by varying the time period between the initial exposure to the stimulus and the test for whether the stimulus is remembered. Or we can explore which aspects of a stimulus are noted and retained by varying which aspects we change during the dishabituation phase. If, for example, a stimulus consists of components A, B, and C, we can test for dishabituation when A alone is changed, when B alone is changed, or when C alone changes. Studies of this sort have revealed both some impressive early competencies in young infants and some important advances in the memory system across the first year or so of life (Courage & Howe, 2004; Rose, Feldman, & Jankowski, 2004).

Other procedures have also contributed to the study of early memory. Consider the arrangement pictured in Figure 12.2. As the figure shows, the ribbon connecting ankle and mobile confers a potential power upon the infant: If the baby kicks, the mobile will jump. Infants as young as 2 months can learn this relation (a form of operant conditioning). Once the response has been established, we can alter the situation in various ways to probe the infant's memory. We can test the duration of memory by reintroducing the mobile after a delay and seeing whether the kick response still occurs. We can test the specificity of memory by presenting new mobiles that vary

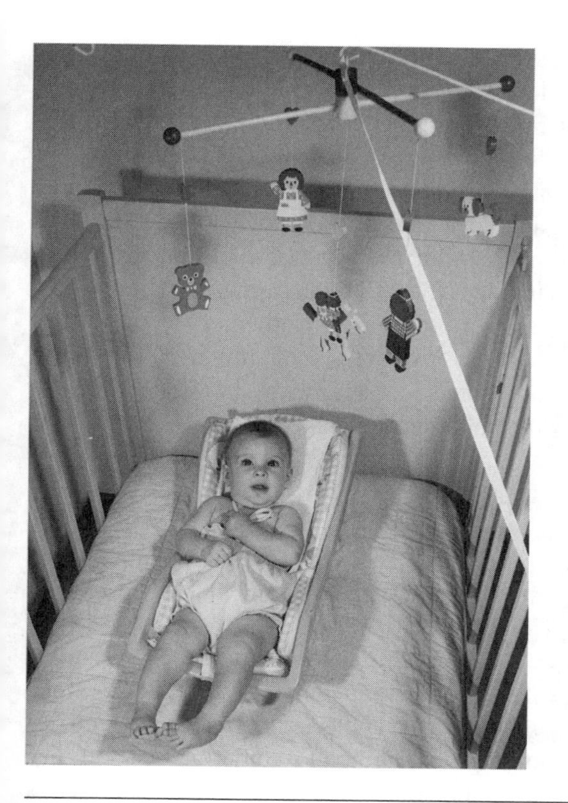

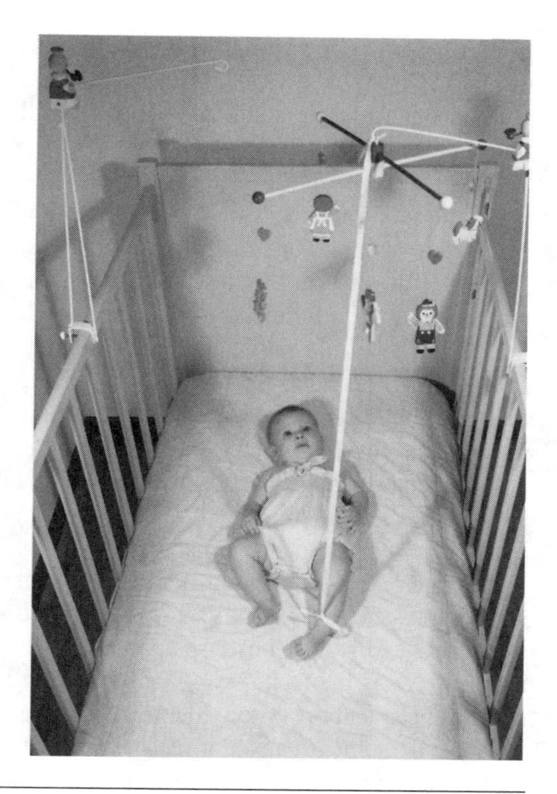

Figure 12.2 An experimental arrangement for studying infants' ability to learn and remember. When the ribbon is attached to the baby's ankle (as in the right-hand photo), kicking the leg makes the mobile above the crib move. Learning is shown by increased kicking whenever the ribbon is attached and the mobile is present.

SOURCE: These photos were made available by Dr. C. K. Rovee-Collier.

in their similarity to the training mobile. These studies, too, reveal both some impressive early abilities and improvements in those abilities across infancy (Rovee-Collier, 1999).

The kind of memory that is most clearly demonstrated in all of the examples discussed thus far is what is called **recognition memory**. Psychology's definition of "recognition" is the same as the everyday, dictionary definition: realizing that something new is the same as something encountered before. An infant who habituates is thus showing recognition, as is an infant who responds to the familiarity of the mother's face. Recognition can be contrasted with another basic form of memory: **recall memory**. Recall refers to the active retrieval of some memory material that is *not* immediately present. A child who relates what happened at his or her birthday party of a week before is demonstrating recall; so too is a child who draws a picture of the party.

Recall is difficult to study in infants, because babies cannot produce the responses (such as verbal reports or drawings) that are used to study recall in older participants. Investigators have therefore had to rely on less direct measures. Probably the most informative of these measures has been **deferred imitation**—that is, imitation of a model that is not immediately present but rather occurred some time in the past. The ability to reproduce behaviors that were seen a day or a week earlier would certainly seem to imply some capacity to recall information over time.

Piaget was the first to study deferred imitation, and he concluded that such imitation emerged at about 18 to 24 months, a finding in keeping with his belief that deferred imitation requires a capacity for representation and recall that is not available until the end of infancy. Later researchers have accepted the theoretical argument but disagreed with respect to the timing. A study by Meltzoff (1988) provided the first challenge to Piaget's account; it showed that 9-month-olds could imitate a novel action (e.g., pushing a button to produce a sound) that they had witnessed 24 hours earlier. Subsequent studies pushed the earliest signs of deferred imitation steadily earlier in development; at present 6 weeks is the youngest age at which success has been demonstrated (Meltzoff & Moore, 1994). Thus, some capacity for recall, if not present at birth, seems to emerge very early.

Infants can imitate and remember not just isolated behaviors but also simple sequences of actions. Eleven-month-olds, for example, can reproduce two-act sequences modeled by an adult (e.g., to make a rattle, first put button in box, then shake box); by 13 months babies can handle sequences with three components (e.g., to make a more complicated rattle, first put ball in larger cup, then invert smaller cup into larger, then shake cups—Bauer & Mandler, 1992). Such memories, moreover, are not necessarily only short-term; 11-month-olds show some retention of modeled sequences across delays as great as 3 months, and by 16 months of age the retention interval has stretched to 6 months (Bauer, 2004). There is even longitudinal evidence that some babies tested first at 11 months of age can remember some aspects of their experience a full year later (McDonough & Mandler, 1994)!

As soon as we move beyond infancy, studies of recall become much more common than studies of recognition. It has been clear for a long time that older children, on the average, recall things better than younger children. This finding emerges in a variety of contexts and for a variety of forms of recall, including more advanced versions of the sort of sequential memory that we just saw is first evident in infancy. The really interesting question is why such improvements in memory occur. I turn next to two kinds of research that have attempted to identify the bases for developmental improvements in memory.

Mnemonic Strategies

The basic idea behind the study of **mnemonic strategies** is that developmental improvements in recall do not result solely—or perhaps at all—from a simple quantitative expansion in the size of the memory "store." The improvements, rather, reflect the greater tendency of older children to *do* something—to utilize some mnemonic strategy—to help themselves remember. Such strategies may come at the time of initial exposure to the material, or during the delay period between exposure and memory test, or at the point of attempting to retrieve the material. These strategies may take a variety of forms: some verbal, some nonverbal, some simple, some complex. Their common property is that they do, at least usually, facilitate memory.

Let us briefly consider what mnemonic strategies look like, before moving on to the question of how to study them. Table 12.2 shows examples of three general classes of mnemonic strategies. The strategies in the table are by no means the only ones that children might use; indeed, we saw an example of another memory strategy—appropriate allocation of study time—in the Dufresne and Kobasigawa (1989) study discussed in chapter 2. Rehearsal, organization, and elaboration are, however, among the strategies that have received the most research attention.

Table 12.2 Examples of Mnemonic Strategies

General strategy	*Experimental task*	*Children's procedure*
Verbal rehearsal	Ten pictures of familiar but unrelated objects are presented one at a time and then removed. The child's task is to recall as many of the pictures as possible.	Single-item rehearsal: Label each picture repeatedly as it is presented—e.g., "apple, apple, apple, flag, flag, flag. . . ." Cumulative rehearsal: Label each picture as it is presented and then rehearse all of the labels to that point—e.g., "apple, apple-flag, apple-flag-moon. . . ."
Elaboration	Twenty pairs of familiar but unrelated words are presented (e.g., "cow-tie," "car-tree"). Subsequently one member of each pair is presented; the child's task is to recall the other member of the pair.	Pictorial elaboration: Create a mental image that relates the members of a pair in some way—e.g., picture to oneself a cow wearing a tie, a car driving up a tree. . . . Verbal elaboration: Create a sentence or phrase that relates the members of a pair in some way—e.g., "the cow wore a tie," "the car drove up the tree. . . ."
Organization	Twenty pictures of familiar objects are presented, drawn from four conceptual categories: animals, vehicles, clothing, furniture. The pictures are not grouped by category during presentation. The child's task is to recall as many of the pictures as possible.	Clustering: Organize the pictures in terms of the four categories, and recall members from the same category together—e.g., "dog, horse, camel, bear, squirrel, car, truck, plane. . . ."

How can strategies be studied? Some strategies are by their nature overt and thus relatively easy to study. Note taking, for example, is a common and easily observable mnemonic strategy. So is asking one's parents for help in remembering something. Perhaps because such external strategies seem so obvious, however, they have not received much attention from developmental psychologists. The interest, rather, has been in more internal, in-the-head strategies such as those in the table—in the kinds of strategies that are necessary precisely when external aids like notes or parents are *not* available. And here, clearly, we run into a measurement problem. How can an in-the-head strategy be measured?

One possibility is to *infer* the use of a strategy from the participant's overt memory performance. Suppose, for example, that we have a memory task that is heavily verbal, that we study verbally mature older children and verbally immature younger children, and that we find that the older children perform better. A reasonable (although of course not certain) inference is that the older children perform better because they are using their verbal skills to help themselves remember. Perhaps, for example, they verbally rehearse the items during the delay period, whereas the younger children do not. Such an inference becomes more certain if we can identify developmental changes not only in the level but in the *pattern* of performance. For example, a classic finding from the adult memory literature concerns the so-called *primacy effect,* which is the tendency for the first items in a stimulus list to be remembered better than later items are remembered. This effect is typically attributed to verbal rehearsal of the early items. When comparable studies are done developmentally, the finding is that older children show the primacy effect; younger children, however, usually do not (e.g., Cole, Frankel, & Sharp, 1971). Such a pattern is consistent with the hypothesis of a developmental increase in verbal rehearsal.

A second general approach is to *induce* the use of the strategy. In this case we do not guess at the presence of the strategy; rather, we instruct children in its use and then observe the effects. We might, for example, tell half of our participants to rehearse the items during the delay period, but give no such instructions to the other, control half. A typical finding is that such instruction is beneficial; children who are helped to use a strategy generally perform better than those who are not. Note that we have here the kind of convergence of evidence that has been discussed at various points. Studies of inferred strategy use suggest that some helpful mnemonic strategy is being used; studies of induced strategy use confirm that the strategy is in fact helpful.

Although the kinds of studies just discussed are informative, they do have their limitations. Studies that infer strategy use have the obvious limitation that the use is inferred; we do not know for certain what the child is doing. Studies that induce strategy use avoid this problem but run into another: Because we have forced the children to use a strategy, we do not know what they would have done on their own.

A breakthrough in the study of children's memory occurred in the 1960s with the discovery of experimental situations in which children would spontaneously produce strategies that were at least somewhat overt and therefore measurable. This discovery made it possible to combine the best elements of the other two approaches: to study strategies that were both spontaneous and observable. Because John Flavell was a pioneering researcher in this area, I will use one of his early studies as an example. It should be stressed, however, that this example is just one of many; both the Flavell research group (e.g., Flavell, Friedrichs, & Hoyt, 1970) and other investigators (e.g., Bjorklund, Coyle, & Gaultney, 1992; Miller, 1990) have been ingenious in devising experimental settings in which strategies can be directly observed.

Flavell, Beach, and Chinsky (1966) presented pictures of seven common objects to kindergarten, second-, and fifth-grade children. On each trial the experimenter pointed in a particular order to a subset of the pictures; the child's task was to recall the designated pictures in the correct order. A delay of 15 seconds intervened between pointing and recall test. The child wore a toy space helmet throughout the study, and during the delay period the visor of the helmet was pulled down, thus ensuring that the child could not see the pictures. The visor had a second purpose as well, which was to allow one of the experimenters to stare at the child's mouth during the delay. This experimenter's job was to record any verbalizations by the child, both overt verbalizations and semiovert ones (the experimenter had been trained to lip-read prior to the study). Of particular interest, of course, were instances of apparent rehearsal.

Children did rehearse, but the probability of rehearsal was strongly tied to age: 17 of 20 fifth-graders showed detectable rehearsal, whereas only 2 of 20 kindergarteners did so.

A subsequent study with a similar procedure (Keeney, Cannizzo, & Flavell, 1967) demonstrated that individual differences in rehearsal within an age group (first grade) correlated with recall performance—that is, children who spontaneously rehearsed showed better recall than children who did not. Both the developmental difference in the tendency to use strategies and the benefit when strategies *are* used are common findings in the memory literature.

In studies of the sort just described, children younger than 5 or 6 often fail to generate mnemonic strategies. Lest it be thought that this is an absolute deficit, consider a study by Wellman, Ritter, and Flavell (1975). Their participants were 3-year-old children, and their experimental procedure is summarized in the following passage:

> I want to tell you a story about this dog. See, here he is on the playground [table top]. He loves to play, he runs, he jumps, . . . but he was playing so hard he got very hungry. So he went to look for some food. When he was looking he went by this dog house, and this dog house, and this dog house, and this dog house [dog is walked by all four cups]. And then he went in this dog house to find some food [dog is hidden]. You know what, I have another toy I could get to help us tell the story. I'll go get it because we need it for the story. (p. 781)

At this point instructions diverged for the two experimental conditions. Children in the Wait condition were told simply to wait with the dog. Children in the Remember condition were told to remember where the dog was. The question, of course, was whether the children who were told to remember would behave differently from those who were told simply to wait. The answer is that they did behave differently, and in a very sensible way: Children instructed to remember were much more likely to spend the delay period with their eyes glued to the critical cup and possibly their fingers touching it as well. These are, it is true, very simple mnemonic strategies, but they *are* strategies, and they are available by age 3. Other studies using a similar hide-and-seek procedure have found the rudiments of strategic behavior in children as young as 18 months (DeLoache, Cassidy, & Brown, 1985). We can see here another example of what has become a recurrent theme in developmental psychology: It is dangerous ever to assert that some ability (in this case, mnemonic strategies) is totally lacking in the young child.

Constructive Memory

Important though they are, strategies do not account for all memory phenomena of interest or for every important developmental change in memory. Consider some of the limitations of the studies of strategic memory. Such studies have typically focused on memory for arbitrary and meaningless material (e.g., lists of unrelated words); clearly, however, much of real-life memory is for meaningful material. Studies of strategic memory have focused on intentional memory; much (perhaps most) of real-life memory, however, is unintentional or incidental, in the sense that we remember things that we never attempted to commit to memory. Finally, studies of strategic memory concern definite and discrete techniques for storing or retrieving the memory material. But not all of the cognitive activities involved in memory can be accounted for in terms of such definite and intentional strategies.

The study of **constructive memory** concerns the effects of the general knowledge system on memory. The basic idea behind this approach is that memory is simply a form of applied cognition, that form that has to do with storing information over time and retrieving information from the past. Like any form of cognition, memory involves action and understanding, not merely passive registration of input. And like any form of cognition, memory shows definite developmental changes as the child's understanding of the world changes.

Let us consider an example. Paris (1975) read the following story to kindergarten through fifth-grade children:

> Linda was playing with her new doll in front of her big red house. Suddenly she heard a strange sound coming from under the porch. It was the flapping of wings. Linda wanted to help so much, but she did not know what to do. She ran inside the house and grabbed a shoe box from the closet. Then Linda looked inside her desk until she found eight sheets of yellow paper. She cut up the paper into little pieces and put them in the bottom of the box. Linda gently picked up the helpless creature and took it with her. Her teacher knew what to do. (p. 233)

Following the story, children were asked eight questions:

1. Was Linda's doll new?
2. Did Linda grab a match box?
3. Was the strange sound coming from under the porch?
4. Was Linda playing behind her house?
5. Did Linda like to take care of animals?
6. Did Linda take what she found to the police station?
7. Did Linda find a frog?
8. Did Linda use a pair of scissors? (p. 233)

Note the basic difference between the first four questions and the last four. The first four concern specific information that is given directly in the story. The last four, however, can be answered only on the basis of inferences that go beyond the information that is explicitly provided. A basic finding—not only from the Paris study but from other such studies as well—is that even young children do make the kinds of inferences that are required by the second set of questions. The tendency to make inferences, as well as the complexity of the inferences that are possible, increases with development. But from early in life, memory—whether for a story, a conversation, an event seen, or whatever—is for meaning and not simply for verbatim detail.

Let us consider another example of this point. We saw earlier that even infants demonstrate some memory for the sequence in which events typically occur. So, of course, do older children—and for events that are a good deal lengthier and more complex than the simple sequences that fall within the span of infant memory. Memories that have to do with the order and the structure of familiar events are referred to as **scripts**. Children form a number of scripts as they develop; among the scripts that have been studied are those for taking a bath, making cookies, going to McDonald's, and having a birthday party.

Scripts are themselves a form of memory; once developed, however, they also influence subsequent memory. Children (and indeed any of us) often make sense of new experiences by relating them to familiar scripts. In general, children remember experiences that preserve the structure of a familiar script better than those that violate a script, and they show better memory for events that are central to a script than for those that are peripheral (Mandler, 1983; McCartney & Nelson, 1981). Children may even rearrange details in their recall to make their experiences fit a script. In one study, for example, preschoolers who heard the phrase "children brought presents" at the end of a story about a birthday party tended to move the phrase to the beginning of the story in their retelling (Hudson & Nelson, 1983). We will encounter such constructive "corrections" again in chapter 13, in the context of a consideration of how children's beliefs about sex differences influence their memory for how boys and girls or men and women behave.

As a final example of the constructive nature of memory, consider the stimuli in Figure 12.3.

Figure 12.3 Meaningful and random configurations of chess pieces. Because they can make use of their knowledge of chess, chess players show better memory for the top array than for the bottom one.

SOURCE: From "Chess Expertise and Memory for Chess Positions in Children and Adults," by W. Schneider, H. Gruber, A. Gold, and K. Opwis, 1993, *Journal of Experimental Child Psychology, 56*, p. 335. Copyright © 1993 by Academic Press. Reprinted by permission from Elsevier.

Imagine that your task is to study each picture for 10 seconds and then attempt to reproduce the configuration from memory. If you do not play chess, the two arrays would probably prove equally difficult; both, after all, contain the same number and the same variety of stimuli. If you do play chess, however, then the top array would almost certainly be easier to remember. It would be easier because the pieces are in positions that could actually occur in a game, whereas those in the bottom array are randomly arranged. Furthermore, if you not only play chess but play it well, then your memory for the top array—as well as the difference in memory for the two arrays—would almost certainly be even greater.

The chess example illustrates the effect of expertise on memory. The term **expertise** refers to organized factual knowledge about some content domain—to what we know about some subject. In general, as in the chess example, when expertise is high memory is high—experts take in and access information more quickly and more effectively than do nonexperts. Because older children and adults possess more expertise for most topics than do young children, this phenomenon provides another explanation for improvements in memory with age. On the other hand, the concept of expertise also leads to an interesting prediction: If we can find situations in which children have more expertise than adults, then the usual developmental differences in memory might be reversed. And this in fact has proved to be the case for a variety of topics, including the memory-for-chess example with which I introduced the concept of expertise (Chi, 1978). Such a finding is instructive, because it indicates that in at least some instances it is expertise, and not other factors associated with age, that is important for memory.

This section has just touched on the many ways in which children's knowledge can affect what they remember. Fuller coverage of such effects—as well as discussion of other bases for developmental changes in memory—can be found in Flavell, Miller, and Miller (2002) and Siegler and Alibali (2005).

Theory of Mind

Methods of Study

Consider the simple scenario depicted in Figure 12.4. To any adult, the answer to the question of where Sally will search for her marble is obvious—in the basket, where she last saw it. She has no way, after all, of knowing that the marble has been moved in her absence. It turns out, however, that this answer is not obvious to many preschool children. Most 3-year-olds indicate that Sally will look in the box. They answer in terms of their own knowledge of the reality of the situation, and not in terms of what Sally could be expected to think.

The Sally/Anne task is an example of research from an exciting new area of study labeled **theory of mind**. Theory of mind has to do with understanding of the mental world—with how children think about phenomena such as desires, intentions, emotions, and (as in the example) beliefs. The interest in such questions is, to be sure, not totally new; the Piagetian approach, in particular, has long encompassed social and mental phenomena within its targets of study. The work on perspective taking discussed earlier in the chapter is in fact a clear forerunner of the contemporary interest in theory of mind. The modern emphasis, however, has gone well beyond these Piagetian beginnings, spawning a host of theories, issues, and research findings that were not available before. It has also led to the creation of a number of new methodologies for studying children's understanding of mental phenomena, and it is the methodological aspect of theory of mind on which I concentrate here.

One such methodology is the one shown in Figure 12.4. The concept at issue in the Sally/Anne task is **false belief**—the realization that it is possible for people to believe something that is not true. Correct response to the task clearly requires this realization; children must set aside their own knowledge of the true location of the marble to realize that Sally would believe something else—would hold a false belief. As already noted, most 3-year-olds fail this task, whereas by age 4 or 5 most children succeed. The false belief task has been of interest because it tells us something very basic about the child's understanding of mental representation. To appreciate the possibility of a false belief requires the understanding that beliefs are simply mental representations, not direct reflections of the world, and as representations they may or may not be true. Thus, what I believe may or may not be the same as what you believe, and what either of us believes may or may not agree with reality. It is this understanding that the young child seems to lack.

There have been two main paradigms for studying false belief. The approach shown in Figure 12.4 is labeled the *unexpected locations* (or *unexpected transfer*) *task* (a procedure invented by Wimmer and Perner, 1983). The other possibility is the *unexpected contents task* (first used by Hogrefe, Wimmer, and Perner, 1986). Suppose that we show the child a candy box and ask what the child thinks is inside. The child replies "candy," at which point we open the box and show that it actually contains pencils. We then close the box back up, bring out a puppet of Ernie from Sesame Street, and pose the following question: "Ernie hasn't seen the inside of the box. What will Ernie think is inside the box before I open it?" You or I (and most 4- and 5-year-olds) would say "candy"—we would realize that Ernie has only the outside of the box as evidence and thus would form a false belief about its contents. Most 3-year-olds say "pencils." As in the locations task, young children are unable to set aside their own knowledge of reality; rather, they assume that everyone else must hold the same true belief that they hold.

The contents task lends itself to another measure as well. Rather than making Ernie the target, we can pose the question in terms of the child's own initial belief. The issue now is whether children realize that they themselves can hold false beliefs. The first part of the procedure remains the same: presentation of the

Figure 12.4 Example of a false belief task. To answer correctly, the child must realize that beliefs are mental representations that may differ from reality.

SOURCE: From *Autism: Explaining the Enigma* (p. 160) by U. Frith, 1989, Oxford: Basil Blackwell. Copyright 1989 by Axel Sheffler. Reprinted by permission.

candy box, elicitation of an initial judgment about the contents, revelation of the true contents. Now, however, the test question becomes the following: "What did you think was inside the box before I opened it?" We might expect this question to be easy; after all, all the child need do is repeat the response from a few seconds before. Three-year-olds, however, are likely to say "pencils." Understanding one's own false belief seems to be just as difficult as understanding false belief in another. The understanding that one's own mental states can change is referred to as **representational change**.

Another of the basic procedures under the theory-of-mind heading is the **appearance-reality task**. This task, too, is directed to an important form of knowledge: the realization that things can appear different from what they really are. You realize, for example, that a straw placed in a glass of water does not really bend upon entering the water—it merely looks bent. Similarly, you understand that a white cutout butterfly placed under a red filter is not really red—it just looks red.

Young children do not at first possess this understanding. The filter task is in fact one common measure, and it generally proceeds as follows (e.g., Flavell, Flavell, & Green, 1983). After some pretraining in the language to be used, the experimenter first presents the butterfly in its nonillusory form, next places the filter over it, and then asks the following two questions: "When you look at the butterfly, does it *look* red or does it *look* white?" "For *real,* is the butterfly *really and truly* white or *really and truly* red?" Correct response requires distinguishing the appearance and the reality, and this is what young children cannot do; most 3-year-olds answer with the appearance in both cases, thus claiming that the butterfly is really, and not just apparently, red. Judging in terms of appearance to the neglect of reality is referred to as a *phenomenalism error.* In other versions of the appearance-reality task, children make the opposite, *realism error:* judging in terms of the reality, to the neglect of the appearance. An often used stimulus in this case is the sponge/rock: a foam rubber sponge that looks for all the world like a gray rock. Once children have had a chance to touch the sponge and learn its true nature, they are asked the usual two questions: What does it look like, and what is it really? Here, remarkably, 3-year-olds tend to claim not only that the object really is a sponge but also that it looks like a sponge. Once they are aware of the reality, they can no longer report the discrepant appearance.

Unlike false belief and appearance-reality, the next methodology to be discussed does not have a single, agreed-upon label. Many investigators have probed what children understand about the *origins of knowledge*—that is, their conceptions of how we form our beliefs about the world (e.g., Miller, Hardin, & Montgomery, 2003; O'Neill, Astington, & Flavell, 1992). Do children appreciate, for example, the central role of perception in belief formation? Can they judge that someone with perceptual access to a stimulus will acquire information about it, whereas someone who lacks access will not? Can they distinguish among different perceptual modalities, judging what it is that can be learned (or that they themselves have learned) through sight as opposed to hearing as opposed to touch? Do they realize that there are sources of knowledge other than perception—logical inference, for example, or communication from others? And can they make all of these judgments not just with respect to themselves but with respect to other people?

As this run-through of (some of) the many possibilities suggests, there are too many different issues here to attempt a summary of findings. I will, however, note one general conclusion from the work on origins of knowledge, a conclusion that parallels findings from the studies of false belief and appearance-reality. The conclusion is that young preschoolers have only a very limited understanding of where beliefs come from, and they often err on tasks that seem transparently obvious to an adult.

They may claim, for example, that they have always known a fact that the experimenter has just taught them (Taylor, Esbensen, & Bennett, 1994), and they may be unable, seconds after learning something, to indicate whether sight, touching, or hearing was the source of the information (O'Neill & Gopnik, 1991). We can see again that there is much to develop with respect to theory-of-mind understanding.

This sampling of tasks hardly exhausts the topics studied under the heading of theory of mind. Researchers have probed, for example, for understanding of various psychological states in addition to thoughts and beliefs—for what children know about emotions (e.g., Lagutta, 2005) or desires (e.g., Moses, Coon, & Wusinich, 2000) or fantasies (e.g., Sharon & Woolley, 2004) or intentions (e.g., Baird & Moses, 2001). They have also explored theory of mind in age periods other than the preschool years, searching for both precursors in infancy (e.g., Legerstee, 2006) and more advanced developments in later childhood (e.g., Bosacki & Astington, 1999). Further discussions of tasks and related findings are available in a number of sources, including Astington (2000), Harris (2006), and Wellman (2002).

Research Issues

Just as my coverage of tasks was selective, so must be the coverage of research issues. I will settle for discussing two questions that have been the focus of much research attention.

The first issue is the same fundamental question addressed with respect to Piaget's studies: the question of assessment. In a sense, the study of false belief has been interesting for the same reason that much of Piagetian research is interesting: It provides surprising, hitherto unsuspected information about young children's cognitive limitations. Just as it is remarkable that an infant should lack object permanence or a preschooler conservation, so is it surprising that a 3-year-old should have no conception of the possibility that a belief can be false. As with Piagetian research, a natural and appropriate response to such a counterintuitive claim is to wonder about the accuracy of the assessment. Do the standard tasks really provide an accurate measure of what young children understand about belief? It is possible that they do not, and for the same general reasons that Piaget's measures have been criticized: The tasks involve language that may confuse a young child, the context for the assessment is strange and at least somewhat unnatural, and the procedures may not optimally engage the child's interest and motivation. Chandler and Hala (1994, p. 412) summarize the criticisms as follows: "These procedures have typically turned upon matters that are often static, hypothetical, third-party, and of no immediate relevance or personal interest to the young subjects in question." Perhaps 3-year-olds would appear more competent if the assessment could somehow be made more "child-friendly."

As with the conservation task, researchers have tried in various ways to probe for earlier competence. Modifications in wording have again been explored. In the locations paradigm, for example, it is possible that young children interpret the standard question as asking where the protagonist *should* search to find the object, or perhaps as asking where the protagonist will *eventually* search to find it; in either case the children would "err" by picking the true location. To test this possibility, Siegal and Beattie (1991) added a potentially clarifying "first" to the question ("Where will Jane look first for her kitten?"). Other researchers have varied the kind of response that is required of the child. It has been suggested that predicting an action based on a false belief (the usual measure) may be more difficult for young children than explaining a false-belief-based action that has already occurred. Several investigators (e.g., Bartsch & Wellman, 1989; Moses & Flavell, 1990), therefore, have contrasted the standard prediction measures with explanation measures, in which the child is asked to interpret a story character's mistaken action (e.g., Jane searching in the wrong place for her kitten). A study by Clements and Perner (1994) reduces the response demands on the child still further. These investigators measured children's looking behavior on a locations task, the attempt being to see whether children might look first at the original, false belief location, even if they were unable to answer the standard verbal question (as indeed many young children did).

Probably the most common procedural modification has been to incorporate *deception* into the false belief assessment. The arguments for doing so are both theoretical and methodological. Theoretically, a full understanding of deception implies an understanding of false belief as well, since the purpose of deception is to implant a false belief in someone else. Methodologically, an assessment based on deception can build upon children's natural experiences with tricks and games, thus adding a degree of familiarity and personal involvement that is lacking with the standard measures. Researchers have explored children's understanding of deception in various ways. Some have added deception to the standard paradigms by enlisting the child's help in "playing a trick" on someone else; the child is then the one to switch the contents of the candy box or move the toy to a new location (e.g., Sullivan & Winner, 1993). Others have devised games in which the child has a chance to deceive a competitor to win a prize—for example, by pointing to the false location for a desired chocolate (e.g., Russell, Mauthner, Sharpe, & Tidswell, 1991). Here, clearly, the motivation to think clearly about the beliefs of another should be maximal.

What have all these modified-assessment studies shown? In answering this question, I will draw from a meta-analysis of false belief studies by Wellman, Cross, and Watson (2001). As with meta-analyses in general (recall the discussion in chapter 8), the Wellman et al. meta-analysis combines and statistically analyzes the results from dozens of different studies. Various "moderator variables," or possible contributors to false belief performance, are explored, including the paradigm used (unexpected locations or unexpected contents), the target for the question (self or other), the wording of the question, and the use of a deceptive context. Variations across studies and relatively early success are sometimes found; in particular, young children sometimes perform better when deception is made part of the assessment. The main conclusion, however, is one of similarity in development across a range of procedures and a range of samples. Wherever and however they are studied, 3-year-olds find it difficult to realize that a belief can be false. Failure to understand false belief, like failure to understand conservation, is a genuine cognitive phenomenon.

The second line of research concerns possible links between theory-of-mind knowledge and social experience. It is certainly plausible that there could be a link, and in both directions—that children's understanding of the mental world can help them interact effectively with other persons, but also that interactions with others can teach children about beliefs and desires and other mental states. This research thus speaks to both the implications and the origins of mental state understanding.

The main approach to the question has been a correlational one. A number of studies have measured both theory-of-mind development and aspects of social experience in preschool samples (e.g., Astington & Jenkins, 1995; Watson, Nixon, Wilson, & Capage, 1999). In general, the relation between the two sets of measures is a positive one, that is, children who are more advanced in theory-of-mind understanding are also more advanced in their social behavior. Astington and Jenkins found, for example, that performance on a battery of false belief tasks was positively related to the level of pretend play with peers. Watson et al. reported a positive relation between false belief understanding and teacher ratings of preschoolers' social skills.

Of course the existence of a correlation between two measures does not tell us the direction of cause and effect. As we saw in chapter 3, one way to make causal inferences more certain is to trace relations over time through longitudinal study. A number of investigators have done so, and the overall results from this research suggest that the cause and effect can flow in both directions (Astington, 2003). In some instances developments in theory of mind appear to pave the way for later advances in social behavior. Jenkins and Astington (2000), for example, found that false belief performance at the initial measurement occasion predicted later competence in pretend play; the reverse relation, however, did not hold. In other instances it is social experience that appears to play the causal role. A program of research by Judy Dunn and associates, for example, has shown that aspects of early family interaction are related to later theory-of-mind understanding. In one study, mothers' talk about feelings when their children were 2 was related to understanding of emotions and beliefs 7 months later (Dunn, Brown, Slomkowski, Tesla, & Youngblade, 1991). In another study, a relatively high degree of family talk about feelings at age 3 predicted children's emotional understanding when they were 6 (Dunn, Brown, & Beardsall, 1991).

The correlational studies are not the only evidence for links between theory of mind and social behavior—or the only evidence that the causal relations between the two domains go in both directions. I mention two other forms of evidence here; Astington (2003) and Hughes and Leekam (2004) provide a fuller discussion.

One kind of evidence concerns the effects of family size. On the average, children with a relatively large number of siblings, perhaps especially older siblings, are faster to master false belief than are children in general (Peterson, 2000; Ruffman, Perner, Naito, Parkin, & Clements, 1998). Children who grow up in large, extended families (Lewis, Freeman, Kyriakidou, Maridaki-Kassotaki, & Berridge, 1996) also show faster than average development. Presumably, the heightened social experience with a variety of social agents gives children a helpful basis for learning about the mental states of others. Whatever the specific explanation, the causal direction for the relation must be a social to cognitive one. It is not plausible that being advanced in theory of mind causes the child to have more siblings.

In contrast, the second sort of evidence indicates a cognitive to social direction. It comes from the phenomenon of *autism*. Autism is a severe disorder, almost certainly biological in origin, that is characterized by a number of abnormalities in development, prominent among which are difficulties in social interaction. From early in life, children with autism show little interest in other people and little ability to form interpersonal relationships. They also show marked deficiencies with regard to theory-of-mind understanding (Baron-Cohen, 1995; Frith, 1989). Even when other aspects of mental functioning are relatively unimpaired, children with autism typically perform very poorly on theory-of-mind tasks. As Baron-Cohen (1995) observed, these children's insensitivity to the thoughts, wishes, and feelings of others is not surprising; they may literally not know that such psychological states exist. The consequences of such "mindblindness" (Baron-Cohen, 1995) are clear testimony to the importance of theory of mind for normal social relations.

Concepts

Infancy

Let us revisit for a moment the work on infant memory. We saw that infants demonstrate recognition memory from early in life, as shown by their differential response to familiar and novel stimuli. Even babies a day or two old, in fact, can recognize stimuli that they have encountered before. Impressive as this ability is, by about 3 months of age babies begin to do something that is more impressive still.

FOCUS ON

Box 12.4. Executive Function

Much research in cognitive development is directed to what children know about particular topics or content areas. This is true for the work under the heading of theory of mind that was just discussed. It is also true for the work on concepts to which we turn next.

One fairly common finding from such research is that what children know or how they reason about one content domain (such as theory of mind) is only weakly related to their knowledge or reasoning about another content domain (such as biological concepts). Thus, development often seems to be *domain-specific*—that is, specific to a particular content area, with only limited generality across areas. Such a conclusion is important, for it stands in contrast to the *domain-general* models that long dominated the field—most notably, Piaget's theory of general stages of development.

Although domain specificity is a common message of recent research, many psychologists are reluctant to abandon the search for general processes that cut across content areas. In recent years a body of research and theorizing has developed that attempts to identify certain basic capacities that may underlie performance on a range of different tasks. The term **executive function** has been coined to refer to such general problem-solving resources. In the words of one summary, executive function "is an umbrella term for all of the complex set of cognitive processes that underlie flexible goal-directed responses to novel or difficult situations" (Hughes & Graham, 2002, p. 131).

What sorts of processes fall under this heading? Let us consider an example that will introduce some of the most often studied ones in early childhood. Figure 12.5 shows the stimuli for the *dimensional card sort task* (Zelazo & Frye, 1998). The child is given a series of cards that vary in both shape and color and is instructed to sort them by one of the two dimensions. If the task is the "color game," for example, then the instructions will be to put the red ones in the box with the red picture and the blue ones in the box with the blue picture. After several such trials the rule changes: Now the task becomes the "shape game," and the instructions are to put the cars in one box and the flowers in the other. Simple though this task seems, it is beyond the capacity of most 3-year-olds. Even when they receive the new instructions at the start of every trial, and even when they themselves succeed in verbalizing the new rule, 3-year-olds continue to sort according to the original rule.

Why should such a seemingly simple task prove so difficult? Success on the card sort task in fact requires a number of component processes, processes that turn out not to be well developed at age 3. The child must possess sufficient *short-term memory* to keep the relevant rule in mind while performing the task. The child must be able to *inhibit* a dominant response in order to make the new response when the rule changes. And the child must possess sufficient *self-awareness* to reflect upon and choose among the rules that have been learned.

As noted, the processes just identified are commonly studied ones in executive function research with children. The card sort task is just one of a variety of assessment procedures that have been developed. Inhibition, for example, has been assessed through a version of a Simon Says game in which the child must respond to commands from "Nice Bear" but refrain from responding when "Nasty Dragon" is the speaker (Carlson & Moses, 2001). Short-term memory has been assessed through tasks of backward digit span or backward word span. Hughes and Graham (2002) provide an overview of measures, including description of several standardized assessment batteries.

Figure 12.5 Dimensional card sort task used in Zelazo and Frye's study of early rule following.

SOURCE: From "Cognitive Complexity and Control: II. The Development of Executive Function in Childhood," by P. D. Zelazo and D. Frye, 1998, *Current Directions in Psychological Science*, 7, 122. Copyright © 1998 by Blackwell Publishing. Reprinted with permission.

What has research on executive function in childhood shown? Two general sets of conclusions have emerged (Hughes, 2002; Hughes, Graham, & Grayson, 2004). The first concerns executive function processes themselves. As the findings from the card sort task suggest, capacities such as short-term memory and inhibition are far from perfectly developed in early childhood. All of the executive function processes that have been studied in childhood improve as children grow older. In addition to developmental differences, there are individual differences—that is, children at any given age vary in capacities such as short-term memory or inhibition. At the extreme, these individual differences extend to the problems that are evident in certain clinical syndromes; in particular, both Attention Deficit Hyperactivity Disorder (ADHD) and autism are characterized by marked deficits in executive function (Hughes, 2002). Indeed, one reason for interest in executive function—in adulthood as well as childhood—has always been the clinical implications of such work.

The second set of conclusions concerns relations of executive function to other cognitive measures. This issue has been examined most fully with respect to performance on theory-of-mind tasks. A large research literature makes clear that there *is* a relation: Children who are more advanced in various areas of executive function are also more advanced in theory-of-mind understanding. This, in fact, is a general finding across a range of cognitive outcomes. Exactly why such relations occur is a matter of dispute among researchers (Moses, 2001; Perner & Lang, 1999). Most, it is probably fair to say, subscribe to some version of a necessary but not sufficient position. By this view, certain executive function processes may be necessary for concepts such as false belief to emerge and be expressed, but domain-specific experiences and knowledge are also necessary. Development, in other words, is *both* domain-general and domain-specific.

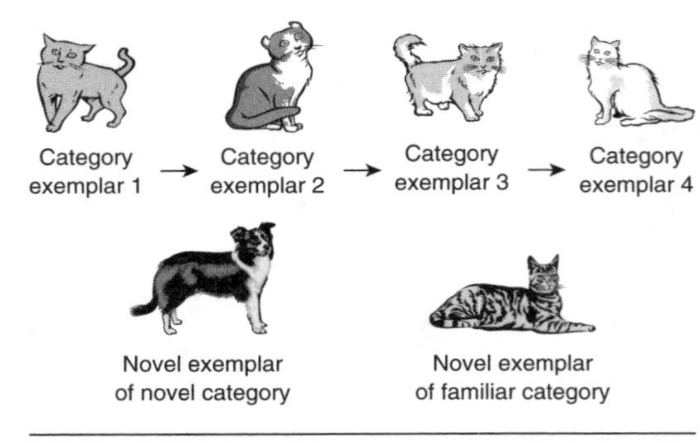

Figure 12.6 Examples of stimuli used in the study of concept formation in infancy

SOURCE: From "Early Category Representation and Concepts" (p. 29), by P. C. Quinn and J. Oates. In J. Oates & A. Grayson (Eds.), *Cognitive and Language Development in Children* (pp. 21–60), Oxford, UK: Blackwell Publishing. Reprinted by permission of Blackwell Publishing, Oxford, UK.

Imagine that we first show the infant a series of cat pictures, such as those in the top part of Figure 12.6. We present one picture at a time, with each picture different from the preceding one. After this initial *familiarization phase,* we run a test using the Fantz preference method. The picture in one window is of a cat that has not been presented before; the picture in the other window is of a dog. Both stimuli, therefore, are new to the baby. The 3-month-old, however, does not treat them as equally new; rather the baby looks more at the dog than at the cat.

What findings such as this demonstrate is that even a young infant is not limited to encoding and remembering only stimuli that have been directly experienced. If this were the case, then the cat in the bottom part of the figure would seem no more familiar than the dog. Rather, from early in life infants go beyond their specific experiences to form *concepts* for familiar stimuli and events. The term concept refers to a mental grouping of different items into a single category on the basis of some underlying similarity—something that makes them all, in a sense, the "same thing." Concepts are a basic way in which we organize and make sense of the world; indeed, without concepts virtually everything we encounter would be experienced as new. Thus, we treat different cats as in some sense the same thing, and so, apparently, do 3-month-olds.

Of course, it may have occurred to you that the cat versus dog comparison is not necessarily clear evidence for the existence of concepts. Perhaps dogs are just more interesting to babies at this age. Or perhaps the specific pictures used provide bases for response other than catness or dogness. These possibilities have also occurred to the investigators who do such research, and they have been careful to build in controls to rule out other explanations for the results. Thus there seems little doubt that 3-month-olds can distinguish cats from dogs. They also can distinguish cats from birds and horses (not, however, from female lions—Quinn, 1999). Moreover, the distinctions that they can make are not limited to biological categories but extend also to human-produced artifacts. Three-month-olds can distinguish animals from furniture, and they can make some differentiations within the category of furniture—can distinguish chairs from couches, beds, and tables, for example (Behl-Chadha, 1996).

The preference method is just one of several techniques that can be used to study concept formation in infancy (Flavell et al., 2002). All are based on the same rationale: See whether the infant responds more similarly to members of a category than to nonmembers, and thus seems in some way to be treating the category members as instances of the same thing. The habituation-dishabituation procedure is another possible source of evidence. We might present a series of cat pictures until interest declines and then test for dishabituation to one of two stimuli: another cat or a dog. Operant conditioning has also been employed. We might use the Rovee-Collier paradigm (described earlier in the chapter) to condition response to a particular mobile, perhaps one covered with cat pictures, then test for generalization to new mobiles that vary in their similarity to the original.

As infants grow older, a wider variety of response measures becomes available. By about a year of age the *sequential touching procedure* can be used. In this case we present an array of objects from different categories (e.g., toy animals, toy vehicles, toy furniture) and record the order in which the infant touches the objects (e.g., Mandler & Bauer, 1988). If the infant shows an above-chance tendency to touch items from the same category in succession—which is, in fact, a typical finding—then we have evidence that the infant considers them as instances of the same thing.

Finally, with slightly older infants it is possible to examine the *inferences* or *inductions* that the infant can draw from knowledge of category membership. In one study, for example, 14-month-olds first imitated various actions by an adult model, such as giving a dog a drink and turning a key in a car door. They were then given a chance to reproduce the actions with new members of the animal and vehicle categories. Even when the new instances were unfamiliar (e.g., an armadillo as the animal), the infants generalized appropriately, performing animal-appropriate activities only with animals and vehicle-appropriate activities only with vehicles (Mandler & McDonough, 1996). The ability to understand and respond appropriately to new instances of a category is a main function that concepts serve. Such findings indicate that children begin to use their concepts adaptively from very early in development.

Older Children

As is always the case, the response measures available to us expand when we move from infants to older children. In particular, it is now possible to convey instructions through words and to receive responses in words. Rather than rely on the indirect measures used with infants, we now can simply ask children to tell us or show us which items in some collection go together and to explain why.

Studies of how children sort items into groups date back more than a century, including some influential work by Piaget (Inhelder & Piaget, 1964). A general conclusion from such studies (a fairly safe conclusion for any aspect of cognitive development) was of early limitations that were gradually overcome with development. In particular, young children often appear perceptually oriented, concentrating on surface features to the neglect of more basic underlying properties. Presented, for example, with a fire engine, car, and apple, most 3- and 4-year-olds group together fire engine and apple, the two red things, rather than fire engine and car, the two vehicles (Tversky, 1985). Note that the perception-oriented nature of the young child is also evident in various findings discussed earlier in the chapter—Piaget's nonconservation phenomenon, for example, or the preschooler's failure to distinguish appearance and reality.

While not negating these early limitations, more recent research suggests a more positive picture of preschoolers' abilities. Table 12.3 shows one example. The problems in the table are from a program of research by Gelman, Markman, and associates (Gelman, 2000; Gelman & Markman, 1986, 1987). The question is how the child will generalize from what is already known to something new, and two possible bases for response are contrasted. If perceptual similarity is taken to be critical, then the new item should be judged to be like the one that it most resembles. This means, for example,

Table 12.3 Sample Items From Gelman and Markman's Studies of Children's Concepts

This bird's legs get cold at night. (picture of flamingo)

This bat's legs stay warm at night. (picture of black bat)

See this bird. Do its legs get cold at night, like this bird, or do its legs stay warm at night, like this bat? (picture of blackbird, looks like the bat) (see pictures)

This fish stays under water to breathe. (picture of tropical fish) This dolphin pops above the water to breathe. (picture of dolphin)
See this fish. Does it breathe under water, like this fish, or does it pop above the water to breathe, like this dolphin? (picture of shark, looks like the dolphin)

This puppy hides bones in the ground. (picture of brown dachshund)
This fox hides food in the ground. (picture of red fox)
See this puppy. Does it hide bones in the ground, like this puppy, or does it hide food in the ground, like this fox? (picture of red dog, looks like the fox)

SOURCE: From "Categories and Induction in Young Children," by S. A. Gelman and E. M. Markman, 1986, *Cognition, 23*, 183–209. Copyright © 1986. Reprinted by permission of Elsevier.

that the blackbird would be expected to have warm legs at night, just like the similar-looking bat. In contrast, if category membership is deemed more important, then the legs would be expected to be cold, just like those of the other bird. The contrast in these tasks is a basic one in studies of children's concepts: between surface similarity and underlying essence as the basis for judging that things are the same.

As noted, much of the research prior to Gelman and Markman's had indicated that young children rely on surface similarity. This was not what Gelman and Markman found. Despite the compelling perceptual cues, most 4-year-olds opted for category membership as the relevant basis for inference. Thus, they judged that the bird's legs would get cold, that the shark would breathe under water, and so forth. A subsequent study using simplified procedures showed that even 2-year-olds had some ability to overlook perceptual appearance in favor of category membership (Gelman & Coley, 1990).

Why the more positive picture of preschoolers' competence in recent research? Two factors are probably important. One is the methods used. As noted, many of the older studies based their conclusions on children's response to explicit instructions to sort items into categories (e.g., "Show me which ones go together"). The Gelman and Markman procedure, in contrast, is tied to a natural, everyday use of concepts—drawing inferences about new instances from what is already known. The grounding of the response measure in this natural function of concepts may be one explanation for the impressive performance in recent research.

The type of concept at issue is probably important as well. Some studies have used arbitrary concepts created on the spot for the purposes of research—for example, the category of blue circles in a study of sorting behavior. There is nothing arbitrary, however, about the concepts that children naturally form—concepts of people, of animals, of vehicles, of food. Such concepts reflect important commonalities among real-life experiences that children extract as they attempt to make sense of the world. The focus on familiar and interesting material may also contribute to the good performance in recent research.

Further evidence for this conclusion comes from perhaps the most active current area in the study of conceptual development: research directed to children's concepts of biology. Again, the main message from the older research literature was one of confusions and limitations in young children's understanding. The best known example of such confusions is Piaget's (1929) concept of **animism**: the tendency to attribute properties of life to nonliving things. The young child who indicates that the sun shines "because it wants to" is engaging in animistic thinking, as is the child who is concerned that a piece of paper will be hurt by being cut. Obviously, such thinking reflects a fundamental gap in biological understanding.

As with concepts in general, contemporary research suggests a more positive picture of young children's competence. This understanding is, to be sure, far from complete, and there are disagreements about exactly how much knowledge to attribute to young children (Astuti, Solomon, & Carey, 2004). Still, when questioned in an optimal fashion, even preschoolers do seem to appreciate some of the basic properties of life. Here are a few examples.

One fundamental property of life is growth. Children as young as 3 or 4 understand some of the basic facts about growth. They realize, for example, that only living things grow, that growth is inevitable (for example, you can't keep a baby pet small and cute just because you want to), and that growth is directional—that is, people, plants, and animals get bigger, not smaller, as they age (Inagaki & Hatano, 1996, 2002; Rosengren, Gelman, Kalish, & McCormick, 1991). They also have some appreciation of the fact that only animals *re*grow—that a cat with a scratch, for example, will eventually heal, but that a car with a scratch will require human intervention (Backschneider, Shatz, & Gelman, 1993).

Whereas growth is characteristic of all life, only animals are capable of self-produced movement. Many other things, of course, do move (cars, bikes, clouds, leaves in the wind, etc.), and young children, as Piaget's studies of animism indicate, are sometimes confused by this fact. Still, even 3-year-olds show some success in judging which things can move by themselves and which things (e.g., statues, plants) cannot (Massey & Gelman, 1988).

One final property of life that has been the subject of much study is inheritance. Living things come from other living things, and they inherit properties both of the species in general and of their own parents in particular. Again, even preschoolers demonstrate some biological knowledge, in this case of origins and kinship. They realize that dogs produce baby dogs, not cats, and that offspring generally resemble their parents (Springer, 1996; Springer & Keil, 1991). By the grade-school years children can also make sensible judgments about which properties are likely to be biological in origin (e.g., physical appearance, dietary preferences) and which are likely to be environmental (e.g., personality traits—Astuti et al., 2004; Springer, 1996).

Summary

This chapter considers five topics that fall under the heading of cognitive development in childhood.

The chapter begins with the historically influential Piagetian approach. Just as is true for infancy, Piaget's research on later childhood is distinguished by its ability to identify interesting and developmentally basic forms of knowledge, all studied through the flexible *clinical method.* Among these forms of knowledge, conservation—the knowledge that quantities remain invariant in the face of perceptual change—has proved especially intriguing. To the discussion of conservation are added descriptions of four other Piagetian concepts: *class inclusion, transitivity, formal-operational reasoning,* and *perspective taking.*

Piaget's work has inspired a host of follow-up studies. A basic question is that of *assessment:* How accurately do our experimental procedures assess the child's abilities? Piaget has long been charged with underestimating the child's ability, primarily because of the heavy verbal emphasis in many Piagetian tasks. That there is some validity to this charge is suggested by a discussion of two sorts of studies: those that have simplified the language involved in Piagetian assessment and those that have increased the naturalness of the assessment situation.

The intelligence test or IQ approach is in many ways quite different from the Piagetian approach. An overview of the differences serves as a lead-in to a discussion of the types of items that appear on two leading IQ tests: the Stanford-Binet and the WISC. These tests are highly standardized instruments whose purpose is to identify individual differences among children. Their content is oriented to the kinds of academic-verbal skills that are important in school. Indeed, it is the predictive power of the tests, including correlations with school performance, that has always served as their chief validity index.

The description of IQ tests is followed by a consideration of some of the theoretical issues that have been of interest in the study of IQ. One is the question of *stability:* Is a child's IQ constant across development, or can the value go up or down? Answering this question requires a longitudinal approach; hence this section of the chapter serves as a reminder of some of the points about longitudinal designs that were made in chapter 3. Another basic question concerns the *determinants of differences* in IQ—specifically, the extent to which such differences are genetic or environmental in origin. Two approaches to this question are reviewed. Twin studies capitalize on the fact that identical twins are more genetically alike than are fraternal twins; greater similarity in IQ for identicals is then taken as evidence for the importance of genetic factors. Adopted child studies offer the opportunity to disentangle the genetic and environmental factors that are confounded in the normal parent-child correlation. The adopted child's IQ can be compared with those of the biological parents and with those of the adoptive parents. Influential though the twin and adopted child studies have been, they do have their limitations, and these limitations are discussed along with the findings.

A topic of longstanding interest in developmental psychology is children's memory. Memory is a basic cognitive process that is present from birth. Memory in early infancy seems to consist mainly of *recognition memory*, which has typically been studied through either the habituation-dishabituation paradigm or the Fantz preference method. Work on deferred imitation suggests that some capacity for *recall memory* emerges by a few weeks of age.

Studies of older children have concentrated on recall memory and on the attempt to explain the developmental improvement in recall across the childhood years. Research on strategies attempts to measure the existence and effects of various *mnemonic strategies* that can be used to aid recall. In some cases it is possible to observe the spontaneous occurrence of the strategy; in other cases it may be necessary either to infer the use of the strategy from overt memory performance or to induce its use experimentally. Research on *constructive memory* examines the effects of the general knowledge system, and of developmental changes in this system, on memory. Several examples of studies of constructive memory are discussed, including the effects of scripts and of expertise on memory.

The term *theory of mind* refers to children's thoughts and beliefs about the mental world. Of special interest has been the concept of *false belief*: the realization that it is possible for people (both others and oneself) to hold beliefs that are false. Another important development is mastery of the *appearance-reality* distinction: the ability to distinguish appearance and reality when the two are in conflict. Several specific methods of study are described not only for false belief and appearance-reality but also for a third focus of theory-of-mind research: children's understanding of the *origins of knowledge*. As is true with Piagetian research, issues of *assessment* are important in studies of theory of mind, and several alternative approaches to the study of false belief are therefore described. Another important issue concerns possible links between theory-of-mind understanding and social experience. A variety of kinds of evidence indicate that there are in fact links and that the causal relations run in both directions.

The chapter concludes with research directed to the formation of *concepts*—that is, mental groupings of items on the basis of some underlying similarity. Infants begin to form simple concepts from early in life, as shown by differential response to members versus nonmembers of particular categories (such as cats and dogs). The specific procedures used to demonstrate such response include the preference method, habituation-dishabituation, sequential touching, and generalization. Such nonverbal procedures are supplanted by verbal techniques in studies with older children, but the rationale remains the same: look for differential response to members of different categories. Recent research indicates that even preschoolers sometimes weight category membership more heavily than perceptual similarity when drawing inferences from what they know. Recent research also indicates that basic forms of biological understanding begin to emerge early in development.

Exercises

1. One of the attractive features of both Piagetian and theory-of-mind tasks is that they are simple to administer and hence lend themselves readily to demonstration projects. The following exercise assumes that you have access to at least one (and preferably several) children with whom you can do simple testing. If the children are between 3 and 5 years old, you should concentrate on theory-of-mind measures; if they are older, you should concentrate on Piagetian tasks. In either case, read both the descriptions in the

textand some of the original sources that are cited, try out some of the tasks yourself, and compare your results with those in the literature. Note: Be sure to obtain the informed consent of both the children and their parents before beginning testing.

2. Follow-up studies of Piaget's final period of development indicate that many adults show less than perfect performance on tasks intended to assess formal-operational thinking. Obtain a copy of the Inhelder and Piaget book cited in the text, reproduce as many of the tasks as you are able, and administer the test battery to a sample of your peers. Do you believe that tasks such as these provide a sufficient measure of the capacity for hypothetical-deductive reasoning? If not, how would you modify or add to the assessment?

3. One criticism of IQ tests has been that they measure only certain kinds of intellectual skills to the neglect of other forms of cognitive competence. One often cited area of neglect is the domain of social or interpersonal intelligence. Imagine that you had the task of devising a test of social intelligence for grade-school children. How would you select items for your measure? How would you validate your test?

4. Think back on your own mnemonic activities of the last week or so. To what extent do they reflect the sorts of processes discussed in the text? Suppose that a researcher wished to study your everyday memory activities. How could the researcher go about collecting evidence?

13

Social Development

Like chapter 12, this chapter concentrates on development during the postinfancy childhood years. The focus now is on the child's social development, a topic at least as large and as methodologically challenging as the topic of how cognition develops. The coverage is again selective, the goal being to consider a few important subjects in some depth, rather than many subjects superficially.

The chapter is divided into four sections. The first three sections are directed to important outcomes of social development. Two such outcomes, attachment and temperament, were discussed in chapter 11. In this chapter I add three others: the development of moral standards and moral behaviors, the development of a self-concept, and the development of sex typing and sex differences.

Identifying and measuring important outcomes speaks to one of the two general issues in the study of social development. The second general issue is the question of determinants—where do these outcomes come from? Why are certain behaviors part of the child's social repertoire, and what is the explanation for developmental or individual differences in them? The final section of the chapter discusses one of the most important and most often studied determinants: parental child rearing.

Moral Development

"Moral development" is potentially a huge topic, encompassing many aspects of the child's development. In fact, as developmental psychologists have studied morality, the topic *is* huge, for many different facets of development are examined under this heading. This first section is consequently the longest one in the chapter.

It is traditional to distinguish three aspects of morality: the behavioral, the emotional, and the cognitive. My organization follows this traditional division, with subsections for each of the three aspects.

Behavior

The behavioral aspect of morality refers to moral action—to behaving in moral, good, socially acceptable ways. Exactly what the criteria are for defining a "moral" behavior is a complex and long-debated question that is well beyond the scope of this chapter. I can note, however, that one component of almost anyone's definition is the notion that the behavior must be at least partly internally generated and not solely in response to immediate external pressures. A child who does not cheat because a teacher is standing over him is not behaving morally (or immorally, for that matter), nor is a child who shares candies with a younger brother because his mother is about to spank him if he refuses to share. On the other hand, a child who is left at least somewhat on his own and *decides* not to cheat or *decides* to share *would* be demonstrating moral behavior.

Independence from one's own self-interest is another common criterion in judging the morality of an action. Behavior is most clearly moral if it is opposed to (or at least not in the service of) one's own self-interests. Thus, a child who shares candy because he hates that kind of candy is not necessarily behaving morally, nor is a child who refrains from cheating because he has no need to cheat. On the other hand, a child who gives away the last piece of a candy that he would dearly love to have himself *is* showing moral behavior.

Note that these definitional points have a methodological corollary. What we need to study are situations, either naturally occurring or experimentally contrived, in which children have an opportunity to produce moral behaviors that do not serve their own interests and are not forced upon them by immediate environmental pressures. I describe a number of such situations as we go.

What specific behaviors fall under the heading of the behavioral aspect of morality? There are a large number of behaviors, and they can be roughly divided into two general classes, classes that correspond to the two examples sketched earlier. In some cases the issue is whether the child can avoid doing something wrong. This is the case, for example, with cheating, a common measure of moral behavior in studies of school-aged children. Cheating is actually a particular instance of a more general construct known as *resistance to temptation.* A resistance-to-temptation test sets up a conflict between some behavior in which the child would like to engage (such as cheating to do better on a task) and some sanction or prohibition against the behavior (such as rules against cheating, or the disapproval of a teacher or parent). The question then is whether the child resists or fails to resist the temptation. A common resistance measure geared to younger children, the "forbidden toys test," is described shortly.

For years the study of moral behavior was oriented mainly to resistance situations and to whether the child could avoid misbehaving. In recent years this negative focus has been complemented by a greatly increased interest in the more positive side of morality: whether the child will not only avoid bad behavior but also actively engage in good behavior. Such active production of socially beneficial behaviors is referred to as **prosocial behavior**. Sharing would be an example of prosocial behavior. So too would showing sympathy for someone in distress or offering to help in some way to alleviate the distress.

Whatever the specific behavior that we decide to study, there are several approaches that we can take in gathering evidence. I discussed these approaches at a general level in chapters 4 and 6; here I consider some specific applications to the measurement of moral development.

One possibility is to go into some naturally occurring "field" setting and collect *naturalistic observations* of the behaviors of interest. A study by Barrett and Yarrow (1977) provides an example. These investigators studied 5- to 8-year-old children who were attending a summer day camp. A time-sampling observational approach was used, with each child observed for a total of eight 15-minute periods. Among the behaviors recorded were instances of prosocial behavior, defined as attempts to fulfill another person's need for physical or emotional support. They include acts of comforting (physically or verbally expressing sympathy or reassurance), sharing (giving materials or work space that one is using or giving a "turn" to another person), and helping (physically assisting or offering physical assistance). (p. 476)

Also recorded were *opportunities* for prosocial behavior—that is, cues from other children that indicated a need for comfort, sharing, or helping.

The obvious strength of an approach like Barrett and Yarrow's was noted in chapter 6. With naturalistic observation, our focus is directly on what we hope to explain: the natural occurrence of behaviors in the natural setting. A further, more specific strength of this particular study is the measurement of opportunities for prosocial behavior as well as actual occurrences of the behavior. A common limitation in naturalistic study is that the eliciting conditions for a behavior may vary across children. Perhaps, for example, a particular child scores low on the comforting measure simply because the child happened to have few opportunities to offer comfort during the observation period. This problem follows from the lack of control in a naturalistic study: We are unable to set up the same environment for all children but rather have to take conditions as they happen to occur. One way around this problem is to measure these conditions for each child and to adjust the analyses accordingly, which is what Barrett and Yarrow did with their measurement of opportunities to be prosocial. Another way, also utilized by Barrett and Yarrow, is to collect a large sample of each child's behavior. The logic in sampling behavior is the same as that discussed earlier for sampling of participants: The larger the sample that we are able to obtain, the more likely it is that chance variations will even out and we will end up with a representative picture.

As we saw earlier, the strengths of the naturalistic approach are balanced by some weaknesses. Ensuring the accuracy of the measurements may be difficult, especially when the observations must be carried out in an uncontrolled field setting. The interobserver reliabilities in Barrett and Yarrow's study were in fact modest, most falling in the .7 and .8 range. Another possible problem concerns effects of the observer on the behavior being observed. It is true that in a free-play school situation, such as that studied by Barrett and Yarrow, the presence of various adults may be common and easily adapted to. Nevertheless, it is wise not to become too sanguine about this issue. Reactive effects from being observed may in fact be especially likely with moral behaviors, a domain in which there is a clear "should" for how one behaves while being watched by adults (i.e., one should share, should offer help, etc.). Prosocial behaviors may be less likely when there is no adult around.

I consider one more example of the naturalistic observation approach. It is an unusually ambitious example, and one that may minimize some of the problems just discussed. Zahn-Waxler, Radke-Yarrow, Wagner, and Chapman (1992) studied responses to distress in others in a sample of 1- to 2-year-olds—in itself an ambitious undertaking, for the toddler is a challenging and understudied research subject. Zahn-Waxler et al. were especially interested in prosocial responses to the other's distress, such as expressing sympathy or offering help. Such responses were studied longitudinally across a 1-year period. Finally, a subset of the observations were made not by the usual research assistant but by the children's own mothers. Prior to the start of the study, the mothers underwent extensive training, during which they learned to make the kinds of observations that the investigators were interested in. The observational method used was a combination of event sampling and narrative record: Whenever an incident of distress to another occurred, the mother dictated a description into a tape recorder, attempting to capture the incident itself, the events preceding it, and the child's response to the distress. Through this approach Zahn-Waxler et al. were able to document early forms of sympathy and prosocial behavior in an age group that had generally been thought to be incapable of such responses.

It is perhaps unnecessary to point out that there are difficulties aplenty in a study like Zahn-Waxler et al.'s, both practical difficulties in securing the necessary cooperation and more substantive difficulties in verifying the accuracy of the mothers' reports. Their approach also has some definite strengths, however, especially in comparison to the usual observational study. Two such strengths are worth noting. First, there is no worry about biasing effects from introducing an observer, for the mother is a natural part of the child's environment. Even though observations are still limited to events that the mother is present to witness, reactivity in the usual sense is not a concern. Note that this is an example of the "participant observation" approach discussed in chapter 4. Second, the breadth of the observations, as well as the resulting quantity of information, is enormously greater than in a typical observational study. We can have considerably more confidence, therefore, that we really have obtained a representative sampling of the child's behavior.

The second general approach to be discussed is *laboratory elicitation* of the behaviors of interest. In this case the locus for our observations is some structured lab environment, an environment designed to permit the occurrence and measurement of the particular moral behavior that we are interested in studying. I consider two examples of this approach, one directed to the negative, avoidance aspect of morality and the other directed to the more positive, prosocial aspect.

As mentioned earlier, a common measure of resistance to temptation with young children is the "forbidden toys test." An early study by Parke (1967) provides a typical example of its use. In Parke's study the participants (first- and second-grade children) were brought individually to a mobile laboratory trailer, where they first spent about 10 minutes drawing. Following the drawing phase, each child was seated at a table containing five toys, and the following instructions were given:

> You can sit here. [The cloth covering the toys is removed. . . .]. Now, these toys have been arranged for someone else, so you'd better not touch them. If you are a good boy (girl) and do not touch the toys, we can play a game together in a little while, but I have forgotten something and have to go into the school and get it. While I am gone, you can look at this book. I'm going to close the door so that no one will bother you. When I come back, I'll knock, so you'll know it's me. (p. 1105)

The child was then left alone for 15 minutes with the toys, during which time the child's behavior was monitored through a one-way mirror. The question, of course, was whether the child would give in and play with the toys, and, if so, how quick and how broad the deviation would be.

In the Parke (1967) study the admonition not to play with the toys came from an experimenter. This is not the only possibility. In some uses of the forbidden toys procedure it is the child's mother who delivers the prohibition (e.g., Aksan & Kochanska, 2005).

The second example of a laboratory study concerns the prosocial behavior of sharing. A study by Barnett, King, and Howard (1979) illustrates a popular procedure for studying sharing. Each of the grade-school participants in their study received 30 prize chips as a reward for answering some questions about what children remember. The chips could be exchanged for prizes at the end of the study once all the children had a chance to participate. Before the child could leave with the chips, however, the experimenter pointed to a nearby donation canister and informed the child that

> there are some other children who go to another school a lot like yours who won't have a chance to be in the study and earn prize chips. Later, if you want, you may share with those children by putting some of your prize chips in the donation can. You don't have to share, but you may if you want. (Barnett et al., 1979, p. 165)

The child was subsequently left alone for 1 minute, during which time he or she could donate or not without being observed. The fact that the canister already contained a number of chips heightened the impression of anonymity. Despite this impression, however, the child's behavior was in fact no freer from scrutiny than that of the children in the forbidden toys study described earlier. By simply counting the number of chips in the canister after each participant, the experimenter was able to measure how much each child had donated.

The laboratory approach has a number of virtues. In comparison with naturalistic observation, a lab assessment is an extremely efficient way of gathering information about a behavior. Rather than wait for perhaps hours for the relevant behavior to occur, we can set up a situation that will elicit the behavior (resisting or not resisting, sharing or not sharing, etc.) within a matter of minutes. We can do so, moreover, in a way that is comparable for all participants. As we saw, lack of comparability in eliciting conditions can be a problem in a purely naturalistic study. And we can do so in a way that permits precise and objective measurement of the behavior. It is quite easy, for example, to record whether and how quickly a child plays with the forbidden toys. It is even easier to record how many chips a child donates.

A further virtue was noted in chapter 6. Once a basic laboratory paradigm has been developed, such as the forbidden toys or donation measure, we can carry out systematic experimental manipulations of a large number of potentially important independent variables. We can see what happens to resistance, for example, if we vary the attractiveness of the forbidden toys, or the strength of the prohibition against playing, or the availability of alternative activities. In a study of sharing we can vary the quantity or the attractiveness of the to-be-shared objects, or the familiarity or deservingness of the recipients of the sharing, or the presence or absence of the experimenter at the time of donation.

The limitations of the laboratory approach can be summarized with the same one-word description used in chapter 6: artificiality. Grusec (1982) summarizes the standard criticisms:

> Is resisting playing with an arbitrarily forbidden toy akin in any way to resisting the temptation to lie or cheat or steal? Is donating tokens just won in a preceding game to unseen poor children at all related to helping a friend who is in real trouble? (p. 259)

The issue is one of external validity—whether the lab situation is similar enough to real life to permit generalization of findings. It is not hard to find possible problems and biases in any specific laboratory measure.

On the other hand, another point made in chapter 6 is also worth reiterating. Lab versus field is a continuum rather than a dichotomy, and experimentally contrived laboratory situations do not have to be radically different from the child's natural experiences. Indeed, a case could be made that moral behavior, more than most topics, lends itself to natural and nonreactive measurement in a laboratory setting. The donation measure is a possible example—a casual and plausible request that is seemingly unrelated to the purposes of the experiment. Similarly, help-giving has been measured by having the experimenter "accidentally" spill a box of tennis balls and then seeing whether the child helps to pick them up. Sympathy has been measured by having the experimenter pinch his or her finger in a drawer and then noting the child's reaction (Yarrow & Waxler, 1976). If skillfully executed (a big "if"), such measures may be quite natural and informative.

I will note finally that it may sometimes be possible to embed the experimental elicitation of a behavior within the natural setting. This was done, for example, in the Zahn-Waxler et al. (1992) study described earlier. In addition to their measures of naturally occurring distress situations, these investigators measured the child's response to simulated distress by the mother—for example, to her apparent pain after banging her ankle. Here we get the twin virtues of "field-field research" discussed in chapter 6: the experimental control that is important for internal validity, coupled with a natural setting to promote external validity.

The final general approach to be considered is the *verbal report* or *rating approach.* Ratings share several characteristics with naturalistic observations. The focus is again on naturally occurring behaviors in the natural setting, as opposed to experimentally elicited behaviors in a laboratory setting. The measurement again requires judgments by a human observer, as opposed to the automatic or essentially automatic recordings that may be made in the lab. Ratings, however, are considerably more global, more evaluative, and more removed from immediate behavior than are observations. With observations the attempt is to capture the specifics of ongoing behavior, generally by means of a precise and objective recording system and highly trained observers. With ratings the attempt is to identify general characteristics of the child—how honest or dishonest, how generous or stingy, and so forth. These characteristics are not direct behavioral measures but are abstracted, usually retrospectively, from a large number of observations made, usually naturally and unsystematically, by someone who knows the child well. This someone is most often a teacher or parent. In studies of older children, it may be other children or even the participants themselves. In any case, our measure is based on what someone who knows the child tells us about the child.

Let us consider a couple of examples. Rutherford and Mussen (1968) used ratings by teachers to assess generosity in a sample of preschool boys. Each teacher was given cards with the names of each boy in her class and was asked to sort the cards into five piles. Pile one was for those "who are among the most generous, least selfish nursery-school boys I have ever known." At the other extreme, pile five was for those "who are among the most selfish, least generous nursery-school boys I have known." In between was pile three: for those "who seem about average in generosity, neither highly generous nor highly selfish in their behavior." This sort of card-sorting procedure is a fairly common method of obtaining ratings. Whether cards are used or not, ratings typically involve placement of the child along some dimension—that is, some evaluation not just of presence or absence but of degree of generosity, degree of honesty, or whatever. Ratings also tend, either explicitly or implicitly, to be relative measures, involving comparison of the child with other children.

As Mussen and Eisenberg-Berg (1977) note, teacher ratings are most likely to be useful with preschool children, with whom teachers have frequent and varied experiences in relatively unstructured situations. In more structured grade-school or high-school classrooms, teachers may lack sufficient experience with their pupils to make many kinds of ratings. By this age, however, peers can often be useful informants. Martin Hoffman, in particular, has used peer ratings in a number of studies to assess aspects of moral development (Hoffman, 1975; Hoffman & Saltzstein, 1967). In one study, for example, Hoffman (1975) measured concern for others by having fifth-grade children nominate the three same-sexed classmates who were most likely to "care about how other kids feel and try not to hurt their feelings" and "to stick up for some kid that the other kids are making fun of or calling names."

Like all measures, ratings have both advantages and disadvantages. On the negative side, the obvious problem with ratings is that they are not direct measures of behavior; rather, they are second-order reports about behavior. There are many reasons why ratings may convey an inaccurate picture of what a child is really like—insufficient or biased opportunities for observation, misunderstanding of instructions, forgetting, willful distortion, stereotyping. Ratings by parents might seem especially susceptible to bias, and there is in fact evidence that parents tend to be unrealistically positive in rating their own children (Seifer, 2005). In addition, ratings, oriented as they are to the abstraction of general characteristics, are poor instruments for the study of the immediate determinants of specific behaviors (recall the trait-state distinction discussed in chapter 4). On the positive side, if skillfully elicited from a genuinely knowledgeable informant, ratings may provide a scope and depth of information about a child that is unavailable with other methods. Laboratory measures are necessarily limited to very brief and possibly atypical samplings of behavior. Observations in the natural setting, the efforts of Zahn-Waxler et al. (1992) notwithstanding, also tend to be limited in the situations and behaviors that can be sampled. With ratings, however, the scope for our conclusions about a child is enormously broader—potentially everything that a parent, teacher, or friend has ever seen the child do.

The most general conclusion to be drawn from this section should be familiar from chapter 6. All methods have their limitations. What we need, therefore, is a convergence of methods—an attack upon the particular research problem (in this case, the development of moral behaviors) through as many different methods as possible.

Emotion

I turn now to the emotional aspect of morality—to how the child feels in morally relevant situations. Of greatest interest historically has been the negative emotion of guilt—that is, unpleasant, self-punitive feelings that occur when one has done something wrong (or perhaps even *thinks* about doing something wrong). More recently, there has been considerable interest in the more positive emotion of empathy—that is, the tendency to share the emotional reactions of others (for example, to feel sad when someone else feels sad).

Emotions are by definition internal phenomena, and as such present a definite methodological challenge. How can we determine what a child is feeling? Until recently, the dominant approach was simply to ask—that is, to elicit some sort of verbal report of emotional reactions across the situations of interest. I begin with this approach, considering first an application to guilt and then an application to empathy. I then move on to some recent attempts to go beyond verbal reports to a more direct measurement of emotional response.

A common method of measuring guilt has been to present stories in which the central character (who is usually similar to the participant) misbehaves in some way. The story is broken off following the misbehavior, and the child is asked to provide an ending. Of particular interest are endings that focus on the story character's emotional reactions, including reactions of guilt. The character's subsequent behaviors may also be of interest; for example, does he run off and hide, confess his wrongdoing, attempt to make restitution, or what? Whatever the particular response ascribed to the character, the assumption is that it tells us something about how the child would feel and act in a similar situation. Because the child is assumed to "project" his or her own emotions onto the character, this kind of measure is called a *projective test.* Box 13.1 provides an example of one such story.

Box 13.1. Example of a Projective Story Used to Assess Guilt

Early one evening Bob and his friend are hurrying along the street on their way to the biggest basketball game of the season. Bob can't wait to see the game. It starts in five minutes, and they don't want to miss any of it. All the kids will be there. On the way they see a little boy wandering around across the street. He seems to be calling out somebody's name. Bob and his friend are the only ones around. They don't know who he is. Bob turns to his friend and says, "Gee, that little kid looks lost. Maybe we ought to go over and help him. It will only take a few minutes." But his friend says, "Come on, let's mind our own business. We don't want to miss any of the game, do we? Besides, his parents will find him after a while and he'll be all right. Come on, are you my friend or aren't you?" Bob finally says, "Okay. I suppose you're right. His folks will find him soon." They get to the game in time and really enjoy it. The next morning Bob goes out to ride his bike. On the way he looks at a newspaper. He notices a picture of the same little boy. The newspaper says that a neighbor lady was taking care of the little boy for the afternoon. She left the four-year-old boy outside a hairdressing shop while she had her hair fixed. She told the little boy to play outside and wait for her. But the little boy started walking around and got lost. Before the neighbor lady could find him, the little boy ran across the street and got hit by a car. The newspaper says he died on the way to the hospital.

SOURCE: From "Conscience, Personality, and Socialization Techniques," by M. L. Hoffman, 1970, Human Development, 13, p. 98. Copyright © 1970 by S. Karger.

The story-completion method has several virtues. It is relatively easy to administer and to score, and it allows us to sample reactions across a wide range of kinds and degrees of transgression, certainly a much wider range than we are likely to be able to sample with any other measurement technique. The most basic criticism of the method, of course, concerns the central assumption underlying its use: the assumption that children respond by projecting their own emotions onto the story character (as opposed, for example, to saying what they think the adult wants to hear). It is worth noting that stories such as that in Box 13.1 can also be used in a more direct, less projective manner by simply asking the children directly how they would think or feel if they were the story protagonist (e.g., Thompson & Hoffman, 1980). Still, the measure remains a verbal report about emotions and not a direct measurement of emotions.

Verbal reports are also common in the study of empathy. Table 13.1 shows one example, a measure developed by Bryant (1982). As can be seen, the questions sample both positive and negative empathic reactions across a number of different situations. Again, the breadth of the assessment is a strength of the verbal-report approach. With young children a simple yes/no response is all that is required. With older children or adolescents a 9-point rating scale can be used to indicate more precisely the degree of agreement or disagreement with a particular statement.

Instruments such as the one in Table 13.1 clearly possess some validity as measures of individual differences in empathy. In particular, scores on such tests show the correlational patterns we would expect of such measures; for example, high empathy is associated with relatively low levels of aggression (Bryant, 1982). Still, such instruments are subject to the same criticism noted earlier for story-completion measures of guilt: They measure verbal reports about emotions and not emotions directly. I turn next to some attempts to devise more direct measures of children's emotional reactions in morally relevant situations.

Table 13.1 The Bryant Empathy Test for Children and Adolescents

Statement	*Response*[a]
1. It makes me sad to see a girl who can't find anyone to play with	(+)
2. People who kiss and hug in public are silly	(–)
3. Boys who cry because they are happy are silly	(–)
4. I really like to watch people open presents, even when I don't get a present myself	(+)
5. Seeing a boy who is crying makes me feel like crying	(+)
6. I get upset when I see a girl being hurt	(+)
7. Even when I don't know why someone is laughing, I laugh too	(+)
8. Sometimes I cry when I watch TV	(+)
9. Girls who cry because they are happy are silly	(–)
10. It's hard for me to see why someone else gets upset	(–)
11. I get upset when I see an animal being hurt	(+)
12. It makes me sad to see a boy who can't find anyone to play with	(+)
13. Some songs make me so sad I feel like crying	(+)
14. I get upset when I see a boy being hurt	(+)
15. Grown-ups sometimes cry even when they have nothing to be sad about	(–)
16. It's silly to treat dogs and cats as though they have feelings like people	(–)
17. I get mad when I see a classmate pretending to need help from the teacher all the time	(–)
18. Kids who have no friends probably don't want any	(–)
19. Seeing a girl who is crying makes me feel like crying	(+)
20. I think it is funny that some people cry during a sad movie or while reading a sad book	(–)
21. I am able to eat all my cookies even when I see someone looking at me wanting one	(–)
22. I don't feel upset when I see a classmate being punished by a teacher for not obeying school rules	(–)

SOURCE: From "An Index of Empathy for Children and Adolescents," by B. K. Bryant, 1982, *Child Development, 53*, p. 416.

a. (+) indicates that an affirmative answer is empathic; (–) indicates that a negative answer is empathic.

Theoretically, guilt is of greatest interest following some misbehavior. What we need, therefore, is a situation in which the child has misbehaved. Rather than wait for misbehavior to occur, several investigators have measured children's responses following *contrived transgressions*—that is, experimentally engineered situations in which children are led to *think* that they have done something wrong.

A study by Kochanska, Gross, Lin, and Nichols (2002) provides an example. Their 2- and 3-year-old participants were presented with an object that the experimenter described as having special value or personal significance—for example, a favorite stuffed animal from her childhood, a toy that she had just assembled herself (the specific objects varied across age). Each of the objects had in fact been rigged so that it fell apart as soon as the child began to handle it. The experimenter responded by saying "Oh, my" in a tone of regret and then sitting silently for 60 seconds, during which time the child's reactions were videotaped for later analysis. This initial period was followed by several standard questions, including "What happened?" "Who did it?" and "Did you do it?" (Given the obvious ethical issues in research of this sort, I should note that the experimenter concluded the procedure by producing an intact version of the object and carefully assuring the child that he or she had not done anything wrong.)

Many of the children did in fact respond in ways suggestive of guilt, averting their gaze, for example, and showing signs of bodily tension (squirming, hunching shoulders, covering face). Such reactions showed some consistency over time and also some expected (although modest) relations to other measures, such as ability to resist temptation. Similar findings have emerged in other research with the contrived-transgressions procedure, including studies with older children (e.g., Dienstbier, 1984).

Recent research on empathy has also moved beyond purely verbal measures. Two alternative indices, in particular, have been explored as possible signs of an empathetic reaction: facial expressions and physiological change.

A study by Eisenberg et al. (1988) illustrates the approach. The preschool and second-grade participants in this study viewed three films in which children were depicted in potentially emotion-arousing circumstances: a boy and girl caught in a loud thunderstorm, a little girl whose pet bird had just died, and a young handicapped girl who was attempting to walk. The children's heart rates were measured during the films, and their facial expressions were videotaped. They were also asked after each film to report how they had felt while viewing the film. The expectation, assuming an empathic response to the characters' plights, was that the first film would evoke feelings of fear or anxiety, whereas the second and third would evoke a feeling of sadness.

This, in fact, is what was found, although somewhat more clearly for the older than for the younger participants. Children reported more fear in response to the first film than for the other two; conversely, reports of sadness were most common for the film about the dead pet. Analysis of the facial expressions painted a similar picture: expressions suggestive of fear or distress in response to the thunderstorm film, expressions of sadness or concern when viewing the other two segments. Finally, heart-rate change was also differential across films: Heart rate tended to accelerate in response to the first film and decelerate in response to the other two. Heart-rate deceleration, as was noted in chapter 11, is a correlate of attention; and Eisenberg et al. argue that its occurrence in response to another's sadness is an indication of an outwardly oriented, truly sympathetic form of response. In support of this interpretation, they have shown in other research (Eisenberg et al., 1989) that heart-rate deceleration when encountering someone in need is associated with an increased willingness to attempt to help.

It should be clear that Eisenberg et al.'s approach to the measurement of empathy embodies the kind of converging operations philosophy that is stressed at various points throughout this book. Self-reports of emotion, although certainly a reasonable source of evidence, may for various reasons be either inaccurate or incomplete. A particular facial expression can clearly be a clue to internal experience, but in themselves facial expressions are unlikely to be sufficient. Any specific pattern of physiological change is even less likely to be free of alternative interpretations. When all three forms of evidence point in the same direction, however, we can be much more confident that we really have succeeded in measuring the target construct.

I have concentrated here on the measurement of the moral emotions of guilt and empathy. A chapter by Eisenberg, Morris, and Spinrad (2005) is a good source with respect to the measurement of emotions more generally.

Cognition

The final topic for this section is the cognitive aspect of morality. The interest now is in how the child reasons about moral issues, and in developmental changes in such reasoning as the child's cognitive abilities mature.

There have been two major bodies of research on the cognitive aspect of morality. I begin with the work carried out by Piaget in the 1920s and since followed up in literally hundreds of studies. I then move on to the other work, a more recent program of research initiated by Lawrence Kohlberg and associates. This section concludes with a brief consideration of several contemporary approaches that extend the Piaget/Kohlberg tradition in some informative ways.

Piaget's work on morality is reported in a single book, *The Moral Judgment of the Child* (Piaget, 1932). As noted, this work was carried out in the 1920s, placing it among the earliest group of Piagetian studies. This placement means that it is not clearly integrated with the later work on cognitive development that was discussed in chapter 12. Nevertheless, many of the basic Piagetian ideas are already evident. These ideas include the belief that the child's moral reasoning is not simply a passive mirror of what parents or society has taught the child but rather reflects the child's own level of cognitive development. They include also the corollary belief that developmental changes in moral reasoning result largely from developmental changes in the child's cognitive abilities. And methodologically, as we will see, they include an emphasis on the flexible "clinical method" of testing as the best way to probe the child's beliefs.

Piaget used two main techniques in his studies of moral reasoning. One technique was to ask children about the rules for games, especially the game of marbles. In Piaget's view, children's games constitute a kind of microcosm of the social world in general, complete with socially transmitted rules, established interpersonal relations, sanctions for deviating from the rules, and so forth. It is for this reason that the study of games can tell us something about the child's level of moral reasoning. Piaget was interested both in the child's adherence to the rules and in his understanding of the origin and nature of rules.[1] The questioning, all conducted in the flexible clinical method style, was directed to points such as the following: What are the rules of the game? Have the rules always been what they are now? Who invented the rules? And could the rules ever be changed?

Box 13.2 shows a small sampling of what Piaget found. As can be seen, there are definite developmental changes in children's conceptions of rules. Younger children tend to view the rules for games as sacred and unchangeable—the rules have always been as they are, having been handed down either from God or the child's father, and they can never be changed. Older children are much more aware that the rules for games are at least somewhat arbitrary and changeable. This developmental shift is, in Piaget's view, part of a much more general shift from a "morality of constraint," or "moral realism," to a "morality of cooperation," or "moral relativism." We will see other examples of this shift shortly.

Box 13.2. Examples of Responses to Piaget's Questions Concerning the Rules for Games

Fal (5) . . . "Long ago when people were beginning to build the town of Neuchatel, did little children play at marbles the way you showed me?—*Yes.*—Always that way?—*Yes.*—How did you get to know the rules?—*When I was quite little my brother showed me. My Daddy showed my brother.*—And how did your daddy know?—*My Daddy just knew. No one told him.*—How did he know?—*No one showed him!*" . . . "Who invented the game of marbles?—*My Daddy did.*"

Stor (7) tells us that children played at marbles before Noah's ark: "How did they play?—*Like we played.*—How did it begin?—*They bought some marbles.*—But how did they learn?—*His daddy taught them.*" Stor invents a new game in the shape of a triangle. He admits that his friends would be glad to play at it, "*but not all of them. Not the big ones.* . . . —Why?—*Because it isn't a game for the big ones.*—Is it as fair a game as the one you showed me?—*No.*—Why?—*Because it isn't a square.*—And if everyone played that way, even the big ones, would it be fair?—*No.*—Why not?—*Because it isn't a square.*"

Malb (12) . . . "Does everyone play the way you showed me?—*Yes.*—And did they play like that long ago?—*No.*—Why not?—*They used different words.*—And how about the rules?—*They didn't use them either, because my father told me he didn't play that way.* . . . —Did they play marbles when your grandfather was little?—*Yes.*—Like they do now?—*Oh, no, different kinds of games.* . . . Could one change the rules?—*Yes.*—Could you?—*Yes, I could make up another game. We were playing at home one evening and we found out a new one* [he shows it to us]. Are these new rules as fair as the others?—*Yes.*—Which is the fairest, the game you showed me first or the one you invented?—*Both the same.*"

SOURCE: From *The Moral Judgment of the Child* (pp. 55, 60, 66-67), by J. Piaget, 1932. New York: Free Press. Copyright 1932 by The Free Press.

Piaget's second general technique for studying morality has had a greater influence on subsequent research. The technique consists of presenting little stories, or **moral dilemmas**, that pose some sort of moral issue that must be judged. The best-known example concerns the issue of whether the morality of a harmful action should be judged "objectively," in terms of the material consequences of the act, or "subjectively," in terms of the intentions behind the act. Box 13.3 shows the five stories that Piaget used to study this issue. His main finding was of a developmental change from an early "objective" focus on consequences to a more mature "subjective" concern with intentions.

To give an idea of some of the other issues examined in Piaget's research, Box 13.4 presents two other examples of Piagetian stories and accompanying responses. The first example concerns the issue of why it is wrong to tell a lie; the second concerns the concept of "immanent justice," or the belief in an automatic system of punishments for bad behavior.

Let us focus on the stories in Box 13.3 and the issue of objective versus subjective morality. Later researchers have found much to criticize in the original Piagetian stories, and it may be a useful exercise to think about possible problems before reading further. I have space, in fact, to list only some of the modifications that have been explored in later research. Some researchers have argued that the order of presentation in Piaget's stories, in which information about consequences always comes last, may bias young children to attend more to consequences than to intentions; they have therefore varied order either within or between

Box 13.3. Stories Used by Piaget to Study Objective Versus Subjective Responsibility

I. A. A little boy who is called John is in his room. He is called to dinner. He goes into the dining room. But behind the door there was a chair, and on the chair there was a tray with fifteen cups on it. John couldn't have known that there was all this behind the door. He goes in, the door knocks against the tray, bang go the fifteen cups and they all get broken!

B. Once there was a little boy whose name was Henry. One day when his mother was out he tried to get some jam out of the cupboard. He climbed up on to a chair and stretched out his arm. But the jam was too high up and he couldn't reach it and have any. But while he was trying to get it he knocked over a cup. The cup fell down and broke.

II. A. There was a little boy called Julian. His father had gone out and Julian thought it would be fun to play with his father's ink-pot. First he played with the pen, and then he made a little blot on the table cloth.

B. A little boy who was called Augustus once noticed that his father's ink-pot was empty. One day that his father was away he thought of filling the ink-pot so as to help his father, and so that he should find it full when he came home. But while he was opening the ink-bottle he made a big blot on the table cloth.

III. A. There was once a little girl who was called Marie. She wanted to give her mother a nice surprise, and cut out a piece of sewing for her. But she didn't know how to use the scissors properly and cut a big hole in her dress.

B. A little girl called Margaret went and took her mother's scissors one day that her mother was out. She played with them for a bit. Then as she didn't know how to use them properly she made a little hole in her dress.

IV. A. Alfred meets a little friend of his who is very poor. This friend tells him that he has had no dinner that day because there was nothing to eat in his home. Then Alfred goes into a baker's shop, and as he has no money, he waits till the baker's back is turned and steals a roll. Then he runs out and gives the roll to his friend.

B. Henriette goes into a shop. She sees a pretty piece of ribbon on a table and thinks to herself that it would look very nice on her dress. So while the shop lady's back is turned (while the shop lady is not looking), she steals the ribbon and runs away at once.

V. A. Albertine had a little friend who kept a bird in a cage. Albertine thought the bird was very unhappy, and she was always asking her friend to let him out. But the friend wouldn't. So one day when her friend wasn't there, Albertine went and stole the bird. She let it fly away and hid the cage in the attic so that the bird should never be shut up in it again.

B. Juliet stole some sweeties from her mother one day that her mother was not there, and she hid and ate them up.

SOURCE: From *The Moral Judgment of the Child* (pp. 122–123), by J. Piaget, 1932. New York: Free Press.

subjects (e.g., Moran & McCullers, 1984). Others have focused on the difficulties that young children may have in understanding and remembering all of the information in pairs of stories that are orally presented. Some have simplified the task by presenting just one story

Box 13.4. Examples of Piagetian Stories and Accompanying Responses for the Concepts of Lying and Immanent Justice

A. "A little boy [or a little girl] goes for a walk in the street and meets a big dog who frightens him very much. So then he goes home and tells his mother he has seen a dog that was as big as a cow."

B. "A child comes home from school and tells his mother that the teacher had given him good marks, but it was not true; the teacher had given him no marks at all, either good or bad. Then his mother was very pleased and rewarded him."

Fel (6) repeats the two stories correctly: "Which of these two children is naughtiest?—*The little girl who said she saw a dog as big as a cow.*—Why is she the naughtiest?—*Because it could never happen.*—Did her mother believe her?—*No because they never are* [dogs as big as cows].—Why did she say that?—*To exaggerate.*—And why did the other one tell a lie?—*Because she wanted to make people believe that she had a good report.*—Did her mother believe her?—Yes.—Which would you punish most if you were the mother?—*The one with the dog because she told the worst lies and was the naughtiest.*"

Arl (10): The naughtiest is the one "*who deceived his mother by saying that the teacher was pleased.*—Why is he the naughtiest?—*Because the mother knows quite well that there aren't any dogs as big as cows. But she believed the child who said the teacher was pleased.*—Why did the child say the dog was as big as the cow?—*To make them believe it. As a joke.*—And why did the other one say that the teacher was pleased?—*Because he had done his work badly.*—Was that a joke?—*No, it is a lie.*—Is a lie the same thing as a joke?—*A lie is worse because it is bigger.*

Once there were two children who were stealing apples in an orchard. Suddenly a policeman comes along and the two children run away. One of them is caught. The other one, going home by a roundabout way, crosses a river on a rotten bridge and falls into the water. Now what do you think? If he had not stolen the apples and had crossed the river on that rotten bridge all the same, would he also have fallen into the water?

Pail (7) . . . "What do you think of that?—*It's fair. It serves him right.*—Why?—*Because he should not have stolen.*—If he had not stolen, would he have fallen into the water?— *No.*— Why?—*Because he would not have done wrong.*—Why did he fall in?—*To punish him.*"

Fran (13) . . . "And if he had not stolen the apples, would he have fallen into the water?—*Yes. If the bridge was going to give way, it would have given way just the same, since it was in bad repair.*"

SOURCE: From *The Moral Judgment of the Child* (pp. 148, 150-151, 157-158, 252, 253-254, 255), by J. Piaget, 1932. New York: Free Press. Copyright 1932 by The Free Press.

with one kind of information at a time and then comparing judgments of naughtiness across the different stories (e.g., Berg-Cross, 1975). Others have replaced the Piagetian oral-presentation format with videotapes of the behaviors to be judged (e.g., Chandler, Greenspan, & Barenboim, 1973). A number of researchers (e.g., Nelson-LeGall, 1985) have noted that Piaget's stories do not clearly distin guish between motive (good or bad purpose? and intentionality (on purpose or accidental?) modifications have been made, therefore, t pull apart these dimensions. Finally, probabl the most important change in later research ha involved an attempt to disentangle the tw bases for response in the original Piagetia stories. Because motive and degree of damage covary in most Piagetian stories, it is impossible to determine exactly what information the child is capable of using.

Three general conclusions can be drawn from this follow-up research (see also Langford, 1995). These conclusions parallel those noted in chapter 12 for follow-up studies of Piaget's work on cognitive development. First, Piaget's methods lead to some underestimation of the young child's abilities, for performance is often more mature with the modified procedures of more recent research. Second, the domain of moral reasoning is more complicated and multidetermined than Piaget envisioned, and all sorts of variables may affect the way a child responds. Finally, whatever the problems in Piaget's studies, there is little doubt about the overall worth of the enterprise, not only with respect to some of the specific findings but also with respect to its pioneering role in establishing the cognitive aspect of morality as an area of study. I turn next to the major contemporary approach to moral reasoning, that of Lawrence Kohlberg.

Like Piaget, Kohlberg based his approach on the participant's response to hypothetical moral dilemmas. Kohlberg was interested in more advanced forms of moral reasoning than was Piaget, however, and the dilemmas that he posed are correspondingly a good deal more complex. There are nine Kohlberg dilemmas in all. Box 13.5 reproduces three of the dilemmas, including the best-known and most often cited one: the Heinz story.

Box 13.5. Examples of Kohlberg's Moral Dilemmas

Dilemma III: In Europe, a woman was near death from a special kind of cancer. There was one drug that the doctors thought might save her. It was a form of radium that a druggist in the same town had recently discovered. The drug was expensive to make, but the druggist was charging 10 times what the drug cost him to make. He paid $200 for the radium and charged $2,000 for a small dose of the drug. The sick woman's husband, Heinz, went to everyone he knew to borrow the money, but he could only get together about $1,000, which is half of what it cost. He told the druggist that his wife was dying and asked him to sell it cheaper or let him pay later. But the druggist said, "No, I discovered the drug and I'm going to make money from it." So Heinz gets desperate and considers breaking into the man's store to steal the drug for his wife. Should Heinz steal the drug? Why or why not?

Dilemma V: In Korea, a company of Marines was greatly outnumbered and was retreating before the enemy. The company had crossed a bridge over a river, but the enemy were mostly still on the other side. If someone went back to the bridge and blew it up, with the head start the rest of the men in the company would have, they could probably then escape. But the man who stayed back to blow up the bridge would probably not be able to escape alive; there would be about 4:1 chance he would be killed. The captain himself is the man who knows best how to lead the retreat. He asks for volunteers, but no one will volunteer. If he goes himself, the men will probably not get back safely and he is the only one who knows how to lead the retreat. Should the captain order a man to go on this very dangerous mission or should he go himself? Why?

Dilemma VIII: In a country in Europe, a poor man named Valjean could find no work, nor could his sister and brother. Without money, he stole food and medicine that they needed. He was captured and sentenced to prison for 6 years. After a couple of years, he escaped from the prison and

(Continued)

(Continued)

went to live in another part of the country under a new name. He saved money and slowly built up a big factory. He gave his workers the highest wages and used most of his profits to build a hospital for people who couldn't afford good medical care. Twenty years had passed when a tailor recognized the factory owner as being Valjean, the escaped convict whom the police had been looking for back in his home town. Should the tailor report Valjean to the police? Would it be right or wrong to keep it quiet? Why?

SOURCE: From "A Longitudinal Study of Moral Judgment," by A. Colby, L. Kohlberg, J. Gibbs, and M. Lieberman, 1983, *Monographs of the Society for Research in Child Development, 48*, pp. 77, 82, and 83. Copyright © 1983 by the Society for Research in Child Development.

The first response elicited following the presentation of a dilemma is the yes/no judgment concerning the morality of the story character's behavior (e.g., Should Heinz steal the drug? Should the captain order a man to go, or go himself?). The real interest, however, is in the reasoning behind this yes/no answer. The "why" question is a first attempt to elicit this reasoning, and the experimenter is then free to follow up on the initial response in a variety of semistandardized ways. In scoring the participant's responses and assigning a stage level, it is the reasoning, and not the yes/no judgment, that is critical.

Most presentations of Kohlberg's theory identify six developmentally ordered stages.[2] The stages, in turn, can be grouped into three developmental levels. Reasoning at the developmentally earliest *preconventional level* is oriented to overt rewards and punishments. An argument at this level against stealing the drug might be "Heinz shouldn't steal the drug because he might get caught and sent to jail." Reasoning at the subsequent *conventional level* shows a concern with society's standards and upholding the letter of the law. An argument at this level against stealing the drug might be "Heinz will know he did wrong after he's punished and sent to jail. He'll always feel guilty for his lawbreaking." Finally, reasoning at the most advanced *postconventional level* reflects internal principles of conscience, principles that in some instances may override society's laws. An argument at this level in favor of stealing the drug might be "If Heinz doesn't steal the drug and lets his wife die, he'd always condemn himself for it afterward. He wouldn't have lived up to his own standards of conscience." It should be noted that the highest stages are unlikely to be found before adolescence and are only moderately common even in adulthood.

Nothing has been said so far about how the participant's answers are scored and assigned to a stage. This issue is difficult to discuss at all briefly, for the Kohlberg scoring system is extremely complex; indeed, it may be *the* most complex scoring system in the psychological literature. I will note, however, that the focus in scoring is not on the specific content of the answer (e.g., whether Heinz should steal the drug) but rather on the level and the structure of the reasoning offered in support of the answer. Furthermore, the stage assignment is based not on a single response but on the pattern of reasoning across several dilemmas and multiple follow-up questions. Though difficult to learn and to apply, the scoring system does show good interrater reliability and moderately good test-retest reliability (Colby & Kohlberg, 1987, Colby et at., 1987).

Given the complexities of the Kohlberg system, it should be noted that two main alternatives to the standard approach have been developed: the Sociomoral Reflection Measure, or SRM (Gibbs, 2003; Gibbs, Basinger, & Fuller, 1992), and the Defining Issues Test, or DIT (Rest, 1979; Rest, Narvaez, Bebeau, & Thoma, 1999). In the SRM, the dilemmas are presented in writing and the participant responds in writing, whereas in the

DIT the participant ranks various issues with regard to how important each is in deciding how the dilemma ought to be resolved. Both measures are less time-consuming than is the standard Kohlberg approach, and each is easier to administer and to score.

Let us move from the method itself to a brief consideration of some of the issues raised by Kohlberg's work. A number of issues have sparked both research and controversy, including the question of the relation between moral reasoning and moral behavior (e.g., Blasi, 1980), the claim of sex differences in moral reasoning (e.g., Gilligan, 1982), and the possibility of moral education programs based on Kohlberg's theory (e.g., Sockett, 1992). I concentrate here, however, on an issue that is central to the theory: the claim that moral reasoning develops through a series of stages.

Two predictions follow from the claim of stages. One is of consistency or concurrence in response. If it makes sense to say that children are "in" a particular stage, then their reasoning should consistently fall within this stage. The second prediction is of invariant sequence. Later stages build upon and are made possible by the stages that precede them. Lower stages should thus always come before higher ones, and no child should ever skip a stage or go backwards.

Research provides fairly good support for both of these predictions (Walker, 2004). Testing the claim of concurrence requires a within-subject approach, in which the same participants respond to a series of moral dilemmas and we can compare their reasoning across the different tasks. Such studies typically reveal substantial but not perfect consistency in stage level. Walker, deVries, and Trevethan (1987), for example, examined response to both the standard Kohlberg dilemmas and a "real-life" dilemma provided by the individual participant. They found that 62% of their sample achieved the same stage classification on the two measures, and that the stages were either identical or adjacent in over 90% of the cases. Similar results emerge when several of the standard dilemmas are compared. Typically, respondents have a modal stage at which 65% to 70% of their responses are classified, with most of the remaining answers falling either one stage above or one stage below the modal stage. Discrepancies of two or more stages are rare (Walker, 1988).

Tests of the claim of invariant sequence require longitudinal study. Kohlberg himself initiated the first such effort: His dissertation study eventually grew into a 20-year-longitudinal project, with participants tested first in the late 1950s when they were between 10 and 16 years old and then five more times at 3- or 4-year intervals across the next 20 years. The final publication from this research (Colby et al., 1983) reports no instances of stage skipping and only a small number of cases of apparent backward movement, a finding that is attributed to measurement error rather than genuine regression. More recent longitudinal research continues to support the claim of invariant sequence (Walker, 1989; Walker & Taylor, 1991).

A further test of the stage claims comes from cross-cultural study. Kohlberg's theory, with its emphasis on the basic cognitive-structural component of morality, predicts a good deal of similarity in moral development across even diverse cultural settings. The theory does allow for differences in the rate of development, or in the final level of development achieved, or in the specific content of some of the answers. What it insists on, however, is that the same basic stages be identifiable in all cultures and that they emerge in the same order in all cultures. Cross-cultural research generally supports these claims (Edwards, 1986; Snarey, 1985). With appropriate adjustments in methodology, the Kohlbergian stages have been found in a wide range of different cultures, the age trends are compatible with the notion of a progression from lower to higher stages, and the data from the subset of longitudinal studies in other cultures support the claim of invariant sequence. Research also suggests, however, that development beyond stage 4 is rare in non-Western societies. At present it is unclear whether this finding reflects a genuine cultural difference in the level of moral reasoning or a failure of the Kohlberg approach to capture forms of moral thought that are important in cultures different from our own.

The final point to be made in this section is somewhat related. Even within our culture, the Piaget and Kohlberg approaches, informative although they clearly are, do not exhaust the domain of moral cognition. Recent years have seen the development of a number of programs of research that fall within the general Piaget-Kohlberg cognitive-structuralist approach but that attempt to explore forms of reasoning not captured by these theorists. Two emphases, in particular, are worth noting.

One concerns the kind of moral content at issue. I noted at the start of the Moral Development section that moral actions divide into two general categories: avoiding bad behavior and producing good behavior. The emphasis in both Piaget and Kohlberg is primarily on reasoning about the first of these categories—on what is sometimes referred to as *justice reasoning*. Work under the heading of *prosocial moral reasoning* speaks to the second category: what sorts of good behavior should be produced when others are in need. The following is an example of one of the dilemmas used:

> A poor farming village named Circleville had a harvest that was just enough to feed the villagers with no extra food left over. Just at that time a nearby town named Larksdale was flooded and all the town's food was ruined, so they had nothing to eat. People in the flooded town of Larksdale asked the poor farmers of Circleville to give them some food. If the farmers did give food to the people of Larksdale, they would go hungry after working so hard all summer for their crops. It would take too long to bring in food from other villages because the roads were bad and they had no airplanes. What should the poor farming village do? (Eisenberg-Berg, 1979, p. 129)

Research with dilemmas of this sort demonstrates that prosocial moral reasoning, like justice reasoning, progresses through various levels of complexity as children develop. It also indicates, however, that many of the important advances in understanding occur by the early grade-school years (Eisenberg, 1982). Such research thus provides a more positive picture of young children's moral reasoning than is true in the Piaget and Kohlberg approaches.

The second emphasis in recent research addresses a basic question: Do children realize that there is a domain of morality within which certain kinds of actions fall? In particular, can they distinguish moral actions from other instances of socially approved behavior, such as social conventions (rules of etiquette, manners of address, etc.)? Do they realize, for example, that actions such as stealing or hitting are bad in any context or any society, but that rules for polite behavior might vary from one setting to another? The answer is that by age 4 or 5 most children do seem to have a basic realization of how moral behaviors differ from other kinds of rules or prescriptions that they have learned to follow (Smetana, 1993; Tisak, 1995). Thus, this work, too, reveals some important early developments not captured by either Piaget or Kohlberg.

Self-Concept

Like morality, self-concept is a large topic with many different facets. As was true with morality, the coverage will be selective, and as with morality, it will follow a commonly accepted division. The division in this case is between the cognitive component of the self-concept and the evaluative component—between what children think or know about the self and what they feel about the self. The first is labeled *self-knowledge* (or *self-awareness*); the second is labeled *self-esteem.*

Self-Knowledge

The developmental study of self-knowledge begins with a very basic question: When are children first aware that they possess a distinct self? A related methodological question immediately follows: Through what kinds of evidence can we infer what the preverbal infant or the scarcely verbal toddler understands about the self? This, it should be clear, is simply a new variant of a question that we have already encountered with respect to numerous aspects of infant development.

Babies do not do anything at birth or during the first few weeks of life that suggests—at least to most developmental psychologists—that they possess any conception of a self. Indeed, in some theoretical formulations, most notably Piaget's, infants begin life with a basic inability to distinguish between self and outer world. Gradually, however, certain behaviors begin to emerge that may signal a dawning, albeit still limited, awareness of the self. Most commonly cited in this regard are various competencies that are grouped under the heading of **personal agency**: the understanding that one can be the cause of events in the world. By a few weeks of life, infants produce behaviors suggestive of a sense of agency, and by mid to late in the first year, they demonstrate such knowledge quite unequivocally. Piaget's studies are one obvious source of evidence; indeed, much of sensorimotor development consists of the increasingly effective ways in which infants act upon the environment to get what they want. So too is the Rovee-Collier research illustrated in Figure 12.2, in which infants learn that they can control the movement of the mobile. And so are a number of phenomena from the study of early social development, a general conclusion from which is that babies become active participants in social interchanges from early in life, emitting various behaviors (cries, gestures, etc.) whose function is to influence the behavior of another.

Personal agency is generally conceptualized as part of the *subjective* dimension of the self—that is, the child's awareness of the self as an active generator of experiences. Also of interest is the *objective* dimension of the self—the self as object of contemplation, or what children think or know *about* the self.

Here, too, the earliest indicators are necessarily nonverbal—various behaviors by which infants or toddlers convey that they are beginning to understand who they are. A basic question is that of **visual self-recognition**: At what point in development can infants first recognize an image of themselves? Suppose, for example, that we place a baby in front of a mirror—will the baby realize that the interesting creature she sees is actually herself? During the first year of life the answer seems clearly to be no. Infants during the first year often react with pleasure to a mirror image, smiling at, reaching toward, and greeting the face that they see; the behaviors, however, are no different from those that they direct toward other infants. It is typically only some time during the second year that we begin to see indications that the infant is beginning to draw a link between image and self. The clearest of these indications comes in response to a measure known as the *rouge test.*

The rouge test was developed independently by Amsterdam (1972) to study infants and by Gallup (1970) to study nonhuman primates. Here I describe the infant version. The test begins with the mother surreptitiously applying a dot of rouge to the child's nose or forehead. The child is then placed in front of a mirror, and his or her reactions are recorded. The question, of course, is whether the child will touch the marked area. If so, then we would have good evidence that the child recognizes the relation between mirror image and self. Prior to about 15 months such self-directed exploration is rare. Touches and other relevant behaviors (e.g., embarrassment at seeing the mark) begin to emerge, however, by about 15 to 18 months, and they are reliably elicited by 24 months (Povinelli, 1995). By this age, then, visual self-recognition is clearly in place.[3]

Other signs of self-understanding emerge either coincident with or soon after success on the rouge test (Stipek, Gralinski, & Kopp, 1990). Two-year-olds can typically recognize themselves not just in mirrors but in photographs or videotapes. Most 2-year-olds have learned their own name and can use it appropriately when referring to the self. First person pronouns—I, me, mine—also begin to appear.

An interesting study by Povinelli and associates (Povinelli, Landau, & Perilloux, 1996) identifies a further early advance in self-knowledge. The 2- to 4-year-old participants first played a simple game, during which the experimenter praised them and patted them on the head at several points. As part of one of the head pats, the experimenter also surreptitiously placed a large colorful sticker on the child's head. A few minutes later the children viewed a videotape of the game, in the course of which they could clearly see the sticker being placed. The question was whether they would reach up and touch or remove the sticker. Although most of the 4-year-olds reached immediately for the sticker, none of the 2-year-olds and only a minority of the 3-year-olds did so. Success on the sticker task requires recognizing the continuity of the self over time—that is, realizing that what happened to the self earlier is relevant to the self's current condition. This sort of extended self-knowledge is apparently more difficult than the immediate self-recognition tapped by the mirror test.

Once language has developed, measures of self-concept tend to work through the medium of words. A basic methodology with older children is the *self-description* approach: Children are asked to provide a description of what they are like (Damon & Hart, 1988; Harter, 1988). Typically, such measures begin with a broad, open-ended question—for example, "Tell me what you are like," or "What kind of person are you?" More specific questions usually follow, with their particular form dependent on the interests of the researcher; among the issues that have been explored are self-concept across different settings and relationships (e.g., "Tell me what you are like with your family," "Tell me what you are like with your friends") and self-concept across time (e.g., "Do you think you'll be the same or different 5 years from now?" "Do you change at all from year to year?").

As the preceding suggests, a variety of issues can be explored with the self-description approach. A basic question, however, is that of developmental change: How does children's thinking about themselves change as they develop? Changes with age do in fact occur, and in general the changes are in line with what we would expect from knowledge of general changes in cognitive development. The self-descriptions of preschool children tend to focus on overt and immediate characteristics: physical attributes, possessions, preferences ("I have red hair." "I like ice cream."). By middle childhood the emphasis has shifted to less tangible characteristics, such as emotions, and to behavioral traits and dispositions ("I'm nice to people." "I'm a good singer."). Finally, by adolescence still higher-order, more abstract characterizations have emerged; general personality attributes now dominate the description, and the picture is considerably more differentiated and multifaceted than was true earlier in development ("I'm concerned about social issues." "I want to be popular but I'm afraid of rejection."). It is worth noting that the same developmental trends are evident in children's descriptions of other people, such as classroom peers (e.g., Livesley & Bromley, 1973).

Self-Esteem

Instruments to measure self-esteem in middle childhood and adolescence are both plentiful and diverse (Coopersmith, 1981; Harter, 1985; Marsh, 1988; Piers, Harris, & Herzberg, 2002; Rosenberg, 1979). Despite their diversity, such instruments share a number of features. One obvious commonality is the focus of interest: Our target now is the evaluative component of the self-system—how children feel, along some positive to negative dimension, about themselves. Also common is the use of a self-report methodology to assess such feelings—on all of these measures conclusions about self-esteem come from children's response to direct questions about their feelings of self-satisfaction and

Table 13.2 Examples of Items From the Coopersmith Self-Esteem Inventory

	Like me	*Unlike me*
2. I'm pretty sure of myself.	______	______
3. I often wish I were someone else.	______	______
17. I'm often sorry for the things I do.	______	______
18. I'm popular with kids my own age.	______	______
35. I'm not doing as well in school as I'd like to.	______	______
36. I can make up my mind and stick to it.	______	______
46. Kids pick on me very often.	______	______
47. My parents understand me.	______	______

SOURCE: From *The Antecedents of Self-Esteem* (pp. 265–266), by S. Coopersmith, 1967, San Francisco: W. H. Freeman and Company.

self-worth. The use of many such questions is a third commonality—self-esteem is a broad target, and we clearly want to sample many rather than few aspects of the child's life.

As the existence of so many instruments suggests, there are also some differences among the various approaches to assessing self-esteem. One basic distinction concerns the contrast between *unidimensional* and *multidimensional* measures. With a unidimensional approach the attempt is to identify a single, overall sense of self-worth—a global self-esteem that cuts across numerous aspects of the respondent's life. Many of the earliest instruments for assessing self-esteem were unidimensional in nature. One influential example is the Coopersmith Self-Esteem Inventory (Coopersmith, 1967), some items from which are shown in Table 13.2. As can be seen, the decision with a unidimensional approach to focus on a single target does not mean a single-item assessment; rather, the Coopersmith Inventory elicits self-evaluations across a range of contexts and life events. There is no attempt, however, to distinguish the different contexts or types of evaluation (e.g., academic versus social); instead, the responses are simply summed to yield one overall score.

A unidimensional measure may be sufficient if people in fact experience just one overarching sense of self-esteem. Perhaps, however, self-esteem is more differentiated. Perhaps, for example, a particular child feels quite good about how he or she is doing in school but not at all good about how he or she is doing in relations with peers. It is to tap such potential distinctions that multidimensional measures were developed.

Table 13.3 presents a sampling of items from one of the most often used multidimensional measures of self-esteem in middle childhood and adolescence: the Harter Self-Perception Profile for Children (Harter, 1985). As the examples suggest, the Self-Perception Profile assesses self-esteem in five specific domains: academic competence, social acceptance, athletic competence, physical appearance, and behavioral conduct. Each of the five dimensions includes six items, and each item yields a score on a 4-point scale. The sum of the six scores for a particular dimension constitutes the measure of self-esteem for that domain.

Must the notion of global self-esteem be abandoned once we move to a multidimensional assessment like the Harter scale? Not necessarily. Harter agrees that global self-worth is a valid and important construct. On the Self-Perception Profile, however, the global score is not simply the sum of responses to the five specific dimensions; rather, global self-worth constitutes a separate, sixth dimension. The final item listed in Table 13.3 is one of the six

Table 13.3 Examples of Items From the Self-Perception Profile for Children

Really true for me	*Sort of true for me*				*Sort of true for me*	*Really true for me*
☐	☐	Some kids feel that they are very good at their school work.	BUT	Other kids worry about whether they can do the school work assigned to them.	☐	☐
☐	☐	Some kids find it hard to make friends.	BUT	Other kids find that it's pretty easy to make friends.	☐	☐
☐	☐	Some kids do very well at all kinds of sports.	BUT	Other kids don't feel that they are very good when it comes to sports.	☐	☐
☐	☐	Some kids are happy with the way they look.	BUT	Other kids are not happy with the way they look.	☐	☐
☐	☐	Some kids often do not like the way they behave.	BUT	Other kids usually like the way they behave.	☐	☐
☐	☐	Some kids are often unhappy with themselves.	BUT	Other kids are pretty pleased with themselves.	☐	☐

SOURCE: From *Manual for the Self-Perception Profile for Children*, by S. Harter, 1985, Denver: University of Denver.

items that are intended to provide a direct measure of global self-esteem.

To many, the idea that self-esteem is multifaceted rather than unidimensional has intuitive appeal. Clearly, however, the validity of an approach like Harter's is an empirical question. Assessing separate domains of self-esteem makes sense only if children do in fact distinguish different domains in their evaluations of the self. Evidence exists to suggest that they do—and not just for the Harter measure but for other multidimensional instruments as well (e.g., Marsh, 1988). The primary validity evidence comes from the kinds of correlational data discussed under the heading of "construct validity" in chapter 4. If self-esteem is really multidimensional, then we would expect that responses to items tapping a particular domain (e.g., academic competence) would correlate positively among themselves—and would correlate more strongly than they do with items directed to a different domain (e.g., athletic competence). This, in fact, is the finding. A basic conclusion from such research, therefore, is that children's (and, for that matter, adults') self-evaluations may differ across different facets of their lives. High or low self-esteem with respect to one aspect of development does not guarantee comparable self-esteem with respect to other aspects.

I noted that the instruments discussed in this section have been directed to middle childhood and adolescence. Can younger children make meaningful self-evaluations? Several assessment instruments and related programs of research suggest that they can.

Figure 13.1 shows one example, a downward extension of the Harter measure labeled the Pictorial Scale of Perceived Competence and Social Acceptance for Young Children (Harter & Pike, 1984). The Pictorial Scale is designed for ages 4 to 7. As the example indicates, the Scale retains the two-phase, 4-point response format of the Self-Perception Profile. It also retains a division into separate dimensions—fewer dimensions, however, than are assessed in older children. The item shown is from the cognitive competence dimension; other domains assessed are physical competence, maternal acceptance, and peer acceptance. A pair of alternatives for the last of these categories, for example, is the following: "This girl has lots of friends to play with. This girl doesn't have very many friends to play with."

The Pictorial Scale is probably the most often used self-esteem measure at the preschool level, but it is not without its critics. The distinctiveness of the different dimensions is not always clear, and there are also doubts about the appropriateness of the measure for low-income children (Fantuzzo, McDermott, Manz, Hampton, & Burdick, 1996). A promising alternative, appropriate for ages 4 and 5, is the Self-Description Questionnaire for Preschoolers, or SDQP,

Figure 13.1 Example of an item from the Pictorial Scale of Perceived Competence and Acceptance for Young Children

SOURCE: From "The Pictorial Scale of Perceived Competence and Acceptance for Young Children," by S. Harter and R. G. Pike, 1984, *Child Development, 55*, p. 1973. Copyright © 1984. Reprinted by permission of Blackwell Publishing.

developed by Marsh and colleagues (Marsh, Ellis, & Craven, 2002). The SDQP assesses six dimensions of self-esteem: physical ability, appearance, peer relations, parent relations, verbal ability, and mathematical ability. In contrast to the Pictorial Scale, the format is strictly verbal: six yes/no questions for each of the six dimensions. An example for the physical dimension is "Can you run fast?" One for the appearance dimension is "Do you like the way you look?" The Marsh et al. report provides some evidence in support of the validity of the measure—in particular, evidence that preschoolers are able to distinguish among the different dimensions. It seems clear both from this and from other research that preschoolers' self-concepts are not yet as consistent or differentiated as those of older children (Davis-Kean & Sandler, 2001; Marsh, Dubus, & Bornholt, 2005). Nevertheless, some ability to make differentiated evaluations of the self emerges early in development.

Sex Differences[4]

Some General Points

As Maccoby and Jacklin (1974) point out, the topic of sex differences is an unusual one in that the majority of findings are incidental to the main purpose of the studies from which they come. That is, when researchers report data on depth perception or conservation or aggression, it is because they set out to study depth perception or conservation or aggression. Data on sex differences, however, may emerge from literally any study on any topic, just as long as the study happens (as do most) to include both sexes.

The incidental nature of much of the data on sex differences means that the database for discussions of the topic is both huge and extremely heterogeneous. This mass of potentially relevant evidence presents a formidable challenge to anyone attempting to survey the literature and distill conclusions from it. How can the results from hundreds of disparate studies be sifted and evaluated?

Box 8.1 introduced the concept of meta-analysis—that is, quantitative approaches to research synthesis that employ various rules to combine and statistically analyze the data from related studies. There is no topic in developmental psychology for which meta-analyses have been either more common or more informative than the topic of sex differences. Such analyses, as I discuss shortly, do not solve all the problems of a complex primary literature—as is always the case, a meta-analysis can be no better than the studies and data that go into it. Still, they do help to make some sense out of what otherwise might be intractable complexity.

What specific topics under the general heading of sex differences have been targets for meta-analysis? Table 13.4 presents some examples; an article by Hyde (2005) provides a fuller overview. Hyde lists 46 meta-analyses spanning dozens of different topics—and her summary is explicitly a partial one! She also comes away with two general conclusions from her review of meta-analyses, conclusions that are worth noting here. The first (which is conveyed in her title: "The Gender Similarities Hypothesis") is that sex differences are both less frequent and smaller in magnitude than is often believed. The second is that the differences that do occur are often context-dependent, appearing under some circumstances but not others. This is also a theme in a major contemporary treatment of gender-role development, Eleanor Maccoby's (1998) *The Two Sexes*.

As noted, meta-analyses do not resolve all of the problems posed by the sex differences literature. Most meta-analyses are limited to the published literature, and hence are potentially susceptible to the "file drawer problem" discussed in Box 8.1—that is, biases in what sorts of findings are likely to be published. Possible publication biases can be argued in either direction. According to Maccoby and Jacklin

Table 13.4 Examples of Meta-Analyses of Sex Differences

Topic	*Source*
Aggressive behavior	Archer (2004)
Computer use	Whitley (1997)
Coping	Tamres, Janicki, & Helgeson (2002)
Facial expression processing	McClure (2000)
Helping behavior	Eagly & Crowley (1986)
Infant activity level	Campbell & Eaton (1999)
Language use	Leaper & Smith (2004)
Mathematics performance	Hyde, Fennema, & Lamon (1990)
Moral orientation	Jaffee & Hyde (2000)
Self-esteem	Kling, Hyde, Showers, & Buswell (1999)
Sexuality	Oliver & Hyde (1993)
Verbal ability	Hyde & Linn (1988)

(1974), the finding of a significant difference is newsworthy; the absence of a difference is much less newsworthy. Researchers tend, therefore, to notice, analyze, and report those cases in which the sexes happen to differ, and to ignore those (much more frequent) cases in which no differences emerge. The result is an inflated picture of the extent to which sex differences actually exist. Block (1976), however, has argued that biases may also work in the opposite direction. As she notes, many researchers regard sex differences as nuisances that are to be ruled out whenever possible. Measures may be selected, therefore, in part precisely because they are already known not to show sex differences. In doubtful cases pilot testing may be used to ensure that the sexes respond equivalently. Even though analyses for sex differences then become trivial, some journals or reviewers require that such analyses be performed and the results reported for any study that includes both sexes. The result, according to Block, is a proliferation of meaningless negative results.

Both kinds of biases just discussed undoubtedly exist; no one, however, knows how widespread they are and which, if either, is more important. This factor contributes to the controversies that have always surrounded the topic of sex differences.

Some of the issues introduced in earlier chapters are also relevant to an evaluation of the sex differences literature. In chapter 5 I discussed the biasing effects that the researcher's expectancies may have on the outcomes of research. As noted then, the obvious way to guard against such bias is to remove the expectancies—to blind the tester or observer with regard to the hypotheses of the study or the group membership of the participant. With many independent variables such blinding is possible. With the variable of sex it generally is not. Sometimes when babies are the targets of study the tester or (more probably) the observer may be kept unaware of the baby's sex. When verbal responses are tape-recorded for later analysis, the person scoring the tapes may not be able to tell whether a boy or girl is talking. If responses are transcribed before scoring, then the sex of participants of any age can usually be disguised. These cases, however, are exceptions to the general rule: Usually the observer who evaluates a behavior, and almost always the tester who elicits the behavior, are aware of the sex of the participant.

There is a further point. I have been discussing the effects of the adult's knowledge of the child's sex. But the converse also holds: The child knows the sex of the adult with whom he or she is interacting. In some cases children respond differently to a male tester than to a female tester. The problematic aspect for interpreting sex differences comes when there is not simply a main effect of sex of tester but an interaction between sex of tester and sex of child. Perhaps, for example, boys respond better to a female tester, whereas girls respond better to a male tester. Note that there *is* a sex difference here; the difference, however (as with all interactions), is more complicated than might at first appear. And it is a difference that might well be misinterpreted by a researcher whose study, as do most, has included only one sex of tester.

The general message of this section can be easily summarized. The determination of which aspects of psychological development show sex differences might seem easy—measures of development already exist, and all that need be done is to apply them to both sexes. For the reasons discussed, the determination is not so easy, and sex differences therefore remains one of the most controversial topics in the field.

Measures of Sex-Role Development

A major point of the preceding section was that one does not need to set out to study sex differences to obtain information about sex differences. But there are, of course, researchers whose primary interest is in the nature and origins of sex differences. There are also a number of measures that have been explicitly designed to provide information about sex-role development. In this section I present a small sampling from the large pool of such measures. Fuller discussions can be found in Beere (1990), Liben and Bigler (2002), and Ruble, Martin, and Berenbaum (2006).

I begin with the historically earliest measures, so-called tests of sex typing. Golombok and Fivush (1994, p. 5) define sex typing as "the extent to which a person conforms to prescribed male and female gender roles." The starting point, therefore, is the assumption that there *are* gender roles and related, on-the-average differences between the sexes—that is, ways of thinking, feeling, or behaving that can be labeled as either "masculine" or "feminine." The goal of a test of sex typing is to capture, within a fairly brief testing period, a child's standing on the masculine-feminine dimension. Such tests can be used to chart developmental changes in degree of sex typing across childhood. They can also be used to look for possible sex differences in sex typing itself—that is, to see whether, at particular points in development, either boys or girls adhere more strongly to their sex-ascribed roles. And they can be used to identify individual differences in degree of sex typing within a sex, differences whose origins can then be explored in further research (e.g., studies of parental child-rearing practices).

Let us consider some examples. Many of the most often used tests of sex typing are self-report measures—that is, we learn about children's activities and preferences by asking them directly about their activities and preferences. Figure 13.2 shows examples of items from one such measure: the Sex Role Learning Index, or SERLI (Edelbrock & Sugawara, 1978). The SERLI is intended to measure both sex-typed preferences and knowledge of sex-role stereotypes in preschool children. The format is partly verbal and partly pictorial. For the knowledge measure the child is shown a series of 10 pictures of common objects, half of which (such as the hammer and nails shown in the figure) are associated with stereotypically masculine activities and half of which (such as a needle and thread) are associated with stereotypically feminine activities. The question in each case is "Who would use a [object name] to [activity name], boys? girls? or both boys and girls?" For the preferences measure, 10 pictures are presented simultaneously; in this case, half depict a stereotypically masculine activity (such as playing baseball), and half depict a stereotypically feminine activity (such as taking care of a baby). There are two such sets of 10, one for childhood activities and one for adult roles and occupations. The question in this case is "If you could do any one of these things, which one would you like to do best?" Once the child makes a choice, that picture is removed, and the question is repeated; in this way a complete ordering of preferences is eventually obtained. Thus, if a little boy, for example, picks all five masculine pictures before any feminine ones, that child would achieve the maximum sex-typing score.

As we have seen before, verbal reports need not be self-reports; we can also elicit information from someone who knows the child. The most common informants in research on sex typing have been parents or teachers. An instrument labeled the Pre-School Activities Inventory (PSAI) provides an example directed (as the name suggests) to sex typing during the preschool period (Golombok & Rust, 1993a, 1993b). The PSAI consists of 24 items distributed across three areas: toys (e.g., dolls, toy cars), activities (e.g., playing house, sports), and personality characteristics (e.g., "likes pretty things," "enjoys rough and tumble play"). As with the SERLI, half of the items in each category are stereotypically masculine, and half are stereotypically feminine. The child's parent or teacher rates the child's standing on each item on a 4-point scale ranging from "never" to "very often." As with the SERLI, therefore, the PSAI provides a range of scores that indicate the extent to which a child's behavior fits the stereotype for his or her sex.

Figure 13.2 Examples of stimuli from the Sex Role Learning Index

SOURCE: From "Acquisition of Sex-Typed Preferences in Preschool-Aged Children," by C. Edelbrock and A. I. Sugawara, 1978, *Developmental Psychology, 14*, p. 616. Copyright © 1978 by the American Psychological Association. Reprinted with permission.

Not all sex-typing assessments are verbal reports. The alternative is to look directly at the behaviors themselves. Such observational measures can be obtained either in the natural setting or in some specially designed laboratory setting. The example described here involves research in the natural setting. Lamb, Easterbrooks, and Holden (1980) carried out observations of a sample of preschoolers across a 9-week span. A time-sampling observational procedure was used, with each child observed for a total of six 10-minute periods. The focus of the observations was sex-typed behavior, and the coding system was therefore divided into two general categories: male-typed activities and female-typed activities. The male-typed activities (as defined from both previous research and adult consensus) included playing with vehicles, climbing, chasing, and wearing male costumes; the female-typed activities included playing with kitchen utensils, art work, doll play, and wearing female costumes. The interest was in whether boys and girls would show average differences across these two categories, and the finding was that they in fact did.

The final example to be considered in this section is a relatively recent addition to the literature, one that is intended to address some limitations in previous measures. Table 13.5 presents a sampling of items from an instrument labeled the COAT (Liben & Bigler, 2002). As can be seen, the measure is broad in scope, assessing children's beliefs about three areas of possible gender differences: occupations, activities, and traits (hence the OAT—C stands for child). In each case the exemplars to be judged, only a small sample of which are shown in the table, divide into three categories (as determined by adult ratings): masculine, feminine, and neutral. The participants (children 11 to 13 years of age) make two sorts of judgments with respect to each item. The "personal" scale assesses their own preferences or attributes; the questions in this case are "How much would you like . . . ?" or "How often do you . . . ?" or "How much like you . . . ?" The "attitudes" scale assesses their beliefs about the appropriateness of the different exemplars for males or females; the question in this case is "Who should . . . ?"

As noted, the goal in developing the COAT was to improve on existing instruments. Three strengths of the measure can be noted. Unlike some instruments, its inclusion of the "personal" and "attitudes" scales provides a clear distinction between children's personal preferences and their general beliefs, as well as a clear comparison between the two. Its sampling of exemplars is much broader than that in most measures, and there is an attempt to equate the desirability of the items across the masculine and feminine categories (one criticism of the SERLI is that the "masculine" activities are on balance more attractive than the "feminine" ones). Finally, most sex-typing measures, including the other three considered in this section, treat "masculine" and "feminine" as opposite ends of a single dimension. That is, a relatively high score on one set of attributes, say those labeled "masculine," automatically guarantees a low score on attributes from the other end of the dimension, in this case so-called "feminine" qualities. It seems possible, however, that some individuals might possess attributes or preferences that combine elements from both of the traditional dimensions—might, for example, be both independent and assertive (typical "masculine" qualities) and compassionate and nurturant (typical "feminine" qualities). The COAT allows for (and in its findings demonstrates) such "androgynous" mixing of attributes.

Several general conclusions from the types of measures described in this section can be noted. First, there *are* on-the-average sex differences on measures such as toy or activity preference. Indeed, it is the fact that such average differences exist that justifies the use of such measures to assess any particular child's degree of sex typing. Second, differences between the

Table 13.5 Examples of Items From the COAT

Masculine	*Feminine*	*Neutral*
Occupation items		
airplane pilot	ballet dancer	artist
auto mechanic	elementary school teacher	baker
doctor	interior decorator	comedian
fire fighter	librarian	elevator operator
school principal	secretary	writer
Activity items		
build with tools	baby-sit	go bowling
go fishing	do gymnastics	go to the beach
play chess	jump rope	listen to music
use a microscope	make jewelry	paint pictures
wash a car	wash clothes	ride a bicycle
Trait items		
adventurous	affectionate	creative
aggressive	emotional	curious
dominant	gentle	friendly
good at math	good at English	good at art
good at sports	neat	truthful

SOURCE: From "The Developmental Course of Gender Differentiation," by L. S. Liben and R. S. Bigler, 2002, *Monographs of the Society for Research in Child Development, 67* (2, Serial No. 269), pp. 114–116. Copyright © 2002. Reprinted with permission from the Society for Research in Child Development.

sexes emerge earlier on measures of naturally occurring play behavior than on verbal instruments like the SERLI or COAT. In some studies, in fact, simple toy and game preferences are evident as early as 1 year of age. Third, sex-typed preferences may emerge somewhat earlier for boys than for girls; on the other hand, girls may outpace boys with respect to *knowledge* of sex-role stereotypes. Finally, there are only very modest correlations between a child's standing on one measure of sex typing and the child's standing on other measures. Thus, children are far from perfectly consistent in the degree to which they exhibit sex-typed attitudes or behaviors.

The final set of measures that we consider parallels approaches discussed in the last part of the Moral Development section. There the interest was in the cognitive side of morality: methods of studying the way that the child reasons about moral issues. Here the focus is on the cognitive side of sex typing: how the child thinks about sex roles and sex differences, as well as how such thinking changes with development.

What sorts of questions fall under the heading of the cognitive aspect of sex typing? The simplest question concerns the child's ability to label the sexes appropriately. At what point in development can a boy accurately label himself as a boy, or a girl label herself as a girl? What about labeling others? Does the toddler know which children in a peer group are boys and which are girls? Suppose that the objects to be labeled are more distinct from the self, perhaps adults rather than other children. Does the young child realize that Daddy is a boy and Mommy is a girl?

Various forms of evidence suggest that the ability to apply gender labels accurately is an early developmental achievement (Fagot & Leinbach, 1993). Labels for gender are common components of most children's early lexicons, and the self label (e.g., "I'm a boy") is typically among the first aspects of the child's emerging self-concept. By 2 1/2 or 3 almost all children can answer correctly when asked directly "Are you a girl or a boy?"(Slaby & Frey, 1975). By this age children are also beginning to label others correctly—for example, to be able to indicate which of a pair of pictures is a boy or a girl, or a man or a woman (Leinbach & Fagot, 1986).

Once we know that a child can label the sexes appropriately, a next question follows naturally: What criteria does the child use in making such distinctions? For adults, of course, the defining criteria are anatomical differences, although in most situations other cues (e.g., hair length, voice, dress) are both more available and quite sufficient. Research suggests that children at first favor the more superficial cues. Thompson and Bentler (1971) examined the relative importance allocated to various cues for sex discrimination in a sample of 4- to 6-year-olds. The stimuli were nude dolls that varied along three dimensions: type of genitals (male or female), body build (a masculine build described as "muscular and sinewy" or a feminine build described as "well proportioned in the breasts and hips"), and hair length (long or short). All possible combinations of each value from each dimension were used, producing eight dolls in all. Each child was shown only one doll and was asked to indicate its sex. The children weighted hair length most heavily, followed by body build. Thus, the doll with male genitals but feminine body and long hair was judged as female by the great majority of children; conversely, the dolls with female genitals but short hair elicited a substantial number of masculine choices.

Studies such as Thompson and Bentler's suggest that the young child's cognitive grasp of gender is less than perfectly formed. This conclusion emerges even more dramatically from research directed to a further aspect of gender understanding: the realization that one's gender is a permanent attribute. Indeed, it was some theorizing by Kohlberg (1966) concerning such **gender constancy** that sparked the contemporary interest in the cognitive side of sex typing. In his 1966 paper Kohlberg argued for two general propositions. The first was that children only gradually come to realize that gender is a permanent quality, immutable in the face of changes in age, volition, or immediate circumstance (e.g., clothing or hair style). The second was that the child's cognitive realization of gender constancy plays a causal role in the development of sex typing and sex differences. It is only when the little boy (for example) realizes that he is a male and will always remain a male that he identifies with the father (who is also male) and takes on masculine preferences and attributes.

Since Kohlberg's original paper, dozens of studies have examined developmental changes in gender constancy. A study by Marcus and Overton (1978) provides one example.

Figure 13.3 Drawings used to assess children's understanding of gender constancy

SOURCE: From "Boy-Girl Identity Task," by W. Emmerich and K. S. Goldman, 1972. In V. Shipman (Ed.), *Disadvantaged Children and Their First School Experiences* (ETS PR 72–20), Princeton, NJ: Educational Testing Service. Copyright © 1972 by the Educational Testing Service. Reprinted with permission.

Figure 13.3 shows two of the stimuli for this study (these stimuli were originally developed by Emmerich and Goldman, 1972). These schematic drawings of a girl and a boy were bound together in a booklet, with one drawing on top of the other. Because the top drawing was cut horizontally across the neck, transformations could be effected by simply turning the top or bottom segment of the top picture. A flip of the bottom part of the girl picture, for example, produced a girl's head perched on a masculinely dressed bottom. A flip of the top part resulted in a boy's head with a skirt beneath.

Five questions were asked of each participant (children between 5 and 7). Starting with the picture that was the same sex as the child, the experimenter produced successive transformations in the figure's hair style, clothing, and both hair style and clothing. Following each change the child was asked whether the figure was still a girl (or boy) or whether the sex had changed. The child was also asked whether the figure would change sex if she (or he) adopted different play interests or "really wanted" to be the other sex.

The procedure just described is one method of studying gender constancy. It is clearly modeled closely after a Piagetian conservation task: In both cases a misleading perceptual transformation is produced, and in both cases the child must look beyond the immediate appearance to recognize the underlying invariance. The other main methodology that has been employed is more purely verbal. Box 13.6 presents an example: a subset of the questions from an interview protocol developed by Slaby and Frey (1975) that has since been used frequently in the study of gender constancy. With the Slaby and Frey procedure there is no misleading appearance to overcome; rather, the modifications remain hypothetical, conveyed through a series of questions that probe understanding of constancy across various times and situations.

Box 13.6. Examples of Questions From the Slaby and Frey Gender Constancy Test

When you were a little baby, were you a little girl or a little boy?
Were you ever a little {opposite sex of first response}?
When you grow up, will you be a mommy or a daddy?
Could you ever be a {opposite sex of first response}?
If you wore {opposite sex of child, i.e. "boys'" or "girls'"} clothes, would you be a girl or a boy?
If you wore {opposite sex of child} clothes, would you be a {opposite sex of first response}?
If you played {opposite sex of child} games, would you be a girl or a boy?
If you played {opposite sex of child} games, would you be a {opposite sex of first response}?
Could you be a {opposite sex of child} if you wanted to be?

SOURCE: Adapted from "Development of Gender Constancy and Selective Attention to Same-Sex Models," by R. G. Slaby and K. D. Frey, 1975, *Child Development, 46*, p. 851.

Several findings from such measures can be briefly noted (Martin, Ruble, & Szkrybalo, 2002). First, Kohlberg was correct that gender constancy is a developmental achievement: Young preschoolers typically fail tasks of gender constancy, and success on such measures emerges some time between 4 and 7. Second, as this wide age range suggests, the method of assessing constancy can make a difference. As we would expect, children typically do better with a purely verbal procedure, such as the Slaby and Frey task, than they do when confronted by a misleading, gender-inconsistent appearance, as in the Marcus and Overton task. Finally—and related to the second of the two claims in Kohlberg's theory—there are only modest and inconsistent relations between understanding of gender constancy and sex-typed behavior (Martin et al., 2002). It appears, then, that constancy contributes to gender-role development but that it is far from a total explanation.

I noted that Kohlberg's (1966) analysis of gender constancy initiated the cognitive approach to the study of sex typing. In recent years the cognitive perspective has been broadened well beyond an emphasis on gender constancy alone. Of particular contemporary interest is the concept of **gender schema** (Martin, 1993). A schema is a mental representation of some familiar class of experiences, and a gender schema is therefore the child's organized set of knowledge and beliefs concerning gender. As this definition suggests, gender schema is a broad notion that encompasses numerous specific forms of thought. It includes mastery of gender labels and their associated criteria, as well as beliefs—at first erroneous and eventually correct—about gender constancy. It includes as well the sort of knowledge of sex-role stereotypes that is tapped by the SERLI and similar instruments—what kinds of toys does each sex prefer, what sorts of personality attributes are typical of boys or of girls? In general, a gender schema includes beliefs about the self and about others, about one's own gender and about the other gender, and about children and adults. It is therefore a much more multifaceted concept than is gender constancy alone, and it has a much more extended developmental history.

It may occur to you that the concept of schema is similar to the notion of script discussed in chapter 12. Scripts are in fact a form of schema—that form that has to do with the structure of familiar events. We saw in chapter 12 that a basic finding with respect to scripts is that a script, once formed, can affect how new information is taken in, interpreted, and remembered. The same point applies to schemas in general, including gender schemas. It has been found, for example, that children show best memory for experiences that are consistent with their gender schemas, and that in some instances they may even distort inconsistent information to make it fit their expectations—recalling, for example, that a doctor was a man and a nurse a woman, when in fact the opposite was the case (Carter & Levy, 1988; Liben & Signorella, 1993). Gender schemas can also affect behavior. Children who are relatively quick to learn gender labels, for example, are more sex-typed in their play than are children who do not yet know the labels (Fagot, 1985). At a general level, therefore, there is support for the second claim in Kohlberg's (1966) cognitive-structuralist theory: How children think about gender affects how they themselves behave, and developmental changes in such thinking contribute to developmental changes in behavior.

Child Rearing

The final section of the chapter moves from outcomes to determinants. Where do the child's social behaviors, both those that we have been considering and others, come from? As noted, I focus on one determinant that has been of considerable theoretical and empirical interest: parents' child-rearing practices. Child rearing, of course, is also a topic of considerable interest to any parent.

I begin with a caveat. Although parents' child-rearing practices may be important for their children's development, they are hardly all-important. Other socialization agents—peers, siblings, grandparents, teachers—may also play important roles. Biological factors contribute as well. Thus this section considers *one* of the explanations for why children turn out the way they do. In doing so, however, we can abstract some general themes that apply to the study of social development more broadly.

Our discussion begins, as does child-rearing research itself, with the most obvious challenge that confronts the researcher of parental practices: How can we figure out what parents do with their children? It is worth thinking for a moment about the difficulties involved in attempts to measure parental behavior. Child rearing is a mostly private enterprise, occurring mainly in the home with parent and child as not only the only participants but also the only witnesses. It is also a very extended enterprise, occurring through thousands of specific interchanges across all the years of childhood. How can researchers dip into this complexity and figure out what a parent typically does with her child?

The general answer to this question should by now sound familiar. We have, as always, three possible methods of study: naturalistic observation, laboratory measurement, and verbal report. I take the possibilities up in that order.

Naturalistic Observation

Our first example under this heading is drawn from perhaps the most influential program of research in the literature on child rearing, some studies carried out by Diana Baumrind some 40 years ago (Baumrind, 1967, 1971). You may recall that the Baumrind research was briefly discussed in chapter 6 in the context of Bronfenbrenner's ecological systems theory and possible variations in parental styles across different cultural contexts. The question we consider now is how the styles were identified in the first place.

The sample for the original Baumrind research was 4-year-old children and their parents. Each parent was interviewed about various child-rearing situations, and the interviews formed part of the basis for the identification of parental styles. The core of the measurement, however, came from direct observation of parent–child interactions. Trained observers visited the homes of the participants on two occasions for several hours at a time, during which they made extensive observations of the interactions between the parents and the child. The visits extended from about an hour before dinner until the child's bedtime, a time period that, in the researchers' words, is "commonly known to produce instances of parent–child divergence and was selected for observation in order to elicit a wide range of critical interactions under maximum stress" (Baumrind & Black, 1967, p. 304). Instances of divergence and stress were in fact common and permitted a special focus on what the researchers labeled "control sequences"—that is, instances in which one family member was attempting to alter or control the behavior of another.

Several hours of detailed observation of family interactions yield a wealth of information about both parent and child behavior. Collecting such data is part of the measurement process, but it is only part. The next step is to make sense of all this information—to abstract general principles and processes from the hundreds of specific interchanges between parent and child. Drawing partly from theory and partly from past research, Baumrind identified 15 "clusters" of parental behavior—that is, sets of interrelated behaviors that appeared to capture important dimensions of child rearing and important differences among parents. Examples of the clusters include Directive versus Nondirective, Firm versus Lax Enforcement Policy, Encourages versus Discourages Independence, and Expresses Punitive versus Nurturant Behavior.

The final step was to determine whether the clusters—which themselves were composites of numerous specific behaviors—could be organized into larger, more general categories. Baumrind's conclusion—based partly on theoretical considerations and partly on various statistical analyses—was that they could be, and the result was the well-known typology of parental styles. In the original work three styles were identified. The *authoritative style* is characterized by firm control in the context of a generally warm and supportive relationship, with an emphasis on reasoning and discussion rather than rigid imposition of parental power. The *authoritarian style* is also high on the dimension of control; it is low in warmth, however, and the strong control is exerted primarily though the greater power of the parent. Finally, the *permissive style* is high in warmth but also, as the name indicates, low in control. (A fourth style, labeled *uninvolved*, was added later; it is low in both warmth and control.)

For a second example of naturalistic study we return to one of the topics discussed in chapter 11. We saw there that infants vary in their security of attachment and that one line of research attempts to identify possible child-rearing contributors to these variations. A general conclusion from such research is that sensitive and responsive forms of caregiving are associated with the development of secure attachment (De Wolff & van IJzendoorn, 1997). The question we consider now is how researchers measure sensitivity of caregiving.

The most common way, dating back to the original research by Mary Ainsworth and colleagues (Ainsworth et al., 1978), is through observations of mother-infant interactions in the home. A study by Isabella (1993) provides a typical example. The mothers and babies in this study were observed a total of nine times: three 30-minute visits, spread across a 2-week period, when the infants were 1, 4, and 9 months old. The instructions to the mother emphasized that the researchers "were interested in learning about her baby's typical daily experiences and thus wished her to go about her routines as if the observers were not present" (Isabella, 1993, p. 608). A running narrative record was kept of the mother-infant interactions, tallying both each partner's behavior and sequences and contingencies between the two. Special attention was paid to behaviors thought to be relevant to attachment, such as signaling behaviors from the infant and vocalizations and stimulation by the mother. These narrative records subsequently formed the basis for classifying the caregiving along 10 rating-scale dimensions; among the dimensions were Sensitivity, Cooperation, Positive Affect, and Appropriateness of Response. The general conclusion—a typical one in this sort of research—was that relatively sensitive and appropriate caregiving heightened the probability of secure attachment.

The strengths and weaknesses of naturalistic observation have been discussed at various points and thus need be only briefly reiterated here. Once again, the great strength of the approach is easy to state: It is the only approach that directly measures what we are interested in, namely the natural occurrence of behavior in the natural setting. The limitations all reflect various obstacles to achieving such naturalistic measurement. Only some families may be willing to let researchers into their homes, and if so the resulting samples may be biased in various ways. Even if the samples are not biased, the sampling of their behavior may be; parents may alter their behavior when they know that a researcher is recording what they do. The sampling of behavior remains in any case a limited one—only a minute proportion of what the parent does with the child, often limited to one particular setting or time period. Because all homes are somewhat different, issues of comparability may arise; different parents may confront different challenges during the periods of observation. Finally, accurately recording and interpreting ongoing behavior is always a challenge, and such difficulties may be especially marked in the uncontrolled setting of the home.

Laboratory Study

The points just made can serve as a transition to the second general approach. Laboratory studies share some of the limitations of naturalistic study. Once again, the range of situations and behaviors that can be sampled is limited, and once again the awareness of being studied may alter how the parent behaves. Lab studies, however, also have some strengths relative to in-the-home efforts. Comparability across parents should not be an issue, since we can set up the identical situation for all parents. Such experimental control makes the laboratory approach a very efficient way to elicit the behaviors of interest; it also allows for systematic manipulation of potentially important factors. Finally, measurement is generally easiest in the controlled setting of the lab; participants can more easily be kept within range, and various technological devices can aid the human eye.

For a first instance of a lab study we can revisit an example from earlier in the book, one that makes the point about measurement especially clearly. One of the examples used in the Observational Methods section of chapter 4 was the Als et al. (1979) study of face-to-face interactions between mothers and their young infants (Table 4.4). As the table indicates, the goal of the research was to record very precise details of facial and bodily movement and vocal expression, with an interest in synchrony and reciprocity between the mother's behavior and that of the infant. Studies of this sort have demonstrated that most mother-infant dyads do establish synchronous and mutually satisfying patterns of interaction from early in infancy, although there are also important individual differences in their success at doing so (Kaye, 1982). Clearly, it would be difficult to obtain measurements of the desired precision in a home setting, let alone to make clear comparisons across mother-infant pairs. Laboratory study is simply a more sensible option for a topic like this.

The lab approach is not limited to infancy or to such micro aspects of development. We saw in the previous section that general parental methods of controlling and disciplining their children can be addressed through naturalistic study, and we will see in the next section that they can also be addressed through verbal report. Laboratory study provides a third general measurement option.

One example comes in the research by Kochanska et al. (2002) discussed under the heading of Moral Emotion. Mothers as well as their 2- and 3-year-old children served as participants for some aspects of this research. The mother's behavior was coded on multiple occasions in two general disciplinary situations: attempts to restrain the child from touching toys or objects that had been designated as off-limits, and attempts to induce the child to put away the toys during clean-up time. Among the forms of control that were coded—and among the forms that in fact proved frequent—were instances of so-called *power assertive* discipline (e.g., threats, spanking, physical control of the child's behavior). Power assertion turned out to be negatively related to guilt—that is, children whose mothers were high in the use of power assertion tended to express less guilt following the contrived transgressions. The results thus showed a nice convergence with those from the Baumrind research, in which high use of power assertion also proved to be a relatively ineffective technique of discipline.

Verbal Report

We have now reached the most frequently employed approach to the measurement of parental behavior. Both observational and laboratory studies do contribute to this literature, as the examples that we have just considered illustrate. In the majority of child-rearing studies, however, researchers learn what parents do by asking what parents do. Because there are at least two parties to any socialization effort, there are two possible classes of informants for such reports: We can ask parents how they treat their children, or we can ask the children themselves. The former, as you might guess, is a good deal more common than the latter. Still, by late childhood or adolescence children become possible sources with respect to what their parents do, and we will consider an example shortly.

In addition to variations in informant, verbal reports can differ along a number of dimensions. They may be elicited through a face-to-face interview or via a questionnaire that the informant completes. They may focus narrowly on one aspect of socialization or attempt to capture parental practices more broadly. They may ask only about current practices, or about practices when the child was younger, or about both. They may employ a closed-choice format in which the respondent chooses among alternatives or elicit open-ended responses that must then subsequently be categorized. And, of course, they may vary in the particular child-rearing situations and parental practices that are the focus of study. Indeed, there have been major changes across the century or so of child-rearing research in how parental practices are conceptualized and which dimensions are emphasized in research (Collins, Maccoby, Steinberg, Hetherington, & Bornstein, 2000; Holden, 1997).

Let us consider a couple of examples. Grusec and Kuczynski (1980) were interested in the consistency of mothers' disciplinary practices across different instances of child disobedience. Is it the case—as many models of socialization have assumed—that mothers have preferred styles of discipline that they apply quite broadly, or are mothers more likely to tailor their response to the specifics of the situation? To find out, Grusec and Kuczynski presented their participants (mothers of 4- to 5- and 7- to 8-year-old children) with 12 common instances of child disobedience—that is, simple scenarios, conveyed via a tape recorder, in which a mother stated a prohibition and a child disobeyed. The scenarios varied in the nature of the prohibition and subsequent disobedience, as well as in the consequences of the child's misbehavior. In some cases, for example, there were no immediate consequences, whereas in others harm ensued to either people or property. An example of a scenario follows.

(Sound of ball bouncing)

Mother: Kim, don't bounce that ball in here. You're too close to that table. That's an expensive vase on there that belonged to your grandmother.

(Sound of ball bouncing)

Mother: Kim, take that ball somewhere else. You're going to knock over that vase.

(Sound of ball bouncing, then loud crash)

The question following each scenario was how the mother would respond if the child were her own. The format was open-ended—that is, rather than choose among alternatives provided by the experimenter, mothers described in their own words what they would do and say. These responses were later coded for the occurrence of different techniques of discipline. Among the categories coded were forms of reasoning with the child—for example, pointing out that the child's behavior would make the mother sad. Also coded were various forms of power-assertive discipline—for example, physically forcing the child to comply with the request, or physically punishing the child following a failure to comply. One finding from the research was that power assertion was a frequently reported form of control. Another finding was that mothers employed a number of different techniques both within and across scenarios, and they were not necessarily consistent from one setting to the next. Thus the specific situation did indeed prove to be an important determinant of how the mother responded.

The second example of a verbal-report approach is drawn from a program of studies by Dornbusch, Lamborn, Steinberg, and associates (Dornbusch et al., 1987; Lamborn et al., 1991; Steinberg, Elmen, & Mounts, 1989). The approach taken in this research contrasts with that in the Grusec and Kuczynski (1980) study in several respects. Rather than having mothers serve as informants, these investigators elicited information about their parents' child-rearing practices from children—adolescents ranging in age from 14 to 18. In contrast to the open-ended response format utilized by Grusec and Kuczynski, the adolescents made their responses by selecting points along various rating scales. Finally, in contrast to the individual interviews of the Grusec and Kuczynski study, data were collected via questionnaires completed in group testing sessions at school.

The starting point for the Dornbusch group's research was the Baumrind work on child rearing discussed earlier in the chapter. As we saw, Baumrind used a combination of naturalistic observation and parental report to identify several general child-rearing styles that varied along the dimensions of warmth, control, and parental involvement. One question underlying the Dornbusch group's research was whether the same styles identified by Baumrind would emerge in adolescents' reports of their parents' rearing practices. The questionnaire items were therefore designed to measure variations along the dimensions of warmth, control, and involvement. Box 13.7 presents some examples of items from one of the studies. A basic conclusion from the research was that the Baumrind typology was indeed applicable to parents of adolescents. A further conclusion concerned correlations between rearing style and various measures of competence and adjustment in the adolescents. In general, the authoritative style proved to be associated with relatively positive outcomes (e.g., good academic performance), whereas the authoritarian style was associated with relatively poor outcomes. These results provided general support for the conclusions that Baumrind had reached about the relative efficacy of different styles of parenting. (Recall, however, a point made in chapter 6: that findings varied to some extent across the different sociocultural groups included in the study.)

Box 13.7. Examples of Items From the Lamborn et al. Adolescent-Report-of-Parenting Scale

Parental Warmth/Involvement

What do you think is usually true or usually false about your father [stepfather, male guardian]? (Response categories are "usually true" and "usually false.")

I can count on him to help me out, if I have some kind of problem.

He keeps pushing me to do my best in whatever I do.

He helps me with my school work if there is something I don't understand.

When he wants me to do something, he explains why.

When you get a poor grade in school, how often do your parents or guardians encourage you to try harder? (Response categories are: "never," "sometimes," and "usually.")

When you get a good grade in school, how often do your parents or guardians praise you? (Response categories are "never," "sometimes," and "usually.")

How much do your parents really know who your friends are? (Response categories are "don't know," "know a little," and "know a lot.")

Parental Strictness/Supervision

In a typical week, what is the latest you can stay out on SCHOOL NIGHTS (Monday-Thursday)? (Response categories are "not allowed out," "before 8:00," "8:00 to 8:59," "9:00 to 9:59," "10:00 to 10:59," "11:00 or later," and "as late as I want.")

In a typical week, what is the latest you can stay out on FRIDAY OR SATURDAY NIGHT? (Response categories are "not allowed out," "before 9:00," "9:00 to 9:59," "10:00 to 10:59," "11:00 to 11:59," "12:00 to 12:59," "1:00 to 1:59," "after 2:00," and "as late as I want.")

My parents know exactly where I am most afternoons after school. (Response categories are "yes" and "no.")

SOURCE: Adapted from "Patterns of Competence and Adjustment Among Adolescents from Authoritative, Authoritarian, Indulgent, and Neglectful Families," by S. D. Lamborn, N. S. Mounts, L. Steinberg, and S. M. Dornbusch, 1991, *Child Development, 62*, pp. 1063–1064.

Let us turn now to an evaluation of the verbal-report approach. As always, the strengths and weaknesses of one approach are clearest when contrasted with the strengths and weaknesses of other approaches. We saw that laboratory studies have the virtues of experimental control and precision of measurement but at the possible cost of artificiality and lack of generalizability. Observational studies tell us what parents actually do in the natural environment, but with an important qualifier: They tell us what parents do when the parents know that they are being watched. In addition, observational measures are dependent on the ability of human observers to form accurate interpretations of complex streams of social behavior. And both laboratory measures and observational measures are limited in the range of situations and behaviors that they encompass.

It is this final variable of breadth of scope that constitutes perhaps the most obvious strength of the verbal-report approach. With verbal reports we are able, within a relatively short time, to gather evidence regarding a wide range of socialization practices, a far wider range—in terms of time periods, environmental settings, and specific behaviors—than could ever be captured in naturalistic observation or laboratory study. This evidence, moreover, concerns the parent's naturally occurring behavior in the natural setting, behavior unaffected by the presence of observers or by the artifices of the lab. Although the direct evidence consists of verbal report rather than behavior, such reports are provided by the people with by far the most knowledge—and in some cases unique knowledge—about parental behavior: either parents themselves or the child targets of the parent's socialization efforts.

The strengths of the verbal-report approach must be set against one obvious and all-important question: Are such reports accurate? The general answer to this question is "sometimes yes and sometimes no." Various kinds of evidence tell us that verbal reports of socialization are not always accurate. Correlations between parents' reports of their socialization practices and direct observations of parental behavior are often modest at best (e.g., Yarrow, Campbell, & Burton, 1968). Studies that solicit information about the same socialization agent from different informants (e.g., ask mother, father, and child about the mother's behavior) also typically report modest correlations (e.g., Gonzalez, Cauce, & Mason, 1996).

Reports may be inaccurate for a number of reasons. In some cases parents may distort their answers, either consciously or unconsciously, to make themselves look better. Such "prideful subject" behavior (see chapter 5) is always a special concern with verbal-report measures. Parents or children may misinterpret questions or use terms in their answers in ways different from the way that the researcher uses the terms. Different parents may have different referent systems or "anchor points." Two mothers, for example, may both describe themselves as "strict" with regard to disobedience; for one mother, however, "strict" may mean occasional admonishments, whereas for the other it may mean invariable physical punishment. Parents may simply forget what it is that they do or used to do with their child. Memory problems are especially likely when the measures are retrospective—that is, concern socialization practices from some earlier period in the child's life. Research has shown that verbal reports that extend back over several years are of very doubtful accuracy (Yarrow, Campbell, & Burton, 1970).

Methods of increasing the accuracy of verbal-report data follow from this list of sources of bias. Accurate reports are more likely if the questions concern contemporary socialization practices than if they concern events from the past. Accuracy is also more likely if the questions are directed to concrete situations and specific behaviors (as in the Grusec and Kuszynski, 1980, study), thus minimizing the need for the respondent to decipher exactly what it is that the researcher is asking. Finally, the inevitable tendency toward positive self-presentation can at least be reduced by couching the interview or questionnaire in as nonevaluative a framework as possible ("no right or wrong answers," "just want to know what parents do," etc.). Given the gaps in our knowledge of child rearing, such wording is in fact reasonably accurate.

Determining Causality

Getting accurate measurements of parental behavior is one major issue in the study of child rearing. The other major issue is the determination of causality. Most child-rearing research has the goal not simply of documenting what parents do but of establishing the effects of different parental practices on important outcomes in the child. This was the case, for example, in all of the programs of research discussed in the preceding sections—the original Baumrind studies, the follow-up of the work by Dornbusch and colleagues, the examinations of maternal sensitivity and infant attachment, and so forth. In each case the measurement of variations among parents was just the first step in the research process. The second step was to relate the variations to the child outcomes of interest.

All of the studies discussed did in fact find relations between parental practices and child outcomes. Why, then, do they not prove that parents' behavior is a causal contributor to how their children develop? They do not prove this because child-rearing studies are inherently *correlational.* What they demonstrate is that some child-rearing practice covaries with some developmental outcome in the child (e.g., relatively frequent use of reasoning with relatively good behavioral morality, relatively sensitive caregiving with security of attachment). But because they lack experimental control, they cannot tell us why the relation comes about. This, of course, was a basic point in the discussion of correlational research in chapter 3.

In child-rearing research there are always a number of possible explanations for a parent-child correlation. One is that the parent's behavior causes the child's behavior—that reasoning promotes moral behavior, that sensitivity leads to secure attachment, and so forth. This, of course, is generally the explanation of greatest interest to the researcher. A second possibility is that the child's behavior causes the parent's behavior. Ever since Bell's (1968) influential paper, the idea that children influence their parents has assumed increasing prominence in socialization research. Perhaps, for example, well-behaved children are easy to reason with, and that is why the correlation between reasoning and good behavior comes about. A third possibility is that causal effects flow in both directions; perhaps over time the parent's behavior influences the child and the child's behavior influences the parent in a complex, reciprocal fashion. A final possibility is that there is no causal relation at all between the particular parental behavior and particular child outcome that we have decided to focus on. The two may both result from some third factor or set of factors, and the relation between the two may therefore be merely statistical and not causal. I will add that a third factor that has received much attention in recent discussions—and that clearly can be important—is shared genes (Rowe, 1994). Parents contribute only some of the child's environment but they contribute *all* of the child's genes, and many parent-child correlations can be interpreted as genetic rather than environmental in origin.

How might a causal role for child rearing be more certainly established? Chapter 3 discussed the issue of correlations and causality at a general level, and an article by Collins et al. (2000) is a good source with respect to child rearing in particular. Table 13.6 summarizes possible sources of evidence (all of which, I might add, *are* found in the child-rearing literature, although not always as strongly or consistently as some advocates of the importance of parenting have claimed). As can be seen, the forms of evidence group into several categories. Some seek to rule out an important third-factor alternative, which is what the adoption-study design does with respect to shared genes. Some track relations between parent and child over time, either on a short-term basis (sequential analysis) or across longer time periods (longitudinal study). Finally, some move beyond correlational analysis to experimental manipulation of aspects of parenting, either controlled-rearing studies with other species, or specific manipulations of hypothesized processes of parenting (analog studies), or broader and more applied interventions with groups of parents (parent training).

Summary

This chapter is divided into two major sections. The first section discusses methods of studying important outcomes in the child's social development. The second section considers one of the determinants of such outcomes: parents' child-rearing practices.

One important set of outcomes consists of the various behaviors, emotions, and cognitions that are grouped under the heading of "moral development." The *behavioral aspect of morality*

Table 13.6 Possible Forms of Evidence for a Causal Role for Parental Factors in Child-Rearing Research

Type of evidence	*Rationale*	*Examples*
Parent-child correlations in adoptive families	The adoption design removes the genetic basis for parent-child relations, making the child-rearing interpretation more plausible.	Bohman (1996), Ge et al. (1996)
Short-term across-time contingencies between parental behavior and child behavior in situations of dyadic interaction (sequential analysis)	Charting the across-time relations permits inferences about the causal direction between parent behavior and child behavior.	Patterson & Cobb (1971)
Long-term across-time relations between parent behaviors and child characteristics assessed during at least two points in development (longitudinal study)	As in the short-term case, charting the across-time relations between parent measures and child measures permits inferences about causal direction.	Steinberg et al. (1994)
Experimental manipulation of rearing conditions in nonhuman species	The experimental manipulation, which is more feasible with nonhuman species, allows clear conclusions about both the nature and direction of causality.	Anisman et al. (1998)
Interventions with human parents	As with animal research, the experimental control (which is necessarily more limited and unidirectional than in animal studies) permits conclusions about causality.	Cowan & Cowan (2002) Forgatch & DeGarmo (1999)
Experimental manipulation of aspects of socialization (e.g., reasoning, punishment) controlled laboratory situation (analog studies)	The experimental control permits the conclusion that practices employed by parents *can* causally affect child behavior.	Bandura et al. (1963) Parke (1981)

includes both the production of positive, prosocial behaviors, such as sharing, and the avoidance of negative, prohibited behaviors, such as cheating. Moral behaviors can be studied in three general ways. With naturalistic observation the attempt is to measure the natural occurrence of behaviors in the natural setting. A laboratory study, in contrast, sets up a structured situation from which the behaviors of interest can be experimentally elicited. The third general approach is to collect ratings of moral behaviors from someone who knows the child well, such as a parent or teacher. Examples are discussed for each of these approaches, followed by a consideration of the strengths and weaknesses of the different methods of study.

Studies of the *emotional aspect of morality* have examined both the negative emotion of guilt and the more positive, prosocial emotion of empathy. A common procedure for studying guilt has been to infer the child's emotions from the endings that the child provides for stories involving transgression, the assumption being that the child "projects" his or her own feelings onto the story character. Verbal measures have also been common in the assessment of empathy, and the text describes one often used example. Also described are more behavioral measures: attempts to infer guilt from the child's responses following a contrived transgression, and attempts to infer empathy from the child's facial and physiological reactions to the plight of another.

Research on the *cognitive aspect of morality* began with the pioneering work of Piaget. Like his research on cognitive development, Piaget's studies of morality stress the flexible clinical method of testing and the identification of qualitatively distinct stages of reasoning as the child matures. In addition to stimulating follow-up study, Piaget's work is important as the forerunner of the major contemporary approach to moral reasoning, that of Kohlberg. Like Piaget, Kohlberg derives stages from children's responses to moral dilemmas; the dilemmas and corresponding stages, however, are a good deal more complex than those found in Piaget. The complexity of the scoring system has in fact always been one issue in evaluating Kohlberg's work. Another issue, central to the theory, concerns the claim that moral reasoning can be characterized by a series of stages. Various methods of testing this claim are discussed. This section concludes with a brief overview of recent work on young children's prosocial moral reasoning.

The second general topic addressed is the development of a *self-concept.* This topic divides, in turn, into two subtopics: a cognitive aspect, or how children think about themselves, and an evaluative aspect, or how children feel about themselves. Research on the cognitive aspect of the self begins very early in development, with a focus on the emergence of personal agency during infancy and of visual self-recognition during toddlerhood. Work on the evaluative component of the self falls under the heading of *self-esteem.* Several self-report instruments for the measurement of self-esteem in childhood are described. Some such instruments are unidimensional, or directed to a single sense of global self-worth; others are multidimensional, or concerned with identifying different dimensions of self-esteem (e.g., academic, physical). Research indicates that even preschool children can make some distinctions in their self-evaluations across different aspects of their lives.

The final major set of outcomes that the chapter considers falls under the heading of *sex typing and sex differences.* Findings regarding sex differences are often incidental to the main purpose of a study, a fact that complicates the interpretation of published sex differences and raises the possibility of biases in what information does get published. The form of research synthesis known as meta-analysis has proved valuable in interpreting the sex differences literature. Biases may also occur at the point of data collection, perhaps especially as a result of the different expectancies that adults hold for boys and girls. The child's knowledge of the adult's sex may also have an effect.

The discussion of general points is followed by a consideration of instruments explicitly designed for the study of sex differences and sex typing. Traditional tests of sex typing, such as the Sex Role Learning Index, attempt to measure the child's adoption of stereotypically masculine or feminine behaviors and preferences. A criticism of such tests is that they treat "masculine" and "feminine" as polar opposites, a limitation that is avoided in a more contemporary measure labeled the COAT. Finally, attention has also been directed to the cognitive side of sex typing and the child's understanding of gender and gender differences. Of interest here are the child's ability to label the sexes appropriately, the criteria that are used in applying labels, the realization that gender is a permanent attribute, and the formation of gender schemas.

The final section of the chapter switches from outcomes to determinants of social development, with a focus on one much studied determinant: parents' child-rearing practices. Like other targets of measurement, child rearing can be studied in three general ways. Two examples of naturalistic observation are described: the influential Baumrind research on parental styles, and measurement of sensitivity of caregiving as a possible contributor to attachment. Laboratory study and verbal reports are the other two measurement options, and examples in these categories are provided as well.

Measuring parental behavior is one of the challenges in child-rearing research; establishing cause-and-effect relations between parental practices and child outcomes is the second major challenge. This challenge arises because child-rearing studies are inherently correlational and thus in the typical case cannot distinguish among three possible explanations for any relations that are found: The parents' practices affect the child, characteristics of the child affect the parent, or some third factor (such as shared genes) accounts for the relations that are found. Various procedures for disentangling causality in child-rearing studies are discussed, included across-time study and experimental manipulation of parental behaviors.

Exercises

1. The text briefly describes a study by Walker et al. in which response to the hypothetical moral dilemmas of Kohlberg was compared with response to real-life moral dilemmas generated by the participants themselves. The wording used by Walker and associates to elicit the real-life dilemmas was as follows:

> Have you ever been in a situation of moral conflict where you had to make a decision about what was right but you weren't sure what to do? Could you describe the situation? What was the conflict for you in that situation? In thinking about what to do, what did you consider? What did you do? Do you think it was the right thing to do? How do you know?

Offer your own answer to the Walker et al. question, and also elicit answers from several of your peers. How do the moral dilemmas that have been important in your lives compare with those studied by Kohlberg?

2. Consider the items from the Lamborn et al. Report-of-Parenting Scale reproduced on page 306. Think about how you might have answered these questions as an adolescent. Think about how your parents might respond if they were asked to provide the same sorts of information—or, if possible, *ask* your parents and compare your perceptions. Do you believe that questions such as these capture important aspects of the parent-child relationship?

3. In the mid-1970s Rheingold and Cook reported a study of sex typing based upon a content analysis of young children's rooms. They were interested in seeing whether girls' rooms, on the average, would contain different toys, decorations, and furnishings than

would boys' rooms—as did in fact prove to be the case. Read the Rheingold and Cook study (the source is *Child Development,* 1975, *46,* 459–463) and then carry out a contemporary replication. Visit as many young children's homes as you are able to gain access to, record the same kinds of information as did Rheingold and Cook, and compare your results with theirs.

4. Discussions of possible effects of the mass media on children's development tend to emphasize TV. Children's books, however, have long been another source of messages and models that may influence how children think and behave. Go to the children's section of your local library and collect a large and random selection of books intended for young children. Perform a content analysis of these books with respect to some topic that interests you, noting both how often the topic arises and how it is treated. The following are some suggested topics:

(a) similarities or differences between the sexes

(b) acts of violence or aggression

(c) depictions of different ethnic groups

(d) depictions of the elderly

Notes

1. The "his" is not sexist in this case. In Switzerland in the 1920s only boys played marbles; hence the participants for this part of Piaget's research were only boys.

2. In Kohlberg's later writings (e.g., Colby, Kohlberg, Gibbs, & Lieberman, 1983) Stages 5 and 6 were combined, resulting in a five-stage rather than a six-stage model.

3. I noted that self-recognition has also been studied in nonhuman primates. These studies have yielded a consistent and interesting pattern of results. Chimpanzees and orangutans are capable of self-recognition; other primates (e.g., all species of monkeys) are not—even if given hundreds of hours of experience with mirrors (Gallup, 1994).

4. In recent years a number of authors have argued that the terms "sex" and "gender" should not be employed interchangeably, but rather should be reserved for different aspects of the phenomena and processes that historically have been grouped under the headings of "sex differences" and "sex-role development." As yet, however, there is no consensus with respect to the desirability of maintaining such a terminological distinction, nor is there agreement about exactly what the "sex"-"gender" contrast should be. This chapter follows the practice of many psychologists and uses the terms as synonyms.

14

Aging

Until recently, the term "developmental psychology" was often merely a synonym for "child psychology." Most psychologists who labeled themselves as "developmental" were really researchers of childhood, and most texts in "developmental" were really texts about children. This state of affairs no longer holds true. Life-span treatments of development, both in general texts (e.g., Cavanaugh & Blanchard-Fields, 2006) and in more advanced treatises (e.g., Baltes, Staudinger, & Lindenberger, 1999), have proliferated. Within this general broadening of scope there has been special interest in the last part of the life span and the changes and challenges of old age. It is to the topic of the elderly years and the psychology of aging that the present chapter is devoted.

A number of reasons can be advanced for an interest in aging. One very simple reason is the increase in both the number and the proportion of elderly people. In 1900 approximately 4% of the population of the United States was aged 65 or older. Today approximately 13% of the population falls into this category, and it is estimated that by the year 2030 the figure will have risen to about 20% (U.S. Bureau of the Census, 2004). Both increases in life expectancy and fluctuations in birth rate have contributed to this trend. The greatest proportional increase, by the way, has come in the age group 80 and older.

A second reason for an interest in aging is similar to one of the reasons for an interest in childhood. Childhood is in many respects a period of vulnerability, a time during which the growing organism is at risk for various problems. One reason to study children is to learn how to prevent these problems. Old age has also been considered a period of vulnerability and risk, a time when abilities decline and losses of various sorts (of health, job, spouse) occur. As we will see, a central question in the study of aging is how accurate this stereotypically negative picture is. Nevertheless, the problems are clearly real for at least some elderly people, and the need to minimize their impact provides a further reason to study aging.

A third reason is more theoretical. "Development" is presumably a lifelong process, and any model of development that stops with adolescence must therefore be incomplete. Within the long stretch from conception to death, it can be argued that special attention should be paid to those parts of the life span that are the locus for definite and major change. Childhood is clearly such a period—a time of frequent, inevitable, and monumental change. Many of the changes, moreover, are directional—a matter of upward movement, of adding on, of growth. Old age, it has been argued, may also be a period of major and at least somewhat inevitable change, but with an important difference: In this case the natural direction for at least some of the changes is negative rather than positive, a matter of losing rather than gaining. Again, the validity of such a negative characterization is very much a matter of dispute. The theoretical importance of the issue is clear, however, and hence the need for empirical study.

A final reason for an interest in aging is perhaps implied by the references in the preceding paragraphs to controversies and uncertainties. Such controversies and uncertainties arise, at least in part, because of the methodological challenges of studying development in old age and of comparing performance across different parts of the life span. Indeed, the psychology of aging illustrates with special clarity the challenges both of doing good research in general and of doing developmental research in particular. It thus constitutes an interesting and informative topic in scientific methodology, even for those who do not have any special professional or personal investment in its study.

It is, of course, the methodological aspects of research on aging that concern us here. The discussion follows the same general organization as chapter 11 on Infancy. The chapter begins with a consideration of general issues in the study of aging. As was true for infancy, these issues are for the most part not unique to the study of any one age group; as just suggested, however, many of them apply with special force to research on aging. The second part of the chapter then takes up some specific research topics in the psychology of aging.

The discussion of general issues will be clearest in the context of a specific example. The example that I use is the study of IQ. IQ is a good example, not only because IQ has been one of the most popular dependent variables in studies of aging but also because many of the methodological issues to be discussed have been most fully explored in the context of IQ. The first part of the chapter, therefore, serves both as an overview of issues and as coverage of one important substantive topic.

General Issues (as Illustrated by the Study of IQ)

Sampling

We begin with the same question that was asked with regard to infancy: How do researchers find samples of elderly people to study? The answer is the same as for infancy: in a variety of ways, few of which fit a textbook model of random sampling. Unlike the researcher of childhood or young adulthood the researcher of aging does not have an institutional setting such as a school or college to draw participants from. Identifying and then randomly sampling from some representative pool of elderly people is therefore difficult. Probably the most common practice has been to solicit participants from various groups or organizations to which elderly people belong, such as religious groups, senior citizens centers, or associations of retired people. Such an initial pool is obviously somewhat biased, because only relatively active and healthy older people are likely to belong to such groups. The sample may then become further biased because only volunteers can be included as participants, and again only relatively competent older people are likely to volunteer for research. The result is that studies of aging have tended to employ somewhat nonrepresentative, positively biased samples of elderly people (Hultsch, MacDonald, Hunter, Maitland, & Dixon, 2002).

Obtaining a representative sample of the elderly is one goal of research on aging. The second goal is to obtain a sample of older people that is comparable to the younger sample with whom the elderly are being compared. Many studies of aging are not concerned simply with documenting levels of performance in older samples; their aim, rather, is to compare performance at older ages with that at younger ages. Because age is the independent variable in such research, it is important that the groups being compared be as similar as possible in every respect except age. As explained in chapter 2, this stricture does not mean that researchers should attempt to rule out all the ways in which 20-year-olds, say, differ from 70-year-olds. But it does mean that they should attempt to rule out any differences that are not naturally associated with age.

Deviations from representativeness in sampling at any one age may bias the comparison of different age groups. The preceding discussion suggested that samples of elderly participants may often be an above-average subset of the elderly population. If no comparable selectivity is exercised in sampling younger age groups, then the age comparison will be biased in favor of the elderly. On the other hand, if the "young adult" group consists of college students (a common practice), or if the elderly participants are drawn from a nursing home, then the bias will go in the opposite direction. And even if the same selection procedures are used at each age, differences in volunteer rate may still lead to noncomparable samples.

Two kinds of confound have been of particular concern in comparisons of young adults and elderly adults. One confound stems from differences in educational level. Imagine a study whose purpose is to compare mean levels of IQ at age 25 and at age 75. We will assume that the researcher solves the sampling problems discussed earlier and is able to obtain representative samples at each age. If the samples are truly representative of their age groups, then they are certain to differ not only in age but also in educational level: On the average, the young adults will be better educated than the older adults. We will have, then, a confounding of age and education. Note the paradox here: Achieving one of the goals in sampling, namely representative samples at each age, ensures some degree of failure with regard to the other goal, comparable samples at the different ages.

The general issue of age by cohort confounds in cross-sectional research was discussed in chapter 3. As noted then, the confounding is a problem only if the variable that is confounded with age bears some plausible relation to the dependent variable being examined. Is it plausible that educational level could relate to IQ? The answer is clearly yes. Indeed, not only is such a relation plausible, but also it has been clearly demonstrated in a number of studies (Gonda, 1980). The standardization data for the WAIS (a commonly used measure of adult IQ), for example, showed a correlation of .68 between educational level and IQ (Wechsler, 1958). A subsequent analysis (Birren & Morrison, 1961) revealed that the relation between IQ and education was substantially stronger than the relation between IQ and age.[1]

What is to be done about this confounding? As suggested in chapter 3, there is no really good solution. It is possible to match mean education level for different age groups through careful selection of participants—for example, by tilting the selection toward unusually well-educated older adults and unusually poorly educated younger adults (e.g., Green, 1969). Such matching typically reduces but does not completely eliminate age differences in IQ. The obvious problem with this procedure is that it creates biased samples that are not representative of the general population. It is also possible to adjust statistically for the differences in education, in particular through a technique called analysis of covariance. Like other forms of statistical adjustment, however, analysis of covariance has a number of limitations, and its application to the age-education confound has been criticized (Storandt & Hudson, 1975). In addition, matching or statistical adjustment can at best equate participants for *quantity* of education, as defined by years of schooling completed. Such procedures do not control either for *recency* of education or for possible changes in the *quality* of education over time.

The issue of education differences is part of the larger issue of cohort effects in cross-sectional designs, a topic that will be returned to shortly when I discuss questions of design. For now, one final point can be made. Whether the age-education confound is worrisome depends on exactly what we want to conclude from research. If our interest is limited to differences between young adults and elderly adults, then there is no problem; it is simply a fact, consistently verified by a large number of studies, that there is an average age difference in IQ. The problem comes when we attempt to move beyond differences to talk about *changes* with age. Cross-sectional studies cannot provide direct evidence that IQ declines with age. Plausible alternative explanations for the superiority of young adults exist, including the possibility of cohort differences in educational opportunity. Indeed, the research just discussed demonstrates that the educational explanation is not merely plausible; differences in education *do* account for at least part of the age difference in IQ.

The second major kind of confound in cross-sectional studies of aging concerns differences in health status. Again, the confound follows naturally from the goal of representative sampling. As people get older, their health, on the average, gets worse. If we achieve representative sampling, therefore, we ensure ourselves of another confound: Our groups will differ not only in age but also in health.[2]

Does health status relate to IQ? A variety of kinds of evidence tell us that it does. As with education, it is possible to produce a rough matching of groups for health status through careful selection of participants. What has been done in particular is to include only exceptionally healthy individuals in samples at the oldest ages. This procedure typically reduces (but again does not eliminate) age differences in IQ. It is also possible to examine IQ as a function of variations in health status within samples of the elderly. The typical finding here is of a clear relation: Relatively healthy older individuals outperform less healthy older individuals. A number of aspects of health can affect performance on IQ tests. Among the factors that at least sometimes (although not necessarily always) show effects are declines in visual and auditory acuity (Baltes & Lindenberger, 1997), hypertension and cardiovascular disease (Schaie, 2005), cerebrovascular disease (Spieth, 1965), anxiety and depression (van Hooren et al., 2005), and overall health status (Rosnick, Small, Graves, & Mortimer, 2004).

In several respects health status is a more complicated issue than is education. First, health status poses a measurement problem that is not present in the case of educational level. How can we know how healthy an elderly participant is? Ideally, what we would like is a physician's assessment based on a broad battery of medical tests. Although some studies do use physicians' reports (e.g., Schaie, 2005), most studies fall well short of this ideal. The most common practice is to base the assessment of health on the participants' own reports of their health. Such self-report measures range from quite simple (e.g., "all participants reported themselves in good health") to considerably more detailed and ambitious. Table 14.1 shows one of the more often used of the self-report measures, the Short-Form Health Survey. The version in the table is the short form of the instrument (SF-12); there is also a longer, 36-item version (SF-36—Ware & Sherbourne, 1992).

As we have seen, self-report measures are always somewhat suspect. It is important to note, therefore, that at least some self-report measures of health status, including the instrument shown in Table 14.1, do correlate with physicians' ratings (Siegler, Bosworth, & Poon, 2003). In addition, such measures may provide information that goes beyond the medical diagnoses that are the physician's province. If our interest is in how health status impacts daily functioning and quality of life, then a self-report measure such as that in Table 14.1 may be more informative than a doctor's report.

Health status is a more complicated variable than education from a conceptual as well as a methodological point of view. It is easy to

Table 14.1. Items From the Short-Form-12 Health Survey

1. In general, would you say your health is:

Excellent	Very good	Good	Fair	Poor
☐	☐	☐	☐	☐

The following item asks about activities you might do during a typical day. Does your health now limit you in these activities? If so, how much?

	Yes, Limited A Lot	Yes, Limited A Little	No, Not Limited At All
2. Moderate activities, such as moving a table, pushing a vacuum cleaner, bowling or playing golf	☐	☐	☐
3. Climbing several flights of stairs	☐	☐	☐

During the past 4 weeks, have you had any of the following problems with your work or other regular daily activities as a result of your physical health?

	YES	NO
4. Accomplished less than you would like.	☐	☐
5. Were limited in the kind of work or other activities	☐	☐

During the past 4 weeks, have you had any of the following problems with your work or daily activities as a result of any emotional problems (such as feeling depressed or anxious)

	YES	NO
6. Accomplished less than you would like	☐	☐
7. Didn't do work or other activities as carefully as usual	☐	☐

8. During the past 4 weeks, how much did pain interfere with your normal work (including both work outside the home and housework)?

Not at all	A little bit	Moderately	Quite a bit	Extremely
☐	☐	☐	☐	☐

These questions are about how you feel and how things have been with you during the past 4 weeks. For each question, please give one answer that comes closest to the way you have been feeling. How much of the time during the past 4 weeks?

	All of the Time	Most of the Time	A Good Bit of the Time	Some of the Time	A Little of the Time	None of the Time
9. Have you felt calm and peaceful?	☐	☐	☐	☐	☐	☐
10. Did you have a lot of energy?	☐	☐	☐	☐	☐	☐
11. Have you felt downhearted and blue?	☐	☐	☐	☐	☐	☐

12. During the past 4 weeks, how much of the time has your physical health or emotional problems interfered with your social activities (like visiting friends, relatives, etc.)?

All of the Time	Most of the Time	Some of the Time	A Little of the Time	None of the Time
☐	☐	☐	☐	☐

SOURCE: From "A 12-Item Short-Form Health Survey: Construction of Scales and Preliminary Tests of Reliability and Validity," by J. E. Ware, Jr., M. Kosinski, and S. Keller, 1996, *Medical Care, 34*, p. 227. Copyright © 1996 by the American Public Health Association. Reprinted with permission of Lippincott, Williams & Wilkins.

imagine—or for that matter to find—cultures or historical periods in which education does not vary with age. This separability of age and education tells us that differences in education are not an intrinsic part of aging—that is, that their covariation in our time and culture *is* a confounding. With health, however, the separability from age is much less clear. Should changes in health, and associated changes in mental performance, be regarded as distinct from aging itself, as "secondary" rather than "primary," as "disease" rather than "normal aging"? Or are declines in health an intrinsic part of the aging process?

How best to conceptualize the relation between health and aging is a complex and much-debated issue that can hardly be resolved here. I will note, however, that whatever the ultimate resolution of the issue, there is little warrant in dismissing health-related declines in IQ as irrelevant or "artifactual." Health *does* vary with age, and IQ *does* vary with health. And even if all declines in IQ turn out to be health-related (which has not been demonstrated so far), the declines would remain just as genuine and important.

A final difference between the education variable and the health variable concerns their role in different designs. Both apply to cross-sectional designs, as the examples just discussed illustrate. Differences in education are generally not a concern in longitudinal designs, as long as the earliest age tested is beyond the age at which formal schooling is completed. Differences in health, in contrast, *are* a concern in longitudinal designs, because health, unlike formal education, does change as people age. Thus health, unlike education, contributes to age differences in both cross-sectional and longitudinal designs.

Let us return for a moment to the general issue of sampling, in this case with regard to longitudinal designs. We saw in chapter 3 that samples for longitudinal studies tend to be somewhat select, given the demands that longitudinal research places on its participants. We saw also that samples for longitudinal studies tend to be limited to members of a single cohort. Both of these factors may affect the conclusions that are drawn about stability or change in IQ with increased age. In addition to possible biases in initial selection of participants, longitudinal research may suffer from selective dropout of participants in the course of the project. In longitudinal studies of IQ it is clear that selective dropout occurs: On the average, participants with relatively low IQs are most likely to be lost from the study. This dropout creates an obvious bias in favor of older age groups: The older we go, the more select the sample becomes.

A dramatic illustration of the phenomenon of selective dropout is provided by Siegler and Botwinick (1979). Siegler and Botwinick's participants were tested first when they were between 65 and 74 years old, and they were then retested up to 10 additional times, depending on availability, over the next 10 years. Some people dropped out after the first test, some after the first two tests, and so on, so that only a small number of participants (8 out of 130) were present for all 11 tests. Figure 14.1 shows the mean scores on the initial test plotted as a function of the number of test sessions for which the participant appeared. Note that scores are almost 30 points lower for people who dropped out after one test than for those who remained available throughout the project. The message is clear: The more test sessions that we attempt in a longitudinal study, the more positively biased our sample becomes.

Why does this pattern occur? There are two general reasons for selective dropout. One is voluntary, having to do with the unwillingness of some people to continue to be tested. On the average, participants with relatively low scores on an initial test are less willing to continue to participate than participants with relatively high scores.

The second basis for selective dropout is death. In longitudinal research with elderly

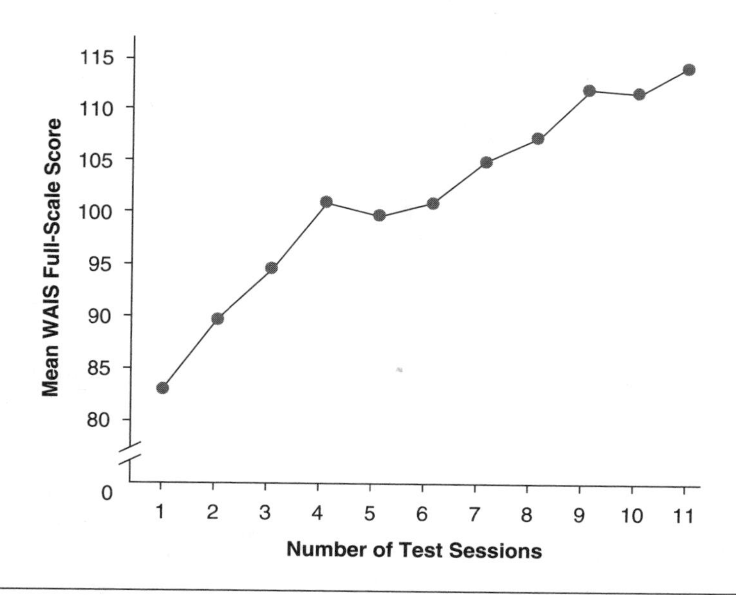

Figure 14.1 Mean IQ scores at initial assessment as a function of the number of longitudinal assessments for which the participant was available for testing

SOURCE: From *Experimental Psychology and Human Aging* (p. 84), by D. H. Kausler, 1982, New York: John Wiley & Sons. Copyright 1982 by John Wiley and Sons. Reprinted with permission. Kausler's figure is adapted from Siegler and Botwinick (1979).

participants, some participants die between one test session and the next. On the average, participants with relatively low test scores are less likely to survive than participants with relatively high test scores (e.g., Botwinick, West, & Storandt, 1978). Thus low IQ turns out to be predictive, in a very rough and imperfect way, of impending death.

A particularly interesting illustration of the relation between IQ and survival is provided by the phenomenon of **terminal drop.** Terminal drop refers to a fairly sharp and abrupt decline in mental abilities in the months or years immediately preceding death. It is associated, presumably, with a general deterioration of functioning that presages death. Terminal drop can be examined in any longitudinal study that includes at least three testing occasions. Imagine a longitudinal study that tests IQ at age 65, again at age 70, and again at age 75. Such a study will have three groups of participants: those who are available for all three tests, those who are available for the first two tests but die before the third, and those who are available for the first test but die before the second. It is the second group that is of interest now. On the average, IQ scores for these individuals show a marked decline from the first test to the second. This pattern is pictured in Figure 14.2. It is this deterioration in performance shortly before death that constitutes terminal drop. (Note that the phenomenon is not always as clear-cut as the figure suggests, and that its generality and importance remain controversial. See Berg, 1996, for a review and discussion.)

It may be helpful to summarize the points that this section has made regarding the relation between health and IQ. There *is* a relation between health and IQ. On the average, healthy people perform better on IQ tests than less healthy people. There is also a relation between health and age. On the average, health

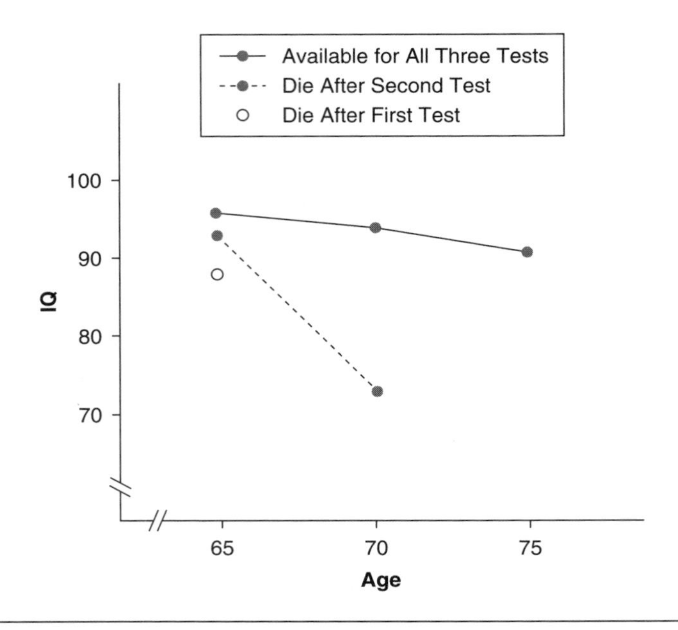

Figure 14.2 Declines in IQ reflective of terminal drop (hypothetical data)

problems become more common as people get older. The relation between health and age has implications for the samples that are examined in aging research. In both cross-sectional and longitudinal designs, samples at the oldest ages are tilted toward relatively healthy individuals, a factor that reduces age differences in IQ. Even with this sampling bias, however, there are still differences in health between samples of different ages, and these differences in health clearly account for at least part of the age difference in IQ.

Measurement

So far I have been discussing findings with regard to IQ and aging without addressing a basic question: How is IQ measured in such research? The present section is directed to this question, as well as some of the more general measurement issues in the study of aging.

As is true in the study of any age group, a variety of different IQ tests have been used in research on aging. Two tests, however, have been most common and most influential. One is the Wechsler Adult Intelligence Scale, or WAIS (Wechsler, 1997). Chapter 12 introduced the two Wechsler tests designed for childhood and presented a sampling of the types of items included on the tests (Table 12.1). Table 14.2 presents a comparable sampling for the adult test. The WAIS, like other Wechsler tests, is an individually administered test that yields an overall estimate of general intelligence. Because the test is divided into a Verbal scale and a Performance scale, it can also yield more differentiated information in the form of separate Verbal and Performance IQs. The Verbal scale and Performance scale are composed of six and five subtests, respectively; thus at the most specific level, age differences on the WAIS can be examined subtest by subtest.

The other commonly used measure in studies of IQ and aging is the Primary Mental Abilities Test, or PMA (Thurstone & Thurstone, 1962). The PMA, unlike the WAIS, can be administered to groups of people at the same time. Like the WAIS, it is divided into subtests. The five PMA subtests are Verbal Meaning, Space, Reasoning, Number, and Word Fluency.

Table 14.2 Simulated Items Similar to Those in the Wechsler Adult Intelligence Scale—Third Edition

Subtest	*Verbal Scale*
Information	What is steam made of? Who wrote Tom Sawyer?
Arithmetic	Three women divided 18 golf balls equally among themselves. How many golf balls did each person receive?
Comprehension	What is the advantage of keeping money in a bank? Why is copper often used in electrical wires?
Subtest	*Performance Scale*
Picture completion: Indicate the missing part.	

SOURCE: *Simulated Items similar to those in the Wechsler Adult Intelligence Scale – Third Edition.*

Examples of the types of items found on several of the subtests are shown in Table 14.3.

What are the general measurement issues that are raised by the study of IQ and aging? One basic issue is *measurement equivalence.* To compare IQ at different ages, whether cross-sectionally or longitudinally, we must have equivalent measures of IQ at the different ages. It is, of course, easy enough to achieve a literal equivalence of measures: All we need do is use the same IQ test at the different ages. The question is whether our test is actually measuring the same thing at different ages, whether we have achieved not only literal but also *functional* equivalence. If the test is less appropriate or valid for one age group than another, then age comparisons are of doubtful significance.

To a good extent, measures that are used in the study of elderly people—not only IQ tests but psychological measures in general—were first developed for the study of younger samples. IQ tests, as we saw, were originally devised to predict school performance in children, and their content was thus oriented to the kinds of academic-verbal skills that are important in school. IQ tests for adults have always had similar academic-pragmatic goals and content. The PMA was developed on samples of college students, and the Army Alpha test was devised to predict the success of young adults in the military. The WAIS, it is true, was standardized on samples that ranged in age from 16 to 74. Even here, however, the content often seems more appropriate for young adults, and the

Table 14.3 Types of Items Included on the Primary Mental Abilities Test (PMA)

Verbal Meaning					
Instructions: Find the word that means the same as the first word in the row.					
ANCIENT	A. dry	B. long	C. happy	D. old	E. sloppy
QUIET	A. blue	B. still	C. tense	D. watery	E. exact
SAFE	A. secure	B. loyal	C. passive	D. young	E. deft
Number					
How do you write in numbers: Eleven thousand and eleven? A. 111 B. 1,111 C. 11,011 D. 110,001 E. 111,011					
1/2 + 1/2 =	A. 1/8	B. 1/4	C. 1/2	D. 1	E. 2
16 x 99 =	A. 154	B. 1,584	C. 1,614	D. 15,084	E. 150,084
Reasoning					
Instructions: Find the letter that follows the last letter in the row.					
c d c d c d	1. c	2. d	3. e	4. f	5. g
a b c a b d a b e a b	1. b	2. c	3. d	4. e	5. f
a m b a n b a o b a p b a	1. m	2. o	3. p	4. q	5. r

SOURCE: From *SRA Primary Mental Abilities* (pp. 1, 6, 11) by L. L. Thurstone and T. G. Thurstone, 1962, Chicago: Science Research Associates. Copyright © 1962 by Science Research Associates, Inc.

validation data have always centered on prediction to academic or occupational contexts that are characteristic of young adulthood.

Note that the issue of measurement equivalence is not limited to the selection of the specific items that go into a test. The issue is a much broader one, for it extends in general to the kinds of settings and contexts within which we assess intellectual performance. To date, these settings have consisted primarily of structured lab environments and highly standardized tests. Such an approach, it can be argued, is biased against the elderly participant, who is many years away from academic contexts and who may therefore be less familiar with and less motivated to respond to such measures than a younger adult. Perhaps the elderly person's competence would appear more impressive if it could be examined in more natural and familiar settings, such as when carrying out a job, shopping for the week's groceries, or explaining the operation of a toy to a grandchild.[3]

Arguments such as those in the preceding paragraphs have led in recent years to a number of attempts to assess more "everyday" intelligence and problem solving in the elderly (Willis, 1996; Willis & Schaie, 1994). These attempts have taken a variety of forms; I briefly describe several examples here and then add some others later in the chapter in the section on Memory.

Table 14.4 presents some items from a measure known as the Observed Tasks of Daily Living, or OTDL (Diehl, Willis, & Schaie, 1995).

Table 14.4 Examples of Items From the Observed Tasks of Daily Living

Baking a Cake in a Microwave Oven

The tester presents a cake mix package with the instructions on the back panel of the package. For this task, the third step of the instructions is relevant: "3. Microwave uncovered on high 2 minutes; rotate pan 1/2 turn. Microwave 2 minutes longer."

The tester provides, along with the cake mix ingredients and the necessary kitchen utensils, the simulated front panel of a microwave oven. This microwave panel had already been used in some previous tasks.

The tester then presents to the participant a 4 × 6 in. (10.2 × 15.2 cm) index card with the following instruction: "Suppose the cake mix is in the baking pan and it is ready to be baked. Show me how you would follow step 3 of the instructions."

The tester records the participant's response in the following way:

Correct Steps	*Participant's Steps*
1. Put the pan in the microwave.	1. ______
2. Touch CLEAR.	2. ______
3. Touch TIME and key in 200.	3. ______
4. Touch START.	4. ______
5. After 2 minutes turn pan 1/2 turn.	5. ______
6. Touch TIME and key in 200.	6. ______
7. Touch START.	7. ______

Loading a Pill Reminder

The tester presents a fictitious person's (named Peggy Wright) medication chart and the medicine bottles for all six medications listed on the chart to the participant. On the chart, the following three medications are marked with an X:

1. Hygroton, 50 mg. Instructions: Take 1 tablet in the morning with food.
2. Capoten, 50 mg. Instructions: Take 1 tablet 3 times a day.
3. Lasix, 40 mg. Instructions: Take 1 tablet every other day from Monday through Friday in the morning.

The tester presents a rectangular pill reminder case with four small compartments (morning, noon, evening, and bedtime) for each day of the week (Sunday through Saturday).

The tester then presents a 4 × 6 in. (10.2 × 15.2 cm) index card with the following instructions: "Mrs. Wright uses this pill reminder so that she does not forget to take her pills. Please fill this timer with the 3 drugs marked on the medication chart."

(Continued)

Table 14.4 (Continued)

Correct Steps	*Participant's Steps*
1. Hygroton all days morning	1. ______
2. Capoten all days morning, noon, and evening	2. ______
3. Lasix Monday, Wednesday, and Friday in the morning	3. ______

Checking Itemized Calls on a Phone Bill

The tester presents a complete monthly phone bill (nine pages) to the participant. All pages are in numerical order and clearly labeled. The tester then presents a 4 × 6 in. (10.2 × 15.2 cm) index card with the following instruction: "According to this bill, on which days were the AT&T long-distance calls to Oregon made? Do not include any calling card calls."

The tester records the participant's response in the following way:

Correct Steps	*Participant's Steps*
1. Goes to page 6 of bill	1.
2. March 4	2.
3. March 11	3.
4. March 23	4.
5. March 24	5.

SOURCE: From "Everyday Problem Solving in Older Adults: Observational Assessment and Cognitive Correlates," by M. Diehl, S. L. Willis, and K. W. Schaie, 1995, *Psychology and Aging, 10*, p. 491.

The OTDL is designed for in-the-home administration and observational assessment: The tester provides props and instructions and records responses, and the quality of the participant's problem solving is later coded along scales that range from complete success to complete failure. The test includes 21 items in all, drawn from three domains: food preparation, medication intake, and telephone use. These domains are commonly sampled ones across a variety of "everyday" instruments; other common sources for tasks are financial management, interpersonal relationships, shopping, and transportation. Indeed, in a recent revision of the OTDL (Diehl et al., 2005), the food preparation domain is replaced by items directed to financial management (e.g., balancing a checkbook, making change).

In addition to variations in content, instruments vary in method of assessment. Posing of actual problems in the home setting (as with the OTDL) is one measurement option. More common—and also perhaps more practical in most contexts—is to assess everyday competence through some sort of paper-and-pencil test. An example in this category is the Everyday Problems Test, or EPT (Marsiske & Willis, 1995). The EPT presents written materials that pose problems across a number of domains of everyday functioning; sample items include reading a medicine label, understanding nutritional information, interpreting traffic laws, and filling out a medical history. Finally, a third general measurement option is to elicit ratings of the participant's ability to cope with various everyday tasks, either self-ratings (e.g., Fillenbaum, 1985) or ratings from someone who knows the individual well enough to judge (e.g., Fillenbaum, 1978).

It is difficult to summarize at all briefly the results from the research on everyday cognition, not only because such research has proliferated in recent years but also because the results may vary across different ways of operationalizing the everyday concept (Allaire & Marsiske, 2002). In some cases performance on such measures shows fairly strong relations to standard indices of ability such as IQ; in other cases the two forms of competence appear independent. Similarly, in some cases success at everyday problem solving declines across adulthood, whereas in other cases performance remains stable well into old age (Berg & Klaczynksi, 1996) Indeed, in some cases, when the tasks are especially familiar to them, older adults outperform younger ones (Artistico, Cervone, & Pezzuti, 2003). Whatever the specific outcomes of any particular research effort, however, the value of the overall enterprise seems clear. Such studies focus on abilities that are important in the daily lives of elderly people, and they can identify both individual differences and forms of competence among the elderly that may be missed with traditional IQ measures.

Let us return now to the domain of IQ. Studies of IQ raise a second measurement issue that is in a sense even more basic than that of measurement equivalence. Intelligence, like any general construct, can be defined and measured in a variety of ways. IQ tests constitute one operational definition of intelligence—hardly the only one and hardly a completely satisfactory one. In addition, "IQ" itself is not a single entity but a diverse conglomeration of different abilities. What we conclude about stability or change in IQ might well depend on the particular aspect of IQ on which we have decided to focus.

Let us briefly consider some data on this latter point. It is worth doing so, for the specific aspect of IQ that is studied turns out to be an important determinant of conclusions about age differences in IQ. Perhaps the best-established finding in this regard concerns the Verbal-Performance distinction on the WAIS. Age differences are much more marked on the Performance scale than on the Verbal scale. Verbal IQ, in fact, often holds up quite well even into fairly advanced old age. Performance IQ, in contrast, is likely to show both earlier and larger decrements. This pattern has been labeled the "classic pattern of aging" (Botwinick, 1984). A typical example of the pattern is shown in Figure 14.3. Note that these data are derived from cross-sectional comparisons of different age groups. The same general Verbal-Performance contrast emerges in longitudinal research, but both the timing and the magnitude of age decrements differ from those observed in cross-sectional studies. I return to this point in the section on Design.

The distinction between Verbal and Performance skills on the WAIS overlaps with another often discussed distinction in work on intelligence and aging: that between crystallized intelligence and fluid intelligence (Horn, 1982). **Crystallized intelligence** refers to forms of knowledge that are gradually accumulated through experience—to all the things that we come to know through years of education and acculturation. Vocabulary is a prototypical

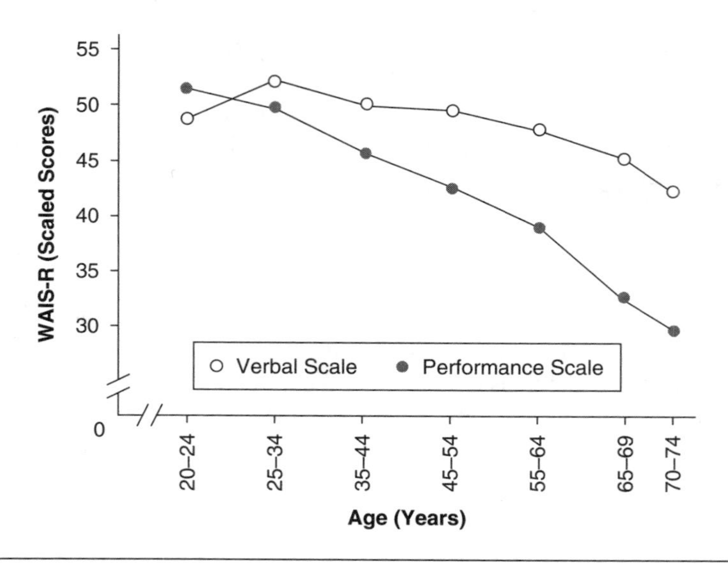

Figure 14.3 WAIS Verbal and Performance scores as a function of age (Verbal scores multiplied by 5/6 to develop a common base with Performance scores)

SOURCE: From *Aging and Behavior* (3rd ed.), by J. Botwinick. Copyright © 1984 by Springer Publishing Company, Inc., New York 10026. Used by permission.

form of crystallized intelligence; so too is the sort of real-world knowledge tapped by the Information subtest of the WAIS (Table 14.2). **Fluid intelligence**, in contrast, has to do with response to novelty and on-the-spot forms of problem-solving—with basic information-processing skills that are relatively independent of experience. Solving analogies or completing puzzles are forms of fluid intelligence, as is solution of the Reasoning items on the PMA (Table 14.3). As people age, it is primarily skills in the fluid category that show declines. Crystallized abilities are maintained well with increasing age; indeed, some forms of crystallized intelligence may increase through much of the adult life span.

One more finding with regard to differential decline is of enough general interest to merit mention. Items on IQ tests can be divided into "speeded" items and "nonspeeded" items. Speeded items are those for which speed of response is important; the answer must be found within a particular time limit, and bonus points may be awarded for an especially quick response. Picture Completion on the WAIS (Table 14.2) is an example of a speeded item. Nonspeeded items are those for which speed of response is not important; participants have as long as they need to arrive at the answer, and quality of response, not speed, is all that is scored. Vocabulary on the WAIS is an example of a nonspeeded item. In general, age differences between young adults and elderly adults are greater on speeded items than on nonspeeded items. Thus, older people seem to have special difficulty in situations that require quick response.

Most of the data concerning the role of speed in IQ performance come from the WAIS, and this situation presents a problem: All of the five Performance items on the WAIS are speeded, whereas only one of the six Verbal items is speeded. There is, then, a confounding between content (Verbal versus Performance) and speed. Given this confound, it is important to note that conclusions about the relation between speed and aging are not limited to the WAIS subtests. Slowing of response with increased age is a general phenomenon that has been demonstrated in a variety of contexts; indeed, a recent summary labels it "one of the most robust findings in the gerontological literature" (Cavanaugh & Blanchard-Fields, 2006, p. 200). It appears, moreover, that this slowing is not purely peripheral in origin—a matter, for example, of decreased sensory acuity or loss of manual dexterity. Evidence indicates, rather, that at least part of the slowing is central in origin, a reflection of general slowing in central nervous system processing that can affect performance on a wide range of tasks (Salthouse, 1996).

We will return to the role of speed of processing in the section on Memory. For now I will mention one more finding from the IQ literature. Earlier I noted that relatively poor performance on IQ tests is a rough predictor of impending mortality. I will add now that difficulty on speeded measures is especially predictive (Deary & Der, 2005; Schaie, 2005).

Design

This section begins with a discussion of the two most common designs in the study of IQ and aging, the cross-sectional and the longitudinal. It then moves on to a consideration of some alternative approaches to the study of changes with age.

Conclusions from cross-sectional studies of IQ and longitudinal studies of IQ are in many respects similar. With both designs, the predominant age effect is a negative one; older individuals perform more poorly than younger ones. With both designs, the same kinds of variables—health status, verbal versus performance content, speed—tend to affect performance. There are, however, also some differences. The major difference is that apparent declines in IQ are much less marked in longitudinal research than in cross-sectional research. Typically, the age at which decrements in IQ first appear is later in longitudinal studies than in cross-sectional studies, and the magnitude of decrements when they do appear is less. Thus, longitudinal research presents a more positive picture of aging than does cross-sectional research.

Why this difference between cross-sectional and longitudinal? The two most obvious explanations have already been discussed, both earlier in this chapter and in chapter 3's general treatment of developmental designs. In cross-sectional studies there is a confounding of age and cohort. Young adults and elderly adults differ not only in age but in generation, and the generational differences (in educational opportunities, for example) seem for the most part to favor the young. Such generational differences might well lead to differences in IQ quite apart from any contribution of age. In longitudinal studies, in contrast, the bias seems to work mainly in the opposite direction. Dropout occurs in longitudinal research, and the dropout tends to be selective, with low-IQ participants becoming progressively less available as the study continues. The result is a positive bias in favor of older age groups.

In addition to selective dropout, a second aspect of longitudinal designs works against decline with age. Participants in a longitudinal study of IQ undergo repeated administrations of the same IQ test. Even though explicit feedback about performance is not provided, the repeated exposure to the same materials might well lead to a positive practice effect. Evidence indicates that such practice effects do occur: People who have already taken an IQ test tend to do better on later administrations of the same test (Rabbitt et al., 2004). This factor, too, acts to minimize decline with age in longitudinal research.

The major alternative to a cross-sectional or longitudinal design is one or more of the *sequential* designs that were introduced in chapter 3. As we saw then, the various sequential designs involve combinations of the simpler cross-sectional, longitudinal, and time-lag approaches. Such designs provide more information than do the simpler approaches, but at the cost of considerably more time and resources. Their use to date has therefore been limited. Here I concentrate on one major sequential study that has been very influential—the Seattle Longitudinal Study by Schaie and associates (Schaie, 1994, 2005).

The Schaie project began in 1956 with testing of 500 adults between the ages of 21 and 70. For purposes of data analysis, the sample (which I will call Sample 1) was organized into seven age groups with age intervals of 7 years. The mean age for the youngest group at the time of initial testing was 25; that for the oldest group was 67. Although several measures were administered, the principal dependent variable was the PMA, with a focus on both individual subtests and composite IQ.

The 1956 testing constituted a cross-sectional study, the results of which were reported by Schaie (1958). Like other cross-sectional research, it painted a negative picture of IQ and aging, with apparent declines in IQ beginning as people reached their 50s. These cross-sectional data, however, were just the starting point for the Schaie research. In 1963 a second testing took place utilizing two distinct samples. One sample was the original one; it consisted of the 303 participants, from the initial 500, who could be located and retested in 1963. The age range for this sample was, of course, now 28 to 74. The second sample (Sample 2) was new: a second independent cross-sectional sample, spanning an age range of 21 to 74. There were 977 participants in this new sample. With the 1963 data, therefore, Schaie and associates had a number of analyses and comparisons available. There were three sets of cross-sectional comparisons spanning most of the adult life span: Sample 1 in 1956 and again, with everyone 7 years older, in 1963, and Sample 2 in 1963. There was a longitudinal comparison, extending 7 years, for Sample 1 in the contrast of 1956 scores with 1963 scores. And there were time-lag comparisons available in the contrast between a particular age group in 1956 and the same age group in 1963.

The 1963 testing, ambitious and challenging though it clearly was, was far from the end of the project. It was, in fact, merely the second "wave" in an effort that now encompasses seven distinct waves, or times of data collection. Schaie and associates have continued to collect data at 7-year intervals, with the most recent reported testing coming in 1998. The general strategy at each testing phase has been the same as that in the original 1963 retesting. First, each wave of data collection includes an attempt to retest, in longitudinal follow-up, as many of the earlier study participants as possible. As might be expected, the number available for retesting goes down as the interval between sessions and the age of the participants go up; by the 1998 testing, only 38 of the original 500 participants were still available. Nevertheless, it is clearly informative, and also quite unusual, to have 42-year longitudinal data for even a subset of one's sample. Second, at each wave of testing a new cross-sectional sample, spanning roughly the ages from 20 to 80, has been added. At the 1998 testing, for example, 719 first-time participants were brought into the study, bringing the total number of participants for the project as a whole to 5676.

A schematic summary may help to sort out the details of this very complex study. Table 14.5 provides such a summary. The numerous cross-sectional, longitudinal, and time-lag comparisons should be derivable from the table. Comparison with Figures 3.2 and 3.3 may help distinguish the particular sequential analyses that are embedded within the overall design. Any pair of rows in the table, for example, provides the basis for a cohort-sequential design—that is, two overlapping longitudinal studies, in which age and cohort can be analyzed as independent variables.

What were the results from all this prodigious effort? To a good extent, the major findings from the project have already been touched upon, since the Schaie research was one

Table 14.5 Summary of the Schaie Sequential Studies: Mean Ages and Sample Sizes at Each Testing Occasion

Time of testing							
	1956	1963	1970	1977	1984	1991	1998
Sample 1	25-67 (500) First Test	32-74 (303) Second Test	39-81(162) Third Test	46-88 (130) Fourth Test	53-95 (97) Fifth Test	60-95 (71) Sixth Test	67-95 (38) Seventh Test
Sample 2		25-74 (997) First Test	32-81 (419) Second Test	39-88 (337) Third Test	46-95 (225) Fourth Test	53-95 (161) Fifth Test	60-95 (111) Sixth Test
Sample 3			25-81 (705) First Test	32-88 (340) Second Test	39-95 (223) Third Test	46-95 (175) Fourth Test	53-95 (127) Fifth Test
Sample 4				25-81 (609) First Test	32-88 (295) Second Test	39-95 (201) Third Test	46-95 (136) Fourth Test
Sample 5					25-81 (629) First Test	32-88 (428) Second Test	39-95 (266) Third Test
Sample 6						25-81 (693) First Test	32-88 (406) Second Test
Sample 7							25-81 (719) First Test

NOTE: Numbers in parentheses indicate sample sizes.

source for the general conclusions about IQ and aging discussed in the preceding pages. One clear finding, for example, concerned the different patterns of change shown by the different abilities assessed by the PMA. In Schaie's (1994, p. 306) words, "there is no uniform pattern of age-related changes in adulthood across all intellectual abilities"; rather, some abilities remain stable or even increase with increased age, whereas others show declines of varying magnitude and beginning at various points in life. Another finding concerned the cross-sectional-longitudinal contrast. As an examination of Table 14.5 should make clear, any age comparison of interest (say 60-year-olds versus 74-year-olds) appears multiple times in the research in both cross-sectional and longitudinal form, and the former set of analyses consistently revealed larger age differences than did the latter. A third conclusion is related, and that is that cohort can play an important role for many of the abilities assessed—certainly more important than had been acknowledged prior to the Schaie research. Although cohort comparisons generally favored the younger generations, it is interesting to note that not all the effects were in this direction. Performance on the number skill subtest, for example, peaked with the 1924 birth cohort and showed progressive declines thereafter.

Box 14.1. Predictors of Maintenance of Intellectual Ability in the Seattle Longitudinal Study

Absence of cardiovascular and other chronic diseases
Living in favorable environmental circumstances
Substantial involvement in activities available in complex and intellectually stimulating environments
Flexible personality style at midlife
Being married to a spouse with a high cognitive status
Maintenance of high levels of perceptual processing speed
Satisfaction with one's life accomplishments in midlife or early old age

SOURCE: Adapted from "The Course of Adult Intellectual Development," by K. W. Schaie, 1994, *American Psychologist, 40*, 304–313.

I will note one more set of findings from the Schaie research. My emphasis so far has been on group-level comparisons—for example, 60-year-olds versus 74-year-olds. A main strength of longitudinal study, however, is that it allows us to move beyond group-level analyses to focus on *individuals* and to trace individual consistency or individual change over time. In a longitudinal study we know not just the average amount of change but also *who* changes—which individuals maintain their abilities and which individuals show declines with age. One of the goals of the Schaie research has been to determine the antecedents of successful aging—that is, to identify factors at midlife that predict maintenance of ability into old age. Box 14.1 summarizes the conclusions that have emerged. We will return to the third of the listed factors, the role of activity in successful aging, in the concluding section of the chapter.

It is probably unnecessary to note that the preceding is a very selective summary of findings from a massive research project; indeed, a number of dependent variables and related theoretical issues have not even been mentioned. Also selective is the treatment of the methodological complexities involved in analyzing sequential designs. In addition to the sources already cited, further discussions of these issues can be found in Donaldson and Horn (1992), Masche and van Dulmen (2004), and Schaie and Caskie (2005). Among the sources for more general treatments of the issues discussed in this section of the chapter are Bergeman and Baker (2005), Hertzog and Dixon (1996), and Salthouse (2000).

With the discussion of general issues as background, we turn now to some specific topics that have been of interest in the study of aging. IQ, of course, is one such topic. I consider next another major approach to the study of cognition in old age, after which the discussion turns to methods of studying personality and social development.

Memory

I noted earlier that IQ has been one of the most popular dependent variables in psychological studies of aging. Memory has been another. Indeed, various forms of memory may be the most common outcome measure in studies of aging.

As with IQ, there is a common stereotype about memory and old age. The stereotype is again one of decline: Old people can no longer

❖FOCUS ON❖

Box 14.2. Wisdom

As we have seen, the dominant stereotype of aging and intelligence is a negative one: Old age is a time when mental abilities decline. There is, however, one exception to this negative characterization, at least in some people's conception of the elderly years. It is the idea that aging brings with it a degree of wisdom beyond that which is typically present earlier in life.

It is not difficult to find anecdotal or historical evidence in support of this idea of the especially wise older individual. Testing the claim more rigorously requires that we solve an obvious challenge: How can we measure wisdom?

Just as there is no single, agreed-upon way to measure other aspects of intellectual functioning, so there is no single approach to the study of wisdom. I concentrate here on what has probably been the most influential approach to date, the "Berlin group" work of Paul Baltes and colleagues (Baltes & Smith, 1990; Baltes & Staudinger, 2000; Kunzmann & Baltes, 2005). Other conceptualizations share some features of the Berlin group approach but also diverge in certain ways. Ardelt (2000, 2004) offers both a critique of the Berlin group's approach and an interesting alternative conceptualization. An edited book by Sternberg and Jordan (2005) provides a more comprehensive overview of different approaches to the study of wisdom.

As chapter 4 indicated, measuring any construct is always a multistep process. A first step is to formulate a conceptual definition of the target for measurement—in the present case, to decide exactly what is meant by "wisdom." Here are two of the definitions offered by the Berlin group: "Wisdom is an expert knowledge system in the fundamental pragmatics of life permitting exceptional insight, judgment, and advice involving complex and uncertain matters of the human condition" (Staudinger & Werner, 2003, p. 590). "Wisdom is defined as *superior knowledge* of how to act for our own good, for the good of others, and the good of the society we live in" (Staudinger & Werner, 2003, p. 590). These definitions contain elements common to any conception of wisdom ("expert," "exceptional," "superior"), but also convey some distinctive aspects of the Berlin group approach. As the references to "others" and "the good of society" suggest, there is a social emphasis; wisdom should not be solely self-directed but should be used for the good of others. Wisdom in this conception is also social in a second sense; it is seen as being embodied not just in individuals but in societies and cultures as a whole—is seen as "a cultural and collective product" (Baltes & Staudinger, 2000, p. 127). There is also a cognitive emphasis in the Berlin group approach ("expert knowledge system," "superior knowledge"), which sets their work in contrast to approaches that treat wisdom more as an aspect of personality than as a body of acquired knowledge (e.g., Ardelt, 2000).

As we have seen, the conceptual definition of a construct must be operationalized—that is, translated into some measurable form. Baltes and colleagues have studied wisdom primarily through response to hypothetical vignettes that pose some complex life problem to be solved. Table 14.6 presents examples of two of the vignettes. The method used is a "think aloud" procedure developed in cognitive psychology—participants are asked to verbalize whatever thoughts they have as they think through the problem, and the scoring of responses takes into account not just the final solution but also the reasoning produced en route. The scoring (which is completed by selected and highly trained raters) evaluates responses along five dimensions: factual knowledge about human nature and the life course; procedural knowledge about how to deal with life problems; contextual knowledge about past, present, and future life contexts; awareness and acceptance of the relativism of different values and goals; and awareness of uncertainty and the limits of one's own knowledge.

(Continued)

(Continued)

Table 14.6 Examples of Vignettes Used to Study Wisdom in the Research by Baltes and Colleagues

Michael, a 28-year-old mechanic with two preschool-aged children, has just learned that the factory in which he is working will close in 3 months. At present, there is no possibility for further employment in this area. His wife has recently returned to her well-paid nursing career. Michael is considering the following options: He can plan to move to another city to seek employment, or he can plan to take full responsibility for child-care and household tasks. Formulate a plan that details what Michael should do and should consider in the next 3 to 5 years. What extra pieces of information are needed?

Joyce, a 60-year-old widow, recently completed a degree in business management and opened her own business. She has been looking forward to this challenge. She has just heard that her son has been left with two small children to care for. Joyce is considering the following options: She could plan to give up her business and live with her son, or she could plan to arrange for financial assistance for her son to cover child-care costs. Formulate a plan that details what Joyce should do and should consider in the next 3 to 5 years. What extra pieces of information are needed?

SOURCE: From "Wisdom-Related Knowledge: Age/Cohort Differences in Response to Life-Planning Problems," by J. Smith and P. B. Baltes, 1990, *Developmental Psychology, 26*, p. 497.

How do we know whether this approach really measures wisdom—that is, what is the evidence for test validity? Baltes and colleagues offer various kinds of evidence. Theoretically, they argue that their emphases and criteria are in agreement with both philosophical writings about wisdom and implicit, layperson conceptions of what it means to be wise. Empirically, they report that the different dimensions in their scoring show the pattern of intercorrelations that would be expected. They also demonstrate that individuals who have been nominated as exceptionally wise do in fact perform better on their measure than does an otherwise comparable group of adults.

What does this work show with respect to the stereotype of the wise older adult with which I began the discussion? I will note first that wisdom, at least as assessed with this approach, is rare at any age; in Kunzmann and Baltes's words (2005, p. 121), "many adults are on their way toward wisdom, but very few people approach a high level." Beyond early adulthood, however, reaching a high level is no more likely at one age than another. Other approaches, I should note, do sometimes find increases in wisdom with age (Ardelt, 2004). And note that even the Berlin group studies show no evidence of a *decline* in wisdom with age; thus, this work provides another counter to the declining-abilities stereotype of old age.

remember things as well as they could when they were younger. As with IQ, there is some truth to the stereotype but also many qualifications and exceptions. The goal of much of the research on memory in the elderly has been to discover the conditions under which memory problems do or do not occur.

Recognition and Recall

One basic distinction in the study of memory at any point in the life span is that between *recognition memory* and *recall memory.* We saw in chapter 12 that recognition memory is both present and powerful from quite early in life and that it is recall memory that shows the more marked developmental changes across the course of childhood.

Recall is also more likely to show changes as people age. A study by Schonfield and Robertson (1966) provides a demonstration. Schonfield and Robertson presented both young adults (ages 20 to 29) and elderly adults (ages 60 to 75) with two lists of 24 words to remember. Memory was subsequently tested in two ways: For one list, participants were asked to say aloud all the words they could remember (thus a recall test), whereas for the other they were presented with five alternatives and asked to circle the one that had been shown (a recognition test). Figure 14.4 shows the results. Not surprisingly, performance was better on recognition than on recall, and this finding held true at both ages. The recognition-recall difference, however, was considerably greater for the elderly than for the young-adult participants. Indeed, on the recognition measure there were no differences at all between the two age groups. Only on recall did young adults outperform older ones. Note that this finding reflects an age by condition interaction: Effects of age varied as a function of the type of memory that was examined.

The Schonfield and Robertson study is often cited, probably both because it was one of the earliest examinations of the recognition-recall issue and because its results were so clear-cut. Since this work, numerous other investigators have also compared recognition and recall in adults of different ages. Similar age by condition interactions are a frequent enough finding to justify the conclusion that difficulties in recall are a more likely accompaniment of old age than are difficulties in recognition (Hess, 2005). On the other hand, the flat age function for recognition reported by Schonfield and Robertson is by no means always obtained. In a number of studies age differences in favor of young adults emerge for recognition as well as recall.

Why might recall and recognition show different developmental patterns as people age? A prior question is why recall and recognition might ever be expected to differ in difficulty. One obvious difference between the two kinds of memory concerns the storage-retrieval distinction. Both kinds of memory require storage of material over time; recall, however, requires a

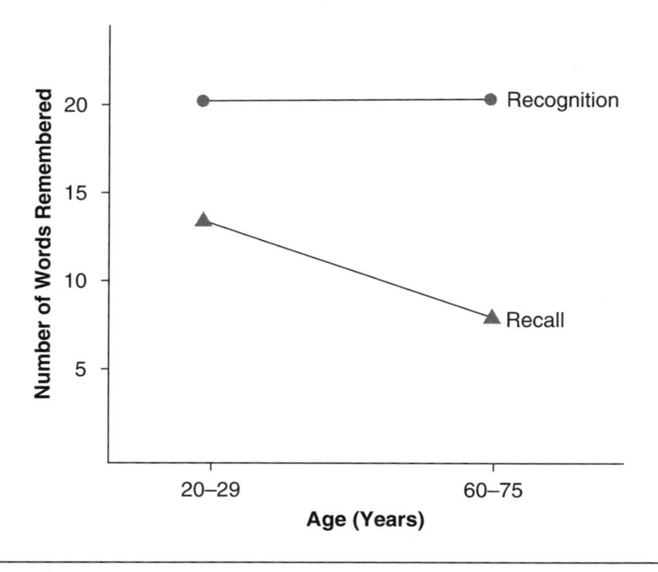

Figure 14.4 Interaction of age and experimental condition in the Schonfield and Robertson study

SOURCE: Adapted from "Memory Storage and Aging," by D. Schonfield and E. A. Robertson, 1966, *Canadian Journal of Psychology, 20*, 228–236.

more active effort at retrieval of the stored material than does recognition, in which the correct answer is present at the time of response. The fact that recognition is relatively impervious to developmental decline suggests, therefore, that the storage component of memory holds up fairly well with age; developmental changes result largely from decrements in retrieval as people get older. Like all generalizations about age differences in memory, this particular generalization has limitations and exceptions (such as the fact that age differences can occur in recognition as well). Nevertheless, it does appear to have some validity as a partial account of age differences in memory.

How else might the distinction between storage and retrieval be studied? Another common approach is the *cued-recall paradigm.* In studies of cued recall, a contrast is drawn between a standard recall test and a recall test on which a hint, or "cue," is provided. Laurence (1967), for example, tested for recall in both young (mean age 20 years) and elderly (mean age 75 years) adults. The stimuli were 36 words drawn from six categories: flowers, trees, birds, formations of nature, vegetables, and countries. The list was presented just once, with the words randomly ordered—that is, not divided by category. Prior to the recall test, however, half of the participants were given cue cards containing the six category names and were told that every word belonged to one of the six categories; the remaining participants received no such hint. The results of this manipulation are shown in Figure 14.5. As can be seen, there was an interaction of age and condition: Older participants benefited much more from the hint than did younger participants.

As with the Schonfield and Robertson study, the early work by Laurence has met a mixed fate in more recent studies. Age by condition interactions of the sort found by Laurence are sometimes but by no means always obtained in research on cued recall; there are a number of instances in which young adults benefit at least as much from a cue as do elderly adults. The general conclusion seems the same as that in the recognition-recall literature. The fact that retrieval cues can be especially helpful for the elderly suggests that older people often learn

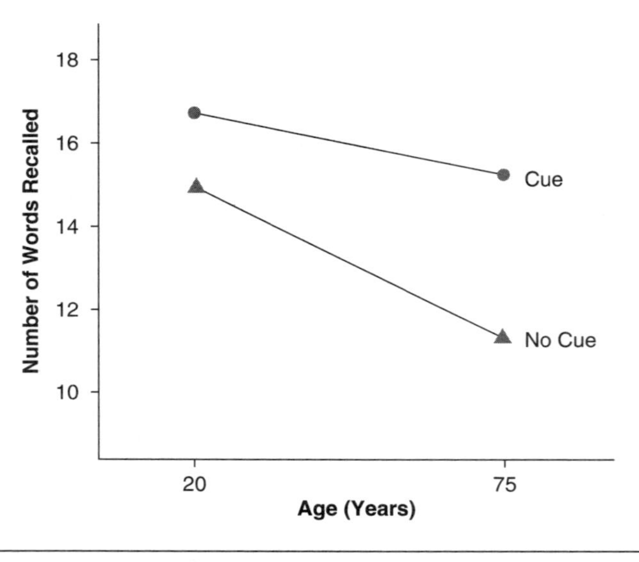

Figure 14.5 Interaction of age and experimental condition in the Laurence study of cued recall

SOURCE: Adapted from "Memory Loss With Age: A Test of Two Strategies for Its Retardation," by M. W. Laurence, 1967, *Psychonomic Science, 9,* pp. 209–210.

and store material successfully but have difficulty in later retrieving the material. The fact that retrieval cues do not always remove young-old differences suggests that there are other factors, in addition to difficulties in retrieval, that contribute to age differences in memory.

I will mention one more source of support for the idea that difficulties in retrieval become more likely with increased age. It comes from a phenomenon known as the *tip of the tongue* experience, or *TOT.* TOT refers to the experience of being certain that one knows a word or a name but simply being unable, at least for the moment, to bring the word or name to mind. You probably have had this experience; research suggests, however, that in 50 or so years you will have it more frequently (MacKay & Abrams, 1996). TOT experiences become more likely with increased age, again suggesting special difficulties with the retrieval component of memory.

Mnemonic Strategies

The argument about the need to be active in retrieving material is part of a more general hypothesis about age differences in memory. The general hypothesis is that elderly people are most likely to have difficulty in situations that require some sort of active processing in order to remember—situations that require *doing* something in order to acquire information in the first place, to maintain information over time, or to retrieve information when it is needed. This argument should have a familiar sound, for it is the same hypothesis that was offered as a partial explanation with respect to changes across childhood—the idea that older children remember better than younger not simply because they have bigger memory "stores" but because they process material in more adaptive ways. In childhood this hypothesis has led to two main lines of research, both of which are also relevant to age differences in adulthood: studies of strategies and studies of constructive memory.

Research on mnemonic strategies in adulthood has followed the same general methodological lines as the research on strategies in childhood that was discussed in chapter 12. In some studies the use or nonuse of strategies is inferred from the overall pattern of memory performance. The organizational strategy of "clustering," for example (Table 12.2), can be inferred from the extent to which the participant brings together conceptually related items in recall. Similarly, rehearsal can be inferred from the existence of a primacy effect in recall—that is, especially good memory for the early items in a list. Another approach is to induce the use of a strategy experimentally. Participants can be instructed to rehearse, for example, and subsequent effects on memory performance can be measured.

To a good extent, conclusions about the role of strategies in memory in old age have already been previewed. Recall that one of the kinds of applied research discussed in chapter 7 concerned mnemonic training programs for elderly individuals. The premise underlying such programs is that some (although of course not all) memory difficulties in old age result because elderly people fail to utilize effective mnemonic techniques. Basic research of the sort sketched in the preceding paragraph supports this premise. Although findings are far from perfectly consistent (see Kausler, 1994, for a review), a variety of kinds of evidence indicate that older adults are often less likely to employ mnemonic strategies spontaneously than are young adults, and also less likely to choose the optimal approach when they do attempt a strategy. Studies in which strategy use is induced—either on a short-term experimental basis or in the more long-term, training sense discussed in chapter 7—are also supportive. As we saw, instruction in strategy use can improve older adults' memory performance.

Why should an elderly adult fail to use a mnemonic strategy when doing so would be helpful? In most instances a literal inability to produce the strategy is not a plausible explanation—most of the strategies in question are simple ones that are well within the capacity of most older adults. One possible explanation is labeled the **disuse hypothesis**. The disuse hypothesis refers to the idea that cognitive skills that are not regularly utilized become harder to access and activate when needed. It is an idea that can be applied quite broadly to a range of cognitive and social abilities in old age. In the case of memory, the argument is that older adults have fewer occasions than do young adults on which they need to commit information to memory, and thus they are at a disadvantage when put in situations that require them to do so. (I should add that not all researchers find the disuse hypothesis a useful explanation for memory difficulties—cf. Light, 1991).

Constructive Memory

In addition to strategies, the other main content area considered in chapter 12 was constructive memory. As noted then, constructive memory concerns the effects of the general knowledge system upon memory—the tendency to fit material in, draw inferences, go beyond the information given, all in an attempt to *understand* what one has experienced.

Memory for meaningful material such as a story or a conversation is a longstanding topic in the adult-memory literature, generally appearing under the labels *discourse memory* or *text memory.* A basic finding from such research concerns similarity in memory across the adult life span. Adults of any age, including the elderly, show better memory for meaningful material than for nonmeaningful, adults of any age draw constructive inferences of various sorts in their attempts to understand such material, and the same sorts of variables (e.g., complexity of text, familiarity of material) tend to affect performance whatever the age of the participant (Johnson, 2003). Nevertheless, there are still situations in which older adults are less likely to show constructive memory than are young adults. We will consider one example here; Kausler (1994) provides a fuller review.

The example on which I focus concerns memory for stories and is thus similar to some of the research discussed in chapter 12. Across a series of experiments, Cohen (1979, 1981) compared young adults' and elderly adults' memory for prose passages of various sorts. Examples of two of the passages used are shown in Box 14.3. The interest was in two kinds of memory. One kind is verbatim recall for information explicitly presented in the story. The first question following each passage taps such verbatim recall. The other is for information that is implicit in the story and hence must be inferred. The second question after each story taps implicit recall.

Box 14.3. Examples of Passages and Questions From Cohen's Research on Constructive Memory

Mrs. Brown goes to the park every afternoon if the weather is fine. She likes to watch the children playing, and she feeds the ducks with bread crusts. She enjoys the walk there and back. For the last three days it has been raining all the time although it's the middle of the summer, and the town is still full of people on holiday.

1. What does Mrs. Brown give the ducks to eat?
2. Did Mrs. Brown go to the park yesterday?

Downstairs there are three rooms; the kitchen, the dining-room, and the sitting-room. The sitting-room is in the front of the house, and the kitchen and dining-room face onto the vegetable garden at the back of the house. The noise of the traffic is very disturbing in the front rooms. Mother is in the kitchen cooking, and Grandfather is reading the paper in the sitting-room. The children are at school and won't be home till tea-time.

1. What is Mother doing?
2. Who is being disturbed by the traffic?

SOURCE: From "Language Comprehension in Old Age," by G. Cohen, 1979, *Cognitive Psychology, 11*, p. 416.

Cohen's basic finding was that age differences were more marked for implicit recall than for explicit recall. Elderly participants did well at understanding and retaining the explicit content of a passage, but they appeared to have difficulty in going beyond the explicit content to infer underlying meanings. And it is, of course, this going beyond what is immediately given that constitutes constructive memory.

Why should there be such age differences in constructive memory? In childhood at least part of the explanation has to do with differences in the knowledge base: Because older children know more than younger children, the kinds of inferences that they can show will differ. In adulthood, however, such differential knowledge seems a less plausible explanation; adults of any age, for example, are certainly capable of drawing the links necessary to answer Cohen's questions. The difference here seems to lie in the tendency to do such linking—to use one's knowledge spontaneously to draw inferences. It is this tendency that apparently decreases in old age.

One explanation of why there should be an age decrement in spontaneous inference implicates speed of processing. We have seen that a slowing of response speed is a typical accompaniment of old age. As applied to text memory, the argument is that older adults' slower rate of processing information makes it difficult for them both to attend to and encode new material and to relate the new material to old, stored material. Because it is the relation between new and old on which inference depends, the result is a decline in constructive memory. This explanation shows a nice compatibility with—and indeed is supported by—other research that demonstrates that slowing of response has negative effects on a variety of kinds of performance in old age.

Everyday Memory

We consider two more topics under the heading of Memory. Thus far the discussion has focused on standard laboratory paradigms that have been common in the study of memory at any age period. The final two approaches to be considered are somewhat more specific to the study of aging; they also (not coincidentally) tend to give a more positive picture of memory in old age.

We saw earlier that the study of "everyday cognition" has become a popular research enterprise. The same can be said for everyday memory. As was true for cognition more generally, a range of content areas and methodological approaches are found in the work on everyday memory. I give two examples here; among the sources for fuller discussions are Hess and Pullen (1996) and Park and Brown (2001).

Spatial memory is one obvious, and popular, subtopic under the heading of everyday memory. All of us utilize forms of spatial memory numerous times every day—following familiar routes, recalling the locations of objects, and so forth. Does such memory decline in old age? We have already seen one laboratory approach to this issue—the Cherry and Park (1993) study introduced in chapter 2. This study yielded two findings that are fairly typical in such research: Older adults showed poorer spatial memory than young adults, and older adults were helped when the task was embedded in a distinctive and familiar context.

Studies of the "everyday" sort take the notion of familiar context considerably further. In these studies the laboratory locus and experimental analogs (e.g., Cherry and Park's model or map) are replaced by naturalistic settings and real-life forms of spatial memory. Shopping at the supermarket, for example, is certainly a familiar real-life task for many people, and it served as the basis for a study by Kirasic (1991). Kirasic's participants were young (mean age = 24 years) and elderly (mean age = 70 years) women, and her experimental tasks involved various forms of spatial cognition that might be applicable while at the grocery store. On one task, for example, participants were asked to plan the most efficient route in order to locate seven items on a shopping list. On another they were asked to rank-order the distances that would be required to reach a set of pictured grocery items. All tasks were administered in two settings: the participants' usual, and thus familiar, supermarket, and an unfamiliar supermarket that they had explored for only 15 minutes.

Several findings emerged. Younger participants did somewhat better than older ones, although the differences were small and occurred on only some tasks. Performance was generally better in the familiar than in the unfamiliar setting, although again the differences were small. More interesting than these main effects was the occurrence of an age by setting interaction, whereby the elderly women benefited more from the familiar context than did the younger women. Figure 14.6 shows this pattern for the route-planning and distance-estimation tasks. Although this finding of special benefit from familiarity for older people does not always obtain (indeed, it held for just two of the four tasks in Kirasic's study), it is a fairly common outcome. And the more general conclusion that emerges from this research is well established: that elderly people may show impressive spatial memory when the assessment focuses on natural activities in natural settings. Among the other settings that have served as locus for such research are office buildings and museums (Uttl & Graf, 1993), familiar neighborhoods (Rabbitt, 1989), and the participants' own homes (West & Walton, 1985).

The second example of an approach to everyday memory is broader in scope; it is also more likely to involve clinical/pragmatic as well as basic-science uses. Recent years have seen the development of a number of self-rating memory scales on which participants can assess their own memory abilities across a range of everyday situations. Table 14.7 shows some of the items from one such measure: the Memory Assessment Clinics Self-Rating Scale, or MAC-S (Crook & Larrabee, 1990; Winterling, Crook, Salama, & Gobert, 1986). In addition to the specific items illustrated in the table, the MAC-S includes several global questions—for example, "Compared to the best your memory has ever been, how would you describe the speed with which you now remember things?" (range from *much slower* to *much faster*).

Instruments such as these lend themselves to a number of issues and uses, of which we will concentrate on two. One question concerns changes with age, and the usual finding is of an increase in self-reported memory problems as people grow older. Thus, instruments such as the MAC-S confirm the common stereotype: Memory problems *are* a common (although of course not inevitable) complaint of old age.

The second question concerns relations to actual memory performance—are people accurate in their assessments of their own mnemonic strengths and weaknesses? As with any self-report measure, the answer turns out

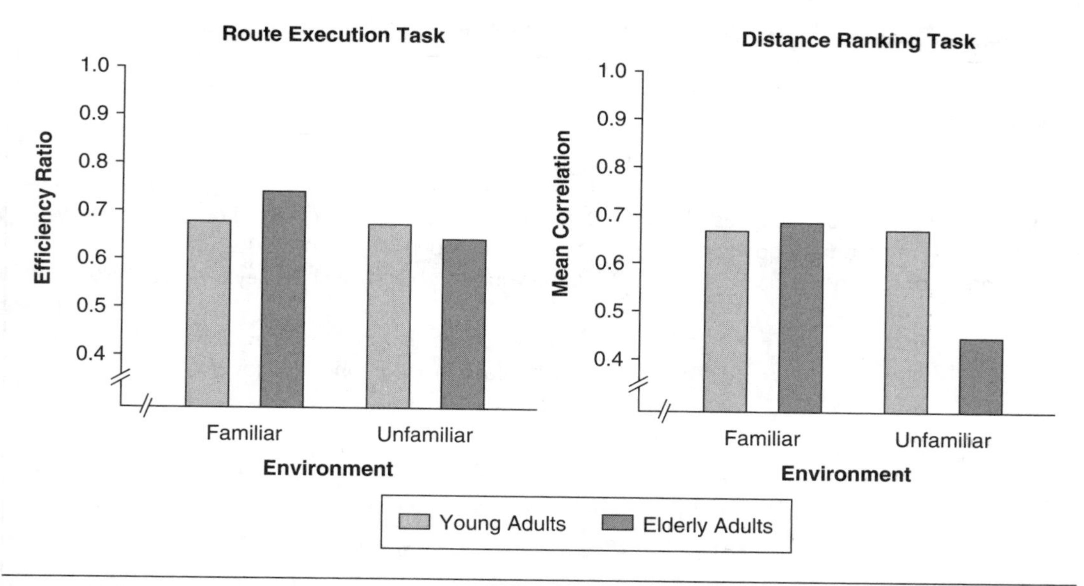

Figure 14.6 Interaction of age and environmental context in the Kirasic study of spatial memory. The dependent variable for the route-execution task was the ratio of selected route to most efficient route; that for the distance-ranking task was the correlation between estimated distance and actual distance. In both cases higher values indicate better performance.

SOURCE: Adapted from "Spatial Cognition and Behavior in Young and Elderly Adults: Implications for Learning New Environments," by K. C. Kirasic, 1991, *Psychology and Aging, 6*, 10–18, p. 14. Copyright © 1991 by the American Psychological Association. Reprinted with permission.

to be "somewhat but not perfectly." Most such measures do show positive correlations with either experimental or clinical indices of memory functioning. The correlations are far from perfect, however, and in some instances depression and other affective disorders are better predictors of self-reported problems than is actual memory performance (Comjis, Deeg, Dik, Twisk, & Jonker, 2002). Note that this finding does not negate the potential importance of memory complaints in an elderly adult's life; it does suggest, however, that such complaints may have various sources.

Remote Memory

There is one final form of memory to be considered. I noted at the outset of the Memory section that the stereotype of memory in old age is one of decline. This stereotype is sometimes qualified, however, by the assertion that really long-term memory may remain remarkably intact in old age, even in the face of declines in memory for recent events. The prototypical example of this modified stereotype is the elderly person who can recall events from childhood in vivid detail yet is unable to remember what was for breakfast that morning. Loss of memory is thus selective, or at least so the argument goes, and selective in surprising ways.

How much truth is there to this notion of the relative intactness of long-term or **remote memory**? The answer is, probably some, but the issue has proved difficult to study, and much of the evidence remains anecdotal. Here we briefly consider several examples of research

Table 14.7 Examples of Items From the Memory Assessment Clinics Self-Rating Scale

Ability scale	
Target of measurement	*Examples*
Ability to remember specific types of information—rated on a 5-point scale from *very poor* to *very good*	Details of family events that occurred during the past year The name of a person just introduced to you Where you put objects (such as keys) in the home or office Meanings of words you once knew fairly well How to reach a geographic location you have visited once or twice
Frequency of occurrence scale	
Target of measurement	*Examples*
Frequency of occurrence of specific memory problems—rated on a 5-point scale from very often to very rarely	Have difficulty recalling a word you wish to use Miss the point someone else is making during a conversation Go into a room to get something and forget what you are after Forget an appointment or other event that is very important to you Fail to recognize people who recognize you

SOURCE: Adapted from "A Self-Rating Scale for Evaluating Memory in Everyday Life," by T. H. Crook and G. J. Larrabee, 1990, *Psychology and Aging, 5,* p. 55.

on remote memory, along with some of the methodological difficulties in doing such research (see also Bahrick, 2000; Erber, 2001).

Bahrick, Bahrick, and Wittlinger (1975) tested high-school graduates ranging in age from 17 to 74 years. The criterion of high-school graduate was important in selecting participants, because the memory examined was memory for one's high-school classmates. Six different memory tests were used, two of recall and four of recognition. The recall tests were to list the names of as many classmates as possible and to indicate the name when presented with a picture of a classmate. The recognition tests were to pick the picture of a classmate from a group of five pictures, to pick the name of a classmate from a group of five names, to match a name with one or several pictures, and to match a picture with one of several names. Obviously, the age of the memory (i.e., time since last exposure to high-school classmates) varies with the age of the participant in this study, varying from extremes of about 3 to 4 months for the youngest participants to better than 50 years for the oldest ones.

Not surprisingly, Bahrick et al. found that young adults were more successful at remembering high-school classmates than were elderly adults. The age differences were especially striking on the two measures of recall. The recognition measures, in contrast, showed both higher overall levels of performance and more similarity in performance across age. Recognition of faces, in particular, was remarkably good even 30 to 40 years after graduation. The strong performance on recognition fits with data discussed earlier regarding the power of recognition memory. The greater decline in recall than in recognition with age fits with other demonstrations of age by type-of-memory interactions. And the fact that some forms of memory are so resilient across such a long period of time provides support for the claim that remote memory can be surprisingly intact in the elderly.

The yearbook study is just one example of research by Bahrick and colleagues on long-term retention of real-world knowledge. Other topics examined include memory for buildings and geographical landmarks (Bahrick, 1983) and memory for Spanish learned in high school or college (Bahrick, 1984). Other research teams have added memory for grade-school and high-school teachers (Schonfield, 1969), memory for song titles (Bartlett & Snelus, 1980), memory for street names (Schmidt, Peeck, Paas, & van Breukelen, 2000), and memory for the content of university courses (Cohen, Stanhope, & Conway, 1992). Like the yearbook study, these studies generally show fairly impressive retention across long intervals, with some but not necessarily large drop-off in performance as people grow older.

The most obvious limitation of studies of this sort, at least with regard to issues of aging, was noted earlier: Age of memory varies with—that is, is confounded with—age of participant. Thus, if we find that older people perform more poorly than younger, we do not know whether to attribute the age difference to a weakening of memory in old age or to the greater difficulty of recapturing a 50-year-old memory than a 5-year-old memory. Probably both factors contribute, but they cannot be disentangled in this kind of research.

How else might we study the notion of remote memory? A second approach has been to examine memory for sociohistorical facts drawn from different historical time periods. The example I use is one of the first such studies; more recent examples include Howes and Katz (1988); Perlmutter, Metzger, Miller, and Nezworski (1980); and Rubin, Rahhal, and Poon (1998).

Botwinick and Storandt (1974) administered a variety of memory tests to adults ranging in age from 20 to 79. Among the tests were questions that dealt with memory for historical facts. Box 14.4 presents the 24 questions that were asked. (For those who wish to test themselves, the answers are included in the Appendix at the end of this chapter.) Because the questions covered events drawn from four different historical time periods, it was possible to examine memory as a function of age of participant, recency of event asked about, and age of participant at the time that the event occurred.

Box 14.4. Items From the Botwinick and Storandt Test of Long-Term Memory

Period: 1950–1969

1. What was the name of the first man to set foot on the moon?
2. What was the name of the man who assassinated Dr. Martin Luther King?
3. In what city was President John F. Kennedy assassinated?
4. In what year did the Russians orbit the first satellite?
5. What was the name of the man who ran against Dwight Eisenhower for president in 1952 and again in 1956?
6. What was the name of the Senator from Wisconsin whose name is associated with congressional investigations of communism in the early 1950s?

(Continued)

(Continued)

Period: 1930–1949

7. What was the name of the man who was elected vice-president in 1948 when Harry Truman was elected president?
8. What was the name of the World War II German general nicknamed "The Desert Fox"?
9. What was the name of the commander of the famous Flying Tigers of World War II?
10. On what date (day, month, year) did the Japanese bomb Pearl Harbor?
11. What was the name of the only president of the United States to be elected to four terms of office?
12. Where (in what state) did the German dirigible, the *Von Hindenburg,* burn and crash?

Period: 1910–1929

13. What was the name of the ship which hit an iceberg and sank on its maiden voyage in 1912?
14. What was the name of the man whose death set off World War I?
15. What was the name of the World War I German flying ace nicknamed "The Red Baron"?
16. What do the initials WCTU stand for?
17. What was the name of the man tried in the famous Monkey Trial of 1925?
18. What was the name of the plane in which Lindbergh flew the Atlantic?

Period: 1890–1909

19. What was the name of the man who discovered the North Pole?
20. In what state did the first legal electrocution for murder in the United States occur?
21. What was the name of the man who became president when President McKinley was assassinated in 1900?
22. Where did the Wright brothers make their first successful flight?
23. What was the name of the boxer who was nicknamed "Gentleman Jim"?
24. In what year did Henry Ford introduce the Model T?

SOURCE: From *Memory, Related Functions, and Age* (pp. 189–191), by J. Botwinick and M. Storandt, 1974. Courtesy of Charles C Thomas, publisher, Springfield, Illinois.

Botwinick and Storandt found no overall age differences in performance on such questions. Thus, 70-year-olds were just as good as 20-year-olds at this kind of remote memory. There was, however, an interesting interaction between age of participant and time period asked about. The interaction reflected the fact that participants of any age were most likely to remember events that occurred when they were 15 to 25 years old. Thus, people in their 50s did best on questions dealing with events from the 1930s and 1940s; people in their 30s were best on events from the 1950s and 1960s. I should add that such age by time period interactions are sometimes but by no means always obtained in other studies of this sort (in fact, the interaction in the Botwinick and Storandt study held only for the male participants). The other studies cited earlier, however, do confirm the most general conclusion from the Botwinick and Storandt research: good memory for sociohistorical facts, with only modest declines, at most, as people age.

The Botwinick and Storandt methodology escapes some of the limitations in the Bahrick studies but runs into some difficulties of its own. When high-school classmates are the stimuli, we can be reasonably certain that the information being asked about was once learned; hence any failures on the recall or recognition tests can be attributed to deficits in memory and not to problems in initial learning. With the kinds of sociohistorical facts examined by Botwinick and Storandt, this equivalence in initial learning is less certain. Perhaps some adults never learned the real name of "The Red Baron." If so, failures to answer correctly can hardly be attributed to failures of memory. In addition, the label "memory" might seem debatable when participants are being asked about events that occurred before their birth. The Botwinick and Storandt questions do not tap the kinds of specific, personal-experience memories that are the focus in the Bahrick et al. study. Instead, they deal more with general information about the world, information that may have been accumulated in a variety of ways.

Whatever their limitations, studies like Botwinick and Storandt's do illustrate some important points about adult memory. Most generally, they provide another demonstration that memory is not necessarily inferior in old age. For some types of questions the elderly may do as well as or even better than young adults. The finding that memory in the general knowledge-about-the-world sense holds up well with age is compatible with a good deal of other research in the aging literature. It fits, for example, with the work on crystallized intelligence discussed earlier in the chapter. Finally, studies such as Botwinick and Storandt's—like the research on everyday memory discussed in the preceding section—suggest that conclusions about memory in the elderly are likely to be more positive when the focus is on meaningful, real-world material, as opposed to arbitrary and unfamiliar laboratory tasks.

Personality and Social Development

The discussion turns now from intelligence to the domain of personality and social functioning. We begin with two examples of approaches to the study of personality in adulthood. We then consider some of the general issues raised by these and similar measures.

Life Satisfaction

One popular topic in the psychology of aging is variously labeled "morale," "subjective well-being," "life satisfaction," or simply "happiness." The question is a basic and very important one: How satisfied is the person with his or her life? A variety of measuring instruments have been devised in an attempt to assess life satisfaction. Here we sample two, beginning with one of the earliest and most influential of such instruments, the Life Satisfaction Index (LSI) of Neugarten, Havighurst, and Tobin (1961).

The Life Satisfaction Index was developed as part of one of the major early studies of aging, the Kansas City Study of Adult Life. The sample for the study consisted of 177 men and women between the ages of 50 and 90. Each participant responded to a series of in-depth interviews, spaced longitudinally across a period of about 2 years. As the following description (Neugarten et al., 1961) suggests, the interviews were quite extensive, surveying a number of aspects of life in middle and old age.

> Included was information on the daily round and the usual weekend-round of activity; other household members; relatives, friends, and neighbors; income and work; religion; voluntary organizations; estimates of the amount of social interaction as compared with the amount at age 45; attitudes toward old age, illness, death, and immortality; questions about loneliness, boredom, anger;

and questions regarding the respondent's role models and his self-image. (p. 136)

The purpose of the interviews was to elicit responses from which various dimensions of life satisfaction might be scored. Eventually, five dimensions proved to be scorable: Zest versus Apathy, Resolution and Fortitude, Congruence between Desired and Achieved Goals, Self-Concept, and Mood Tone. For each dimension, scoring was on a 5-point scale, with 5 representing the positive end of the dimension and 1 the negative end.

The interview method, while perhaps necessary for the initial delineation of dimensions, is obviously a time-consuming approach to individual assessment. Neugarten et al.'s next step, therefore, was to attempt to develop a briefer technique for eliciting the same information. The result was the LSI: a self-report questionnaire that requires participants simply to agree or disagree with a series of 20 statements. Table 14.8 presents a sampling of the items, along with an indication of the choice that would result in the most positive score on the various dimensions.

Table 14.8 Items From the Life Satisfaction Index

Here are some statements about life in general that people feel differently about. Would you read each statement on the list, and if you agree with it, put a check mark in the space under "AGREE." If you do not agree with a statement, put a check mark in the space under "DISAGREE." If you are not sure one way or the other, put a check mark in the space under "?" PLEASE BE SURE TO ANSWER EVERY QUESTION ON THE LIST.

(Key: score 1 point for each response marked X.)

	Agree	*Disagree*	*?*
1. As I grow older, things seem better than I thought they would be.	X	—	—
2. I have gotten more of the breaks in life than most of the people I know.	X	—	—
3. This is the dreariest time of my life.	—	X	—
4. I am just as happy as when I was younger.	X	—	—
5. My life could be happier than it is now.	—	X	—
6. These are the best years of my life.	X	—	—
7. Most of the things I do are boring or monotonous.	—	X	—
8. I expect some interesting and pleasant things to happen to me in the future.	X	—	—
9. The things I do are as interesting to me as they ever were.	X	—	—
10. I feel old and somewhat tired.	—	X	—

SOURCE: From "The Measurement of Life Satisfaction," by B. L. Neugarten, R. J. Havighurst, and S. S. Tobin, 1961, *Journal of Gerontology, 16*, p. 141.

As noted, the LSI is intended to measure life satisfaction in old age. Clearly, however, life satisfaction is not an issue that is specific to the last part of the life span. Furthermore, any attempt to chart stability or change in life satisfaction across the adult years requires an instrument that is appropriate for different age groups. A number of such instruments and related programs of research have emerged in recent years. I present one of them here; for another major approach, see Ryff (1989, 1995).

One measure that spans the adult years is the Satisfaction with Life Scale, or SWLS (Diener, Emmons, Larsen, & Griffin, 1985). The SWLS is a brief instrument and thus easy to administer; the five items that make up the test are reproduced in Table 14.9. Despite the brevity of the assessment, various forms of evidence support the validity of the SWLS as a measure of subjective well-being in adults of various ages (Myers & Diener, 1995; Pavot, Diener, Colvin, & Sandvik, 1991). The test has been shown, for example, to correlate with other, lengthier assessments of life satisfaction. It also correlates with other sources of information about the individual's morale—reports from friends, for example, or clinical assessments.

What have measures such as the LSI and the SWLS revealed about life satisfaction in old age? We can ask first the major developmental question: Are there changes with age in life satisfaction? Answering this question is complicated by the fact that the great majority of studies are cross-sectional; hence apparent changes with age may reflect cohort differences rather than actual change as people get older. It is also complicated by the fact that results are somewhat inconsistent, varying across samples and across different dimensions of life satisfaction. Some studies do find an on-the-average negative relation between age and life satisfaction, at least for individuals 60 or older. Edwards and Klemmack (1973), for example, reported a correlation of −.14 between age and LSI scores in a sample of 507 middle-aged and elderly adults. On the other hand, many other studies have found no relation at all between age and life satisfaction; indeed, this seems to be the usual conclusion in the more recent generation of research (Myers & Diener, 1995). And even when effects do emerge, the relation between age and morale is modest at best. Thus, the presence of substantial individual differences among the elderly is again a noteworthy point. There are plenty of elderly people who feel quite satisfied with their lives.

The results just discussed lead to a natural next question: What are the determinants of developmental or individual differences in life satisfaction? Much of the research with

Table 14.9 Satisfaction With Life Scale

Items	*Instructions*
1. In most ways my life is close to my ideal. 2. The conditions of my life are excellent. 3. I am satisfied with my life. 4. So far I have gotten the important things I want in life. 5. If I could live my life over, I would change almost nothing.	"Below are five statements with which you may agree or disagree. Using the 1-7 scale below, indicate your agreement by placing the appropriate number on the line preceding that item. Please be open and honest in your responding." Scale from 1 (*disagree*) to 7 (*strongly agree*).

SOURCE: From "The Satisfaction with Life Scale," by E. Diener, R. A. Emmons, R. J. Larsen, and S. Griffin, 1985, *Journal of Personality Assessment, 49*, p. 72. Copyright © 1985 by Lawrence Erlbaum Associates, Inc. Reprinted with permission.

instruments such as the LSI and SWLS has been directed to this issue. A number of correlates have been identified, most of which are quite predictable (Diener, 1984; Myers & Diener, 1995; Ryff, 1995). Health, for example, is positively related to life satisfaction; on the average, relatively healthy people report greater satisfaction than less healthy people. Marital status is also related; life satisfaction scores are higher for married people than for individuals who have been either widowed or divorced. Income also shows modest correlations, in the expected direction, with life satisfaction. Finally, variables of a more behavioral or psychological sort also relate. Perhaps the most interesting finding concerns the relation between activity and life satisfaction. A number of studies have demonstrated that the relation is a positive one—that is, that elderly people who remain active report greater life satisfaction than those whose activity level has decreased. We will return to this finding later.

Measures of Personality

Large and important though the target may be, life satisfaction constitutes just one aspect of adult personality. The measures to which I turn now are broader in scope, for their goal is to capture all of the various facets that go into defining personality and individual differences in personality. Needless to say, there are disagreements about how best to conceptualize such a broad target, as well as a corresponding diversity in the assessment instruments available. Aldwin and Levenson (1994) provide a good overview of some of the most often used measures in the study of aging. The example on which I concentrate is the chief instrument in one influential program of research; it is the NEO Personality Inventory (NEO PI-R), developed by Costa and McCrae (1992).

The NEO PI-R can be administered in either a self-report or third-person format. In either case, the instrument consists of 240 statements descriptive of aspects of adult personality. Table 14.10 presents a sampling of items, worded for the self-report version. The participant responds to each item on a 5-point scale that ranges from *strongly disagree* to *strongly agree.*

It should be clear from even this brief sampling that the NEO PI-R provides considerable information about adult personality. But what is done with all this information—that is, how do we move beyond the 240 individual statements to draw more general conclusions about personality? In addressing this issue, Costa and McCrae, like the developers of many other personality tests, made use of the statistical technique of factor analysis. The goal of factor analysis is to determine the number of distinct components or "factors," that can be identified in the measurement of some target construct. The approach (which is a good deal more complicated than I attempt to describe here) involves analysis of the patterns of correlations among a set of assessment items, with items that show relatively high intercorrelations conceptualized as measures of the same underlying factor.

Using both previous research and theory and factor analysis of their own test battery, Costa and McCrae concluded that five factors are sufficient to capture most individual differences in personality, factors that have since come to be called "The Big Five." The first three of these factors to be identified provided the "NEO" part of the test's name: Neuroticism, Extraversion, and Openness. Two other factors were subsequently added: Agreeableness and Conscientiousness. Table 14.10 shows some of the items that go into the measurement of each factor. (Note that in the actual test administration the items are not labeled and blocked as in the table; rather, the items that make up a factor are randomly dispersed throughout the assessment.)

As always, the existence of an assessment instrument is merely the starting point for research. The NEO PI-R and other personality inventories have been directed to a number of research issues. I focus here on one basic

Table 14.10 Examples of Items From the NEO Personality Inventory

Domain	*Items*
Neuroticism	I often feel tense and jittery. I often get angry at the way people treat me. I rarely feel lonely or blue. It's often hard for me to make up my mind.
Extraversion	I really like most people I meet. I shy away from crowds of people. I have often been a leader of groups I have belonged to. When I do things, I do them vigorously.
Openness	I have a very active imagination. I experience a wide range of emotions or feelings. I'm pretty set in my ways. I often enjoy playing with theories or abstract ideas.
Agreeableness	I believe that most people are basically well-intentioned. Sometimes I trick people into doing what I want. I would rather cooperate with others than compete with them. I'd rather not talk about myself and my achievements.
Conscientiousness	I tend to be somewhat fastidious or exacting. I try to perform all the tasks assigned to me conscientiously. I have trouble making myself do what I should. I rarely make hasty decisions.

SOURCE: From Costa, Jr., P. T., and McCrae, R. R., *Manual for the Revised NEO Personality Inventory (NEO-PI-R) and NEO Five-Factor Inventory* (NEO-FFI), copyright © 1992 by Psychological Assessment Resources, Inc. Reprinted by permission.

question in the study of aging, and that is the issue of the stability of personality across the life span. Can we assume that the particular constellation of traits that defines an individual's personality at age 20 will still be the same at 40 or 60 or 80? Or do our personalities change as we age? And are the personalities of older people in general different from those of younger people?

The research by Costa and McCrae, utilizing both the NEO PI-R and its predecessors, is one major source of evidence on this question (Costa & McCrae, 1989; McCrae & Costa, 1990, 1994; McCrae et al., 1999; Terracciano, McCrae, Brant, & Costa, 2005). Their studies include both cross-sectional comparisons of adults of different ages and longitudinal examinations of the same individuals as they age. Both sorts of evidence lead them to a strong position with respect to the stability of personality traits. Cross-sectionally, they report some changes in the average level of certain traits up until about age 30 but little evidence for systematic change beyond this point. Thus 70-year-olds, with minor exceptions, show the same patterns of response to the NEO PI-R as do 30-year-olds. Longitudinally, their research indicates remarkable stability of personality across the adult years, with correlations in the range of .60 to .80 even across intervals as great as 30 years.

In their words, "Stability appears to characterize all five of the major domains of personality. . . . [Evidence] suggests that an adult's personality profile as a whole will change little over time" (McCrae & Costa, 1994, p. 173).

Having emphasized the Costa and McCrae program of research, I should note that not everyone subscribes to as strong a position on the stability issue as do they (Caspi & Roberts, 2001; Helson, Kwan, John, & Jones, 2002; Roberts, Walton, & Viechtbauer, 2006; Srivastava, John, Gosling, & Potter, 2003). The evidence on which they rely can be interpreted in different ways, and there are also indications that other approaches to the study of personality may reveal more change than does the trait approach embodied in the NEO PI-R. Nevertheless, it seems fair to say that their work has persuaded most students of aging that personality is a good deal more stable than was once believed. It also serves to counteract another stereotype of aging, in this case with regard to personality in the elderly. Some people undoubtedly do become more rigid, more withdrawn, more depressed, or whatever as they grow old. These, however, are individual and exceptional changes, not the general pattern.

Some General Points

Let us turn now to some of the general methodological issues that are raised by this brief sampling of personality measures. One issue is that of measurement equivalence. As has been stressed repeatedly, conclusions about age differences or age changes are valid only if the measures used are equally appropriate for the age groups being compared. As is true in the study of intelligence, many of the personality tests that are employed with the elderly were originally developed with younger samples. Even measures that were devised with life-span research in mind often seem to have content that is more appropriate for younger participants. And even if the content is age-appropriate, test-taking factors (e.g., difficulty in filling out a lengthy questionnaire) may affect the responses of elderly individuals.

A second issue concerns the general approach that is taken to the assessment of personality. Throughout this book we have seen three general approaches that can be taken in the study of any behavioral domain: direct observation of the behaviors of interest in the natural setting, experimental elicitation of the behaviors in some structured laboratory setting, and ratings or verbal reports about typical behavior from someone who knows the participant. The measures considered in this section clearly fall under this third heading of ratings or verbal reports. The rating approach has been by far the most common in the assessment of adult personality. Exceptions do exist. Experimental approaches have occasionally been used in the study of some topics; examples include conformity (e.g., Klein, 1972), rigidity (e.g., Ohta, 1981), and cautiousness (e.g., Okun & Elias, 1977). Observational data, although rare, do exist for some topics—for example, observational assessment of the elderly residents of a nursing home (Martino-Salzman, Blasch, Morris, & McNeal, 1991). Nevertheless, the bulk of what we know about personality and aging is derived from rating measures. The earlier discussions of the pluses and minuses of such measures should therefore be kept in mind when evaluating this literature.

There is a further point. The measures that I have reviewed are not only verbal reports; in most instances they are *self*-reports. The source for information about the participant is the participant himself or herself. All of the problems and biases that can accompany self-report measures are therefore applicable to these tests. A review by Lawton, Whelihan, and Belsky (1980) provides a helpful discussion both of test-taking biases and of ways to minimize the biases when doing personality assessment with the elderly. As Lawton et al. make clear, such biases (e.g., evaluation apprehension, response sets, misunderstanding of instructions, fatigue) can affect not only conclusions about the elderly but also conclusions about age differences in personality. An apparent difference in depression, for example, might reflect an age-related difference in the willingness to admit negative emotions rather than a genuine personality difference between young and old.

The final issue concerns the correlational nature of much of the work that has been considered. The core relation of interest, that between personality and age, is, of course, correlational. Beyond this, many of the research programs that I have discussed rely heavily on the establishment of correlations. Consider the question of the determinants of life satisfaction. Does health make a contribution? Does a relatively high level of activity? The answer is based on correlations. What we are interested in are cause-and-effect relations (e.g., activity as a cause of satisfaction), yet what we are working with are nonexperimental designs that do not permit the establishment of causality. As is always true with correlations, several causal possibilities exist. In the case of the activity-satisfaction correlation, it is quite possible that maintaining an active lifestyle contributes to life satisfaction. But it is also possible that the causal direction is the reverse: that relatively high life satisfaction helps people to maintain an active lifestyle. And, of course, it is possible that the cause and effect flow in both directions: Staying active promotes happiness, and happiness promotes staying active.

Both the limitations of correlational research and ways to overcome these limitations were discussed in chapter 3. As we saw then, there is no point in bemoaning the correlational nature of many research domains, for correlations are sometimes the best that we can do. This is clearly the case for many of the predictors of life satisfaction. We can hardly carry out an experimental study in which we induce ill health or divorce in an elderly sample in order to determine the effects on their subsequent life satisfaction. We can, however, do several things that can move us closer to causality. One is to use various statistical methods (such as the partial correlation technique discussed in chapter 3) to control for potential confounding factors and specify more exactly the relation between the variables of interest. We might ask, for example, what happens to the relation between marital status and morale if we control for socioeconomic status. Such statistical controls are common for the topics that have been considered. Another possibility is to trace the patterns of correlation over time, making use of the fact that causes must precede their effects. A demonstration, for example, that life satisfaction declines in the time period following a divorce suggests a more certain causal relation than does a simple one-point-in-time correlation. Similarly, a demonstration that a relatively high level of activity in early old age relates to relatively good functioning in later old age—which is in fact the outcome in a number of research programs (e.g., McAuley et al., 2005; Schaie, 2005; Schooler & Mulatu, 2001)—provides evidence that activity can in fact play a causal role. We can see here once again the value of longitudinal research for the study of central issues in developmental psychology.

Summary

The first part of this chapter is devoted to general issues in the study of aging. It also provides an overview of research on one extensively studied topic: stability or change in IQ as people age.

Research on aging faces the same methodological challenges as does developmental research in general, but often in heightened form. *Sampling* of participants has two general goals: achieving representative samples at each age and achieving comparable samples across different ages. Two kinds of confounds in particular have been of concern in comparisons of young adults and elderly adults. One is a confounding of age and education: On the average, young adults are better educated than are elderly adults. The other is a confounding of age and health: On the average, young adults are healthier than are elderly adults. Both factors contribute to the average IQ decrement in old age.

The discussion of the health variable leads to a general consideration of sampling problems in longitudinal research. *Selective dropout* is one possible problem. In studies of IQ, the relatively low IQ participants are most likely to be lost from the study. Such dropout has two bases: voluntary withdrawals from the study and involuntary withdrawals for reasons of health. The result is a positive bias in favor of the elderly.

The discussion turns next to issues of *measurement.* The two most commonly used instruments in the study of IQ and aging have been the WAIS and the PMA. Both tests are divided into subscales, and the probability of age differences varies across different subscales. Age differences are less likely on measures of verbal intelligence than on measures of performance intelligence, on measures of crystallized intelligence than on measures of fluid intelligence, and on nonspeeded measures than on speeded measures. Also, age differences may be less likely on measures of so-called everyday cognition than they are on standard IQ assessments. The discussion of alternatives to IQ raises an issue that is central to any comparison of young and old: the problem of *measurement equivalence.* The tests and contexts used to assess the elderly often seem more appropriate for young adults, and this factor is another probable contributor to age differences in performance.

The first section of the chapter concludes with a consideration of issues of *design. Cross-sectional* designs involve a confounding of age and cohort; *longitudinal* designs involve a confounding of age and time of measurement. These various confounds, together with selective dropout in longitudinal studies, provide an explanation for the often-reported finding that age decrements are greater in cross-sectional research than in longitudinal research. An attempt to overcome the limitations of simpler designs is found in the *sequential* studies of Schaie and associates. Sequential designs involve combinations of the simpler designs and hence provide more opportunity to disentangle the contributions of age, time, and cohort.

The second part of the chapter addresses selected research topics in the psychology of aging. Memory has been one popular topic. As with IQ, the probability of age differences in memory depends on a number of factors. Although exceptions exist, memory decrements in old age are less likely on *recognition* than on *recall,* and they are less likely on tasks of *cued-recall* than on noncued measures. Both of these findings suggest that elderly adults may have special difficulty in retrieving information from storage. More generally, it has been suggested that elderly adults have difficulty in situations that require the use of *mnemonic strategies.* Evidence from a variety of paradigms provides some support for this notion. Evidence also indicates that *constructive memory* may decline in old age, perhaps because of a general slowdown in information processing. Finally, a more positive picture of memory and aging emerges from two lines of research: that concerned with *everyday memory,* and that concerned with the intactness of really long-term or *remote memory.*

This chapter concludes with a consideration of methods for studying personality and social relations in the elderly. Several sample tests are described, along with illustrative findings from their use. The LSI and SWLS are among the instruments developed to assess *life satisfaction* or subjective well-being in adulthood. The NEO PI-R is broader in scope; it is one of a number of verbal-report (usually self-report) inventories whose purpose is to measure individual differences in adult personality. A central question addressed with both sorts of measure is that of stability or change as people age. Although change certainly can occur, research suggests that most aspects of personality are stable across the adult years.

The presentation of specific tests is followed by a discussion of general issues in the study of personality in the elderly. One issue is measurement equivalence: Are the tests equally appropriate for the different age groups being compared? Another issue is validity: Do the tests in fact measure what they claim to measure? A particular concern with regard to validity is the self-report nature of many measures. The final issue considered is the correlational nature of many research programs in the study of aging. Such correlational designs cannot conclusively establish the cause-and-effect links (e.g., between activity and morale) that have been of interest in such research.

Exercises

1. The main emphasis of the Schaie sequential research has been on intellectual performance. Suppose that you had the opportunity to carry out a comparable sequential study of some outcome other than IQ. What dependent variable would you pick, and why? What would you expect to find with respect to the relative contributions of age, cohort, and time of measurement?

2. As the chapter notes, negative stereotypes about aging are common. These stereotypes have themselves been the subject of research, the question being how strongly particular beliefs are held by participants of different ages. Among the examples of such research are articles by Austin (*Gerontologist,* 1985, *25,* 431–434), Braithwaite et al. (*Australian Psychologist,* 1993, *28,* 9–15), and Seccombe and Ishii-Kuntz (*Gerontologist,* 1991, *31,* 527–533); a number of additional sources are cited in a book edited by Nelson, *Ageism: Stereotyping and Prejudice Against Older Persons* (2004, Cambridge: MIT Press). Find a perceptions-of-aging measure that you like and administer it to members of as many different age groups as you are able to obtain. Try in particular to include at least one elderly participant; if the measure is age-appropriate, it would also be interesting to include children.

3. The issue of measurement equivalence is a recurring one throughout the chapter. Imagine that you had the task of constructing a "tasks of daily living" measure that would be an appropriate assessment for you and your peers. What sorts of tasks would you include, and how similar would your instrument be to the measure summarized in Table 14.4? Do you think that it would be possible to devise a test of this sort that would be equally valid for 20-year-olds and 70-year-olds?

4. Table 14.6 presents two of the vignettes used in the Berlin group studies of wisdom. One finding from such research is that participants sometimes perform best when the vignette reflects issues common to their own age group—thus, in the examples, the Michael story for young adults and the Joyce story for older adults. Construct a vignette to assess wisdom that presents a dilemma with which a college student might have to deal. Administer all three vignettes to several of your peers and compare the results. (Information about how to score responses can be found in Smith and Baltes, 1990.)

Notes

1. I should forestall any confusion about the sense in which IQ might vary with age. All IQ tests are constructed to yield the same mean IQ across all the ages for which the test is appropriate. Thus IQ, by definition, does not vary with age. To achieve this constancy, however, different norms and standards are used for different age groups. On the WAIS, for example, a 70-year-old needs fewer correct responses to achieve a score of 100 than does a 25-year-old. In the research that we are considering here, these age adjustments are removed; thus what is being compared is raw score performance.

2. Any fuller discussion of health changes with age is beyond my scope. For those who are interested, however, I will note that a primary source of information is the Baltimore Longitudinal Study of Aging, begun in 1958 and now the nation's longest running study of aging (www.grc.nia.nih.gov/branches/blsa/blsa.htm).

3. A program of research by May and colleagues (May, Hasher, & Foong, 2005; May, Hasher, & Stoltzfus, 1993; Yoon, May, & Hasher, 2000) provides an interesting example of a feature of laboratory research that may hinder the performance of elderly participants. The starting point for the research was evidence suggesting that bodily or "circadian" rhythms vary with age, with most older adults experiencing optimal arousal in the morning and most young adults favoring afternoon or evening. Across a series of studies May and associates have demonstrated that performance on many (although not all) cognitive tasks varies with time of day, with older adults generally doing best in the morning and younger adults doing best in the afternoon and evening. The implications seem clear: If (as informal evidence suggests) experimental sessions tend to be scheduled at times unfavorable to elderly adults, then the result will be a magnification of age differences.

Appendix

Answers to the Botwinick and Storandt questions.

1. Neil Armstrong
2. James Earl Ray
3. Dallas
4. 1957
5. Adlai Stevenson
6. Joseph McCarthy
7. Alben Barkley
8. Erwin Rommel
9. Claire Chennault
10. December 7, 1941
11. Franklin Roosevelt
12. New Jersey
13. *Titanic*
14. Archduke Ferdinand
15. Baron Manfred Von Richthofen
16. Women's Christian Temperance Union
17. John T. Scopes
18. *Spirit of St. Louis*
19. Robert E. Peary
20. New York
21. Theodore Roosevelt
22. Kitty Hawk
23. James Corbett
24. 1908

Glossary

alpha level Probability value (e.g., .05) below which results are considered statistically significant—indicates the probability of Type 1 error.

appearance-reality distinction Understanding of the distinction between how objects appear and what they really are.

archival data Already collected and available information about people that can be used for research purposes.

attachment Emotional bond, typically first formed in infancy, between infant and caregiver.

Attachment Q-Set Ratings method for the assessment of attachment, based upon the rater's judgment of the extent to which attributes indicative of attachment are characteristic of the child.

baby biography Method of study in which a parent makes observations of the development of his or her own children.

baseline Initial level of a target behavior prior to an experimental intervention.

between-subject design Experimental design in which different participants are assigned to the different levels of the independent variable.

blinding Withholding of potentially biasing information from a tester or an observer.

carryover effect Tendency for response to one task or condition to be affected by a preceding task or condition.

ceiling effect Performance on a dependent variable that is at or close to the maximum value possible, thus reducing the probability of finding differences between groups.

central tendency Score that indicates the dominant pattern of response in a sample.

chronosystem Temporal and historical change as a context for development—a component of Bronfenbrenner's ecological systems theory.

class inclusion Knowledge that a subclass cannot be larger than the superordinate class that contains it.

clinical method Flexible, semistandardized method of posing problems and questioning children used by Piaget.

cohort Group of people who share particular experiences and characteristics—most often defined by year of birth.

concept Mental grouping of items on the basis of some underlying similarity or set of similarities.

conditioned head turning Method for the study of perceptual ability in infancy—infers detection and discrimination of stimuli from the presence or absence of a conditioned head turn response when the stimuli appear.

confidence interval Statistical concept that expresses the range of possible values within which the true population value can be assumed to fall with a given probability.

confidentiality Principle of ethics that states that information obtained in research must not be made public in a way that could harm or embarrass participants.

confounding Unintended conjunction of two potentially important variables.

consensual drift Tendency for observers who are frequently paired to begin to err in the same way.

conservation Knowledge that the quantitative properties of an object or collection of objects are not altered by a change in perceptual appearance.

construct validity (of studies) Accuracy of the theoretical interpretation of the results of research.

construct validity (of tests) Form of test validity based on the extent to which a variety of forms of evidence support the theoretical interpretation of the attribute being measured.

constructive memory Effects of the general knowledge system upon memory.

content validity Form of test validity based on the adequacy with which the test items represent the attribute being measured.

convenience sampling Procedure for selection of research participants in which selection is based largely on availability or cooperation.

converging operations Use of a variety of different methods in studying a particular topic.

correlation statistic An index of the strength and direction of the relation between two variables.

correlational research Form of research in which there is no control of an independent variable but rather examination of possible relations between two or more measured dependent variables—also labeled *nonexperimental research.*

counterbalancing In within-subject designs, distribution of the tasks or conditions across the various ordinal positions of presentation.

criterion validity Form of test validity based on the correlation of test scores with some external criterion of the attribute being measured.

cross-sectional design Design for age comparisons in which different participants are studied at the different ages, all at the same point in time.

crystallized intelligence Forms of knowledge that are accumulated through experience.

debriefing Ethical principle that states that participants should receive a postexperiment explanation of the true purposes of the research when the study has involved incomplete disclosure or deception.

deception Provision of deliberately incorrect information about a study.

demand effect The tendency of research participants to respond in ways that are intended to confirm the experimenter's hypothesis.

dependent variable Variable that the researcher measures in response to variations in the independent variable.

descriptive statistics Statistics whose purpose is to organize and summarize data.

diffusion Unintended spread of a treatment effect from an experimental group to an untreated control group.

dishabituation Recovery of the orienting response when a habituated stimulus changes.

disuse hypothesis Theory that cognitive skills that are not regularly used become harder to access when needed.

effect size Measure of the magnitude of the relation between the independent and dependent variables.

ethnographic research Description and interpretation of the central practices of some cultural group—a method of study used in qualitative research.

evaluation apprehension effect The tendency of research participants to respond in ways that are intended to maximize the experimenter's positive evaluation of them.

event sampling Form of observational measurement in which the observer begins recording when the behavior of interest occurs and continues recording for the duration of the behavior.

exosystem Social systems that can affect an individual but in which the individual does not participate directly—the third of the four layers of context in Bronfenbrenner's ecological systems theory.

experience sampling method Procedure for collecting data about everyday experiences in the natural setting—participants provide self-reports of ongoing activities when signaled to do so.

experimenter expectancy effect Self-fulfilling and biasing influence of the experimenter's expectancies upon the outcomes of research.

expertise Organized factual knowledge with respect to some content domain.

exploratory research Research whose purpose is to identify the phenomena of interest in a new area of study; characterized by a relatively flexible, nonstandardized method of study.

external validity Accuracy with which the results of research can be generalized.

eye-movement recording Method for the study of visual ability in infancy—measures the sequence of fixations as the infant scans a visual stimulus.

factor Another term for independent variable.

false belief Realization that people can hold beliefs that are not true.

floor effect Performance on a dependent variable that is at or close to the minimum value possible, thus reducing the probability of finding differences between groups.

fluid intelligence Basic problem-solving abilities that are largely independent of experience.

forensic developmental psychology Subfield of developmental psychology directed to the study of children as witnesses, especially in cases of suspected abuse.

functional imaging techniques Procedures (such as positive emission tomography and functional magnetic resonance imaging) for measuring brain activity during cognitive processing.

gender constancy The understanding that gender is a permanent attribute.

habituation Decline in the orienting response as an initially novel stimulus is repeated and becomes familiar.

history Potentially important events occurring between early and later measurements in addition to the independent variables being studied.

incomplete disclosure Withholding of information about a study.

independent review Principle of ethics that states that all research with human participants must undergo an independent review for ethical standards.

independent variable Variable that the researcher controls through manipulation or selection in order to examine effects on the dependent variable.

inferential statistics Statistics whose purpose is to determine whether observed differences among groups are greater than would be expected by chance.

informed consent Agreement to participate in research following receipt of information about what participation will entail.

institutional review board Organization responsible for evaluating the ethics of proposed research prior to the start of data collection.

instrumentation Unintended changes in experimenters, observers, or measuring instruments across the course of a study.

interaction Research outcome in which the effect that one independent variable has on the dependent variable varies across the levels of another independent variable.

internal consistency reliability Form of reliability based on consistency of response across the different items of a test.

internal validity Accuracy of conclusions concerning cause-and-effect relations between the variables of a study.

intersubject communication Unintended and potentially biasing communication between research participants.

interval scale Scale of measurement in which the numbers assigned represent equal quantitative intervals.

levels Values assumed by the independent variables in a study.

life course Elder's term for the sequence of socially defined, age-graded roles and events that people experience throughout life.

longitudinal design Design for age comparisons in which the same participants are studied at different ages over time.

macrosystem Culture or subculture in which the individual lives—the fourth of the four layers of context in Bronfenbrenner's ecological systems theory.

main effect Research outcome in which an independent variable has a direct effect on a dependent variable that is independent of other independent variables in the study.

matched-groups design Form of between-subject design in which participants are matched on a potentially important characteristic or characteristics prior to assignment to experimental conditions.

maturation Naturally occurring changes in the participants as a function of the passage of time during the study.

mean Arithmetical average of a set of scores—a measure of central tendency.

measurement equivalence Comparability of procedures and measurements across the groups being compared.

median Midmost value in a set of scores—a measure of central tendency.

mesosystem Interrelations among an individual's microsystems—the second of the four layers of context in Bronfenbrenner's ecological systems theory.

meta-analysis A method of reviewing the research literature that uses statistical procedures to establish the existence and the size of effects.

microgenetic method A research method in which a sample of individuals are observed repeatedly at closely spaced intervals in an attempt to identify processes of developmental change.

microsystem Environmental system that is closest to the individual—the first of the four layers of context in Bronfenbrenner's ecological systems theory.

mnemonic strategies Techniques (such as rehearsal or organization) that people use in an attempt to remember something.

mode Most common value in a set of scores—a measure of central tendency.

molar observational system System for observing behavior in which the categories are relatively global and interpretive.

molecular observational system System for observing behavior in which the categories are relatively fine-grained and specific.

mono-method bias Bias resulting from the use of a single methodology to examine possible relations between the independent and dependent variables.

mono-operation bias Bias resulting from the use of a single operationalization of either the independent or dependent variable.

moral dilemmas Stories used by Piaget and others to assess levels of moral reasoning.

narrative record Form of observational measurement in which the observer records a continuous descriptive account of the behaviors of interest.

narrative research Collection of participants' stories about their experiences—a method of study used in qualitative research.

nominal scale Scale of measurement in which the numbers assigned are merely labels with no quantitative significance.

nonparametric tests Tests of statistical inference whose validity does not depend on assumptions about the distributions of scores in the target population.

normal distribution Bell-shaped distribution of scores in which the mean, median, and mode are synonymous.

null hypothesis The assumption, adopted for purposes of statistical inference, that there is no true relation between the variables being examined.

object concept Knowledge that objects have a permanent existence that is independent of our perceptual contact with them.

observer bias Tendency of observers to perceive and record behavior in a biased fashion that conforms to their preexisting expectations.

observer drift Tendency for observers to become less reliable once they are no longer being monitored.

observer influence Biasing effects of the presence of the observer upon the behavior being observed.

operational definition Definition of a variable in terms of the operations used to produce or measure the variable.

order effect Tendency for response to change in a systematic fashion from early in an experimental session to later in the session.

ordinal scale Scale of measurement in which the numbers assigned represent a rank ordering in terms of quantity.

orienting response Natural attentional response to new stimuli.

oversampling Procedure for selection of research participants in which members of a group are selected at a rate greater than their proportion in the target population, the goal being to ensure that a sufficient number are included in the final sample.

parametric tests Tests of statistical inference whose validity depends on particular assumptions about the distribution of scores in the target population.

partial correlation technique Procedure for statistically removing the contribution of a potentially important third factor from the correlation between two variables.

participant observation Form of observational research in which the observations are made by someone who is already part of the setting.

personal agency The understanding that one can be the cause of events.

population The total set of people, observations, or events within some domain.

power Probability that a statistical test will result in a correct rejection of the null hypothesis.

preference method Method for the study of visual ability in infancy—measures the infant's attention to two simultaneously presented visual stimuli.

primary variance Differences attributable to the independent variables in a study.

prosocial behavior Aspect of behavioral morality that comprises the production of socially beneficial behaviors such as sharing and helping.

random assignment Procedure for assigning participants to experimental conditions in which each participant has an equal chance of being assigned to each condition.

random sampling Procedure for selection of research participants in which each member of the target population has an equal chance of being selected.

ratio scale Scale of measurement in which the numbers assigned represent equal quantitative intervals and the scale includes a true zero point.

reactivity Unintended effects of the experimental arrangements upon participants' responses.

recall memory Retrieval of some past stimulus or event that is not perceptually present.

recognition memory Realization that some perceptually present stimulus or event has been encountered before.

regression toward the mean Tendency of initially extreme scores to move toward the group mean upon retesting.

reliability Consistency or repeatability of measurement.

remote memory Memory for events from the distant past.

replication Duplication of the essential elements of a previous study, the goal being to determine whether the same results are obtained.

representational change The understanding that one's own mental states can change.

response set Preexisting, biased form of response that is independent of task content.

robustness Degree to which an inferential statistical test is unaffected by violations of its underlying mathematical assumptions.

sample Subset of a target population that is selected for study.

sampling Selection of a sample of research participants from a larger population.

scripts Representations of the typical order and structure of familiar events.

selection bias Assignment of initially non-equivalent participants to the groups being compared.

selective dropout Nonrandom, systematically biased loss of participants in the course of the study.

self-report measures Measures on which conclusions about participants' characteristics are based on the participants' own direct reports of their characteristics.

sequential design Design in which longitudinal, cross-sectional, and time-lag components are combined in an attempt to distinguish effects of age, cohort, and time of measurement.

standard deviation Square root of the variance—a measure of variability.

standardization Making all aspects of the experimental procedure the same for all participants within a particular experimental condition.

state Level of physiological and behavioral arousal—ranges from quiet sleep to intense crying.

statistical conclusion validity Accuracy of the statistical conclusions drawn from the analysis of data.

Strange Situation Structured laboratory method for the assessment of attachment—measures the infant's response to a series of separations and reunions with mother and stranger.

stratified sampling Procedure for selection of research participants in which members of different groups are selected at rates that reflect their proportions in the target population.

subject variable Independent variable whose levels reflect preexisting, inherent differences among participants, such as age or sex.

terminal drop Decline in mental abilities in the time period immediately preceding death.

test validity Accuracy with which a test measures what it is intended to measure.

testing Effects of taking a test upon performance on a later test.

test-retest reliability Form of reliability based on consistency of scores across two administrations of the same test.

theory of mind Thoughts and beliefs about the mental world.

time sampling Form of observational measurement in which the behaviors of interest are recording during preselected (usually brief) units of time.

time-lag design Design in which participants of the same age but different cohorts are studied at different times of measurement.

time-series design Form of within-subject design in which the dependent variable is measured on multiple occasions in response to the presence or absence of an experimental treatment.

transitivity Ability to combine relations logically to deduce necessary conclusions—for example, if $A > B$ and $B > C$, then $A > C$.

Type 1 error False rejection of a true null hypothesis.

Type 2 error Failure to reject a false null hypothesis.

validity Accuracy of the conclusions drawn from research.

variability Degree of dispersion in a set of scores.

variance Average of the squared deviation scores when each score in a group is subtracted from the group mean—a measure of variability.

violation-of-expectation method Method of study in cognitive development—measures the infant's or child's reaction to an apparent violation of a logical or physical law.

visual cliff Method for the study of depth perception in infancy—measures the infant's response to the apparent drop-off from a glass-covered table.

visual self-recognition Infant's ability to recognize himself or herself, typically inferred from response to a mirror image.

withholding treatment Decision not to provide a potentially beneficial experimental treatment to a subset of research participants.

within-subject design Experimental design in which the same participants are assigned to the different levels of the independent variable.

References

Abramovitch, R., Freedman, J. L. Henry, K., & Van Brunschot, C. (1995). Children's capacity to agree to psychological research: Knowledge of risks and benefits and voluntariness. *Ethics and Behavior, 5,* 25–48.

Abramovitch, R., Freedman, J. L., Thoden, K., & Nikolich, C. (1991). Children's capacity to consent to participation in psychological research: Empirical findings. *Child Development, 62,* 1100–1109.

Acredolo, L. P., & Hake, J. L. (1982). Infant perception. In B. B. Wolman (Ed.), *Handbook of developmental psychology* (pp. 244–283). Englewood Cliffs, NJ: Prentice-Hall.

Adams, R. J. (1995). Further exploration of human neonatal chromatic-achromatic discrimination. *Journal of Experimental Child Psychology, 60,* 344–360.

Ainsworth, M. D. S., Blehar, M. C., Waters, E., & Wall, S. (1978). *Patterns of attachment: A psychological study of the Strange Situation.* Hillsdale, NJ: Lawrence Erlbaum.

Aksan, N., & Kochanska, G. (2005). Conscience in childhood: Old questions, new answers. *Developmental Psychology, 41,* 506–516.

Aldwin, C. M., & Levenson, M. R. (1994). Aging and personality assessment. *Annual Review of Gerontology and Geriatrics, 14,* 182–209.

Allaire, J. C., & Marsiske, M. (2002). Well- and ill-defined measures of everyday cognition: Relationship to older adults' intellectual ability and functional status. *Psychology and Aging, 17,* 101–115.

Als, H., Tronick, E., & Brazelton, T. B. (1979). Analysis of face-to-face interaction in infant-adult dyads. In M. E. Lamb, S. J. Suomi, & G. R. Stephenson (Eds.), *Social interaction analysis: Methodological issues* (pp. 33–76). Madison: The University of Wisconsin Press.

American Psychological Association. (2001). *Publication manual of the American Psychological Association* (5th ed.). Washington, DC: Author.

American Psychological Association. (2002). Ethical principles of psychologists and code of conduct. *American Psychologist, 57,* 1060–1073.

Amsterdam, B. (1972). Mirror self-image reactions before age 2. *Developmental Psychobiology, 5,* 297–305.

Anders, T. F., & Roffwarg, H. P. (1973). The effects of selective interruption and deprivation of sleep on the human newborn. *Developmental Psychobiology, 6,* 77–89.

Anderson, C. A., Berkowitz, L., Donnerstein, E., Huesmann, L. R., Johnson, J. D., Linz, D., et al. (2003). The influence of media violence on youth. *Psychological Science in the Public Interest, 4,* 81–110.

Anderson, C. A., Lindsay, J. J., & Bushman, B. J. (1999). Research in the psychological laboratory:Truth or triviality? *Current Directions in Psychological Science, 8,* 3–9.

Anisman, H., Zaharia, M. D., Meaney, M. J., & Merali, Z. (1998). Do early-life events permanently alter behavioral and hormonal responses to stressors? *International Journal of Developmental Neuroscience, 16,* 149–164.

Applebaum, M. I., & McCall, R. B. (1983). Design and analysis in developmental psychology. In P. H. Mussen (Series Ed.) & W. Kessen (Vol. Ed.), *Handbook of child psychology: Vol. 1. History, theory, and methods* (4th ed., pp. 415–476). New York: Wiley.

Archer, J. (2004). Sex differences in aggression in real-world settings: A meta-analytic review. *Review of General Psychology, 8,* 291–322.

Ardelt, M. (2000). Antecedents and effects of wisdom in old age. *Research on Aging, 22,* 360–394.

Ardelt, M. (2004). Wisdom as expert knowledge system: A critical review of a contemporary operationalization of an ancient concept. *Human Development, 47,* 257–285.

Artistico, D., Cervone, D., & Pezzuti, L. (2003). Perceived self-efficacy and everyday problem solving among young and older adults. *Psychology and Aging, 18,* 68–79.

Astington, J. W. (Ed.). (2000). *Minds in the making.* Malden, MA: Blackwell.

Astington, J. W. (2003). Sometimes necessary, never sufficient: False belief-understanding and social competence. In B. Repacholi & V. Slaughter (Eds.), *Individual differences in theory of mind* (pp. 13–38). New York: Psychology Press.

Astington, J. W., & Jenkins, J. M. (1995). Theory of mind development and social understanding. *Cognition and Emotion, 9,* 151–165.

Astuti, R., Solomon, G. E. A., & Carey, S. (2004). Constraints on conceptual development. *Monographs of the Society for Research in Child Development, 69*(3, Serial No. 277).

Backschneider, A. G., Shatz, M., & Gelman, S. A. (1993). Preschoolers' ability to distinguish living kinds as a function of regrowth. *Child Development, 64,* 1242–1257.

Bahrick, H. P. (1983). The cognitive map of a city—50 years of learning and memory. In G. Bower (Ed.), *The psychology of learning and motivation: Advances in research and theory* (Vol. 17, pp. 125–163). New York: Academic Press.

Bahrick, H. P. (1984). Semantic memory content in permastore: Fifty years of memory for Spanish learned in school. *Journal of Experimental Psychology: General, 113,* 1–26.

Bahrick, H. P. (2000). Long-term maintenance of knowledge. In E. Tulving & F. I. M. Craik (Eds.), *Oxford handbook of memory* (pp. 411–425). Oxford, England: Oxford University Press.

Bahrick, H. P., Bahrick, P. O., & Wittlinger, R. P. (1975). Fifty years of memory for names and faces: A cross-sectional approach. *Journal of Experimental Psychology: General, 104,* 54–75.

Baillargeon, R. (1987). Object permanence in 3.5- and 4.5-month-old infants. *Developmental Psychology, 23,* 655–664.

Baillargeon, R. (2004). Infants' reasoning about hidden objects: Evidence for event-general and event-specific expectations. *Developmental Science, 7,* 391–414.

Baird, J. A., & Moses, L. J. (2001). Do preschoolers appreciate that identical actions may be motivated by different intentions? *Journal of Cognition and Development, 2,* 413–448.

Bakeman, R., Deckner, D. F., & Quera, V. (2005). Analysis of behavioral streams. In D. Teti (Ed.), *Handbook of research methods in developmental science* (pp. 394-420). Malden, MA: Blackwell Publishers.

Baltes, P. B., & Lindenberger, U. (1997). Emergence of a powerful connection between sensory and cognitive functions across the adult life span: A new window to the study of cognitive aging? *Psychology and Aging, 12,* 12–21.

Baltes, P. B., & Smith, J. (1990). The psychology of wisdom and its ontogenesis. In R. J. Sternberg (Ed.), *Wisdom: Its nature, origins, and development* (pp. 87–120). New York: Cambridge University Press.

Baltes, P. B., & Staudinger, U. M. (2000). Wisdom: A metaheuristic (pragmatic) to orchestrate mind and virtue toward excellence. *American Psychologist, 55,* 122–136.

Baltes, P. B., Staudinger, U. M., & Lindenberger, U. (1999). Lifespan psychology: Theory and application to intellectual functioning. *Annual Review of Psychology, 50,* 471–507.

Bandura, A., Ross, D., & Ross, S. A. (1963). Imitation of film-mediated aggressive models. *Journal of Abnormal and Social Psychology, 66,* 3–11.

Barber, T. X. (1976). *Pitfalls in human research: Ten pivotal points.* New York: Pergamon Press.

Barber, T. X., & Silver, M. J. (1968). Fact, fiction, and the experimenter bias effect. *Psychological Bulletin Monograph Supplement, 70,* 1–29.

Barker, R. G., & Wright, H. F. (1951). *One boy's day.* New York: Harper & Row.

Barlow, D. H., & Hersen, M. (1984). *Single case experimental designs.* Elmsford, NY: Pergamon Press.

Barnett, M. A., King, L. M., & Howard, J. A. (1979). Inducing affect about self or other: Effects on generosity in children. *Developmental Psychology, 15,* 164–167.

Baron-Cohen, S. (1995). *Mindblindness: An essay on autism and theory of mind.* Cambridge, MA: MIT Press.

Barrett, D. E., & Yarrow, M. R. (1977). Prosocial behavior, social inferential ability, and assertiveness in children. *Child Development, 48,* 475–481.

Bartlett, J. C., & Snelus, P. (1980). Lifespan memory for popular songs. *American Journal of Psychology, 93,* 551–560.

Bartsch, K., & Wellman, H. M. (1989). Young children's attribution of action to beliefs and desires. *Child Development, 60,* 946–964.

Bauer, P. J. (2004). Getting explicit memory off the ground: Steps toward construction of a neuro-developmental account of changes in the first two years of life. *Developmental Review, 24,* 347–373.

Bauer, P. J., & Mandler, J. M. (1992). Putting the horse before the cart: The use of temporal order in recall of events by one-year-old children. *Developmental Psychology, 28,* 441–452.

Baumrind, D. (1967). Child care practices anteceding three patterns of preschool behavior. *Genetic Psychology Monographs, 75,* 43–88.

Baumrind, D. (1971). Current patterns of parental authority. *Developmental Psychology Monograph, 4,* 1–103.

Baumrind, D. (1985). Research using intentional deception: Ethical issues revisited. *American Psychologist, 40,* 165–174.

Baumrind, D. (1989). Rearing competent children. In W. Damon (Ed.), *Child development today and tomorrow* (pp. 349–378). San Francisco: Jossey-Bass.

Baumrind, D., & Black, A. E. (1967). Socialization practices associated with dimensions of competence in preschool boys and girls. *Child Development, 38,* 291–327.

Bayley, N. (2005). *Bayley Scales of Infant and Toddler Development* (3rd ed.). San Antonio, TX: Harcourt Assessment.

Beere, C. A. (1990). *Gender roles: A handbook of tests and measures.* New York: Greenwood.

Behl-Chadha, G. (1996). Basic-level and superordinate-like categorical representations in infancy. *Cognition, 60,* 65–141.

Bell, R. Q. (1968). A reinterpretation of the direction of effects of socialization. *Psychological Review, 75,* 81–95.

Belsky, J. (2001). Developmental risks (still) associated with early child care. *Journal of Child Psychology and Psychiatry, 42,* 845–859.

Belsky, J. (2002). Quantity counts: Amount of child care and children's socioemotional development. *Journal of Developmental and Behavioral Pediatrics, 23,* 167–170.

Belsky, J., & Braungart, J. (1991). Are insecure-avoidant infants with extensive day-care experience less stressed by and more independent in the Strange Situation? *Child Development, 62,* 567–571.

Belsky, J., & Rovine, M. J. (1990). Q-set security and first-year nonmaternal care. In K. McCartney (Ed.), *New directions for child development: No. 49. Child care and maternal employment: A social ecology approach* (pp. 7–22). San Francisco: Jossey-Bass.

Bem, D. J. (1995). Writing a review article for *Psychological Bulletin. Psychological Bulletin, 118,* 172–177.

Bem, D. J. (2004). Writing the empirical journal article. In J. M. Darley, M. P. Zanna, & H. L. Roediger (Eds.), *The compleat academic: A career guide* (2nd ed., pp. 185–219). Washington, DC: American Psychological Association.

Berg, C. A., & Klaczynski, P. A. (1996). Practical intelligence and problem solving: Searching for perspectives. In F. Blanchard-Fields & T. M. Hess (Eds.), *Perspectives on cognitive change in adulthood and aging* (pp. 323–357). New York: McGraw-Hill.

Berg, K. M., Berg, W. K., & Graham, F. K. (1971). Infant heart rate response as a function of stimulus and state. *Psychophysiology, 8,* 30–44.

Berg, S. (1996). Aging, behavior, and terminal decline. In J. E. Birren & K. W. Schaie (Eds.), *Handbook of the psychology of aging* (4th ed., pp. 323-337). San Diego, CA: Academic Press.

Berg, W. K., & Berg, K. M. (1987). Psychophysiological development in infancy: State, startle, and attention. In J. D. Osofsky (Ed.), *Handbook of infant development* (2nd ed., pp. 238–317). New York: Wiley.

Berg-Cross, L. G. (1975). Intentionality, degree of damage, and moral judgments. *Child Development, 46,* 970–974.

Bergeman, C. S., & Baker, S. M. (Eds.). (2005). *Methodological issues in aging research.* Mahwah, NJ: Lawrence Erlbaum.

Berkowitz, L., & Donnerstein, E. (1982). External validity is more than skin deep. *American Psychologist, 37,* 245–257.

Bernstein, T. M. (1971). *Miss Thistlebottom's hobgoblins: The careful writer's guide to the taboos, bugbears, and outmoded rules of English usage.* New York: Farrar, Strauss and Giroux.

Bertenthal, B. I., & Campos, J. J. (1990). A systems approach to the organizing effects of self-produced locomotion during infancy. In C. Rovee-Collier (Ed.), *Advances in infancy research* (Vol. 6, pp. 1–60). Norwood, NJ: Ablex.

Bertenthal, B. I., Campos, J. J., & Barrett, K. C. (1984). Self-produced locomotion: An organizer of emotional, cognitive, and social development in infancy. In R. Emde & R. Harmon (Eds.), *Continuities and discontinuities in development* (pp. 175–210). New York: Plenum.

Birren, J. E., & Morrison, D. F. (1961). Analysis of the WAIS subtests in relation to age and education. *Journal of Gerontology, 16,* 363–369.

Bjorklund, D. F. (Ed.). (2000). *False memory creation in children and adults.* Mahwah, NJ: Lawrence Erlbaum.

Bjorklund, D. F., Coyle, T. R., & Gaultney, J. F. (1992). Developmental differences in the acquisition and maintenance of an organizational strategy: Evidence for the utilization deficiency hypothesis. *Journal of Experimental Child Psychology, 54,* 434–448.

Blasi, A. (1980). Bridging moral cognition and moral action. *Psychological Bulletin, 88,* 593–637.

Block, J. H. (1976). Issues, problems, and pitfalls in assessing sex differences: A critical review of *The psychology of sex differences. Merrill-Palmer Quarterly, 22,* 285–308.

Bogartz, R., Shinskey, J. L., & Schilling, T. H. (2000). Object permanence in five-and-a-half-month-old infants? *Infancy, 1,* 403–428.

Bohlin, G., Hagekull, B., & Rydell, A. (2000). Attachment and social functioning: A longitudinal study from infancy to middle childhood. *Social Development, 9,* 24–39.

Bohman, M. (1996). Predispositions to criminality: Swedish adoption studies in retrospect. In G. R. Bock & J. A. Goode (Eds.), *Genetics of criminal and antisocial behavior* (pp. 99–114). Chichester, UK: Wiley.

Borsboom, D., Mellenbergh, G. J., & van Heerden, J. (2004). The concept of validity. *Psychological Review, 111,* 1061–1071.

Bosacki, S., & Astington, J. W. (1999). Theory of mind in preadolescence: Relations between social understanding and social competence. *Social Development, 8,* 237–255.

Bottoms, B. L., Goodman, G. S., Schwartz-Kenney, B. M., Sachsenmaier, T., & Thomas, S. (1990, March). *Keeping secrets: Implications for children's testimony.* Paper presented at the meeting of the American Psychology and Law Society, Williamsburg, VA.

Botwinick, J. (1984). *Aging and behavior* (3rd ed.). New York: Springer.

Botwinick, J., & Storandt, M. (1974). *Memory, related functions, and age.* Springfield, IL: Charles C. Thomas.

Botwinick, J., West, R. L., & Storandt, M. (1978). Predicting death from behavioral test performance. *Journal of Gerontology, 33,* 755–762.

Bouchard, T. J., Jr. (1997). IQ similarity in twins reared apart: Findings and responses to critics. In R. J. Sternberg & E. L. Grigorenko (Eds.), *Intelligence, heredity, and environment* (pp. 126–160). New York: Cambridge University Press.

Bower, T. G. R. (1966). The visual world of infants. *Scientific American, 215,* 90–92.

Brewer, B. W., Scherzer, C. B., Van Raalte, J. L., Petitpas, A. J., & Andersen, M. B. (2001). The elements of (APA) style: A survey of psychology journal editors. *American Psychologist, 56,* 266–267.

Brewer, J., & Hunter, A. (1989). *Multimethod research: A synthesis of styles.* Newbury Park, CA: Sage.

Bridgman, P. W. (1927). *The logic of modern physics.* New York: Macmillan.

Brody, G. H., Stoneman, Z., & Wheatley, P. (1984). Peer interaction in the presence and absence of observers. *Child Development, 55,* 1425–1428.

Bronfenbrenner, U. (1977). Toward an experimental ecology of human development. *American Psychologist, 32,* 513–531.

Bronfenbrenner, U. (1979). *The ecology of human development.* Cambridge, MA: Harvard University Press.

Bronfenbrenner, U. (1989). Ecological systems theory. In R. Vasta (Ed.), *Annals of child development: Vol. 6. Six theories of child development: Revised formulations and current issues* (pp. 187–249). Greenwich, CT: JAI Press.

Bronfenbrenner, U. (1993). The ecology of cognitive development: Research models and fugitive findings. In R. H. Wozniak & K. W. Fischer (Eds.), *Development in context* (pp. 3–44). Hillsdale, NJ: Lawrence Erlbaum.

Bronfenbrenner, U., & Evans, G. W. (2000). Developmental science in the 21st century: Emerging questions, theoretical models, research designs, and empirical findings. *Social Development, 9,* 115–125.

Bronfenbrenner, U., & Morris, P. A. (1998). The ecology of developmental processes. In R. M. Lerner (Vol. Ed.) and W. Damon (Series Ed.), *Handbook of child psychology: Vol. 1. Theoretical models of human development* (5th ed., pp. 993–1028). New York: Wiley.

Brooks, P. H., & Kendall, E. D. (1982). Working with children. In R. Vasta (Ed.), *Strategies and techniques of child study* (pp. 325–343). New York: Academic Press.

Brooks-Gunn, J., Phelps, E., & Elder, G. H., Jr. (1991). Studying lives through time: Secondary data analyses in developmental psychology. *Developmental Psychology, 27,* 899–910.

Brown, R. (1973). *A first language: The early stages.* Cambridge, MA: Harvard University Press.

Bruck, M., & Ceci, S. J. (2004). Forensic developmental psychology. *Current Directions in Psychological Science, 13,* 229–232.

Bruzzese, J. M., & Fisher, C. B. (2003). Assessing and enhancing the research consent capacity of children and youth. *Applied Developmental Science, 7,* 13–26.

Bryant, B. K. (1982). An index of empathy for children and adolescents. *Child Development, 53,* 412–425.

Bullock, M. (1995, July/August). What's so special about a longitudinal study? *Psychological Science Agenda, 8,* 9–10.

Buss, A. H., & Plomin, R. (1984). *Temperament: Early developing personality traits.* Hillsdale, NJ: Lawrence Erlbaum.

Cairns, R. B., Cairns, B. D., Neckerman, H. J., Ferguson, L. L., & Gariepy, J. L. (1989). Growth and aggression: 1. Childhood to early adolescence. *Developmental Psychology, 25,* 320–330.

Camic, P. N., Rhodes, J. E., & Yardley, L. (Eds.). (2003). *Qualitative research in psychology.* Washington, DC: American Psychological Association.

Camp, C. J. (1998). Memory interventions and pathological older adults. In R. Schulz, G. Maddox, & M. P. Lawton (Eds.), *Annotated review of gerontology and geriatrics* (Vol. 18, pp. 155–189). New York: Springer.

Campbell, D. T., & Stanley, J. C. (1966). *Experimental and quasi-experimental designs for research.* Chicago: Rand McNally.

Campbell, D. W., & Eaton, W. O. (1999). Sex differences in the activity level of infants. *Infant and Child Development, 8,* 1–17.

Canadian Psychological Association. (2001). *Canadian Code of Ethics for Psychologists* (3rd ed.). Ottawa, Ontario: Author.

Carlson, S. M., & Moses, L. J. (2001). Individual differences in inhibitory control and children's theory of mind. *Child Development, 72,* 1032–1053.

Carter, D. B., & Levy, G. D. (1988). Cognitive aspects of early sex-role development: The influence of gender schemas on preschoolers' memories and preferences for sex-typed toys and activities. *Child Development, 59,* 782–792.

Casey, B. J., & de Haan, M. (Eds.). (2002). Imaging techniques and their application to developmental science [Special issue]. *Developmental Science, 5*(3).

Caspi, A., & Roberts, B. W. (2001). Personality development across the life course: The argument for change and continuity. *Psychological Inquiry, 12,* 49–66.

Cavanaugh, J. C., & Blanchard-Fields., F. (2006). *Adult development and aging* (5th ed.). Belmont, CA: Wadsworth.

Ceci, S. J., & Bruck, M. (1998). Children's testimony. In W. Damon (Series Ed.) & I. E. Sigel & K. A. Renninger (Vol. Eds.), *Handbook of child psychology: Vol. 4. Child psychology in practice* (5th ed., pp. 713–774). New York: Wiley.

Chandler, M., & Chapman, M. (Eds.). (1991). *Criteria for competence: Controversies in the conceptualization and assessment of children's abilities.* Hillsdale, NJ: Lawrence Erlbaum.

Chandler, M., Greenspan, S., & Barenboim, C. (1973). Judgments of intentionality in response to videotaped and verbally presented moral dilemmas: The medium is the message. *Child Development, 44,* 315–320.

Chandler, M., & Hala, S. (1994). The role of personal involvement in the assessment of early false belief skills. In C. Lewis & P. Mitchell (Eds.), *Children's early understanding of mind* (pp. 403–425). Hillsdale, NJ: Lawrence Erlbaum.

Charmaz, K. (2005). Grounded theory. In J. A. Smith (Ed.), *Qualitative psychology* (pp. 81–110). Thousand Oaks, CA: Sage.

Chase-Lansdale, P. L., Mott, F. L., Brooks-Gunn, J., & Phillips, D. A. (1991). Children of the National Longitudinal Survey of Youth: A unique research opportunity. *Developmental Psychology, 27,* 918–931.

Chen, Z., & Siegler, R. S. (2000). Across the great divide: Bridging the gap between understanding of toddlers' and older children's thinking. *Monographs of the Society for Research in Child Development, 65*(2, Serial No. 261).

Cherry, K. E., & Park, D. C. (1993). Individual difference and contextual variables influence spatial memory in younger and older adults. *Psychology and Aging, 8,* 517–526.

Chi, M. T. H. (1978). Knowledge structures and memory development. In R. S. Siegler (Ed.), *Children's thinking: What develops?* (pp. 73–96). Hillsdale, NJ Lawrence Erlbaum.

Cho, G. E., & Miller, P. J. (2004). Personal storytelling Working-class and middle-class mothers in comparative perspective. In M. Farr (Ed.), *Ethnolinguistic Chicago: Language and literacy in the city's neighborhoods* (pp. 79–101). Mahwah, NJ: Lawrence Erlbaum.

Christensen, T. M., Barrett, L. F., Bliss-Moreau, E., Lebo, K., & Kaschub, C. (2003). A practical guide to experience-sampling procedures. *Journal of Happiness Studies, 4,* 53–78.

Cillessen, A. H. N., & Bellmore, A. D. (2002). Social skills and interpersonal perception in early and middle childhood. In P. K. Smith & C. H. Hart (Eds.), *Blackwell handbook of childhood social development* (pp. 355–374). Malden, MA: Blackwell Publishers.

Clarke-Stewart, K. (1989). Infant day care: Maligned or malignant? *American Psychologist, 44,* 266–273.

Clearfield, M. W., & Westfahl, S. M. (2006). Familiarization in infants' perception of addition problems. *Journal of Cognition and Development, 7,* 27–43.

Clements, W. A., & Perner, J. (1994). Implicit understanding of belief. *Cognitive Development, 9,* 377– 395.

Cliff, N. (1993). What is and isn't measurement. In G. Keren & C. Lewis (Eds.), *A handbook for data analysis in the behavioral sciences: Methodological issues* (pp. 59–93). Hillsdale, NJ: Lawrence Erlbaum.

Cohen, G. (1979). Language comprehension in old age. *Cognitive Psychology, 11,* 412–429.

Cohen, G. (1981). Inferential reasoning in old age. *Cognition, 9,* 59–72.

Cohen, G., Stanhope, N., & Conway, M. A. (1992). Age differences in the retention of knowledge by young and elderly students. *British Journal of Developmental Psychology, 10,* 153–164.

Cohen, J. (1960). A coefficient of agreement for nominal scales. *Educational and Psychological Measurement, 20,* 37–46.

Cohen, J. (1977). *Statistical power analysis for the behavioral sciences* (Rev. ed.). New York: Academic Press.

Cohen, J., Cohen, P., West, S. G., & Aiken, L. S. (2003). *Applied multiple regression/correlation analysis for the behavioral sciences* (3rd ed.). Mahwah, NJ: Lawrence Erlbaum.

Cohen, L. B., & Cashon, C. H. (2006). Infant cognition. In W. Damon and R. M. Lerner (Series Eds.) & D. Kuhn & R. S. Siegler (Vol. Eds.), *Handbook of child psychology: Vol. 2. Cognition, perception, and language* (6th ed., pp. 214–251). New York: Wiley.

Cohn, L. D., & Becker, B. J. (2003). How meta-analysis increases statistical power. *Psychological Methods, 8,* 243–253.

Colby, A., & Kohlberg, L. (1987). *The measurement of moral judgment: Vol. 1. Theoretical foundations and research validation.* New York: Cambridge University Press.

Colby, A., Kohlberg, L., Gibbs, J., Candee, D., Hewer, A., Power, C., & Speicher-Dubin, B. (1987). *The measurement of moral judgment: Vol. 2. Standard issue scoring manual.* New York: Cambridge University Press.

Colby, A., Kohlberg, L., Gibbs, J., & Lieberman, M. (1983). A longitudinal study of moral judgment. *Monographs of the Society for Research in Child Development, 48*(1–2, Serial No. 200).

Cole, M., Frankel, F., & Sharp, D. (1971). Development of free recall learning in children. *Developmental Psychology, 4,* 109–123.

Collins, W. A., Maccoby, E. E., Steinberg, L., Hetherington, E. M., & Bornstein, M. H. (2000). Contemporary research on parenting: The case for nature *and* nurture. *American Psychologist, 55,* 218–232.

Comjis, H. C., Deeg, D. J. H., Dik, M. G., Twisk, J. W. R., & Jonker, C. (2002). Memory complaints: The association with psycho-affective and health problems and the role of personality characteristics: A 6-year follow-up study. *Journal of Affective Disorders, 72,* 157–166.

Committee for Ethical Conduct in Child Development Research. (1990, Winter). Report from the Committee for Ethical Conduct in Child Development Research. *SRCD Newsletter,* pp. 5–7.

Connell, J. P., & Tanaka, J. S. (Eds.). (1987). Introduction to the special section on structural equation modeling. *Child Development, 58,* 2–3.

Cook, T. D., & Campbell, D. T. (1979). *Quasi-Experimentation.* Boston: Houghton Mifflin.

Cooke, R. A. (1982). The ethics and regulation of research involving children. In B. B. Wolman (Ed.), *Handbook of developmental psychology* (pp. 149–172). Englewood Cliffs, NJ: Prentice-Hall.

Coopersmith, S. (1967). *The antecedents of self-esteem.* San Francisco: W. H. Freeman.

Coopersmith, S. (1981). *Coopersmith Self-Esteem Inventory.* Palo Alto, CA: Consulting Psychologists Press.

Costa, P. T., Jr., & McCrae, R. R. (1989). Personality continuity and the changes of adult life. In M. Storandt & G. R. VandenBos (Eds.), *The adult years: Continuity and change* (pp. 45–77). Washington, DC: American Psychological Association.

Costa, P. T., Jr., & McCrae, R. R. (1992). *Manual for Revised NEO Personality Inventory (NEO PI-R) and NEO Five Factor Inventory (NEO-FFI).* Odessa, FL: Psychological Assessment Resources.

Courage, M. L., & Howe, M. L. (2004). Advances in early memory development research: Insights about the dark side of the moon. *Developmental Review, 24,* 6–32.

Cowan, P. A., & Cowan, C. P. (2002). What an intervention design reveals about how parents affect their children's academic achievement and behavior problems. In J. Borkowski, S. Landesman-Ramey, & M. Bristol (Eds.), *Parenting and the child's world* (pp. 75–98). Mahwah, NJ: Lawrence Erlbaum.

Coyle, T. R., & Bjorklund, D. F. (1997). Age differences in, and consequences of, multiple- and variable-strategy use on a multitrial sort-recall task. *Developmental Psychology, 33,* 372–380.

Creswell, J. W. (2003). *Research design: Qualitative, quantitative, and mixed methods approaches* (2nd ed.). Thousand Oaks, CA: Sage.

Creswell, J. W., & Maietta, R. C. (2002). Qualitative research. In D. C. Miller & N. J. Salkind (Eds.), *Handbook of research design and social measurement* (6th ed., pp. 143–184). Thousand Oaks, CA: Sage.

Cronbach, L. J. (1990). *Essentials of psychological testing* (5th ed.). New York: Harper & Row.

Crook, T. H., & Larrabee, G. J. (1990). A self-rating scale for evaluating memory in everyday life. *Psychology and Aging, 5,* 48–57.

Csikszentmihalyi, M., & Larson, R. (1984). *Being adolescent.* New York: Basic Books.

Csikszentmihalyi, M., & Schneider, B. (Eds.). (2001). Conditions for optimal development in adolescence: An experiential approach [Special issue]. *Applied Developmental Science, 5*(3).

Damon, W., & Hart, D. (1988). *Self-understanding in childhood and adolescence.* New York: Cambridge University Press.

Darwin, C. (1877). Biographical sketch of an infant. *Mind, 2,* 285–294.

Davis, K. F., Parker, K. P., & Montgomery, G. L. (2004). Sleep in infants and young children: Part 1: Normal sleep. *Journal of Pediatric Health Care, 18,* 56–71.

Davis-Kean, P. E., & Sandler, H. M. (2001). A meta-analysis of measures of self-esteem for young children: A framework for future measures. *Child Development, 72,* 887–906.

Dawe, H. C. (1934). An analysis of two hundred quarrels of preschool children. *Child Development, 5,* 139–157.

Deary, I. J., & Der, G. (2005). Reaction time explains IQ's association with death. *Psychological Science, 16,* 64–69.

de Haan, M., & Thomas, K. M. (2002). Applications of ERP and fMRI techniques to developmental science. *Developmental Science, 5,* 335–343.

DeLoache, J. S., Cassidy, D. J., & Brown, A. L. (1985). Precursors of mnemonic strategies in very young children's memory. *Child Development, 56,* 125–137.

Denzin, N. K., & Lincoln, Y. S. (Eds.). (2000). *The handbook of qualitative research.* Thousand Oaks, CA: Sage.

Deur, J. L., & Parke, R. D. (1970). Effects of inconsistent punishment on aggression in children. *Developmental Psychology, 2,* 403–411.

De Wolff, M., & van IJzendoorn, M. H. (1997). Sensitivity and attachment: A meta-analysis on parental antecedents of infant attachment. *Child Development, 68,* 571–591.

Diehl, M., Willis, S. L., & Schaie, K. W. (1995). Everyday problem solving in older adults: Observational assessment and cognitive correlates. *Psychology and Aging, 10,* 478–491.

Diehl, M. K., Marsiske, M., Horgas, A. L., Rosenberg, A., Saczynski, J. S., & Willis, S. L. (2005). The Revised Observed Tasks of Daily Living: A performance-based assessment of everyday problem solving in older adults. *Journal of Applied Gerontology, 24,* 211–230.

Diener, E. (1984). Subjective well-being. *Psychological Bulletin, 107,* 542–575.

Diener, E., Emmons, R. A., Larsen, R. J., & Griffin, S. (1985). The Satisfaction With Life Scale. *Journal of Personality Assessment, 49,* 71–75.

Dienstbier, R. A. (1984). The role of emotion in moral socialization. In C. E. Izard, J. Kagan, & R. B. Zajonc (Eds.), *Emotions, cognition, and behavior* (pp. 484–514). New York: Cambridge University Press.

Donaldson, G., & Horn, J. L. (1992). Age, cohort, and time developmental muddles: Easy in practice hard in theory. *Experimental Aging Research, 18* 213–222.

Dornbusch, S. M., Ritter, P. L., Leiderman, P. H., Roberts, D. F., & Fraleigh, M. J. (1987). The relation of parenting style to adolescent school performance. *Child Development, 58,* 1244–1257.

Dufresne, A., & Kobasigawa, A. (1989). Children's spontaneous allocation of study time: Differential and sufficient aspects. *Journal of Experimental Child Psychology, 47,* 274–296.

Dunn, J., Brown, J., & Beardsall, L. (1991). Family talk about feeling states and children's late understanding of others' emotions. *Developmental Psychology, 27,* 448–455.

Dunn, J., Brown, J., Slomkowski, C., Tesla, C., & Youngblade, L. (1991). Young children's understanding of other people's feelings and beliefs: Individual differences and their antecedents. *Child Development, 62,* 1352–1366.

Dunn, L. M., & Dunn, L. M. (2007). *Peabody Picture Vocabulary Test* (4th ed.). Circle Pines, MN: American Guidance Service.

Eagly, A. H., & Crowley, M. (1986). Gender and helping behavior: A meta-analytic review of the social psychological literature. *Psychological Bulletin, 100,* 283–308.

Edelbrock, C., & Sugawara, A. I. (1978). Acquisition of sex-typed preferences in preschool-aged children. *Developmental Psychology, 14,* 614–623.

Edwards, C. P. (1986). Cross-cultural research on Kohlberg's stages: The basis for consensus. In S. Modgil & C. Modgil (Eds.), *Lawrence Kohlberg: Consensus and controversy* (pp. 419–430). Philadelphia: Falmer.

Edwards, J., & Klemmack, D. (1973). Correlates of life satisfaction: A reexamination. *Journal of Gerontology, 28,* 497–502.

Eid, M, & Diener, E. (Eds.). (2006). *Handbook of multimethod measurement in psychology.* Washington, DC: American Psychological Association.

Eisen, M., Quas, J., & Goodman, G. S. (Eds.) (2002). *Memory and suggestibility in the forensic interview.* Mahwah, NJ: Lawrence Erlbaum.

Eisenberg, N. (1982). The development of reasoning regarding prosocial behavior. In N. Eisenberg (Ed.), *The development of prosocial behavior* (pp. 219–249). New York: Academic Press.

Eisenberg, N., Fabes, R. A., Bustamante, D., Mathy, R. M., Miller, P. A., & Lindholm, E. (1988). Differentiation of vicariously induced emotional reactions in children. *Developmental Psychology, 24,* 237–246.

Eisenberg, N., Fabes, R. A., Miller, P. A., Fultz, J., Shell, R., Mathy, R. M., & Reno, R. R. (1989). Relation of sympathy and personal distress to prosocial behavior: A multimethod study. *Journal of Personality and Social Psychology, 57,* 55–66.

Eisenberg, N., Morris, A. S., & Spinrad, T. L (2005). Emotion-related regulation: The construct and its measurement. In D. M. Teti (Ed.), *Handbook of research methods in developmental science* (pp. 423–442). Malden, MA: Blackwell Publishers.

Eisenberg-Berg, N. (1979). Development of children's prosocial moral judgment. *Developmental Psychology, 15,* 128–137.

Elder, G. H., Jr. (1999). *Children of the great depression* (25th anniversary ed.). Boulder, CO: Westview Press.

Elder, G. H., Jr., & Caspi, A. (1990). Studying lives in a changing society: Sociological and personological explanations. In A. I. Rabin, R. A. Zucker, & S. Frank (Eds.), *Studying persons and lives* (pp. 201–247). New York: Springer.

Elder, G. H., Jr., Pavalko, E. K., & Clipp, E. C. (1993). *Working with archival data.* Newbury Park, CA: Sage.

Emmerich, W., & Goldman, K. S. (1972). Boy-girl identity task. In V. Shipman (Ed.), *Disadvantaged children and their first school experiences* (ETS PR-7220). Princeton, NJ: Educational Testing Service.

Emond, R. (2005). Ethnographic research methods with young people. In S. Greene & D. Hogan (Eds.), *Researching children's experience* (pp. 123–137). London: Sage.

Engel, S. (2005). Narrative analysis of children's experience. In S. Greene & D. Hogan (Eds.), *Researching children's experience* (pp. 199–216). London: Sage.

Erber, J. T. (2001). Remote memory: In G. L. Maddox (Ed.), *The encyclopedia of aging* (3rd ed., pp. 873–875). New York: Springer.

Fagan, J. F., III, & Detterman, D. H. (1992). The Fagan Test of Infant Intelligence: A technical summary. *Journal of Applied Developmental Psychology, 10,* 173–193.

Fagan, J. F., III, & Shepherd, P. A. (1986). *The Fagan Test of Infant Intelligence: Training manual.* Cleveland: Infantest Corporation.

Fagot, B. I. (1985). Changes in thinking about early sex role development. *Developmental Review, 5,* 83–98.

Fagot, B. I., & Leinbach, M. D. (1993). Gender-role development in young children: From discrimination to labeling. *Developmental Review, 13,* 205–224.

Fantuzzo, J. W., McDermott, P. A., Manz, P. H., Hampton, V. R., & Burdick, N. A. (1996). The Pictorial Scale of Perceived Competence and Social Acceptance: Does it work with low-income urban children? *Child Development, 67,* 1071–1084.

Fantz, R. L. (1961). The origin of form perception. *Scientific American, 204,* 66–72.

Ferguson, R. P., & Bray, N. W. (1976). Component processes of an overt rehearsal strategy in young children. *Journal of Experimental Child Psychology, 21,* 490–506.

Feshbach, S., & Singer, R. D. (1971). *Television and aggression: An experimental field study.* San Francisco: Jossey-Bass.

Fidler, F., Thomason, N., Cumming, G., Finch, S., & Leeman, J. (2004). Editors can lead researchers to confidence intervals, but can't make them think. *Psychological Science, 15,* 119–126.

Field, J. (1977). Coordination of vision and prehension in young infants. *Child Development, 48,* 97–103.

Field, T. (1982). Infancy. In R. Vasta (Ed.), *Strategies and techniques of child study* (pp. 13–48). New York: Academic Press.

Fillenbaum, G. G. (1978). Reliability and validity of the OARS Multidimensional Functional Assessment Questionnaire. In Duke University Center for the Study of Aging (Eds.), *Multidimensional functional assessment: The OARS methodology* (2nd ed., pp. 20–28). Durham, NC: Duke University.

Fillenbaum, G. G. (1985). Screening the elderly: A brief instrumental activities of daily living measure. *Journal of the American Geriatrics Society, 33,* 698–706.

Finders, M. J. (1996). *Just girls.* New York: Teachers College Press.

Fisher, C. B. (2003). *Decoding the Ethics Code: A practical guide for psychologists.* Thousand Oaks, CA: Sage.

Fisher, C. B., Hoagwood, K., Boyce, C., Duster, T., Grisso, T., Levine, R. J., et al. (2002). Research ethics for mental health science involving ethnic minority children and youth. *American Psychologist, 57,* 1024–1040.

Fisher, C. B., & Lerner, R. M. (Eds.). (1994a). *Applied developmental psychology.* New York: McGraw-Hill.

Fisher, C. B., & Lerner, R. M. (1994b). Foundations of applied developmental psychology. In C. B. Fisher & R. M. Lerner (Eds.), *Applied developmental psychology* (pp. 3–20). New York: McGraw-Hill.

Fisher, C. B., & Lerner, R. M. (Eds.). (2005). *Encyclopedia of applied developmental science.* Thousand Oaks, CA: Sage.

Fisher, C. B., & Tryon, W. W. (Eds.). (1990). *Ethics in applied developmental psychology: Emerging issues in an emerging field.* Norwood, NJ: Ablex.

Flavell, J. H. (1963). *The developmental psychology of Jean Piaget.* Princeton, NJ: Van Nostrand.

Flavell, J. H. (1985). *Cognitive development* (2nd ed.). Englewood Cliffs, NJ: Prentice Hall.

Flavell, J. H. (1992). Perspectives on perspective taking. In H. Beilin & P. Pufall (Eds.), *Piaget's theory: Prospects and possibilities* (pp. 107–139). Hillsdale, NJ: Lawrence Erlbaum.

Flavell, J. H., Beach, D. H., & Chinsky, J. M. (1966). Spontaneous verbal rehearsal in memory task as a function of age. *Child Development, 37,* 283–299.

Flavell, J. H., Botkin, P. T., Fry, C. L., Jr., Wright, J. W., & Jarvis, P. E. (1968). *The development of role-taking and communication skills in children.* New York: Wiley.

Flavell, J. H., Flavell, E. R., & Green, F. L. (1983). Development of the appearance-reality distinction. *Cognitive Psychology, 15,* 95–120.

Flavell, J. H., Friedrichs, A. G., & Hoyt, J. (1970). Developmental changes in memorization processes. *Cognitive Psychology, 1,* 324–340.

Flavell, J. H., Miller, P. H., & Miller, S. A. (2002). *Cognitive development* (4th ed.). Englewood Cliffs, NJ: Prentice-Hall.

Flynn, J. R. (1988). IQ gains over time: Toward finding the causes. In U. Neisser (Ed.), *The rising curve: Long-term gains in IQ and related measures* (pp. 25–66). Washington, DC: American Psychological Association.

Follett, W. (1966). *Modern American usage: A guide.* New York: Hill & Wang.

Forgatch, M. S., & DeGarmo, M. S. (1999). Parenting through change: An effective prevention program for single mothers. *Journal of Consulting and Clinical Psychology, 67,* 711–724.

Fowler, H. G. (1996). *The new Fowler's modern English usage* (3rd ed.). New York: Oxford University Press.

Frith, U. (1989). *Autism: Explaining the enigma.* Oxford, England: Basil Blackwell.

Gallup, G. G., Jr. (1970). Chimpanzees: Self-recognition. *Science, 167,* 86–87.

Gallup, G. G., Jr. (1994). Self-recognition: Research strategies and experimental design. In S. Parker, R. W. Mitchell, & M. L. Boccia (Eds.), *Self-awareness in animals and humans* (pp. 35–50). New York: Cambridge University Press.

Garbarino, J. (1982). Sociocultural risk: Dangers to competence. In C. B. Kopp and J. B. Krakow (Eds.),

The child: Development in a social context (pp. 631–685). Reading, MA: Addison-Wesley.

Ge, X., Conger, R., Cadoret, R., Neiderhiser, J., Yates, W., Troughton, E., et al. (1996). The developmental interface between nature and nurture: A mutual influence model of child antisocial behavior and parent behavior. *Developmental Psychology, 32,* 547–589.

Gelman, R. (1991). Epigenetic foundations of knowledge structures: Initial and transcendent constructions. In S. Carey & R. Gelman (Eds.), *The epigenesis of mind: Essays on biology and cognition* (pp. 293–322). Hillsdale, NJ: Lawrence Erlbaum.

Gelman, S. A. (2000). The role of essentialism in children's concepts. In H. W. Reese (Ed.), *Advances in child development and behavior* (Vol. 27, pp. 55–98). San Diego, CA: Academic Press.

Gelman, S. A., & Coley, J. D. (1990). The importance of knowing a dodo is a bird: Categories and induction in 2-year-old children. *Developmental Psychology, 26,* 796–804.

Gelman, S. A., & Markman, E. M. (1986). Categories and induction in young children. *Cognition, 23,* 183–209.

Gelman, S. A., & Markman, E. M. (1987). Young children's inductions from natural kinds: The role of categories and experience. *Child Development, 58,* 1532–1541.

Gibbons, J. D. (1993). *Nonparametric statistics: An introduction.* Newbury Park, CA: Sage.

Gibbs, J. C. (2003). *Moral development and reality: Beyond the theories of Kohlberg and Hoffman.* Thousand Oaks, CA: Sage.

Gibbs, J. C., Basinger, K. S., & Fuller, D. (1992). *Moral maturity: Measuring the development of sociomoral reflection.* Hillsdale, NJ: Lawrence Erlbaum.

Gilligan, C. (1982). *In a different voice: Psychological theory and women's development.* Cambridge, MA: Harvard University Press.

Ginsburg, H., & Opper, S. (1988). *Piaget's theory of intellectual development: An introduction* (3rd ed.). Englewood Cliffs, NJ: Prentice-Hall.

Giorgi, A., & Giorgi, B. (2005). Phenomenology. In J. A. Smith (Ed.), *Qualitative psychology* (pp. 25–50). Thousand Oaks, CA: Sage.

Gliner, J. A., & Morgan, G. A. (2000). *Research methods in applied settings.* Mahwah, NJ: Lawrence Erlbaum.

Goldman, B. A., & Mitchell, D. F. (2003). *Directory of unpublished experimental mental measures* (Vol. 8). Washington, DC: American Psychological Association.

Goldsmith, H. H., Buss, K. A., & Lemery, K. S. (1997). Toddler and childhood temperament: Expanded content, stronger genetic evidence, new evidence for the importance of environment. *Developmental Psychology, 33,* 891–905.

Goldsmith, H. H., & Rothbart, M. K. (1991). Contemporary instruments for assessing early temperament by questionnaire and in the laboratory. In J. Strelan & A. Angleitner (Eds.), *Explorations in temperament: International perspectives on theory and measurement* (pp. 249–272). New York: Plenum Press.

Goldsmith, H. H., & Rothbart, M. K. (1992). *The Laboratory Temperament Assessment Battery.* Eugene, OR: Personality Development Laboratory.

Goldstein, A. O., Frasier, P., Curtis, P., Reid, A., & Kreher, N. E. (1996). Consent form readability in university sponsored research. *Journal of Family Practice, 42,* 606–611.

Golombok, S., & Fivush, R. (1994). *Gender development.* Cambridge, UK: Cambridge University Press.

Golombok, S., & Rust, J. (1993a). The measurement of gender role behavior in pre-school children: A research note. *Journal of Child Psychology and Psychiatry, 34,* 805–811.

Golombok, S., & Rust, J. (1993b). The Pre-School Activities Inventory: A standardized assessment of gender role in children. *Psychological Assessment, 5,* 131–136.

Gonda, J. (1980). Relationship between formal education and cognitive functioning: A historical perspective. *Educational Gerontology, 5,* 283-291.

Gonzalez, N., Cauce, A., & Mason, C. (1996). Interobserver agreement in the assessment of parental behavior and parent-adolescent conflict: African American mothers, daughters, and independent observers. *Child Development, 67,* 1483–1498.

Goodman, G. S., Hirschmann, J. E., Hepps, D., & Rudy, L. (1991). Children's memory for stressful events. *Merrill-Palmer Quarterly, 37,* 109–158.

Goodman, G. S., Pyle Taub, S., Jones, D. P. H., England, P., Port, L. K., Rudy, L., et al. (1992). Testifying in criminal court. *Monographs of the Society for Research in Child Development, 57*(5, Serial No. 229).

Gottman, J. M. (Ed.). (1995). *The analysis of change.* Mahwah, NJ: Lawrence Erlbaum.

Green, R. F. (1969). Age-intelligence relationship between ages sixteen and sixty-four: A rising trend. *Developmental Psychology, 1,* 618–627.

Greene, S., & Hogan, D. (Eds.). (2005). *Researching children's experience.* London: Sage.

Gruen, G. E. (1965). Experiences affecting the development of number conservation. *Child Development, 36,* 963–979.

Grusec, J. E. (1982). Prosocial behavior and self-control. In R. Vasta (Ed.), *Strategies and techniques of child study* (pp. 245–272). New York: Academic Press.

Grusec, J. E., & Kuczynski, L. (1980). Direction of effect in socialization: A comparison of the parent's versus the child's behavior as determinants of disciplinary techniques. *Developmental Psychology, 16,* 1–9.

Haith, M. M., & Benson, J. B. (1998). Infant cognition. In W. Damon (Series Ed.) & D. Kuhn & R. S. Siegler (Vol. Eds.), *Handbook of child psychology: Vol. 2. Cognition, perception, and language* (5th ed., pp. 199–254). New York: Wiley.

Hall, R. V., Fox, R., Willard, D., Goldsmith, L., Emerson, M., Owen, M., et al. (1971). The teacher as observer and experimenter in the modification of disputing and talking-out behaviors. *Journal of Applied Behavior Analysis, 4,* 141–149.

Harlow, H. F. (1958). The nature of love. *American Psychologist, 13,* 673–685.

Harlow, H. F. (1962). Fundamental principles for preparing psychology journal articles. *Journal of Comparative and Physiological Psychology, 55,* 893–896.

Harlow, L. L., Mulaik, S. A., & Steiger, J. S. (Eds.). (1997). *What if there were no significance tests?* Mahwah, NJ: Lawrence Erlbaum.

Harris, P. L. (2006). Social cognition. In W. Damon & R. M. Lerner (Series Eds.) & D. Kuhn & R. S. Siegler (Vol. Eds.), *Handbook of child psychology: Vol. 2. Cognition, perception, and language* (6th ed., pp. 811–858). New York: Wiley.

Harter, S. (1985). *The Self-Perception Profile for Children.* Denver, CO: University of Denver.

Harter, S. (1988). Developmental processes in the construction of the self. In T. D. Yawkey & J. E. Johnson (Eds.), *Integrative processes in socialization: Early to middle childhood* (pp. 45–78). Hillsdale, NJ: Lawrence Erlbaum.

Harter, S., & Pike, R. (1984). The Pictorial Scale of Perceived Competence and Social Acceptance for Young Children. *Child Development, 55,* 1969–1982.

Hartmann, D. P. (2005). Assessing growth in longitudinal investigations: Selected measurement and design issues. In D. M. Teti (Ed.), *Handbook of research methods in developmental science* (pp. 319-339). Malden, MA: Blackwell Publishers.

Hartmann, D. P., Barrios, B. A., & Wood, D. D. (2004). Principles of behavior observation. In S. N. Haynes & E. M. Heiby (Eds.), *Comprehensive handbook of psychological assessment: Vol. 3. Behavioral assessment* (pp. 108–127). New York: Wiley.

Hartmann, D. P., & Pelzel, K. E. (2005). Design, measurement, and analysis in developmental research. In M. H. Bornstein & M. E. Lamb (Eds.), *Developmental psychology: An advanced textbook* (5th ed., pp. 103–184). Mahwah, NJ: Lawrence Erlbaum.

Hartmann, D. P., & Wood, D. W. (1990). Observational methods. In A. S. Bellack, M. Hersen, & A. E. Kazdin (Eds.), *International handbook of behavior modification and therapy* (2nd ed., pp. 107–138). New York: Plenum.

Hays, W. L. (1981). *Statistics* (3rd ed.). New York: Holt, Rinehart & Winston.

Hektner, J., & Csikszentmihalyi, M. (2002). The experience sampling method: Measuring the context and content of lives. In R. B. Bechtel & A. Churchman (Eds.), *Handbook of environmental psychology* (pp. 233–243). New York: Wiley.

Helson, R., Kwan, V. S. Y., John, O. P., & Jones, C. (2002). The growing evidence for personality change in adulthood: Findings from research with personality inventories. *Journal of Research in Personality, 36,* 287–306.

Hertzog, C., & Dixon, R. A. (1996). Methodological issues in research on cognition and aging. In F. Blanchard-Fields & T. M. Hess (Eds.), *Perspectives on cognitive change in adulthood and aging* (pp. 66–121). New York: McGraw-Hill.

Hertzog, C., & Nesselroade, J. R. (2003). Assessing psychological change in adulthood: An overview of methodological issues. *Psychology and Aging, 18,* 639–657.

Hess, T. M. (2005). Memory and aging in context. *Psychological Bulletin, 131,* 383–406.

Hess, T. M., & Pullen, S. M. (1996). Memory in context. In F. Blanchard-Fields & T. M. Hess (Eds.), *Perspectives on cognitive change in adulthood and aging* (pp. 387–427). New York: McGraw-Hill.

Hey, V. (1997). *The company she keeps: An ethnography of girls' friendships.* Buckingham, PA: Open University Press.

Hill, M. (2005). Ethical considerations in researching children's experiences. In S. Greene & D. Hogan

(Eds.), *Researching children's experience* (pp. 61–86). London: Sage.

Hoffman, M. L. (1975). Sex differences in moral internalization and values. *Journal of Personality and Social Psychology, 32,* 720–729.

Hoffman, M. L., & Saltzstein, H. D. (1967). Parent discipline and the child's moral development. *Journal of Personality and Social Psychology, 5,* 45–57.

Hogrefe, G. J., Wimmer, H., & Perner, J. (1986). Ignorance versus false belief: A developmental lag in attribution of epistemic states. *Child Development, 57,* 567–582.

Holden, G. W. (1997). *Parents and the dynamics of child rearing.* Boulder, CO: Westview Press.

Holmes, A., & Teti, D. M. (2005). Developmental science and the experimental method. In D. M. Teti (Ed.), *Handbook of research methods in developmental science* (pp. 66–80). Malden, MA: Blackwell Publishers.

Hood, B. M. (2004). Is looking good enough or does it beggar belief? *Developmental Science, 7,* 415–417.

Horka, S., & Farrow, B. (1970). A methodological note on intersubject communication as a contaminating factor in psychological experiments. *Journal of Experimental Child Psychology, 10,* 363–366.

Horn, J. L. (1982). The theory of fluid and crystallized intelligence in relation to concepts of cognitive psychology and aging in adulthood. In F. I. M. Craik & S. Trehub (Eds.), *Aging and cognitive processes* (pp. 237–278). New York: Plenum.

Howes, J. L., & Katz, A. N. (1988). Assessing remote memory with an improved public events questionnaire. *Psychology and Aging, 3,* 142–150.

Hudson, J. A., & Nelson, K. (1983). Effects of script structure on children's story recall. *Developmental Psychology, 19,* 625–635.

Hughes, C. (2002). Executive functions and development: Emerging themes. *Infant and Child Development, 11,* 201–209.

Hughes, C., & Graham, A. (2002). Measuring executive functions in childhood: Problems and solutions. *Child and Adolescent Mental Health, 7,* 131–142.

Hughes, C., Graham, A., & Grayson, A. (2004). Executive functions in childhood: Development and disorder. In J. Oates & A. Grayson (Eds.), *Cognitive and language development in children* (pp. 205–230). Oxford, UK: Blackwell Publishing.

Hughes, C., & Leekam, S. (2004). What are the links between theory of mind and social relations? Review, reflections and new directions for studies of typical and atypical development. *Social Development, 13,* 590–619.

Hultsch, D. F., MacDonald, S. W., Hunter, M. A., Maitland, S. B., & Dixon, R. A. (2002). Sampling and generalisability in developmental research: Comparison of random and convenience samples of older adults. *International Journal of Behavioral Development, 26,* 345–359.

Hurley, J. C., & Underwood, M. K (2002). Children's understanding of their research rights before and after debriefing: Informed assent, confidentiality, and stopping participation. *Child Development, 73,*132–143.

Hyde, J. S. (2005). The gender similarities hypothesis. *American Psychologist, 60,* 581–592.

Hyde, J. S., Fennema, E., & Lamon, S. (1990). Gender differences in mathematics performance: A meta-analysis. *Psychological Bulletin, 107,* 139–155.

Hyde, J. S., & Linn, M. C. (1988). Gender differences in verbal ability: A meta-analysis. *Psychological Bulletin, 104,* 53–69.

Inagaki, K., & Hatano, G. (1996). Young children's recognition of commonalities between animals and plants. *Child Development, 67,* 2823–2840.

Inagaki, K., & Hatano, G. (2002). *Young children's naïve thinking about the biological world.* New York: Psychology Press.

Inhelder, B., & Piaget, J. (1958). *The growth of logical thinking from childhood to adolescence.* New York: Basic Books.

Inhelder, B., & Piaget, J. (1964). *The early growth of logic in the child.* London: Routledge & Kegan Paul.

Isabella, R. A. (1993). Origins of attachment: Maternal interactive behavior across the first year. *Child Development, 64,* 605–621.

Izard, C. E. (1989). *The Maximally Discriminative Facial Movement Coding System* (MAX) (Rev. ed.). Newark: University of Delaware, Information Technologies and University Media Services.

Jacobsen, T., & Hofmann, V. (1997). Children's attachment representations: Longitudinal relations to school behavior and academic competency in middle childhood and adolescence. *Developmental Psychology, 33,* 703–710.

Jaffee, S., & Hyde, J. S. (2000). Gender differences in moral orientation: A meta-analysis. *Psychological Bulletin, 126,* 703–726.

Jahromi, L. B., Putnam, S. P., & Stifter, C. A. (2004). Maternal regulation of infant reactivity from 2 to 6 months. *Developmental Psychology, 40,* 477–487.

Jenkins, J. M. & Astington, J. W. (2000). Theory of mind and social behavior: Causal models tested in a longitudinal study. *Merrill-Palmer Quarterly, 46,* 203–220.

Jensen, A. R. (1981). *Straight talk about mental tests.* New York: Free Press.

Johnson, R. E. (2003). Aging and the remembering of text. *Developmental Review, 23,* 261–346.

Jordan, T. E. (1994). The arrow of time: Longitudinal study and its applications. *Genetic, Social, and General Psychology Monographs, 120,* 469–531.

Josephson, W. L. (1987). Television violence and children's aggression: Testing the priming, social script, and disinhibition predictions. *Journal of Personality and Social Psychology, 53,* 882–890.

Kagan, J. (1964). American longitudinal research on psychological development. *Child Development, 35,* 1–32.

Kagan, J. (1998). Biology and the child. In W. Damon (Series Ed.) & N. Eisenberg (Vol. Ed.), *Handbook of child psychology: Vol. 3. Social, emotional, and personality development* (5th ed., pp. 177–236). New York: Wiley.

Kagan, J., Reznick, J. S., & Gibbons, J. (1989). Inhibited and uninhibited types of children. *Child Development, 60,* 838–845.

Kaiser, M. K., McCloskey, M., & Proffitt, D. R. (1986). Development of intuitive theories of motion: Curvilinear motion in the absence of external forces. *Developmental Psychology, 22,* 67–71.

Kausler, D. H. (1994). *Learning and memory in normal aging.* New York: Academic Press.

Kaveck, M. (2004). Predicting later IQ from infant visual habituation and dishabituation: A meta-analysis. *Journal of Applied Developmental Psychology, 25,* 369–393.

Kaye, K. (1982). *The mental and social life of babies.* Chicago: University of Chicago Press.

Kazdin, A. E. (1998). *Research design in clinical psychology* (3rd ed.). Boston: Allyn and Bacon.

Keeney, T., Cannizzo, S. R., & Flavell, J. H. (1967). Spontaneous and induced verbal rehearsal in a recall task. *Child Development, 38,* 953–966.

Kellman, P. J., & Arterberry, M. E. (2006). Infant visual perception. In W. Damon & R. M. Lerner (Series Eds.) & D. Kuhn & R. S. Siegler (Vol. Eds.), *Handbook of child psychology: Vol. 2. Cognition, perception, and language* (6th ed., pp. 109–160). New York: Wiley.

Kent, R. N., O'Leary, K. D., Diament, C., & Dietz, A. (1974). Expectation biases in observational evaluation of therapeutic change. *Journal of Consulting and Clinical Psychology, 42,* 774–780.

Keppel, G. (1991). *Design and analysis: A researcher's handbook* (3rd ed.). Englewood Cliffs, NJ: Prentice-Hall.

Kerlinger, F. N., & Lee, H. B. (2000). *Foundations of behavioral research* (4th ed.). Fort Worth, TX: Harcourt College Publishers.

Killeen, P. R. (2005). An alternative to null-hypothesis significance tests. *Psychological Science, 16,* 345–353.

Kirasic, K. C. (1991). Spatial cognition and behavior in young and elderly adults: Implications for learning new environments. *Psychology and Aging, 6,* 10–18.

Kirk, R. E. (2001). Promoting good statistical practices: Some suggestions. *Educational and Psychological Measurement, 61,* 213–218.

Klein, R. L. (1972). Age, sex, and task difficulty as predictors of social conformity. *Journal of Gerontology, 27,* 229–236.

Kline, R. B. (2004). *Beyond significance testing.* Washington, DC: American Psychological Association.

Kling, K. C., Hyde, J. S., Showers, C. J., & Buswell. B. N. (1999). Gender differences in self-esteem: A meta-analysis. *Psychological Bulletin, 125,* 470–500.

Kochanska, G., Gross, J. N., Lin, M., & Nichols, K. E. (2002). Guilt in young children: Development, determinants, and relations with a broader system of standards. *Child Development, 73,* 461–482.

Kohlberg, L. (1966). A cognitive-developmental analysis of children's sex-role concepts and attitudes. In E. Maccoby (Ed.), *The development of sex differences* (pp. 82–173). Stanford, CA: Stanford University Press.

Korner, A. F., & Thoman, E. B. (1970). Visual alertness in neonates as evoked by maternal care. *Journal of Experimental Child Psychology, 10,* 67–78.

Kuhn, D. (1995). Microgenetic study of change: What has it told us? *Psychological Science, 6,* 133–139.

Kunzmann, U., & Baltes, P. B. (2005). The psychology of wisdom: Theoretical and empirical challenges. In R. J. Sternberg & J. Jordan (Eds.), *A handbook of wisdom* (pp. 110–135). New York: Cambridge University Press.

LaGreca, A. M. (Ed.). (1990). *Through the eyes of the child: Obtaining self-reports from children and adolescents.* Needham Heights, MA: Allyn & Bacon.

Lagutta, K. H. (2005). When you shouldn't do what you want to do: Young children's understanding of desires, rules, and emotions. *Child Development, 76,* 713–733.

Lamb, M. E. (1976). Twelve-month-olds and their parents: Interaction in a laboratory playroom. *Developmental Psychology, 12,* 237–244.

Lamb, M. E. (1998). Nonparental child care. In W. Damon (Series Ed.) & I. E. Sigel & K. A. Renninger (Vol. Eds.), *Handbook of child psychology: Vol. 4. Child psychology in practice* (5th ed., pp. 73–133). New York: Wiley.

Lamb, M. E., Bornstein, M. H., & Teti, D. M. (2002). *Development in infancy* (4th ed.). Mahwah, NJ: Lawrence Erlbaum.

Lamb, M. E., Easterbrooks, M. A., & Holden, G. W. (1980). Reinforcement and punishment among preschoolers: Characteristics, effects, and correlates. *Child Development, 51,* 1230–1236.

Lamb, M. E., & Thierry, K. L. (2005). Understanding children's testimony regarding their alleged abuse: Contributions of field and laboratory analog research. In D. M. Teti (Ed.), *Handbook of research methods in developmental science* (pp. 489–508). Malden, MA: Blackwell Publishing.

Lamborn, S. D., Mounts, N. S., Steinberg, L., & Dornbusch, S. M. (1991). Patterns of competence and adjustment among adolescents from authoritative, authoritarian, indulgent, and neglectful families. *Child Development, 62,* 1049–1065.

Langford, P. E. (1995). *Approaches to the development of moral reasoning.* Hove, UK: Lawrence Erlbaum.

Larson, R., & Seepersad, S. (2003). Adolescents' leisure time in the United States: Partying, sports, and the American Experiment. In S. Verma & R. Larson (Eds.), *New directions for child and adolescent development: No. 99. Examining adolescent leisure time across cultures* (pp. 53–64). San Francisco: Jossey-Bass.

Laurence, M. W. (1967). Memory loss with age: A test of two strategies for its retardation. *Psychonomic Science, 9,* 209–210.

Lavelli, M., Pantoja, A. P. F., Hsu, H., Messinger, D., & Fogel, A. (2005). Using microgenetic designs to study change processes. In D. M. Teti (Ed.), *Handbook of research methods in developmental science* (pp. 40–65). Malden, MA: Blackwell Publishers.

Lawton, M. P., Whelihan, W. M., & Belsky, J. K. (1980). Personality tests and their uses with older adults. In J. E. Birren & R. B. Sloane (Eds.), *Handbook of mental health and aging* (pp. 537–553). Englewood Cliffs, NJ: PrenticeHall.

Leaper, C., & Smith, T. E. (2004). A meta-analytic review of gender variations in children's language use: Talkativeness, affiliative speech, and assertive speech. *Developmental Psychology, 40,* 993–1027.

Legerstee, M. (2006). *Infants' sense of people.* New York: Cambridge University Press.

Leinbach, M. D., & Fagot, B. I. (1986). Acquisition of gender labeling: A test for toddlers. *Sex Roles, 15,* 655–666.

Leong, F. T., & Muccio, D. J. (2006). Finding a research topic. In F. T. Leong & J. T. Austin (Eds.), *The psychology research handbook* (2nd ed., pp. 23–40). Thousand Oaks, CA: Sage.

Lerner, R. M., Jacobs, F., & Wertlieb, D. (Eds.). (2003). *Handbook of applied developmental science.* Thousand Oaks, CA: Sage.

Levin, J. R. (1985). Some methodological and statistical "bugs" in research on children's learning. In M. Pressley & C. J. Brainerd (Eds.), *Cognitive learning and memory in children: Progress in cognitive development research* (pp. 205–233). New York: Springer-Verlag.

Lewis, C., Freeman, N. H., Kryiakidou, C., Maridaki-Kassotaki, K., & Berridge, D. M. (1996). Social influences on false belief access: Specific sibling influences or general apprenticeship? *Child Development, 67,* 2930–2947.

Lewis, M., & Johnson, N. (1971). What's thrown out with the bath water: A baby? *Child Development, 42,* 1053–1055.

Liben, L. S., & Bigler, R. S. (2002). The developmental course of gender differentiation. *Monographs of the Society for Research in Child Development, 67*(2, Serial No. 269).

Liben, L. S., & Signorella, M. L. (1993). Gender-schematic processing in children: The role of initial interpretations of stimuli. *Developmental Psychology, 29,* 141–149.

Liebert, R. M., & Baron, R. A. (1972). Some immediate effects of televised violence on children's behavior. *Developmental Psychology, 6,* 469–475.

Light, L. (1991). Memory and aging: Four hypotheses in search of data. *Annual Review of Psychology, 42,* 333–376.

Light, P. H., Buckingham, N., & Robbins, A. H. (1979). The conservation task as an interactional setting.

British Journal of Educational Psychology, 49, 304–310.

Lipsitt, L. P. (1992). Discussion: The Bayley Scales of Infant Development: Issues of prediction and outcome revisited. In C. Rovee-Collier & L. P. Lipsitt (Eds.), *Advances in infancy research* (Vol. 7, pp. 239–245). Norwood, NJ: Ablex.

Livesley, W. J., & Bromley, D. B. (1973). *Person perception in childhood and adolescence.* London: Wiley.

MacCallum, R. C., & Austin, J. T. (2000). Applications of structural equation modeling in psychological research. *Annual Review of Psychology, 51,* 201–226.

Maccoby, E. E. (1998). *The two sexes: Growing up apart, coming together.* Cambridge, MA: Harvard University Press.

Maccoby, E. E., & Jacklin, C. N. (1974). *The psychology of sex differences.* Stanford, CA: Stanford University Press.

MacKay, D. G., & Abrams, L. (1996). Language, memory, and aging: Distributed deficits and the structure of new-versus-old connections. In J. E. Birren & K. W. Schaie (Eds.), *Handbook of the psychology of aging* (4th ed., pp. 251–265). San Diego, CA: Academic Press.

Main, M. (2000). The organized categories of infant, child, and adult attachment: Flexible vs. inflexible attention under attachment-related stress. *Journal of the American Psychoanalytic Association, 48,* 1055–1096.

Main, M., & Solomon, J. (1991). Procedures for identifying infants as disorganized/disoriented during the Ainsworth Strange Situation. In M. Greenberg, D. Cicchetti, & E. M. Cummings (Eds.), *Attachment in the preschool years: Theory, research, and intervention* (pp. 121–160). Chicago: University of Chicago Press.

Mandler, J. M. (1983). Representation. In P. H. Mussen (Series Ed.) & J. H. Flavell & E. M. Markman (Vol. Eds.), *Handbook of child psychology: Vol. 3. Cognitive development* (4th ed., pp. 420–494). New York: Wiley.

Mandler, J. M., & Bauer, P. J. (1988). The cradle of categorization: Is the basic level basic? *Cognitive Development, 3,* 247–264.

Mandler, J. M., & McDonough, L. (1996). Drinking and driving don't mix: Inductive generalization in infancy. *Cognition, 59,* 307–335.

Mann, J., Have, J., Plunkett, J., & Meisels, S. (1991). Time sampling: A methodological critique. *Child Development, 62,* 227–241.

Marascuilo, L. A., & McSweeney, M. (1977). *Nonparametric and distribution free methods for the social sciences.* Monterey, CA: Brooks/Cole.

Marcus, D. E., & Overton, W. F. (1978). The development of cognitive gender constancy and sex role preferences. *Child Development, 49,* 434–444.

Marsh, H. W. (1988). *Self-Description Questionnaire, I.* San Antonio, TX: The Psychological Corporation.

Marsh, H. W. Debus, R., & Bornholt, L. (2005). Validating young children's self-concept responses: Methodological ways and means to understand their responses. In D. M. Teti (Ed.), *Handbook of research methods in developmental science* (pp. 138–160). Malden, MA: Blackwell Publishers.

Marsh, H. W., Ellis, L., & Craven, R. G. (2002). How do preschool children feel about themselves? Unraveling measurement and multidimensional self-concept structure. *Developmental Psychology, 38,* 376–393.

Marsiske, M., & Willis, S. L. (1995). Dimensionality of everyday problem solving in older adults. *Psychology and Aging, 10,* 269–283.

Martin, C. L. (1993). New directions for investigating children's gender knowledge. *Developmental Review, 13,* 184–204.

Martin, C. L., Ruble, D., & Szkrybalo, J. (2002). Cognitive theories of early gender development. *Psychological Bulletin, 128,* 903–933.

Martino-Salzman, D., Blasch, B. B., Morris, R. D., & McNeal, L. W. (1991). Travel behavior of nursing home residents perceived as wanderers and nonwanderers. *Gerontologist, 31,* 666–672.

Masche, J. G., & van Dulmen, M. H. M. (2004). Advances in disentangling age, cohort, and time effects: No quadrature of the circle, but a help. *Developmental Review, 24,* 322–342.

Massey, C. M., & Gelman, R. (1988). Preschoolers' ability to decide whether a photographed unfamiliar object can move itself. *Developmental Psychology, 24,* 307–317.

May, C. P., Hasher, L., & Foong, N. (2005). Implicit memory, age, and time of day: Paradoxical priming effects. *Psychological Science, 16,* 96–100.

May, C. P., Hasher, L., & Stolzfus, E. R. (1993). Optimal time of day and the magnitude of age differences in memory. *Psychological Science, 4,* 326–330.

McAuley, E., Elavsky, S., Motl, R. W., Konopack, J. F., Hu L., & Marquez, D. X. (2005). Physical activity, self-efficacy, and self-esteem: Longitudinal relationships in older adults. *Journals of Gerontology: Psychological Sciences, 60B,* P268-P275.

McCall, R. B. (1977). Challenges to a science of developmental psychology. *Child Development, 48,* 333–344.

McCall, R. B., & Green, B. L. (2004). Beyond the methodological gold standards of behavioral research: Considerations for practice and policy. *SRCD Social Policy Report, 18,* 2.

McCall, R. B., & Mash, C. W. (1994). Infant cognition and its relation to mature intelligence. In R. Vasta (Ed.), *Annals of child development* (Vol. 10, pp. 27–56). London: Kingsley.

McCartney, K., & Nelson, K. (1981). Children's use of scripts in story recall. *Discourse Processes, 4,* 59–70.

McClowry, S. G. (1995). The development of the School-Age Temperament Inventory. *Merrill-Palmer Quarterly,* 41, 272–285.

McClure, E. B. (2000). A meta-analytic review of sex differences in facial expression processing and their development in infants, children, and adolescents. *Psychological Bulletin, 126,* 424–453.

McCrae, R. R., & Costa, P. T., Jr. (1990). *Personality in adulthood.* New York: Guilford.

McCrae, R. R., & Costa, P. T., Jr. (1994). The stability of personality: Observations and evaluations. *Current Directions in Psychological Science, 3,* 173–175.

McCrae, R. R., Costa, P. T., Jr., Lima, M., Simoes, A., Ostendorf, F., Angleitner, A., et al. (1999). Age differences in personality across the adult life span: Parallels in five cultures. *Developmental Psychology, 35,* 466–477.

McCrink, K., & Wynn, K. (2004). Large-number addition and subtraction by 9-month-old infants. *Psychological Science, 15,* 776–781.

McDonough, L., & Mandler, J. M. (1994). Very-long-term recall in infants: Infantile amnesia reconsidered. *Memory, 2,* 339–352.

McGarrigle, J., & Donaldson, M. (1974). Conservation accidents. *Cognition, 3,* 341–350.

McGrath, J. E., & Johnson, B. A. (2003). Methodology makes meaning: How both qualitative and quantitative paradigms shape evidence and its interpretation. In P. M. Camic, J. E. Rhodes, & L. Yardley (Eds.), *Qualitative research in psychology* (pp. 31–48). Washington, DC: American Psychological Association.

McGue, M., Bouchard, T. J., Jr., Iacono, W. G., & Lykken, D. T. (1993). Behavior genetics of cognitive ability: A life-span perspective. In R. Plomin & G. E. McClearn (Eds.), *Nature, nurture, and psychology* (pp. 59–76). Washington, DC: American Psychological Association.

McGuire, W. J. (1997). Creative hypothesis generating in psychology: Some useful heuristics. *Annual Review of Psychology, 48,* 1–30.

McLeod, J. M., Atkin, C. K., & Chaffee, S. H. (1972). Adolescents, parents, and television use: Adolescent self-report measures from Maryland and Wisconsin samples. In G. A. Comstock & E. A. Rubinstein (Eds.), *Television and social behavior: Vol. 3. Television and adolescent aggressiveness* (pp. 173–238). Washington, DC: U.S. Government Printing Office.

Meltzoff, A. N. (1988). Infant imitation and memory: Nine-month-olds in immediate and deferred tests. *Child Development, 59,* 217–225.

Meltzoff, A. N., & Moore, M. K. (1994). Imitation, memory, and the representation of persons. *Infant Behavior and Development, 17,* 83–99.

Mertens, D. M. (2005). *Research and evaluation in education and psychology* (2nd ed.). Thousand Oaks, CA: Sage.

Messick, S. (1983). Assessment of children. In P. H. Mussen (Series Ed.) & W. Kessen (Vol. Ed.), *Handbook of child psychology: Vol. 1. History, theory, and methods* (4th ed., pp. 477–526). New York: Wiley.

Michell, J. (1986). Measurement scales and statistics: A clash of paradigms. *Psychological Bulletin, 100,* 398-407.

Miller, P. H. (1990). The development of strategies of selective attention. In D. F. Bjorklund (Ed.), *Children's strategies: Contemporary views of cognitive development* (pp. 157–184). Hillsdale, NJ: Lawrence Erlbaum.

Miller, P. H., & Coyle, T. R. (1999). Developmental change: Lessons from microgenesis. In E. Scholnick, K. Nelson, S. A. Gelman, & P. H. Miller (Eds.), *Conceptual development: Piaget's legacy* (pp. 209–239). Mahwah, NJ: Lawrence Erlbaum.

Miller, P. H., & Scholnick, E. K. (Eds.). (2000). *Toward a feminist developmental psychology.* New York: Routledge.

Miller, P. J., Cho, G. E., & Bracey, J. R. (2005). Working-class children's experience through the prism of personal storytelling. *Human Development, 48,* 115–135.

Miller, P. J., Fung, H., & Mintz, J. (1996). Self-construction through narrative practices: A Chinese and American comparison of early socialization. *Ethos, 24,* 237–280.

Miller, P. J., Hengst, J. A., & Wang, S. (2003). Ethnographic methods: Applications from developmental cultural psychology. In P. M. Camic, J. E. Rhodes, & L. Yardley (Eds.), *Qualitative research in psychology* (pp. 219–242). Washington, DC: American Psychological Association.

Miller, P. J., Wiley, A., Fung, H., & Liang, C. (1997). Personal storytelling as a medium of socialization in Chinese and American families. *Child Development, 68,* 557–568.

Miller, S. A. (1976a). Nonverbal assessment of conservation of number. *Child Development, 47,* 722–728.

Miller, S. A. (1976b). Nonverbal assessment of Piagetian concepts. *Psychological Bulletin, 83,* 405–430.

Miller, S. A. (1977). A disconfirmation of the quantitative identity-quantitative equivalence sequence. *Journal of Experimental Child Psychology, 24,* 180–189.

Miller, S. A. (1982). Cognitive development: A Piagetian perspective. In R. Vasta (Ed.), *Strategies and techniques of child study* (pp. 161–207). New York: Academic Press.

Miller, S. A., Hardin, C. A., & Montgomery, D. E. (2003). Young children's understanding of the conditions for knowledge acquisition. *Journal of Cognition and Development, 4,* 325–356.

Mitchell, S. K. (1979). Interobserver agreement, reliability, and generalizability of data collected in observational studies. *Psychological Bulletin, 86,* 376–390.

Moen, P., Elder, G. H., & Luscher, K. (Eds.). (1995). *Examining lives in context: Perspectives on the ecology of human development.* Washington, DC: American Psychological Association.

Mook, D. G. (1983). In defense of external invalidity. *American Psychologist, 38,* 379–387.

Moran, J. D., III, & McCullers, J. C. (1984). The effects of recency and story content on children's moral judgments. *Journal of Experimental Child Psychology, 38,* 447–455.

Moses, L. J. (2001). Executive accounts of theory-of-mind development. *Child Development, 72,* 688–690.

Moses, L. J., Coon, J. A., & Wusinich, N. (2000). Young children's understanding of desire formation. *Developmental Psychology, 36,* 77–90.

Moses, L. J., & Flavell, J. H. (1990). Inferring false beliefs from actions and reactions. *Child Development, 61,* 929–945.

Murray, M. (2003). Narrative psychology and narrative analysis. In P. M. Camic, J. E. Rhodes, & L. Yardley (Eds.), *Qualitative research in psychology* (pp. 95–112). Washington, DC: American Psychological Association.

Murray, M. (2005). Narrative psychology. In J. A. Smith (Ed.), *Qualitative psychology* (pp. 111–131). Thousand Oaks, CA: Sage.

Mussen, P. H., & Eisenberg-Berg, N. (1977). *Roots o caring, sharing, and helping: The development o prosocial behavior in children.* San Francisco W. H. Freeman.

Myers, D. G., & Diener, E. (1995). Who is happy *Psychological Science, 6,* 10–19.

Nannis, E. D. (1991). Children's understanding of thei participation in psychological research: Implication for issues of assent and consent. *Canadian Journal o Behavioural Science, 23,* 133–141.

Neale, J. M., & Liebert, R. M. (1986). *Science and behav ior: An introduction to methods of research* (3r ed.). Englewood Cliffs, NJ: Prentice-Hall.

Nelson, C. A., & Luciano, M. (Eds.). (2001). *Handbook o developmental cognitive neuroscience.* Cambridge MA: MIT Press.

Nelson, K. (2003). Narrative and self, myth and mem ory: Emergence of the cultural self. In R. Fivush & C. A. Haden (Eds.), *Autobiographical memory an the construction of a narrative self* (pp. 3–28) Mahwah, NJ: Lawrence Erlbaum.

Nelson-LeGall, S. (1985). Motive-outcome matchin and outcome foreseeability: Effects on attribution o intentionality and moral judgments. *Developmenta Psychology, 21,* 332–337.

Neugarten, B. L., Havighurst, R. J., & Tobin, S. S. (1961) The measurement of life satisfaction. *Journal c Gerontology, 16,* 134–143.

NICHD Early Child Care Research Network. (1997) The effects of infant child care on infant-mothe attachment security: Results of the NICHD Study o Early Child Care. *Child Development, 68,* 860–879.

NICHD Early Child Care Research Network. (2003) Does amount of time spent in child care predic socio-emotional adjustment during the transitio to kindergarten? *Child Development, 74,* 976–1005

NICHD Early Child Care Research Network. (2004). Typ of child care and children's development at 54 month *Early Childhood Research Quarterly, 19,* 203–230.

Nickerson, R. S. (2000). Null hypothesis significanc testing: A review of an old and continuing contro versy. *Psychological Methods, 5,* 241–301.

Nunnally, J. C. (1978). *Psychometric theory* (2nd ed.). New York: McGraw-Hill.

Odom, S. L., & Ogawa, I. (1992). Direct observation of young children's social interaction with peers: A review of methodology. *Behavioral Assessment, 14,* 407–441.

Ogloff, J. R. P., & Otto, R. K. (1991). Are research participants truly informed? Readability of informed consent forms used in research. *Ethics and Behavior, 1,* 239–252.

Ohta, R. J. (1981). Spatial problem-solving: The response selection tendencies of young and elderly adults. *Experimental Aging Research, 1,* 81–84.

Okun, M. A., & Elias, C. S. (1977). Cautiousness in adulthood as a function of age and payoff structure. *Journal of Gerontology, 32,* 451–455.

Oliver, M. B., & Hyde, J. S. (1993). Gender differences in sexuality: A meta-analysis. *Psychological Bulletin, 114,* 29–51.

O'Neill, D. K., Astington, J. W., & Flavell, J. H. (1992). Young children's understanding of the role that sensory experiences play in knowledge acquisition. *Child Development, 63,* 474–490.

O'Neill, D. K., & Gopnik, A. (1991). Young children's ability to identify the sources of their beliefs. *Developmental Psychology, 27,* 390–397.

Orne, M. T. (1962). On the social psychology of the psychological experiment: With particular reference to demand characteristics and their implications. *American Psychologist, 17,* 776–783.

Paris, S. G. (1975). Integration and inference in children's comprehension and memory. In F. Restle, R. Shiffrin, J. Castellan, H. Lindman, & D. Pisoni (Eds.), *Cognitive theory* (Vol. 1, pp. 223–246). Hillsdale, NJ: Lawrence Erlbaum.

Park, D. C., & Brown, S. C. (2001). Everyday memory and aging. In G. L. Maddox (Ed.), *The encyclopedia of aging* (3rd ed., pp. 363–365). New York: Springer.

Parke, R. D. (1967). Nurturance, nurturance withdrawal, and resistance to deviation. *Child Development, 38,* 1101–1110.

Parke, R. D. (1979). Interactional designs. In R. B. Cairns (Ed.), *The analysis of social interactions: Methods, issues, and illustrations* (pp. 15–35). Hillsdale, NJ: Lawrence Erlbaum.

Parke, R. D. (1981). Some effects of punishment on children's behavior—revisited. In E. M. Hetherington & R. D. Parke (Eds.), *Contemporary readings in child psychology* (2nd ed., pp. 176–188). New York: McGraw-Hill.

Parke, R. D., Berkowitz, L., Leyens, J. P., West, S. G., & Sebastian, R. J. (1977). Some effects of violent and nonviolent movies on the behavior of juvenile delinquents. In L. Berkowitz (Ed.), *Advances in experimental social psychology* (Vol. 10, pp. 135–172). New York: Academic Press.

Parke, R. D., & Buriel, R. (1998). Socialization in the family: Ethnic and ecological perspectives. In W. Damon (Series Ed.) & N. Eisenberg (Vol. Eds.), *Handbook of child psychology: Vol. 3. Social, emotional, and personality development* (5th ed., pp. 463–552). New York: Wiley.

Parrott, L., III (1999). *How to write psychology papers* (2nd ed.). New York: Longman.

Passarotti, A. M., Paul, B. M., Bussiere, J. R., Buxton, R. B., Wong, E. C., & Stiles, J. (2003). The development of face and location processing: An FMRI study. *Developmental Science, 6,* 100–117.

Patterson, C. J., & Carter, D. B. (1979). Attentional determinants of children's self-control in waiting and working situations. *Child Development, 50,* 272–275.

Patterson, G. R., & Cobb, J. A. (1971). A dyadic analysis of "aggressive" behaviors. In J. P. Hill (Ed.), *Minnesota symposia on child psychology* (Vol. 5, pp. 72–129). Minneapolis: University of Minnesota Press.

Pavot, W., Diener, E., Colvin, C. R., & Sandvik, E. (1991). Further validation of the Satisfaction with Life Scale: Evidence for the cross-method convergence of well-being measures. *Journal of Personality Assessment, 57,* 149–161.

Pedhazur, E. J. (1997). *Multiple regression in behavioral research* (3rd ed.). Belmont, CA: Wadsworth.

Pedhazur, E. J., & Schmelkin, L. P. (1991). *Measurement, design, and analysis.* Hillsdale, NJ: Lawrence Erlbaum.

Pellegrini, A. D. (2004). *Observing children in their natural worlds* (2nd ed.). Mahwah, NJ: Lawrence Erlbaum.

Pence, A. R. (Ed.). (1988). *Ecological research with children and families: From concepts to methodology.* New York: Teachers College Press.

Perlmutter, M., Metzger, R., Miller, K., & Nezworski, T. (1980). Memory of historical events. *Experimental Aging Research, 6,* 47–60.

Perner, J., & Lang, B. (1999). Development of theory of mind and executive control. *Trends in Cognitive Science, 3,* 337–344.

Peters, D. P. (1991). The influence of stress and arousal on the child witness. In J. Doris (Ed.),

The suggestibility of children's recollections (pp. 60–76). Washington, DC: American Psychological Association.

Peterson, C. C. (2000). Kindred spirits: Influences on siblings' perspectives on theory of mind. *Cognitive Development, 15,* 435–455.

Piaget, J. (1926). *The language and thought of the child.* New York: Harcourt Brace.

Piaget, J. (1929). *The child's conception of the world.* London: Routledge & Kegan Paul.

Piaget, J. (1932). *The moral judgment of the child.* London: Routledge & Kegan Paul.

Piaget, J. (1951). *Play, dreams, and imitation in childhood.* New York: Norton.

Piaget, J. (1952). *The origins of intelligence in children.* New York: International Universities Press.

Piaget, J. (1954). *The construction of reality in the child.* New York: Basic Books.

Piaget, J., & Inhelder, B. (1973). *Memory and intelligence.* New York: Basic Books.

Piaget, J., & Szeminska, A. (1952). *The child's conception of number.* Atlantic Highlands, NJ: Humanities.

Piers, E. V., Harris, D. B., & Herzberg, D. S. (2002). *Piers-Harris Self-Concept Scale for Children* (2nd ed.). Los Angeles: Western Psychological Services.

Pipe, M., Lamb, M. E., Orbach, Y., & Esplin, P. W. (2004). Recent research on children's testimony about experienced and witnessed event. *Developmental Review, 24,* 440–468.

Plake, B. S., Impara, J. C., & Spies, R. A. (2003). *The fifteenth mental measurements yearbook.* Lincoln: University of Nebraska Press.

Plomin, R. (1990). *Nature and nurture.* Belmont, CA: Wadsworth.

Pollio, H. R., Graves, T. R., & Arfken, M. (2006). Qualitative methods. In F. T. L. Leong & J. T. Austin (Eds.), *The psychology research handbook* (2nd ed., pp. 254–274). Thousand Oaks, CA: Sage.

Poole, D. A., & Lamb, M. E. (1998). *Investigative interviews of children: A guide for helping professionals.* Washington, DC: American Psychological Association.

Poole, D. A., & Lindsay, D. S. (2002). Children's suggestibility in the forensic context. In M. L. Eisen, J. A. Quas, & G. S. Goodman (Eds.), *Memory and suggestibility in the forensic context* (pp. 355–381). Mahwah, NJ: Lawrence Erlbaum.

Porges, S. W. (1979). Developmental designs for infancy research. In J. D. Osofsky (Ed.), *Handbook of infant development* (pp. 742–765). New York: Wiley.

Povinelli, D. J. (1995). The unduplicated self. In P. Rochat (Ed.), *The self in early infancy* (pp. 161–192). Amsterdam: North-Holland-Elsevier.

Povinelli, D. J., Landau, K. R., & Perilloux, H. K. (1996). Self-recognition in young children using delayed versus live feedback: Evidence of a developmental asynchrony. *Child Development, 67,* 1540–1554.

Pressley, M. (1992). How *not* to study strategy development. *American Psychologist, 47,* 1240–1241.

Quas, J. A., Goodman, G. S., Bidrose, S., Pipe, M., Craw, S., & Ablin, D. S. (1999). Emotion and memory: Children's long-term remembering, forgetting, and suggestibility. *Journal of Experimental Child Psychology, 72,* 235–270.

Quinn, P. C. (1999). Development of recognition and categorization of objects and their spatial relations in young infants. In L. Balter & C. S. Tamis-Monda (Eds.), *Child psychology: A handbook of contemporary issues* (pp. 85–115). Philadelphia: Psychology Press.

Rabbitt, P. (1989). Inner-city decay? Age changes in structure and process in recall of familiar topographical information. In L. W. Poon, D. C. Rubin, & B. A. Wilson (Eds.), *Everyday cognition in adulthood and late life* (pp. 284–299). New York: Cambridge University Press.

Rabbitt, P., Diggle, P., Holland, F., & McInnes, L. (2004). Practice and dropout effects during a 17-year longitudinal study of cognitive aging. *Journals of Gerontology: Psychological Sciences, 59B,* P84-P97.

Reed, J. G., & Baxter, P. M. (2006). Bibliographic research. In F. T. L. Leong & J. T. Austin (Eds.), *The psychology research handbook* (2nd ed., pp. 41–58). Thousand Oaks, CA: Sage.

Reid, J. B. (1970). Reliability assessment of observation data: A possible methodological problem. *Child Development, 41,* 1143–1150.

Reisinger, H. S. (2004). Counting apples as oranges: Epidemiology and ethnography in adolescent substance abuse treatment. *Qualitative Health Research, 14,* 241–258.

Rest, J. R. (1979). *Development in judging moral issues.* Minneapolis: University of Minnesota Press.

Rest, J., Narvaez, D., Bebeau, M. J., & Thoma, S. J. (1999). *Postconventional moral thinking.* Mahwah, NJ: Lawrence Erlbaum.

Rheingold, H. L. (1982). Ethics as an integral part of research in child development. In R. Vasta (Ed.), *Strategies and techniques of child study* (pp. 305–325). New York: Academic Press.

Richardson, G. A., & McCluskey, K. A. (1983). Subject loss in infancy research: How biasing is it? *Infant Behavior and Development, 6,* 235–239.

Roberts, B. W., Walton, K. E., & Viechtbauer, W. (2006). Patterns of mean-level change in personality traits across the life course: A meta-analysis of longitudinal studies. *Psychological Bulletin, 132,* 1–25.

Roid, G. (2003). *Stanford-Binet Intelligence Scales* (5th ed.). Chicago: Riverside Publishing.

Rose, S. A., & Blank, M. (1974). The potency of context in children's cognition: An illustration through conservation. *Child Development, 45,* 499–502.

Rose, S. A., Feldman, J. F., & Jankowski, J. J. (2004). Infant visual recognition memory. *Developmental Review, 24,* 74–100.

Rose, S. A., Schmidt, K., & Bridger, W. H. (1978). Changes in tactile responsivity during sleep in the human newborn infant. *Developmental Psychology, 14,* 163–172.

Rosenberg, M. (1979). *Conceiving the self.* New York: Basic Books.

Rosenberg, M. J. (1965). When dissonance fails: On eliminating evaluation apprehension from attitude measurement. *Journal of Personality and Social Psychology, 1,* 28–42.

Rosengren, K. S., Gelman, S. A., Kalish, C. W., & McCormick, M. (1991). As time goes by: Children's early understanding of growth in animals. *Child Development, 62,* 1302–1320.

Rosenstein, D., & Oster, H. (1988). Differential facial responses to four basic tastes in newborns. *Child Development, 59,* 1555–1568.

Rosenthal, R. (1968). Experimenter expectancy and the reassuring nature of the null hypothesis decision procedure. *Psychological Bulletin Monograph Supplement, 70,* 30–47.

Rosenthal, R. (1976). *Experimenter effects in behavioral research* (Enl. ed.). New York: Halsted Press.

Rosenthal, R. (1994). Parametric measures of effect size. In H. Cooper & L. V. Hedges (Eds.), *The handbook of research synthesis* (pp. 231–244). New York: Russell Sage Foundation.

Rosenthal, R. (2002). Covert communication in classrooms, clinics, courtrooms, and cubicles. *American Psychologist, 57,* 839–849.

Rosenthal, R., & DiMatteo, M. R. (2001). Meta-Analysis: Recent developments in quantitative methods for literature reviews. *Annual Review of Psychology, 52,* 59–82.

Rosnick, C. B., Small, B. J., Graves, A. B., & Mortimer, J. A. (2004). The association between health and cognitive performance in a population-based study of older adults: The Charlotte County Healthy Aging Study (CCHAS). *Aging Neuropsychology and Cognition, 11,* 89–99.

Rosnow, R. L., & Rosenthal, R. (1989). Statistical procedures and the justification of knowledge in psychological science. *American Psychologist, 44,* 1276–1284.

Rosnow, R. L., & Rosenthal, R. (1995). "Some things you learn aren't so": Cohen's paradox, Asch's paradigm, and the interpretation of interaction. *Psychological Science, 6,* 3–9.

Rosnow, R. L., & Rosnow, M. (2006). *Writing papers in psychology: A student guide to research reports, essays, proposals, posters, and brief reports* (7th ed.). Belmont, CA: Thomson/Wadsworth.

Ross, J. B., & McLaughlin, M. M. (Eds.). (1949). *The portable medieval reader.* New York: Viking Press.

Rothbart, M. K. (1986). A psychobiological approach to the study of temperament. In G. A. Kohnstamm (Ed.), *Temperament discussed: Temperament and development in infancy and childhood* (pp. 63–72). Lisse, Netherlands: Swets & Zeitlinger Publishers.

Rothbart, M. K., Ahadi, S. A., Hersey, K. L., & Fisher, P. (2001). Investigations of temperament at three to seven years: The Children's Behavior Questionnaire. *Child Development, 72,* 1394–1408.

Rothbart, M. K., & Bates, J. E. (2006). Temperament. In W. Damon & R. M. Lerner (Series Eds.) & N. Eisenberg (Vol. Ed.), *Handbook of child psychology: Vol. 3. Social, emotional, and personality development* (6th ed., pp. 99–166). New York: Wiley.

Rothbart, M. K., Derryberry, D., & Hershey, K. (2000). Stability of temperament in childhood: Laboratory infant assessment to parent report at seven years. In V. J. Molfese & D. L. Molfese (Eds.), *Temperament and personality development across the life span* (pp. 85–119). Mahwah, NJ: Lawrence Erlbaum.

Rovee-Collier, C. K. (1999). The development of infant memory. *Current Directions in Psychological Science, 8,* 80–85.

Rowe, D. (1994). *The limits of family influence: Genes, experience, and behavior.* New York: Guilford Press.

Rubin, D. C., Rahhal, T. A., & Poon, L. W. (1998). Things learned in early adulthood are remembered best. *Memory and Cognition, 26,* 3–19.

Ruble, D. N., Martin, C. L., & Berenbaum, S. A. (2006). Gender development. In W. Damon & R. M. Lerner

(Series Eds.) & N. Eisenberg (Vol. Ed.), *Handbook of child psychology: Vol. 3. Social, emotional, and personality development* (6th ed., pp. 858–932). New York: Wiley.

Rudy, L., & Goodman, G. S. (1991). Effects of participation on children's reports: Implications for children's testimony. *Developmental Psychology, 27,* 527–538.

Ruffman, T., Perner, J., Naito, M., Parkin, L., & Clements, W. A. (1998). Older (but not younger) siblings facilitate false belief understanding. *Developmental Psychology, 34,* 161–174.

Ruhland, R., & van Geert, P. (1998). Jumping into syntax: Transitions in the development of closed class words. *British Journal of Developmental Psychology, 16,* 65–95.

Russell, A., Russell, G., & Midwinter, D. (1992). Observer influences on mothers and fathers: Self-reported influence during a home observation. *Merrill-Palmer Quarterly, 38,* 263–283.

Russell, J., Mauthner, N., Sharpe, S., & Tidswell, T. (1991). The "windows task" as a measure of strategic deception in preschoolers and autistic subjects. *British Journal of Developmental Psychology, 9,* 331–349.

Rutherford, E., & Mussen, P. H. (1968). Generosity in nursery school boys. *Child Development, 39,* 755–765.

Ryff, C. D. (1989). Happiness is everything, or is it? Explorations on the meaning of psychological well-being. *Journal of Personality and Social Psychology, 57,* 1069–1081.

Ryff, C. D. (1995). Psychological well-being in adult life. *Current Directions in Psychological Science, 4,* 99–103.

Sackett, G. P., Ruppenthal, G. C., & Gluck, J. (1978). Introduction: An overview of methodological and statistical problems in observational research. In G. P. Sackett (Ed.), *Observing behavior: Vol. 2. Data collection and analysis methods* (pp. 1–14). Baltimore: University Park Press.

Saffran, J. R., Werker, J. E., & Werner, L. A. (2006). The infant's auditory world: Hearing, speech, and the beginnings of language. In W. Damon & R. M. Lerner (Series Eds.) & D. Kuhn & R. S. Siegler (Vol. Eds.), *Handbook of child psychology: Vol. 2. Cognition, perception, and language* (6th ed., pp. 58–108). New York: Wiley.

Salthouse, T. A. (1996). The processing-speed theory of adult age differences in cognition. *Psychological Review, 103,* 403–428.

Salthouse, T. A. (2000). Pressing issues in cognitive aging. In D. C. Park & N. Schwarz (Eds.), *Cognitive aging: A primer* (pp. 43–54). New York: Psychology Press.

Sanson, A., Hemphill, S. A., & Smart, D. (2002). Temperament in social development. In P. K. Smith & C. H. Hart (Eds.), *Blackwell handbook of childhood social development* (pp. 97–116). Malden, MA: Blackwell Publishers.

Saywitz, K., Goodman, G., Nicholas, G., & Moan, S. (1991). Children's memory of a physical examination involving genital touch: Implications for reports of child sexual abuse. *Journal of Consulting and Clinical Psychology, 5,* 682–691.

Schaie, K. W. (1958). Rigidity-flexibility and intelligence: A cross-sectional study of the adult life-span from 20 to 70. *Psychological Monographs, 7*[illegible] (9, Whole Number 462).

Schaie, K. W. (1994). The course of adult intellectual development. *American Psychologist, 49,* 304–313.

Schaie, K. W. (2005). *Developmental influences on adult intelligence: The Seattle Longitudinal Study.* New York: Oxford University Press.

Schaie, K. W., & Caskie, G. I. L. (2005). Methodological issues in aging research. In D. M. Teti (Ed.), *Handbook of research methods in developmental science* (pp. 21–39). Malden, MA: Blackwell Publishers.

Schmidt, F. L., & Hunter, J. E. (2003). Meta-analysis. In J. A. Schinka & W. F. Velicer (Eds.), *Handbook of psychology: Vol. 2. Research methods in psychology* (pp. 533–554). New York: Wiley.

Schmidt, H. G., Peeck, V. H., Paas, F., & van Breukelen, G. J. (2000). Remembering the street names of one's childhood neighbourhood: A study of very long term retention. *Memory, 8,* 37–49.

Schneider, B. H., Atkinson, L., & Tardif, C. (2001). Child–parent attachment and children's peer relations: A quantitative review. *Developmental Psychology, 37,* 86–100.

Schonfield, A. E. D. (1969). *In search of early memories.* Paper presented at the International Congress of Gerontology, Washington, DC.

Schonfield, D., & Robertson, E. A. (1966). Memory storage and aging. *Canadian Journal of Psychology, 2*[illegible] 228–236.

Schooler, C., & Mulatu, M. S. (2001). The reciprocal effects of leisure time activities and intellectual functioning in older adults: A longitudinal analysis. *Psychology and Aging, 16,* 446–482.

Segal, N. (1999). *Entwined lives.* New York: Dutton.

Seifer, R. (2005). Who should collect our data: Parents or trained observers? In D. M. Teti (Ed.), *Handbook of research methods in developmental science* (pp. 123–137). Malden, MA: Blackwell Publishers.

Seitz V. (1984). Methodology. In M. H. Bornstein & M. E. Lamb (Eds.), *Developmental psychology: An advanced textbook* (pp. 37–79). Hillsdale, NJ: Lawrence Erlbaum.

Shadish, W. R., Cook, T. D., & Campbell, D. T. (2002). *Experimental and quasi-experimental designs for generalized causal inference.* Boston: Houghton Mifflin.

Sharon, T., & Woolley, J. D. (2004). Do monsters dream? Young children's understanding of the fantasy/real distinction. *British Journal of Developmental Psychology, 22,* 293–310.

Sharpe, D. (1997). Of apples and oranges, file drawers and garbage: Why validity issues in meta-analysis will not go away. *Clinical Psychology Review, 17,* 881–901.

Sheikh, J. I., Hill, R. D., & Yesavage, J. (1986). Long-term efficacy of cognitive training for age-associated memory impairment: A six-month follow-up study. *Developmental Neuropsychology, 2,* 413- 421.

Shweder, R. A., Goodnow, J., Hatano, G., LeVine, R. A., Markus, H., & Miller, P. J. (1998). The cultural psychology of development: One mind, many mentalities. In W. Damon (Series Ed.) & R. M Lerner (Vol. Ed.), *Handbook of child psychology: Vol. 1. Theoretical models of human development* (5th ed., pp. 865–937). New York: Wiley.

Sieber, J. E. (1992). *Planning ethically responsible research: A guide for students and internal review boards.* Newbury Park, CA: Sage.

Siegal, M. (1991). *Knowing children: Experiments in conversation and cognition.* Hillsdale, NJ: Lawrence Erlbaum.

Siegal, M., & Beattie, K. (1991). Where to look first for children's knowledge of false beliefs. *Cognition, 38,* 1–12.

Siegel, S., & Castellan, N. J., Jr. (1988). *Nonparametric statistics for the behavioral sciences* (2nd ed.). New York: McGraw-Hill.

Siegler, I. C., Bosworth, H., & Poon, L. (2003). Disease, health, and aging. In R. M. Lerner, M. A. Easterbrooks, & J. Mistry (Eds.), *Handbook of psychology: Vol. 6. Developmental psychology* (pp. 423–442). New York: Wiley.

Siegler, I. C., & Botwinick, J. (1979). A long-term longitudinal study of intellectual ability of older adults: The matter of selective subject attrition. *Journal of Gerontology, 34,* 242–245.

Siegler, R. S. (1996). *Emerging minds: The process of change in children's thinking.* New York: Oxford University Press.

Siegler, R. S., & Alibali, M. W. (2005). *Children's thinking* (4th ed.). Upper Saddle River, NJ: Prentice Hall.

Siegler, R. S., & Crowley, K. (1992). Microgenetic methods revisited. *American Psychologist, 47,* 1241–1243.

Siegler, R. S., & Jenkins. E. (1989). *How children discover new strategies.* Hillsdale, NJ: Lawrence Erlbaum.

Sigel, I. E., & Renninger, K. A. (Eds.). (2006). *Handbook of child psychology: Vol. 4. Child psychology in practice* (6th ed.). New York: Wiley.

Silverman, I. (1977). *The human subject in the psychological laboratory*. New York: Pergamon.

Slabach, E. H., Morrow, J., & Wachs, T. D. (1991). Questionnaire measurement of infant and child temperament: Current status and future directions. In S. Strelan & A. Angleitner (Eds.), *Explorations in temperament: International perspectives on theory and measurement* (pp. 205–234). New York: Plenum Press.

Slaby, R. G., & Frey, K. S. (1975). Development of gender constancy and selective attention to same-sex models. *Child Development, 46,* 849–856.

Slater, A., Mattock, A., & Brown, E. (1990). Size constancy at birth: Newborn infant's responses to retinal and real size. *Journal of Experimental Child Psychology, 49,* 314–322.

Smetana, J. G. (1993). Understanding of social rules. In M. Bennett (Ed.), *The child as psychologist* (pp. 111–141). London: Guilford Press.

Smith, I. D. (1968). The effects of training procedures upon the acquisition of conservation of weight. *Child Development, 39,* 515–526.

Smith, J., & Baltes, P. B. (1990). Wisdom-related knowledge: Age/cohort differences in response to life-planning problems. *Developmental Psychology, 26,* 494–505.

Smith, J. A. (Ed.). (2005). *Qualitative psychology.* Thousand Oaks, CA: Sage.

Snarey, J. R. (1985). Cross-cultural universality of social-moral development: A critical review of Kohlbergian research. *Psychological Bulletin, 97,* 202–232.

Sockett, H. (1992). The moral aspects of the curriculum. In P. W. Jackson (Ed.), *Handbook of research on curriculum* (pp. 543–569). New York: Macmillan.

Spieth, W. (1965). Slowness of task performance and cardiovascular disease. In A. T. Welford & J. E. Birren (Eds.), *Behavior, aging, and the nervous system* (pp. 366–400). Springfield, IL: Charles C. Thomas.

Springer, K. (1996). Young children's understanding of a biological basis for parent-offspring relations. *Child Development, 67,* 2841–2856.

Springer, K., & Keil, F. C. (1991). Early differentiation of causal mechanisms appropriate to biological and nonbiological kinds. *Child Development, 62,* 767–781.

Srivastava, S., John, O., Gosling, S., & Potter, J. (2003). Development of personality in early and middle adulthood: Set like plaster or persistent change? *Journal of Personality and Social Psychology, 84,* 1041–1053.

Sroufe, L. A. (1983). Infant-caregiver attachment and patterns of adaptation in preschool: The roots of maladaptation and competence. In M. Perlmutter (Ed.), *Minnesota symposia on child psychology: Vol. 16. Development and policy concerning children with special needs* (pp. 41–83). Hillsdale, NJ: Lawrence Erlbaum.

Sroufe, L. A., Waters, E., & Matas, L. (1974). Contextual determinants of infant affective response. In M. Lewis & L. A. Rosenblum (Eds.), *The origins of fear* (pp. 49–72). New York: Wiley.

Stake, R. E. (2000). Case studies. In N. K. Denzin & Y. S. Lincoln (Eds.), *The handbook of qualitative research* (pp. 435–454). Thousand Oaks, CA: Sage.

Stanley, B. H., Sieber, J. E., & Melton, G. B. (Eds.). (1996). *Research ethics: A psychological approach.* Lincoln: University of Nebraska Press.

Staudinger, U. M., & Werner, I. (2003). Wisdom: Its social nature and lifespan development. In J. Valsiner & K. J. Connolly (Eds.), *Handbook of developmental psychology* (pp. 584–600). Thousand Oaks, CA: Sage.

Steinberg, L., Elmen, J. D., & Mounts, N. S. (1989). Authoritative parenting, psychosocial maturity, and academic success among adolescents. *Child Development, 60,* 1424–1436.

Steinberg, L., Lamborn, S., Darling, J., Mounts, N., & Dornbusch, S. (1994). Over-time changes in adjustment and competence among adolescents from authoritative, authoritarian, indulgent, and neglectful families. *Child Development, 65,* 754–770.

Steiner, J. E. (1979). Human facial expressions in response to taste and smell stimulation. In H. W. Reese & L. P. Lipsitt (Eds.), *Advances in child development and behavior* (Vol. 13, pp. 257–295). New York: Academic Press.

Sternberg, R. J. (Ed.). (2000). *Getting published in psychology journals.* New York: Cambridge University Press.

Sternberg, R. J. (2003). *The psychologist's companion: A guide to scientific writing for students and researchers* (4th ed.). New York: Cambridge University Press.

Sternberg, R. J., & Jordan, J. (Eds.). (2005). *A handbook of wisdom.* New York: Cambridge University Press.

Stevens, S. S. (1968). Measurement, statistics, and the schemapiric view. *Science, 161,* 849–856.

Stipek, D. J., Gralinski, J. H., & Kopp, C. B. (1990). Self-concept development in the toddler years. *Developmental Psychology, 26,* 972–977.

Storandt, M., & Hudson, W. (1975). Misuse of analysis of co-variance in aging research and some partial solutions. *Experimental Aging Research, 1,* 121–125.

Strunk, W., Jr., & White, E. B. (2000). *The elements of style* (4th ed.). New York: Longman.

Sullivan, K., & Winner, E. (1993). Three-year-olds' understanding of mental states: The influence of trickery. *Journal of Experimental Child Psychology, 56,* 135–148.

Tamres, L. K., Janicki, D., & Helgeson, V. S. (2002). Sex differences in coping behavior: A meta-analytic review and examination of relative coping. *Personality and Social Psychology Review, 6,* 2–30.

Tannenbaum, A. S., & Cooke, R. A. (1977). Research involving children. In National Commission for the Protection of Human Subjects of Biomedical and Behavioral Research (Eds.), *Appendix to report and recommendations on research involving children* (pp. 1-1-1-129). Washington, DC: U.S. Government Printing Office.

Taplin, P. S., & Reid, J. B. (1973). Effects of instructional set and experimenter influence on observer reliability. *Child Development, 44,* 547–554.

Tatsuoka, M. (1993). Effect size. In G. Keren & C. Lewis (Eds.), *A handbook for data analysis in the behavioral sciences: Methodological issues* (pp. 461–479). Hillsdale, NJ: Lawrence Erlbaum.

Taylor, M., Esbensen, B. M., & Bennett, R. T. (1994). Children's understanding of knowledge acquisition:

The tendency for children to report that they have always known what they have just learned. *Child Development, 65,* 1581–1604.

Taylor, M. J., & Baldeweg, T. (2002). Application of EEG, ERP and intracranial recordings to the investigation of cognitive functions in children. *Developmental Science, 5,* 318–334.

Terracciano, A., McCrae, R. R., Brant, L. J., & Costa, P. T., Jr. (2005). Hierarchical linear modeling of the NEO-PI-R scales in the Baltimore Longitudinal Study of Aging. *Psychology and Aging, 20,* 493–506.

Teti, D. M. (Ed.). (2005). *Handbook of research methods in developmental psychology.* Malden, MA: Blackwell Publishers.

Thomas, A., & Chess, S. (1977). *Temperament and development.* New York: Brunner/Mazel.

Thomas, A., Chess, S., & Birch, H. (1968). *Temperament and behavior disorders in children.* New York: New York University Press.

Thompson, R. A. (1990). Vulnerability in research: A developmental perspective on research risk. *Child Development, 61,* 1–16.

Thompson, R. A. (2000). The legacy of early attachments. *Child Development, 71,* 145–152.

Thompson, R. A., & Hoffman, M. L. (1980). Empathy and the development of guilt in children. *Developmental Psychology, 16,* 155–156.

Thompson, S. K., & Bentler, P. M. (1971). The priority of cues in sex discriminations by children and adults. *Developmental Psychology, 5,* 181–185.

Thorne, B. (1993). *Gender play: Girls and boys in school.* New Brunswick, NJ: Rutgers University Press.

Thurstone, L. L., & Thurstone, T. G. (1962). *SRA Primary Mental Abilities.* Chicago: Science Research Associates.

Tisak, M. S. (1995). Domains of social reasoning and beyond. In R. Vasta (Ed.), *Annals of child development* (Vol. 11, pp. 95–130). Philadelphia: Jessica Kingsley Publishers.

Tobey, A. E., & Goodman, G. S. (1992). Children's eyewitness memory: Effects of participation and forensic context. *Child Abuse and Neglect, 16,* 779–796.

Tomlinson-Keasey, C. (1993). Opportunities and challenges posed by archival data sets. In D. C. Funder & R. D. Parke (Eds.), *Studying lives through time* (pp. 65–92). Washington, DC: American Psychological Association.

Trials of war criminals before the Nuremberg military tribunals, U.S. vs. Karl Brandt (Vol. 2). (1949). Washington, DC: U.S. Government Printing Office.

Turkheimer, E. (1991). Individual and group differences in adoption studies of IQ. *Psychological Bulletin, 110,* 392–405.

Tversky, B. (1985). Development of taxonomic organization of named and pictured categories. *Developmental Psychology, 21,* 1111–1119.

Tymchuk, A. J. (1991). Assent processes. In B. Stanley & J. E. Sieber (Eds.), *Social research on children and adolescents: Ethical issues* (pp. 128–139). Newbury Park, CA: Sage.

Ullman, J. R., & Bentler, P. M. (2003). Structural equation modeling. In J. A. Schinka & W. F. Velicer (Eds.). (2003). *Handbook of psychology: Vol. 2. Research methods in psychology* (pp. 607–634). New York: Wiley.

Underwood, M. K., Coie, J. D., & Herbsman, C. R. (1992). Display rules for anger and aggression in school-age children. *Child Development, 63,* 366–380.

U.S. Bureau of the Census. (2004). *Population projections.* Washington, DC: U.S. Government Printing Office.

Uttl, B., & Graf, P. (1993). Episodic spatial memory in adulthood. *Psychology and Aging, 8,* 257–273.

van Hooren, S. A. H., Valentijn, S. A. M., Bosma, H., Ponds, R. W. H. M., van Boxtel, M. P. J., & Jolles, J. (2005). Relation between health status and cognitive functioning: A 6-year follow-up of the Maastricht Aging Study. *Journals of Gerontology: Psychological Sciences, 60B,* P57–60.

van IJzendoorn, M. H. (1995). Adult attachment representations, parental responsiveness, and infant attachment: A meta-analysis on the predictive validity of the Adult Attachment Interview. *Psychological Bulletin, 117,* 387–403.

van IJzendoorn, M. H., Juffer, F., & Poelhuis, C. W. K. (2005). Adoption and cognitive development: A meta-analytic comparison of adopted and non-adopted children's IQ and school performance. *Psychological Bulletin, 131,* 301–316.

van IJzendoorn, M. H., & Sagi, A. (1999). Cross-cultural patterns of attachment: Universal and contextual dimension. In J. Cassidy & P. R. Shaver (Eds.), *Handbook of attachment: Theory, research, and clinical applications* (pp. 713–734). New York: Guilford.

van IJzendoorn, M. H., Vereijken, C. M. J. L, Bakermanns-Kraneburg, J., & Riksen-Walraven, J. M. (2004). Assessing attachment security with the Attachment Q-Set: Meta-analytic evidence for the validity of the observer AQS. *Child Development, 75,* 1188–1213.

Vaughn, B. E., & Bost, K. K. (1999). Attachment and temperament: Redundant, independent, or interacting influences on interpersonal adaptation and personality development? In J. Cassidy & P. R. Shaver (Eds.), *Handbook of attachment: Theory, research, and clinical applications* (pp. 265–286). New York: Guilford.

Velicer, W. F., & Fava, J. L. (2003). Time series analysis. In J. A. Schinka & W. F. Velicer (Eds.). (2003). *Handbook of psychology: Vol. 2. Research methods in psychology* (pp. 581–606). New York: Wiley.

Voyat, G. E. (1982). *Piaget systematized.* Hillsdale, NJ: Lawrence Erlbaum.

Wachs, T. D., & Bates, J. E. (2001). Temperament. In G. Bremmer & A. Fogel (Eds.), *Blackwell handbook of infant development* (pp. 465–501). Malden, MA: Blackwell.

Wachs, T. D., Pollitt, E., Cueto, S., & Jacoby, E. (2004). Structure and cross-contextual stability of neonatal temperament. *Infant Behavior and Development, 27,* 382–396.

Walk, R. D., & Gibson, E. J. (1961). A comparative and analytical study of visual depth perception. *Psychological Monographs, 75*(15, Whole No. 519).

Walker, L. J. (1988). The development of moral reasoning. In R. Vasta (Ed.), *Annals of child development* (Vol. 5, pp. 33–78). Greenwich, CT: JAI Press.

Walker, L. J. (1989). A longitudinal study of moral reasoning. *Child Development, 60,* 157–166.

Walker, L. J. (2004). Progress and prospects in the psychology of moral development. *Merrill-Palmer Quarterly, 50,* 546–557.

Walker, L. J., deVries, B., & Trevethan, S. D. (1987). Moral stages and moral orientations in real-life and hypothetical dilemmas. *Child Development, 58,* 842–848.

Walker, L. J., & Taylor, J. H. (1991). Stage transitions in moral reasoning: A longitudinal study of developmental processes. *Developmental Psychology, 27,* 330–337.

Ware, J. J., & Sherbourne, C. D. (1992). The MOS 36-item short-form health survey (SF-36). I. Conceptual framework and item selection. *Medical Care, 30,* 473–483.

Waters, E. (1995). The Attachment Q-Set (version 3.0). In E. Waters, B. E. Vaughn, G. Posada, & K. Kondo-Ikemura (Eds.), Caregiving, cultural, and cognitive perspectives on secure-base behavior and working models (pp. 234–246). *Monographs of the Society for Research in Child Development, 60*(2–3, Serial No. 244).

Waters, E., & Deane, K. E. (1985). Defining and assessing individual differences in attachment relationship: Q-methodology and the organization of behavior in infancy and early childhood. In I. Bretherton & E. Waters (Eds.), Growing points of attachment theory and research (pp. 41–65). *Monographs of the Society for Research in Child Development, 50*(1–2, Serial No. 209).

Watson, A. C., Nixon, C. L., Wilson, A., & Capage, L. (1999). Social interaction skills and theory of mind in young children. *Developmental Psychology, 35,* 386–391.

Webb, E. J., Campbell, D. T., Schwartz, R. D., & Sechrest, L. (2000). *Unobtrusive measures* (Rev. ed.). Thousand Oaks, CA: Sage.

Wechsler, D. (1958). *The measurement and appraisal of adult intelligence* (4th ed.). Baltimore: Williams & Wilkins.

Wechsler, D. (1997). *Wechsler Adult Intelligence Scale* (3rd ed.). San Antonio, TX: Harcourt Assessment.

Wechsler, D. (2002). *Wechsler Preschool and Primary Scale of Intelligence* (3rd ed.). San Antonio, TX: Harcourt Assessment.

Wechsler, D. (2003). *Wechsler Intelligence Scale for Children* (4th ed.). San Antonio, TX: Harcourt Assessment.

Weizmann, F., Cohen, L. B., & Pratt, R. J. (1971). Novelty, familiarity, and the development of infant attention. *Developmental Psychology, 4,* 149–154.

Wellman, H. M. (2002). Understanding the psychological world. In U. Goswami (Ed.), *Blackwell handbook of childhood cognitive development* (pp. 167–187). Malden, MA: Blackwell.

Wellman, H. M., Cross, D., & Watson, J. (2001). Meta-analysis of theory-of-mind development: The truth about false belief. *Child Development, 72,* 655–684.

Wellman, H. M., Ritter, K., & Flavell, J. H. (1975). Deliberate memory behavior in the delayed reactions of very young children. *Developmental Psychology, 11,* 780–787.

West, R. L. (1995). Compensatory strategies for age-associated memory impairment. In A. D. Baddely, B. A. Wilson, & F. N. Watts (Eds.), *Handbook of memory disorders* (pp. 481–500). New York: Wiley.

West, R. L., & Walton, M. (1985, March). *Practical memory functioning in the elderly.* Paper presented at the meeting of the Second National Forum on Research in Aging, Lincoln, NE.

Wetherford, M. J., & Cohen, L. B. (1973). Developmental changes in infant visual preferences for novelty and familiarity. *Child Development, 44,* 416–424.

White, T. L., Leichtman, M. D., & Ceci, S. J. (1997). The good, the bad, and the ugly: Accuracy, inaccuracy, and elaboration in preschoolers' reports about a past event. *Applied Cognitive Psychology, 11,* S37-S54.

Whitehurst, G. J., & Lonigan, C. J. (2001). Emergent literacy: Development from prereaders to readers. In S. B. Nueman & D. K. Dickinson (Eds.), *Handbook of early literacy research* (Vol. 1, pp. 11–29). New York: Guilford.

Whitley, B. E. (1997). Gender differences in computer-related attitudes and behavior: A meta-analysis. *Computers in Human Behavior, 13,* 1–22.

Willis, S. L. (1996). Everyday problem solving. In J. E. Birren & K. W. Schaie (Eds.), *Handbook of the psychology of aging* (4th ed., pp. 287–307). San Diego, CA: Academic Press.

Willis, S. L. (2001). Methodological issues in behavioral intervention research with the elderly. In J. E. Birren & K. W. Schaie (Eds.), *Handbook of the psychology of aging* (5th ed., pp. 78–108). San Diego, CA: Academic Press.

Willis, S. L., & Schaie, K. W. (1994). Assessing everyday competence in the elderly. In C. B. Fisher & R. M. Lerner (Eds.), *Applied developmental psychology* (pp. 339–372). New York: McGraw-Hill.

Wilkinson, L., and the Task Force on Statistical Inference (1999). Statistical methods in psychology journals: Guidelines and explanations. *American Psychologist, 54,* 594–604.

Wimmer, H., & Perner, J. (1983). Beliefs about beliefs: Representation and constraining function of wrong beliefs in young children's understanding of deception. *Cognition, 13,* 103–128.

Windle, M., & Lerner, R. M. (1986). Reassessing the dimensions of temperament individuality across the life span: The revised Dimensions of Temperament Survey (DTS-R). *Journal of Adolescent Research, 1,* 213–230.

Winterling, D., Crook, T. H., Salama, M., & Gobert, J. (1986). A self-rating scale for assessing memory loss. In A. Bes, J. Cahn, S. Hoyer, J. P. Marc-Vergnes, & H. M. Wisniewski (Eds.), *Senile dementias: Early detection* (pp. 482–486). London: John Libbey Eurotext.

Wolff, P. H. (1966). The causes, controls, and organization of behavior in the neonate. *Psychological Issues, 5,* 7–11.

World Medical Association. (1964). Declaration of Helsinki. In H. K. Beecher (Ed.), *Research and the individual: Human studies* (pp. 277–278). Boston: Little, Brown.

Xu, F., Spelke, E. S., & Goddard, S. (2005). Number sense in human infants. *Developmental Science, 8,* 88–101.

Yarrow, M. R., Campbell, J. D., & Burton, R. V. (1968). *Child rearing: An inquiry into research and methods.* San Francisco: Jossey-Bass.

Yarrow, M. R., Campbell, J. D., & Burton, R. V. (1970). Recollections of childhood: A study of the retrospective method. *Monographs of the Society for Research in Child Development, 35*(5, Serial No. 138).

Yarrow, M. R., & Waxler, C. Z. (1976). Dimensions and correlates of prosocial behavior in young children. *Child Development, 47,* 118–125.

Yarrow, M. R., & Waxler, C. Z. (1979). Observing interaction: A confrontation with methodology. In R. B. Cairns (Ed.), *The analysis of social interactions: Methods, issues, and illustrations* (pp. 37–65). Hillsdale, NJ: Lawrence Erlbaum.

Yin, R. K. (2003). *Case study research: Design and methods* (3rd ed.). Thousand Oaks, CA: Sage.

Yonas, A. (1981). Infants' responses to optical information for collision. In R. N. Aslin, J. R. Alberts, & M. R. Peterson (Eds.), *Development of perception: Psychobiological perspectives: Vol. 2. The visual system* (pp. 313–334). New York: Academic Press.

Yoon, C., Hasher, L., Feinberg, F., Rahhal, T. A., & Winocur, G. (2000). Cross-cultural differences in memory: The role of culture-based stereotypes about aging. *Psychology and Aging, 15,* 694–704.

Yoon, C., May, C. P., & Hasher, L. (2000). Aging, circadian arousal patterns, and cognition. In D. C. Park & N. Schwartz (Eds.), *Cognitive aging: A primer* (pp. 151–171). Philadelphia: Psychology Press.

Zahn-Waxler, C., Radke-Yarrow, M., Wagner, E., & Chapman, M. (1992). Development of concern for others. *Developmental Psychology, 28,* 126–136.

Zaitzow, B. H., & Fields, C. B. (2006). Archival data sets: Revisiting issues and considerations. In F. T. L. Leong & J. T. Austin (Eds.), *The psychology research handbook* (2nd ed., pp. 251–261). Thousand Oaks, CA: Sage.

Zelazo, P. D., & Frye, D. (1998). Cognitive complexity and control: II. The development of executive function in childhood. *Current Directions in Psychological Science, 7,* 121–126.

Zigler, E. F., & Finn-Stevenson, M. (1999). Applied developmental psychology. In M. H. Bornstein & M. E. Lamb (Eds.), *Developmental psychology: An advanced textbook* (4th ed., pp. 555–598). Mahwah, NJ: Lawrence Erlbaum.

Zigler, E. F., & Muenchow, S. (1992). *Head Start: The inside story of America's most successful educational experiment.* New York: Basic Books.

Zigler, E. F., & Styfco, S. (Eds.). (1993). *Head Start and beyond: A national plan for extended childhood intervention.* New Haven, CT: Yale University Press.

Author Index

Subject Index

About the Author

Scott A. Miller (Ph.D., Child Development, University of Minnesota, 1971) is Professor of Psychology at the University of Florida, where he regularly teaches survey courses in developmental psychology at the undergraduate and graduate levels and a course on research methods in developmental psychology. He is a Fellow in the American Psychological Association Division 7, Developmental Psychology, and in the Society for Research in Child Development. His areas of research are cognitive development, social cognition, and Piaget. He has authored numerous research articles in such journals as *Child Development,* the *Journal of Experimental Child Psychology, Cognitive Development,* the *Journal of Cognition & Development,* and the *British Journal of Developmental Psychology.* He is also an experienced textbook writer. In addition to *Developmental Research Methods,* he is a coauthor of Flavell, Miller, and Miller's fourth edition of *Cognitive Development* (2002) and Vasta, Miller, and Miller's fourth edition of *Child Psychology* (2004).

About the Author

Scott A. Miller (Ph.D., Child Development, University of Minnesota, 1971) is Professor of Psychology at the University of Florida, where he regularly teaches survey courses in developmental psychology at the undergraduate and graduate levels and a course on research methods in developmental psychology. He is a Fellow in the American Psychological Association Division 7, Developmental Psychology, and in the Society for Research in Child Development. His areas of research are cognitive development, social cognition, and Piaget. He has authored numerous research articles in such journals as *Child Development,* the *Journal of Experimental Child Psychology, Cognitive Development,* the *Journal of Cognition & Development,* and the *British Journal of Developmental Psychology.* He is also an experienced textbook writer. In addition to *Developmental Research Methods,* he is a coauthor of Flavell, Miller, and Miller's fourth edition of *Cognitive Development* (2002) and Vasta, Miller, and Miller's fourth edition of *Child Psychology* (2004).